AGRICULTURAL LAND REFORM IN SOUTH AFRICA

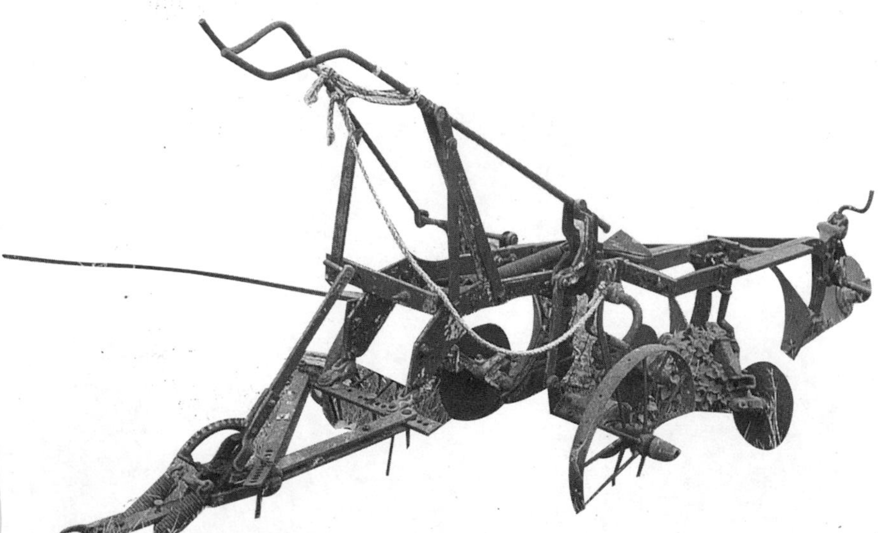

Agricultural Land Reform in South Africa

Policies, Markets and Mechanisms

GENERAL EDITORS:

Johan van Zyl
Johann Kirsten
Hans P. Binswanger

CAPE TOWN
OXFORD UNIVERSITY PRESS
1996

OXFORD UNIVERSITY PRESS
Walton Street, Oxford OX2 6DP, United Kingdom

Oxford New York
Athens Bangkok Bombay
Calcutta Cape Town Dar es Salaam Delhi
Florence Hong Kong Istanbul Karachi
Kuala Lumpur Madras Madrid Melbourne
Mexico City Nairobi Paris Singapore
Taipei Tokyo Toronto

and associated companies in
Berlin Ibadan

AGRICULTURAL LAND REFORM IN SOUTH AFRICA: POLICIES, MARKETS AND MECHANISMS
ISBN 0 19 571385 0

First published 1996

Editor: Judith Marsden
Designer: Mark Standley
Cover designer: Mark Standley
Indexer: Ethné Clark

Published by Oxford University Press Southern Africa,
Harrington House, Barrack Street, Cape Town 8001, South Africa

Text set in 10 on 12 pt Times New Roman by RHT desktop publishing cc
Reproduction by RHT desktop publishing cc
Cover reproduction by RJH Graphic Reproductions
Printed and bound by ABC Book Printers, Epping

Acknowledgements: The authors and publishers would like to thank Mark van Aardt, Photographic Enterprises and Mark Standley for the use of their photographs.

PREFACE

The land issue in South Africa has been debated intensively over the past few decades. Discussions were usually characterized by widely diverse points of view among the various concerned parties. The earlier academic debates on the land question focused on a number of wide-ranging issues, including traditional or freehold tenure, ownership or tenancy, and how to redistribute land.

After 1990, the discussions on land became much more focused in anticipation of, and in preparation for, democratic rule. Conferences, seminars and workshops were held across the country and overseas in an effort to find common ground on how to approach the difficult and often emotional task of land reform. At first, the initiatives were ad hoc in nature but, after 1992, they gained momentum under the general guidance of the Land and Agricultural Policy Centre (LAPC). Major initiatives were the Land Options Conference held in October 1993 in Johannesburg and numerous discussions at grass roots level to popularize approaches to land reform.

With the advent of democratic elections in April 1994, land reform obtained a central place in the Reconstruction and Development Programme (RDP) of the new Government of National Unity. Under the RDP, the aim is to redistribute 30 per cent of white-owned farm land within the first 5 years of democratic rule.

This book supports the notion of substantial land reform and advocates a market-assisted approach to achieve this goal. In this respect, the book provides the theoretical underpinnings of a major land reform initiative in South Africa by providing lessons based on international experience, analyzing the South African policy and legal environment, and basing proposals on these realities.

The idea for this book was borne out of the realization of the need to publish the theoretical research done for the preparation, execution and management of a land reform programme in South Africa. From the South African perspective, it is important to disseminate the theoretical principles and international experiences underlying the findings and results of the large research effort to design land reform options for South Africa. This has not been adequately addressed before. In addition, several new insights have emerged, and the process has proceeded far beyond the initial proposals expressed at the 'Options Conference' of the LAPC held from October 12–15, 1993. The research and lessons emerging from the land reform process need to be freely available in order to obtain the transparency required to make the process a success.

Interest from the international community on the developments in South Africa, specifically related to land reform, is already huge and growing. Dissemination of the results and process will also stimulate this debate at the international level. In addition, this book could make a positive contribution to, and form an important part of, the curricula of universities. In this respect, the programme advocated in South Africa contributes to existing knowledge and is really shifting the frontier on this important issue.

The South African land reform programme can also act as an important experience to international development aid organizations and the donor community on the complexity of the issues involved, how to tackle the process, and the pitfalls and problems which need to be managed. Many countries are involved in similar land reforms or, at least, the same ideas are being debated in a wide variety of areas, including Brazil, Colombia, Poland, and many more. All these countries can benefit tremendously from the South African experience.

The purpose of the book is to address these needs, namely to:

- document the research, proposals and experiences in addressing South African land reform with a strong emphasis on the theoretical underpinnings;
- contribute to the stock of knowledge on the subject;
- increase the dissemination of the information and transparency of the land reform process;
- stimulate the debate on the issues involved at both the national and international levels;
- serve as a resource for similar future programmes.

The book is intended for South African and international academics and students, development economists and other development practitioners, government officials, and policy and other decision-makers. We trust that it will bring a clearer understanding of the theoretical arguments for land reform and the principles underlying a market-assisted redistribution process. In addition, we intend it to stimulate further debate on the mechanisms for land reform in South Africa.

The book consists of six major sections, each with a number of chapters: setting the scene; learning from experience; agricultural policies, law and land reform; options for land reform; implementing the programme; and managing the process. The first part, *Setting the Scene*, consists of four chapters that give a general introduction; it describes the five broad patterns of rural development, discusses the evolution of the South African agrarian structure, and analyzes the legacy of history and current options for South African land policy. Part II, summarized in three chapters, addresses international experience and provides the characteristics and performance of settlement programmes, the political implications of alternative models of land reform and compensation, and rural development and poverty reduction. This is followed by Part III, consisting of six chapters, which describes the legal and policy environment for land reform. These chapters deal with the legal environment, the contemporary agricultural policy environment, natural resource issues, the farm size-efficiency relationship, the land market and agricultural growth linkages. Part IV on *Options for Land Reform* consists of four chapters: an overview chapter, one on restitution, one on how market-assisted rural land reform will work, and the last on rural livelihoods and the cost of rural restructuring. Part V comprises three chapters on *Implementing the Programme*; these address the selection of beneficiaries, the provision of agricultural support, and the restructuring of rural finance and land reform financing mechanisms. The last section, Part VI, deals with the land reform process; its five chapters cover the general process followed, the pilot programme, a benefit-cost analysis of redistribution, managing the environment and managing the political process.

Finally, it remains for us to thank all the contributors to this volume. This initiative was launched at short notice at the end of November 1994. Also, we want to thank all those people who contributed to discussions and commented on previous drafts. Many of the ideas in this book come from these sources.

* * *

THE CONTRIBUTORS

MICHAEL ALIBER is an agricultural economist with the Land and Agriculture Policy Centre in Johannesburg.

HANS BINSWANGER is the Senior Policy Adviser in the Agriculture and Natural Resources Department of the World Bank.

DANIEL BROMLEY is chairman and the Anderson-Bascom Professor in the Department of Agricultural Economics at the University of Wisconsin-Madison.

ROBERT CHRISTIANSEN was an economist in the Agricultural Division of the Southern Africa Department of the World Bank, and was task manager for agriculture and natural resource issues in South Africa. At present, he is in the African Technical Department.

GERHARD COETZEE is a senior policy analyst in the Centre for Policy and Information Analysis at the Development Bank of Southern Africa, and specializes in rural finance.

DAVID COOPER is the Director of the Land and Agriculture Policy Centre, a policy think-tank in Johannesburg with close links to the Department of Land Affairs of the new Government of National Unity in South Africa.

KLAUS DEININGER was a Young Professional in the World Bank. At present, he is an agricultural economist in the Policy Research Department of the World Bank.

MIKE DE KLERK is Senior Lecturer in the School of Economics at the University of Cape Town.

SALIEM FAKIR is responsible for the Land and Agricultural Policy Centre's research programme on environmental issues in South Africa.

CHRISTINA GOLINO is with the Development Bank of Southern Africa and is the Divisional Manager: Rural and Agricultural Development for the Northern Region.

STEPHEN HOBSON was with the Development Bank of Southern Africa, but has subsequently joined the Small Farmer Development Corporation in Cape Town.

BILL KINSEY is with the Centre for the Study of African Economies at the University of Oxford.

JOHANN KIRSTEN is Senior Lecturer in the Department of

Agricultural Economics, Extension and Rural Development at the University of Pretoria.

HEINZ KLUG is with the Law School of the University of the Witwatersrand in Johannesburg.

SUSAN LUND is Deputy Director-General in the Department of Land Affairs.

MASIPHULA MBONGWA is with the Centre for Policy and Information Analysis at the Development Bank of Southern Africa.

CRAIG McKENZIE co-ordinates the Environmental Research Programme in the Centre for Policy and Information Analysis at the Development Bank of Southern Africa.

KGOTOKI NHLAPO is a private consultant working on rural finance issues.

BONGIWE NJOBE-MBULI was a senior lecturer in the Department of Agricultural Economics, Extension and Rural Development at the University of Pretoria. At present, she is Deputy Director-General in the national Department of Agriculture in Pretoria.

KATRINA TREU is an operations analyst in the Agricultural Division of the Southern Africa Department of the World Bank.

ROGIER VAN DEN BRINK is a country economist in the Agricultural Division of the Southern Africa Department of the World Bank.

JOHAN VAN ROOYEN is Professor in Agricultural Economics and Director of the School of Agriculture and Rural Development at the University of Pretoria.

HERMAN VAN SCHALKWYK is a lecturer in the Department of Agricultural Economics, Extension and Rural Development at the University of Pretoria.

JOHAN VAN ZYL is Professor in Agricultural Economics and Dean of the Faculty of Biological and Agricultural Sciences at the University of Pretoria. He is currently on sabbatical with the Agricultural and Natural Resources Department of the World Bank in Washington, D.C.

NICK VINK was a Divisional Manager in the Centre for Policy and Information Analysis at the Development Bank of Southern Africa. At present he is head of the Department of Agricultural Economics at the University of Stellenbosch.

CONTENTS

LIST OF FIGURES

Setting the Scene

PART I

1

Introduction

Johan van Zyl, Johann Kirsten and Hans Binswanger

South Africa has experienced far-reaching changes in the political sphere resulting in the new democracy that ended the apartheid era and opened the way for resuming a full role within the international community. It is generally accepted that equality should be the guiding principle in the political and social spheres of the new South African society. The consensus also favours equality of opportunity in the economic sphere.

In the agricultural context, equality of opportunity is probably most important where it concerns the ownership of land. Thus, the redistribution of land is at the centre of the debate. This book advocates substantial land reform. Building on international experience, it proposes a methodology and mechanism for land reform based on utilizing the land market. It also outlines the early steps of the land reform programme.

Why land reform?

Why is land reform necessary in countries like South Africa? The motivation rests on: increased efficiency, increased growth and poverty reduction.

The most obvious motivation for land reform is the unsustainability – from a political, social, economic and equity point of view – of the present distribution of the ownership of agricultural land. The South African agricultural sector is dominated by large farms that are owned and operated by a small number of individuals or companies. As a result of a history of distortion (discussed in Chapters 3 and 9), this minority owns a startling 86 per cent of South Africa's agricultural land.[1]

1 South Africa is not unique in this respect, but is in a similar position to countries such as Brazil, Colombia, Guatemala and others which are characterized by extreme inequality in land ownership. In Brazil, for example, the majority (63 per cent) of rural inhabitants are smallholders who collectively own only 8,4 per cent of the total agricultural land, while the large landowners (25,5 per cent) own 62 per cent of agricultural land. As a result, these countries presently grapple with similar issues to South Africa.

The equity and equality case for land redistribution rests on the history of racial and economic policies which favoured white commercial farmers relative to the landless and smallholders. The majority of the poor in South Africa reside in rural areas, often with no or limited access to land. Land reform could therefore be one strategy to alleviate the present crushing poverty situation in South Africa. It should be kept in mind that the relationship between rural poverty and access to land is complex. Many factors are involved, including differences in land quality, the availability of complementary inputs, access to credit, and markets and opportunities for off-farm employment. When land quality is poor and access to water, inputs and markets is limited, access to land in itself does not significantly reduce poverty.

In South Africa, land reform also has an ethical and political–economic dimension. While equity, or equality of opportunity, seems to be a principle that can easily be applied, it requires favouring A over B. In South Africa, the status quo is a consequence of exclusion and oppression of one group by another. Future equality of opportunity starts from the present distribution of rights, privileges, wealth and power. To accept the present distribution of ownership of wealth and assets as a starting point, is to accept the past policies from which they have been derived. The principle of equality of opportunity thus questions the status quo with respect to the distribution of wealth and economic power and, at the very least, suggests a comprehensive affirmative action approach to ensure a more equal starting point.

The efficiency argument for land reform is that the redistribution of agricultural land to smallholders will increase, or certainly not reduce, total factor productivity and efficiency in the longer term. In most of the developing world there exists an inverse relationship between farm size and efficiency. Once a small minimum size is exceeded, family farms – relying primarily on family labour, are generally more productive than larger farms relying primarily on hired labour. They also generate more employment. The evidence for the inverse relationship has been challenged by partial analysis, often relating output only to land, rather than to all inputs. But the relevant efficiency indicators are total factor productivity across different farm sizes, fully accounting for differences in factor quality, or profit, net of family labour, per unit of capital invested. When using these measures, the inverse relationship has been confirmed in many studies, and especially in this volume, for South Africa.

Land reform will increase employment and self-employment for two reasons. First, smaller farms are more labour intensive than larger mechanized farms and livestock ranches. Second, smaller farms create greater linkages with the non-farm economy. The existence of farm–non-farm growth linkages provides an important argument for the implementation of land reform in South Africa. Evidence (discussed in Chapter 13) shows that areas with relatively equal land and farm income distributions have a higher share of non-farm consumption produced locally and more labour intensively. Thus, depending on the strength of these linkages, the multiplier effect on the incomes of the poor (unskilled labourers and smallholders) will be greater, the more equal the land distribution. It is therefore argued that land reform will result in considerable gains in employment and growth in the off-farm and non-farm rural sectors.

Each of these arguments provides a good case for redistributive land reform. Taken together, the argument for redistribution of agricultural land is strong.

Large-scale commercial agriculture in South Africa

Commercial agriculture in South Africa is widely regarded as a highly sophisticated and successful sector. In its recent study of South African agriculture, the World Bank (*1994*), summarizing a large volume of work conducted mainly by South Africans over a number of years, reached the conclusion that the performance of South Africa's commercial agricultural sector is deceptive in that it successfully conveys the appearance of efficiency. The sector has followed a pattern of growth that is far from normal due to the distortions prevailing during a long history of persistent government intervention in its favour.[2] This has encouraged excessive farm size growth and mechanization, and excessive shedding of labour. While these decisions were privately profitable and technically efficient, they reduced economic efficiency. Several analysts conclude that South African agriculture makes sub-optimal use of its most abundant resource – labour. Subsidized low interest rates and various tax breaks encouraged the excessive substitution of capital for labour.

In sharp contrast, apartheid policies have resulted in the concentration of about eight million blacks on 13 per cent of agricultural land –

2 Some of these distortions are not particular to agriculture, but characterize the entire economy, e.g. sluggish growth in total factor productivity and the high capital-intensity of production in the presence of widespread unemployment. Nonetheless, it appears to be the case that agriculture has produced distortions that have been extremely far-reaching.

primarily in the former homelands. Combined with inadequate access to markets, inadequate infrastructure and support services, these policies caused the virtual elimination of small-scale black agriculture. This has prevented the development of a viable, employment-intensive rural economy centred on agriculture. Thus, the usual vibrant and wide range of informal business activities created through forward and backward linkages of agricultural development have never emerged in South Africa's rural economy.

The agricultural strategy pursued in South Africa has achieved its two main objectives – food self-sufficiency and acceptable income levels for white farmers. It has done so by distorting the policy environment and causing society to pay an enormous financial and social cost. Society has been deprived of a large contribution to GDP that a more efficient and dynamic agricultural sector could have provided. At this point, the agricultural sector is in a state of crisis. Nearly a decade ago, the financial cost of the agricultural strategy became sufficiently burdensome to prompt the introduction of a limited liberalization. This policy change, along with adverse weather conditions, had the effect of exposing farmers to a more exacting environment. As a result, profits declined and debt increased. Many parts of large-scale commercial agriculture are not viable under the new policy environment.

The combination of economic inefficiency in the large-scale sector, the inequitable distribution of land and the new democracy has produced a politically unsustainable situation that threatens the future viability of the entire economy. Despite the enormous difficulty of land reform and settlement processes, failure to execute a major land reform in countries with highly dualistic farm size structures, or delayed implementation of such reforms and continued neglect of native peasant sectors, seems to have had far more adverse consequences than the relatively minor risks associated with the process of land reform.

Therefore, we reach the conclusion that there are few development options for agriculture and the rural economy. South Africa needs solutions which will allow it to act quickly and cheaply. A successful strategy for the growth and development of the rural economy will require at least three elements (*World Bank, 1994*):

1. **The removal of current distortions in white commercial agriculture** to increase competition and induce a shift towards more employment-intensive forms of production, processing and marketing. The process of ongoing liberalization would include: (a) further reform of the input and output marketing system; (b) efforts to reduce the

concentration in the agro-processing sector; (c) revision of land sub-division guidelines; and, (d) restructuring of the present agricultural credit system. Continued liberalization along these lines is likely to result in more bankruptcies among large-scale farms, the expansion of small-scale farming (especially near urban areas), expansion of the horticultural sector, and contraction of grain and livestock production.

2. **The development of a new type of commercial, small-scale agriculture** centred on the family farm to further increase employment intensity and efficiency in agriculture. This development strategy would also feature upgrading of agricultural support services and investing in an improved physical and social infrastructure in the former homelands. This has the advantage of continuing the process of policy liberalization and of concentrating public sector resources on some of the most obvious victims of apartheid.
3. **A fundamental institutional restructuring in order to support the new vision:** on the one hand, a down-sized and employment-intensive white farming sub-sector and, on the other hand, an emerging commercial small-scale farming sub-sector. In addition, this option supports redistribution of land in the large farm sector. A redistribution of agricultural land would achieve three critical objectives: (a) to reduce the uncertainty experienced by current owners, thereby encouraging those who continue farming to invest; (b) to address the present inequitable distribution of land use; and, (c) to encourage, if the redistribution mechanism is properly designed, those with an interest in land use to gain access to and use land efficiently. This latter objective would lead to a more dynamic rural economy and to greater employment and income generation among low-income groups than would either of the first two strategic options.

Agricultural policies, land reform and rural development: the international experience

A central theme of the book is the relation between land reform and the policy environment, both macro-economic and sectoral. Binswanger's Chapter 2 introduces this theme by showing how poor agricultural and rural development performance has retarded economic growth and increased poverty, both rural and urban. From the 'painful lessons' of rural development in many countries across the world has emerged a broad professional consensus. This argues that foreign exchange, trade and taxation policies should not discriminate against agriculture; that an open economy, employment intensive and small farmer oriented strategy is both economically efficient and most likely to reduce poverty; that providing privileges or reducing competition in output, input and credit markets is costly to consumers and taxpayers and ends up hurting small farmers and the rural poor; and finally, that in the case of unequal land distribution, substantial land reform is required.

Chapter 3 by Mbongwa *et al* illustrates the South African example of some of the 'painful lessons', and shows the devastating effects of continued and forceful depression of the profitability of small-scale agriculture through discriminatory policies and by providing privileges to white farmers.

One of the most important questions in the debate about a future land policy in South Africa is whether the current land distribution provides a good basis for future growth of output and employment in the agricultural sector, and for a gradual elimination of the inequalities in access to land, or whether land reform and resettlement are necessary to achieve these goals. These issues depend critically on the nature of economies of scale in agriculture. Binswanger and Deininger (Chapter 4) review these issues based on worldwide evidence, and then examine the congruence with, or divergence from, international experience of the specific history of South African agriculture. They make a strong plea in favour of a rapid and large-scale land reform programme, and sketch out the mechanisms by which such a task could be accomplished using judicial and market-assisted processes for the transfer and resettlement of land.

Kinsey and Binswanger review the worldwide experience with land settlement in Chapter 5. A number of links between core characteristics of resettlement and specific outcomes appear to be fairly well established. An understanding of these links is useful for avoiding costly mistakes in the initial design and establishment of settlements.

Chapter 6 focuses on the administrative cost and political economy implications of alternative land reform processes and modes of compensation. Expropriation is one of the alternatives that can be considered, but Binswanger argues against this option mainly because expropriation is likely to be a lengthy and protracted process, especially when just compensation at below market prices is contemplated and expropriation judgements are appealed in the higher level courts. Not only does this increase the administrative costs, but also delays benefits from accruing to beneficiaries. Reliance on the willing-buyer – willing-seller model will greatly reduce the political opposition to land reform, and may increase its political sustainability over the decade or so required to implement any sizeable programme.

Binswanger's following chapter discusses the failure of integrated rural development and indicates how countries, other donors and multilateral lenders, as well as the World Bank, have tried to find alternatives to integrated rural development. Most of these initiatives deal with administrative and/or fiscal decentralization, and greater involvement of beneficiaries. One of the key elements of the new consensus on rural development and poverty reduction is substantial prior or concurrent effort at land reform in a decentralized and participatory fashion based on a market-assisted approach.

Against this background, and based on international experience, it is argued that market-assisted land reform programmes will have great advantages. The need for reliance on market mechanisms stems from the observed weaknesses of non-market oriented programmes that typically vest too much control in public sector bureaucracies. These public sector bureaucracies develop their own set of interests that are in conflict with the rapid redistribution of land. Nonetheless, a well-functioning land market is not a sufficient condition for large mechanized and relatively inefficient farms to be subdivided into smaller family-type farms, specifically where economic and institutional distortions favour large farms over small farms. Therefore, non-market interventions in the form of grants are necessary to ensure successful implementation of any land reform programme. Executing land reform through grants or vouchers to beneficiary groups who buy from willing sellers obviates the need for a land reform/settlement agency and the chances for bureaucratic rent-seeking are therefore less. Beneficiaries are free to choose the land in the market and to do with it what they wish, rather than having to follow the guidelines of an agency.

Does the South African reality allow for a market-assisted land reform?

One important element of the land reform process in South Africa is the restitution of land rights. This will take place through a land claims process as described in Chapter 15. The majority of the chapters in this volume deal with the land redistribution aspect of land reform. As motivated earlier, a market-assisted land reform process is considered to bring about a successful redistribution of agricultural land. A number of legal and policy related questions, specifically applied to South Africa, are important when considering a market-assisted land reform programme. This is discussed by the contributors in Parts III and IV of the book. Taken together, these chapters debate the implementation and likely success of market-assisted land reform in South Africa.

The history of land reform around the world demonstrates that land invasions, which governments then normalize through legal processes of expropriation and allocation, have been the most common and effective processes of land reform. Given the potential of violent conflict inherent in the process of land invasions, a legal framework must aim to reduce the likelihood of this alternative. In this light, the question can be asked whether the present legal framework in South Africa is conducive to land redistribution?

Heinz Klug argues in Chapter 8 that the characteristics of the current framework for South African land law undermine its legitimacy and have profound consequences for the establishment of a functional system of land law. Freehold is privileged; land law is fragmented in different parts of the country; there is an inadequate system for recording land rights; bureaucratic discretion exists over land rights and the disposition of land claims, and racial and gender stratification of inheritance in land rights are present. While the restitution element of land reform has largely been taken care of through the adoption of the Restitution of Land Rights Act in November 1994, there is still a need to develop comprehensive land legislation to facilitate the land redistribution process. For example, the Act on the Subdivision of Agricultural Land has to be abolished. The elements of comprehensive land legislation will include:

- the creation of a single national legislative framework for land rights and administration;
- provisions enabling the state to redistribute land;
- new rules to regulate the tension between ownership rights and illegal use of land based on necessity;
- the creation of a viable system of land administration at the district level which may be coupled with the recognition of land rights.

The success of market-assisted land reform relies on a policy environment which does not favour larger-scale farmers. Fortunately, today it can be stated that the policy environment in South Africa is conducive to market-assisted land reform. The policy environment has become much more liberalized since the early 1980s, with most of the privileges to large farms having been abolished, or in the process of being abolished (Chapter 9). This has paved the way for better functioning markets with fewer distortions, and a general environment that is conducive to market-assisted land reform.

International experience on the relationship between farm size and efficiency shows that economies of scale in agricultural production are often real but 'false' in the sense that they are the result of market and policy distortions favouring large farms over smaller family-type farms. In the absence of policy distortions and market failures, small family-type farms are more efficient than, and superior to, large mechanized farms. Analysis presented in Chapter 11 – using a variety of analytical approaches, including econometric and non-parametric – shows that the South African experience closely correlates with observations elsewhere in the world. The analyses provide clear answers to the issues in question: *from an efficiency point of view, farms should be smaller in the white commercial sector and larger in the former homeland sector.* Land reform will therefore result in relatively large efficiency gains. In addition, it will substantially increase on-farm employment as smaller farms generally use more labour-intensive production techniques.

Based on international experience and the earlier arguments, this book suggests a design for land reform in South Africa that relies as much as possible on the existing land market. The question is, therefore, can the land market handle a land reform?

An analysis of the underlying factors driving land price changes in South Africa, reported in Chapter 12, provides a comprehensive framework for determining the relative importance of factors driving farmland prices over the past two decades. The results of the analysis clearly demonstrate that *the agricultural land market is not only active enough, but also stable enough to be used as a transfer mechanism for substantial amounts of agricultural land* to the people disadvantaged and excluded by past policies. In addition, the current relatively small difference between the market price of land and the capitalized value of farm profits enhances repayment ability since buyers of land will now find it easier to repay a loan from the productive capacity of the land itself. *It also firmly establishes a market-assisted approach as a real and workable option for land reform.*

Given the necessity and the potential benefits of land reform, Chapters 14, 15 and 16 discuss the options and detail for land reform. Chapter 14 is concerned with the broad issues of rural restructuring and land redistribution options. The chapter advocates a combination of restoration and redistribution, and examines options for how such a process of land reform might work in South Africa. In the light of this favoured approach to land reform in South Africa, individuals (or groups) that wish to gain access to land have three choices: (a) seek land through the restoration process as outlined in Chapter 15; (b) acquire land through the redistribution channel; or, (c) purchase land without assistance from the programme. Individuals (or groups) that qualify would submit an application to the appropriate institution (i.e. a land claims court, a land commission or a land committee).

Chapter 14 provides another well-supported case as to why a market-assisted redistribution process is favoured in South Africa. In this process, the beneficiaries would be responsible for identifying a piece of land, but would be eligible for assistance in this activity. Once a suitable piece of land is identified, financing could be arranged from three possible sources: (a) the grant component available from the programme; (b) the beneficiaries' own resources; and, (c) a bank loan. The grant elements of the programme are essential in order to accomplish a redistribution of assets and to ensure that beneficiaries emerge from the programme with a net increase in their asset position. A matching grant scheme that forces participants to use some of their own resources in order to gain access to land will help to assist in self-selection of beneficiaries.

Van Zyl and Binswanger argue in Chapter 16 that the current debt crisis in commercial agriculture, following market liberalization and the withdrawal of other privileges to white large-scale farmers, can be managed by linking it to a market-assisted land reform process. Agricultural credit provided or guaranteed by the government should be used as an instrument to correct some of the imbalances caused by previous policies rather than to perpetuate their consequences. It can be an extremely powerful tool in achieving the objectives of the Reconstruction and Development Programme (RDP). Existing credit lines can be retargeted and restructured to benefit new small-scale family-type black farmers. Calling-up of bad and overdue debt could be used to start the land reform process.

Within this context, the basic premise of this book is that *managing the agricultural debt crisis, increasing efficiency, improving food security and addressing some of the racial imbalances in South Africa's farm sector, can be done by facilitating a market-assisted land reform process rather than a much more costly, inefficient and inequitable blanket debt relief programme. It will also have the added advantages of increasing employment at a low cost and adding to the rural safety-net.*

Accepting the fact that a market-assisted land reform process can successfully be implemented in South Africa, the impact on livelihoods and costs needs to be determined. Chapter 17 calculates the impact of land reform on rural livelihoods, and also estimates the fiscal costs of a nationwide land reform programme. The main hypothesis emerging from the analysis in this chapter is that large numbers of poor, but often potentially commercial, farm households can obtain reasonably attractive household and employment levels through fiscally affordable land redistribution. This process can be stable, sustainable, consistent with urban, rural and household food security, and can help to restrain the speed of migration from rural to urban areas. Results from four representative models indicate that when land moves from large to small holdings, and a more dynamic rural and peri-urban economy emerges, there is a substantial net increase in the number of livelihoods.

The total five year fiscal cost of a nationwide land reform programme is estimated at R22,1 billion. Farm capital costs represent 59 per cent of total costs, and beneficiaries would be responsible for about 29 per cent of total costs. This implies a total fiscal cost of roughly R15 000 per net livelihood. These figures translate into the creation of more than 600 000 additional net full-time equivalent employment opportunities at a fiscal cost of approximately R35 000 per additional job.

It is assumed that 6 per cent of the land of the commercial farm sector could be transferred in every year of the programme. This would lead to a total transfer of 30 per cent of the medium and high quality land of the commercial farm sector over five years, with a total of around 1,5 million net livelihoods created.

Implementing the land reform programme

The ultimate success of a land reform programme in South Africa should be tested against its ability to address equity in land distribution and livelihood upgrading, reduction of poverty, creation of rural employment and income-generating opportunities, inter alia raising the number of successful black agricultural producers and enhancing overall productivity, whilst maintaining sustainable natural resource management and utilization.

At its heart, this programme should be aimed at redressing the impact of past wrongs. In this way, potential participants would clearly come from black rural society which has been denied entitlements in the past in terms of apartheid ideology and laws. However, it is also recognized that a programme of this nature cannot and will not make an agricultural producer out of every participant, although it should go a long way in creating a viable rural economy within which agriculture and the related linkages can develop. Ultimately, there will be critical, difficult and also emotive choices that will need to be made about the selection of beneficiaries and the focus of such a programme. The inability to make timely decisions, in the long term, may result in politically motivated and potentially 'inappropriate' criteria, or the lack of effective criteria being used to establish a viable rural economy.

In their chapter, Van Rooyen and Njobe-Mbuli propose a possible set of criteria for selection to participate in the land reform programme. In the interest of attaining optimum productivity in land-use, and recognizing that apartheid policies resulted in structural poverty, it becomes important to offer special assistance in accessing appropriate services to the victims of apartheid. The proposed selection system is extended to apply to these criteria so as to facilitate the determination of the level of assistance. Important attributes to be tested at this level include health, age, net worth, education levels, gender, previous experience in farming, entrepreneurial skills and managerial aptitude.

Land is just one of the factors of production. Land reform will therefore have to be complemented by reorientation of support services towards beneficiaries, particularly research, extension and information, credit, input provision and output markets (see Chapter 19). The restructuring of rural finance will also be important to enable potential beneficiaries to access credit to purchase land through the market-assisted process. Without the proper functioning of the market for agricultural credit, the proposed land reform process will be doomed to failure.

Against this background, Coetzee, Mbongwa and Nhlapo (Chapter 20) provide a framework for a new rural financial structure based on certain guidelines. A state financial assistance structure to ensure access to financial services in rural areas is proposed. The land reform mechanism should link into these mechanisms. No justification can be found for a separate financing structure just for the land reform programme and the land reform beneficiaries. Only in the case of the disbursement of grants may there be a temporary role for District Offices to act as a financial mechanism. The importance of programming development activities at the local level and of programming macro-level activities has also been argued. Proposals were made for an interim structure to accommodate the urgent needs for the land reform programme.

The role of the informal sector and savings mobilization are extremely important in this regard. The importance of access to information cannot be emphasized enough. Land reform beneficiaries must have information on their rights and what is being offered in the programme. Financial intermediaries must have information on which to base decisions. Without information, the market-assisted approach would not be implementable.

The implementation of a nationwide land reform programme was initiated in February 1995 with the launch of a Land Reform Pilot Programme in each of the nine provinces. An overview of the Pilot Programme is provided in Chapter 22. The Land Reform Pilot Programme is an effort to translate the outcomes of an extensive South African debate on land reform policy options into an implementable strategy for redistribution. The piloting process is one of learning by doing, while establishing a facilitative role for the state in a manner that is both replicable and affordable. As Sue Lund elaborates in her chapter, the Pilot Programme places strong emphasis on shaping the nature of government intervention for land reform, and on building rural capacity to plan and manage the expenditure of state resources for land purchase according to locally devised and negotiated solutions.

The effects of land reform

Although increasing food production, and therefore a policy of food self-sufficiency, will not automatically ensure that people are food secure and have enough to eat, the South African Government for many years pursued an agricultural policy which had food self-sufficiency as a major objective. Although this objective was realized to a large extent, many people in South Africa are still food insecure. Estimations are that nearly 50 per cent of the population live below the poverty line, the large majority of whom are black and rurally-based. In this context, national food self-sufficiency has little policy relevance. Appropriate policy should rather address the real and most pressing problem in rural South Africa, namely that of crushing poverty and resultant food insecurity. Land reform is one important and powerful tool in alleviating these problems since it addresses both sides of the hunger equation, namely availability of and ability to buy food. The first premise is that increasing food production, storage and trade can assure food availability, but this will not automatically ensure that all people have access to a well-balanced diet or an end to hunger. The second premise is that, because poverty is a central cause of hunger and malnutrition, special efforts are needed to help increase the access and entitlement to food. The conclusion is that rural restructuring and specifically *land reform is necessary, but not sufficient, to improve the food security situation of the poor* who constitute the majority of the rural population. Even if this negatively affects food production in the short-term, there is enough 'surplus capacity' to ensure that the food needs of the urban population are not placed at risk. In addition, the extra rural livelihoods created will allow many households to obtain food entitlements.

Improved food security is one of the benefits of land reform which Michael Aliber has included in his estimation (Chapter 23) of the benefits and costs of the land reform programme. The benefit-cost analysis of land redistribution helps answer the broad question of whether land redistribution is in society's best interest. It simultaneously may help to answer more specific – but equally important – questions: (a) who gains from land redistribution, who loses, and by how much?; (b) is the target beneficiary group really made better off?; (c) what is the cost to the taxpayer?; and, (d) what are the unseen social effects (benefits and costs) that accrue to third parties? The calculations of the benefits and costs of land reform provided in Chapter 23 indicate that land reform will yield substantial advantages to the beneficiaries, to society and to government.

Land reform will address some of the most pressing natural resource issues in rural South Africa by alleviating some of the population pressure in the former homelands. In addition, it will also have positive impacts on harmful practices, for example maize monoculture, in the white commercial areas (see Chapters 10 and 24). Poverty poses the most serious environmental threat to the less developed areas of South Africa. The millions of people who live near the subsistence minimum will over-exploit natural resources to survive. The eradication of poverty, perhaps through improved access to land, could be one of the important ways to prevent further environmental degradation. Once the issues of rights to natural resources have been adequately addressed, other issues could be addressed in the process of formulating a coherent environmental policy.

Finally, land reform will have a profound effect on the structure of South African agriculture. In the final chapter, Johan van Zyl presents his vision for South Africa's agriculture sector in the aftermath of a land reform programme and a process of rural restructuring. This vision should be to increase efficiency, equity and employment-intensity in agriculture by moving towards a more diversified farm structure centred around competitive commercial, owner-operated, family farms. These farms should not be dependent on subsidies and government support for their sustainability, but should be supported and serviced primarily by the private sector. The role of government should be to establish a comprehensive legal, institutional and policy framework which will ensure a level playing field for all players. This framework will include increased reliance on markets, privatization, deconcentration and decentralization.

Conclusion

In summary, this book advocates that a land reform programme should avoid: the maintenance of privileges to the farm sector, most of which are captured by the large-scale mechanized farms, for example, tax shelters and credit subsidies; general debt bail-outs and blanket debt relief to farmers; confiscation, expropriation or any other administrative/legal 'cheap option' to get land; land acquisition by state or local land reform or administration bureaucracies; administrative beneficiary selection; and settling land reform beneficiaries on low quality land.

Desirable design features include: making the law consistent with the objectives and process of the reform; the use of a market-assisted approach, involving willing buyers and willing sellers, with targeted financial assistance to poor buyers; the use of self-selection by communities and individuals through the rules and incentive structures of the programme; the use of grants to target the poor and to provide appropriate self-selection incentives; limiting restrictions on the beneficiaries, for instance, on renting or selling the farm, or choosing alternative enterprise and community models; involving beneficiaries in the planning of infrastructure and services; making use of a decentralized institutional structure; decentralizing supervision; and emphasizing ex-post supervision rather than ex-ante approval at each step, combined with suitable penalties.

2

Patterns of rural development: painful lessons

Hans Binswanger[1]

Introduction

Some countries have stumbled onto successful agricultural and rural development strategies as if by accident (but usually as the consequence of bitter political and/or military conflict), and have been fortunate enough to maintain them. Others had to experience dramatic failures and painful learning from their own and other's experiences before they abandoned inadequate policies. Some stubbornly continue to cling to misguided prescriptions of the past, supported by political inequities which do not provide adequate voice to large segments of their poor or rural populations. Severe external shocks and adverse policies of developed countries have often contributed to their unsatisfactory performance. Poor agricultural and rural development performance has retarded economic growth and increased poverty, both rural and urban.

In this chapter, the theories and misconceptions which have provided the intellectual and ideological underpinnings of misguided agricultural and rural development policies are reviewed. This is followed by a discussion of the actual major misguided policies supported by these intellectual traditions which either singly, or in combination, have been shown to reduce agricultural growth and to harm the welfare of the rural poor.

The next section proposes a (non-exclusive) classification of countries into five groups, distinguished by their adoption or non-adoption of these misguided policies, and discusses their growth, rural development and poverty reduction performance. In the process, some issues of degradation of the agricultural resource base are briefly considered.

1 This chapter is largely based on the Simon Brand Address delivered by the author on 19 September 1994 at the 32nd Annual Meeting of the Agricultural Economics Association of South Africa in Pretoria, South Africa. These remarks are not an official position of the World Bank but represent the author's own views.

Agriculture in early development theories

Backwardness, low supply response and low productivity

During the early days of 'development economics' in the 1950s, rural poverty was often explained by the 'backwardness' of traditional smallholder agriculture. The sector was considered to have almost no potential for development. In his seminal work on *Transforming Traditional Agriculture*, Theodore Schultz characterizes the prevailing views of that time. The doctrinal answers run as follows:

- the opportunity for growth from agriculture is among the least attractive of the sources of growth;
- agriculture can provide a substantial part of the capital that is required to mount industrialization in poor countries;
- it also can provide an unlimited supply of labour for industry;
- it can even provide much labour at zero opportunity costs because a considerable part of the labour force in agriculture is redundant in the sense that its marginal productivity is zero;
- farmers are not responsive to normal economic incentives but instead often respond perversely, with the implication that the supply curve of farm products is backward sloping;
- large farms are required in order to produce farm products at minimum costs.

In addition, international commodity markets for agricultural goods were regarded as hostile, exposing countries which relied on them for growth to undue risks. Agriculture could be taxed with little adverse consequence for economic growth or poverty reduction.

The solution to rural poverty is in agricultural modernization and urban growth. It is therefore not surprising that the solution to the reduction in rural poverty was seen almost universally as being associated with urban growth and rural–urban migration. Rural population and employment should decline, if not in absolute terms, at least as a proportion of the labour force. In Colombia, for example, Blakemore and Smith (*1975*) summarized these views as follows:

> ... during the 1940s and 1950s ... there emerged a school of opinion, voiced most strongly by Lauchlin Currie, that suggested that the best method of increasing the efficiency of Colombian agriculture would be to allow fewer, more technically sophisticated farmers, using all

> the technological aids available, to farm land vacated by the majority of what can only be considered as inefficient agriculturists. These would be permitted, or should be persuaded, to migrate to the towns.

These views have been thoroughly discredited by research, much of it by agricultural economists. Yet they also provided the ideological justification for patterns of agricultural policies and programmes which have been highly detrimental to rural populations, especially the poor.

The key misconceptions

Underestimating supply response

A vast literature demonstrated early that individual crop supply response is highly elastic since individual crops can expand by withdrawing factors of production – land, capital and labour – from other agricultural sectors. Aggregate supply response, however, is highly inelastic in the short run because technical change, investment and changes in migration patterns are required to bring about such a response. As Mundlak and his various proponents have shown, in the long run these factor reallocations do take place leading to a large long run supply response of the sector (*Binswanger et al, 1987*).

Believing in the superiority of large farms

Communist countries as well as many market economies have paid an enormous price for assuming – without much empirical evidence – that large farms are more efficient than small ones. Large farmers are generally deeply committed to dynamic agricultural development and technological sophistication. Their farms are often well-managed with technically efficient, high volumes of output, yet their economic costs of production exceed those of smaller enterprises relying primarily on family labour. Their production is capital intensive and they generate very little employment. Because small farms have less wealth and/or access to credit markets, they use an input mix which relies much more on labour than capital, and thereby generates far more employment and self-employment than their large counterparts.

In communist countries, the belief in the superiority of large-scale farming was one of the motives for the failed collectivization of agriculture. In market economies, it led to costly differential policies and programmes in their favour, including subsidies to inputs, investments and marketing, and especially credit.

Genuine economies of scale only exist in the so-called plantation crops: sugarcane, bananas for export, tea, and oilpalm. They do not arise from economies of scale in the farming enterprise, but rather in processing or shipping which, for these commodities, must be co-ordinated with the harvesting of the crop. Yet even these economies of scale can be circumvented by contract farming, as with world class competitors in sugar; Thailand for example.

In other commodities, technical economies of scale exist where large machines must be owned by farmers themselves, and in information gathering about technology, marketing and credit. These technical economies of scale are offset by the superior incentive of family labour on farms which rely primarily on family members, rather than on hired labour. Thus, in most commodities, economies of scale are exhausted by the farm size which one family can handle, relying primarily on family labour. The larger the size, the larger the opportunity cost of the family labour and management skills outside agriculture; therefore optimal farm sizes tend to grow with economic development (*Binswanger, Deininger & Feder, 1993*). But even in the developed world, much of the observed growth of family farms is fuelled by distortions such as tax incentives to mechanization.

This does not mean that small farm size or fragmentation could never be a constraint to agricultural growth. But recent experience in China, Vietnam and Albania suggests that these fears are vastly exaggerated. China's agricultural growth has broken all world records for the last 16 years. Yet the average farm size is half a hectare, fragmented into an average of nine plots.

Assuming technological backwardness

The spectacular adoption of green revolution technologies by smallholders in most of Asia, Latin America, North Africa and selected sub-Saharan countries, has long dispelled notions that smallholders are necessarily 'backward'. Smallholders rarely lag more than a few years behind their larger counterparts in technology adoption. Not only have they adopted divisible technology such as fertilizers, new seed or pesticides, they have also adopted machines through rental arrangements and markets where economies of scale are pronounced.

Ignoring women farmers and workers

In various parts of Africa, women head the majority of farm households. If account is taken of women farmers who form part of extended households whose husbands or partners work in urban areas or in mines, the percentage of female farmers becomes even larger. In countries such as Colombia, up to a third of rural households are female-headed. In Asia, where it is more rare for women to farm on their own account, they often provide as much farm labour as men, either as unpaid family workers or as hired workers. The full implications of these facts for the design of agricultural and rural development strategies, and for the design of support service and credit programmes, have only entered into the development debate during the last two decades, and are yet to be fully translated into practice.

Ignoring rural non-farm linkages

Dynamic, smallholder-based agricultural growth has fuelled non-farm activities and employment through forward, backward and consumer-demand linkages. Consumer demand for rural goods, housing and services tends to play a major part. The linkages have been strong in Asia and Latin America where taxation of agriculture has been less than in Africa, and where rural incomes and technology are higher (*Haggblade & Hazell, 1989*).

Underestimating the poverty reduction impact of agricultural growth

That agricultural growth can have a major impact on poverty reduction has been amply demonstrated. It reduces consumer prices of non-tradable or semi-tradable foods (unless their markets are heavily protected or monopolized, of course). It can generate rapid growth of rural employment and self-employment in rural areas. The corresponding tightening of the labour market raises rural wages and has spill-over effects to urban informal sector wages. The modest government support which smallholders require in research, extension, infrastructure and marketing can usually produce many jobs at a low budget cost per job (unless, of course, the support is diverted to large-scale enterprises).

A recent study of India quantifies the contribution of rural and urban consumption growth to poverty reduction in India during the 1951–1990 period (*Ravallion & Datt, 1994*). Growth in each of the sectors reduced poverty within each sector. But rural growth had a much stronger effect on reducing urban poverty than the other way around. Rural growth clearly tended to reduce national poverty. The same could not be shown for urban growth due to the sector's smaller share of the poor, and its relatively low intra- and inter-sectoral poverty reduction effects.

The misguided policies

These and other misconceptions have provided the intellectual justification for a number of misguided policies which are discussed below. The mix of these policies varied a great deal across countries and over time. Some countries have consistently applied all of them, while others have used only a portion, or abandoned some or all of them in the course of the last 30 years. Their respective performances are discussed in the following section.

Excessive agricultural taxation

In a comparative study, Maurice Schiff and Alberto Valdes compare agricultural policy patterns for 18 developing countries from 1960 to about 1983 (*1992*). Krueger (*1992*) provides a political economy interpretation of the policy patterns found. Over time, interventions varied widely across countries, and within countries by commodity and sometimes by region. Nevertheless some patterns emerged:

- All but two countries (Taiwan and Portugal) taxed their agricultural sectors over the 25 year period studied.
- Taxation was most severe in the three African countries studied (Ivory Coast, Ghana and Zambia) where, on average, about half the value added was extracted by policies.
- A middle group taxed agriculture to the extent of 30 to 40 per cent (Argentina, Colombia, Dominican Republic, Egypt, Morocco, Pakistan, Philippines, Sri Lanka, Thailand and Turkey).
- Low taxation between 8 and 22 per cent was applied by Brazil, Chile and Malaysia.
- On average, three quarters of the taxation was derived from indirect measures, overvalued exchange rates and industrial protection.
- Only one fourth of the taxation was derived from direct agricultural policy measures.
- Direct interventions tended to tax export commodities, both food and non-food, where countries had strong comparative advantage.
- Direct protection, on the other hand, was concentrated on importable food commodities with little or no comparative advantage.
- The countries with the highest degree of discrimination against agriculture had the lowest rates of economic growth, and vice versa.
- These protection patterns cannot be explained by the intention of policy makers to increase foreign exchange earnings, or to provide cheap food for industrial development or to poor urban consumers.

While many governments raised revenues through direct taxation of agriculture, the taxation was highly inefficient.

- The only policy goal consistently achieved by these policies was some stabilization of domestic (relative to world) market prices.
- Direct support to agriculture through subsidized credit, infrastructure investment, research and extension, etc, did not compensate for the losses of the agricultural producers on account of heavy taxation.

Urban bias in development expenditures

Urban bias of policies was not confined to industrial protection and agricultural taxation. Michael Lipton (*1977*) documents that many countries had a pervasive urban bias in public expenditures on productive infrastructure and social services. These are reflected in the poorer health and nutrition status of rural compared to urban populations, and in lower educational attainment in virtually all developing countries. Many compensatory social programmes, especially food subsidies, have also been concentrated on urban areas in countries as diverse as India, Brazil, Zambia and Mexico. An ironic implication is that hunger and malnutrition are often the most severe where food is being produced, namely in the rural areas.

Compensating inefficient food sectors in the name of national self-sufficiency

In developed and developing countries alike, farmer interests have used national food security objectives to justify direct interventions in favour of agriculture, some with more, others with less success. Interestingly, it is often importable commodities with little comparative advantage which have received the highest direct protection, such as maize in Mexico and wheat in Guatemala, Nigeria and Brazil. Such policies make food more expensive for all net buyers of food, including many rural poor. They may reduce rather than increase household food security.

Compensating the rural elites

Politically articulated farmer groups have also been able to obtain producer subsidies in the form of subsidized or not repaid credit, irrigation investment without cost recovery, agricultural extension, parastatal marketing, and sometimes monopoly marketing or trading rights. They also often obtained tax privileges in income, land and capital gains taxation.

While insufficient to offset the losses the farm sectors suffered through indirect and direct taxation, the compensatory programmes were usually heavily concentrated towards large farmers. The concentration is partly explained by the fact that virtually all these supports are proportional to business volumes. The lower administrative costs of dealing with large farmers rather than small ones in credit or extension provides additional explanation. But extra concentration towards the large farmers is explained by their better ability to manipulate the administrative systems which provide the support. In extreme cases, privileged farmer groups were able to monopolize either production or marketing of certain commodities.

Many of these compensatory policies drive land prices above the capitalized value of farm profits because they provide the better-off with additional income from farming, such as tax shelters or access to subsidized credit. Such capitalization of non-farm benefits of land ownership makes it unfeasible for the poor to buy land, even if they were offered credit at market rates similar to those available to large enterprises. The policies thus contribute to the concentration of land ownership far beyond what would be warranted on account of economies of scale.

Misguided land policies

Policies for the allocation of land rights and for land titling have often led to the elimination of the rights of previous occupants of land and their eviction, explaining much of the concentration of ownership in Southern Africa and Latin America. Subdivision acts and other restrictions to ownership transfer have been used in Brazil, the US and Southern Africa to prevent the purchase of land by politically emancipated black populations.

The inability of the poor to ascend the tenancy ladder through the purchase of land has been aggravated in many countries by the destruction of the lower rungs of the tenancy ladder through misguided land and labour policies. The most common pattern is the removal of the option of long-term tenancy of small farms from large owners. In some cases, tenancy or 'indirect exploitation' is prohibited outright (South Africa or Colombia). In other cases, such as India or Brazil, ceilings on land rental or crop shares and/or the acquisition of permanent rights of long-term tenants make tenancy unattractive to owners under all but the most informal and short-term arrangements.

In these cases, landowners have reacted either by mechanizing their farms – usually with the help of credit subsidies, or by converting their crop farms into extensive livestock ranches, freeing them of tenants or large long-term hired work forces. The effect has been the conversion of farm tenants/workers into rural or semi-urban proletariats who derive a precarious existence as seasonal harvest workers. The lowest rung of the tenancy ladder, namely, the ability to acquire farming skills by participation in the entire production cycle as a worker, has therefore also been eliminated.

Elimination of the access to land through rental or sales markets leaves the poor with few land access options. They can intensify farming on good land in their holdings through labour-intensive land investments, giving rise to remarkably productive and sustainable farming systems all over the world. They can farm marginal lands in their holdings or on mountain sides, usually with disastrous consequences since such land rarely warrants sufficient investment in erosion protection. They can migrate to frontiers, but in many countries such as Colombia or Brazil, most fertile frontier land is already allocated to large-scale farmers by the time it receives an infrastructure. Much of the resulting smallholder migration may therefore be to marginal frontier areas with adverse environmental consequences. These are some of the linkages between failed policies, poverty and environmental destruction.

Protectionism in the developed world

Farmers in the developed world receive enormous protection against international competition. Protection is most extreme in countries with the least comparative advantage in agriculture such as Norway, Switzerland and Japan. The European Community's Common Agricultural Policy is a sophisticated machinery to protect their producers. Countries with strong comparative advantage such as New Zealand and Australia provide relatively little protection to their own producers. The US provides heavy protection to sectors which face stiff international competition from the tropics or sub-tropics such as sugar, tobacco, peanuts and citrus. Many surplus producers use the international market for costly disposal of commodities, such as European sugar. Increasingly they restrict the production of these commodities to reduce fiscal costs. Others, like the US, subsidize export credit and shipping costs. Only a small fraction of the surpluses is provided in the form of concessionary food aid.

The price depressing effects of these policies have been evaluated in a number of major studies which have concluded that, across agricultural commodities, they depress world market prices by between 9 and 16 per cent. Negative impacts are most pronounced for dairy products, followed by beef, wheat, rice and coarse cereals – in that order. A tabulation done by Gardner (*1989*) summarizes these studies in the footnote below.[2]

Many developing countries which are net importers of food benefit in the aggregate from these policies so that, as a whole, the gains for the developing world outweigh the losses. Those who lose are farmer groups who are net sellers of food and rural areas in the aggregate. Losses transmit themselves to rural workers through direct and indirect employment effects. Undoubtedly these developed country policies have contributed to rural poverty in the developing world.

Under the recently concluded Uruguay Round Agreement, agriculture for the first time has come under GATT rules. The agreement includes:

- the conversion of all existing non-tariff barriers into bound tariffs, and agreements to reduce them over time;
- guaranteeing minimum access quotas to guard against the impact of initially high tariff rates;
- restrictions on expenditures on subsidizing exports and on quantities of exports subsidized;
- limits on expenditures on domestic support policies which are heavily trade-distorting (i.e. those outside the 'green box').

2 Gardner, 1989, p. 367, provides the following summary table:

Simulated results of OECD trade liberalization: percentage change in selected world prices

	Tyers and Anderson	IIASA	OECD	USDA-ERS
Wheat	10	18	–1	29
Rice	11	21	1	32
Coarse grains	3	11	–3	23
Beef	27	17	15	17
Dairy products	61	31	44	53
Mean of Agricultural Commodities	16	9	–	10

Note:
Results for 10 per cent *ad valorem* liberalization multiplied by 10. No mean price effect reported by OECD.

These are welcome steps which, in the long run, will reduce the negative impact of the policies of developed countries on international prices and trading opportunities. The International Agricultural Trade Research Consortium (IATRC) has evaluated these policies and concludes:

> The extent to which the Agreement will lead to greater market access, curb export subsidies and modify domestic policies in the next few years can only be determined from a detailed inspection of the Schedule of Commitments made by the individual countries. Paradoxically, the immediate impact on national policies is likely in most cases to be small. Many countries have been engaged in a process of reducing government support to agriculture, and making such support more closely targeted to needs, in advance of the outcome of the Round. Policy reforms in the EU, Canada, Sweden, Australia and New Zealand, along with much of Latin America, have been strongly influenced by the negotiations in the Uruguay Round. The Agreement thus takes on the task of supporting and locking-in such reforms, and encouraging them in other countries. (*IATRC, 1994:iii.*)[3]

3 The Consortium makes the following points:

- The most far-reaching element is a change in the rules regarding market access. With very few exceptions, all participating countries have agreed to convert all existing non-tariff barriers (along with unbound tariffs) into bound tariffs, as well as tariffs already bound earlier, according to Schedules included as a part of the Agreement.
- 'Tariffication' will impose changes in import policies for a number of countries. Canada will replace import quotas for dairy and poultry products with tariffs, initially at a high level. The European Union will replace its variable levy with tariffs, though a maximum duty-paid price for cereals has been negotiated which puts a limit on the tariff charged ... The US will forgo the use of Section 22 import quotas and the negotiation of voluntary export restraint agreements with beef suppliers ... Japan and Korea have been allowed to delay tariffication in the case of rice for the next few years.
- It provides in cases of tariffication for 'minimum access opportunities' to guard against the impact of high initial tariff rates ...
- ... Under the Agreement, countries accept commitments on reducing expenditure on export subsidies as well as on the quantity of subsidized exports. This will limit export subsidies by the EU and other countries for such products as wheat, dairy products and beef, and should lead to firmer world market prices in these commodities.
- It also sets rules and commitments for domestic support policies. It defines a set of policies which are deemed to be less trade-distorting than others, and allocates them

Five patterns of development

Countries differ greatly in the extent to which they have implemented the misguided policies discussed above and to which they have been affected by developed country's protectionist policies. Depending on these factors, five broad paths of agricultural and rural development across which most countries cluster can be distinguished.

A few countries which either did not use or abandoned the implied policy mixes early, such as Taiwan, Indonesia, Malaysia, Thailand and China, have had spectacular successes in agricultural growth and rural poverty reduction. Others, such as India or Kenya, have abandoned parts (but not all) of the implied policy prescriptions, and at least have provided substantial support to their smallholder sectors. These countries have had acceptable rural growth rates and some rural poverty reduction. Others, for example, Brazil, Colombia and South Africa, have heavily subsidized their commercial farm sectors and experienced rapid agricultural 'modernization'. In the process, they have prematurely evicted workers from their agricultural sectors, only to find them piling up in rural and urban slums. In most of Eastern Europe and Central Asia, a disastrous pursuit of large-scale and collective farming has led to an inefficient and fiscally unsustainable farm sector which has contributed to the downfall of the communist regimes. They are a long way from realizing the growth and employment potential of an efficient private farm sector based on family farms, competitive private marketing and other agricultural services. The countries where discredited policy recommendations have not been abandoned until very recently, for example, many African countries and Argentina, have experienced spectacular failures of rapid growth in agriculture with catastrophic results for the rural poor.

Successful rural development and poverty reduction

Countries in this cluster include South Korea and Taiwan which, after early land reform efforts, consistently implemented policies and

to a 'green box' which is broadly immune to challenge. Other policies not sheltered in this way are subject to reduction through a limit on the total support given by domestic subsidies and administered prices.

- Along with the provisions on domestic and trade policies in the Agreement, participants also concluded an Agreement on Sanitary and Phytosanitary Measures (the SPS Agreement). The goal was to make it easier to distinguish between genuine health and safety issues and disguised protection. (*International Agricultural Trade Consortium, 1994:iii–iv.*)

programmes to support smallholder agriculture. More recently, they have been joined by Indonesia and Malaysia and, in 1978, by China when it abandoned excessive agricultural taxation and farm collectives and, since then, broke all world records in rural growth and poverty reduction.

All these countries either did not have large-scale ownership holdings or reformed these into owner-operated family farms, with the exception of the plantation crops discussed above. Even in these crops, they pursued smallholder strategies based on contract farming and the smallholder share of these crops has steadily increased. They invested heavily in agricultural infrastructure or, in the case of China, inherited the investment from the period of collectivization. They also invested heavily in agricultural technology for smallholders and largely refrained from heavily subsidizing credit. They have either not taxed or only lightly taxed their agricultural sectors through indirect or direct taxation.

These countries have experienced rapid agricultural growth, modernization of technology and sharp reductions in rural poverty.

Agricultural stagnation

At the opposite extreme are countries such as Argentina, Zambia, Tanzania and Guinea which have heavily discriminated against their rural sectors, primarily through overvalued exchange rates and industrial protection. They have either not provided much government support to their sectors, as in the case of Argentina, or provided it via parastatal investments which were often unsuccessful and sometimes became fiscal drains and instruments of further taxation of agriculture. Public investment for agriculture was inadequate for both their commercial farm sectors and their smallholder sectors. The agricultural and rural sectors of these countries have not grown fast or have stagnated, leading to increasing poverty, especially in rural areas.

Modest taxation and heavy public investment

A third group of countries – India, Mexico, Kenya and the Philippines – has discriminated against agriculture through indirect and direct taxation, but has attempted to compensate for the discrimination with heavy investment in road and irrigation infrastructure, research and extension, and subsidies to credit, electricity, irrigation water and fertilizers. They have directed substantial support to the larger commercial farmers and to their smallholder sectors, and have implemented extensive but not radical land reform efforts.

These countries have experienced adequate agricultural growth based largely on Green Revolution Technologies, and their food supplies have grown faster than population growth rates. They have also seen some reduction in rural poverty. However, the quality of their public expenditures has steadily eroded as a consequence of rent-seeking farmer groups and bureaucrats, shifting increasingly to untargeted credit and input subsidies at the expense of investment. The quality of government expenditures has also suffered from a shift to wage and salary costs in the large parastatal sectors serving agriculture. In some countries growth rates have faltered as a consequence.

All these countries have started to reduce indirect and direct agricultural taxation, and to reform their public expenditure policies for agriculture and rural development, with Mexico being the most advanced in its reform effort. Their growth and poverty reduction rates have not yet started to accelerate.

State farms and collectives with state monopoly of services

This pattern of development was based on a belief in economies of scale in farming and the intention to avoid private sector involvement in marketing and services. The sectors built on this model have been technologically backward and inefficient, with excessive dependence on fossil fuels and imported feedgrain for overly large livestock sectors. They have contributed to huge unsustainable government deficits which have accelerated the downfall of the regimes.

The abandonment of this model has led to spectacular agricultural and rural growth in China, Vietnam and Albania. But, in most of Eastern Europe and Central Asia, the transformation to private sector farming and competitive private sector services and marketing has only just begun. It is by no means clear that family-based farming will be favoured, or that the monopolies in marketing, input supply and credit will give way to more competitive structures within which an efficient family-based farm sector can flourish.

Premature expulsion of labour

A group of countries including Brazil, Colombia, Guatemala and South Africa, inherited unequal land ownership distributions at the end of World War II, and did little to change them through land reform. These countries imposed relatively modest taxation on their agricultural sectors and concentrated the bulk of their public sector support on their large-scale farming sectors. This support took the form of subsidized farm credit, infrastructure investment without cost recovery, and assistance in marketing through parastatals or statutory monopoly rights. Most Central American countries pursued similar policies until the early 1980s when they led to civil war.

These countries successfully abolished 'feudal' land-labour relations in their large-scale farming sectors and modernized them into large-scale commercial farms, although at a heavy fiscal cost. They have fostered a dynamic, technologically sophisticated and politically articulated class of commercial farm owners. At the same time, they have seen most of their smallholder sectors decay and sink deeper into poverty, except where special rural development efforts were made. Rural labour forces have been largely excluded from participation in the modernization process and have converted to slum dwellers in rural towns and the cities.

The fiscal burden of supporting the technologically sophisticated but economically inefficient large-scale farming sectors has proved to be unsustainable in all of these countries. Credit and other subsidies have been substantially reduced, tax privileges are threatened, and protection at the border or through monopolies is being eliminated. Farm land prices have declined. A new effort at macroeconomic adjustment in Brazil is likely to lead to a wave of farm bankruptcies. In Colombia, bankruptcies could be stemmed only through blanket debt relief and backsliding on agricultural policy reforms which simply prolonged the lives of these subsidy-dependent farms.

The promised elimination of rural poverty through migration to the cities has not materialized. Today these countries have two to three times as many poor rural people as in 1950. Much of the massive population transfer to the cities has been into the informal sector and into slum areas. Most of these countries are characterized by high rates of rural violence and urban crime. And many of these countries have undergone decades of land-related uprisings and land invasions.

Many smallholders who cannot access land through purchase, rental or land reform turn to environmentally destructive cultivation of marginal land or to petty production and trading activities; some turn to crime.

Summary

Development professionals have examined the development theories, measured the impacts of various development policies, and painstakingly sifted through the experiences of many countries in an effort to fathom what makes the difference between success and failure. While there are many areas where knowledge is still limited, a fairly clear picture has emerged. This picture is constantly being blurred and obfuscated in heated debates, usually with the purpose of defending the vested interests of specific farmer or producer groups, or of bureaucrats and politicians, confusing the novice and well-meaning observers and policy makers. The broad professional consensus includes the following key elements:

- The foreign exchange, trade and taxation regime should not discriminate against agriculture, but should tax it lightly, preferably using the same progressivity and instruments as for the urban economy.
- An open economy and an employment intensive, small farmer oriented strategy is economically efficient and most likely to reduce rural and urban poverty.
- Providing privileges or reducing competition in output, input and credit markets is costly to consumers and taxpayers and ends up hurting small farmers and the rural poor, even if such an effect is unintended.
- The strategy of unequal land redistribution requires a substantial prior or concurrent effort at land reform. Constraining land rental and insisting on expropriation without compensation has a perverse impact on the rural poor. Centralized ministries or parastatal bureaucracies are not good at implementing land reforms.

These lessons have been acquired at an enormous cost for poor and nonpoor rural populations of the developing world. They have been synthesized with painstaking effort by our profession and others. It is politically difficult to change the misguided policies. Yet the lessons can only be ignored at a high social cost – for economic growth, for the fiscal balance, for agricultural resources, for the rural and urban poor, and for social peace.

References

Binswanger, Hans. 1989. The Policy Response of Agriculture. *Supplement to the World Policy Economic Review and the World Bank Research Observer*, Proceedings of the World Bank Conference on Development Economics.

Binswanger, Hans, and Deininger, Klaus. 1994. World Bank Land Policy: Evolution and Current Challenges. *Paper prepared for the World Bank Agricultural Symposium, January 1994.* Washington, D.C.

Binswanger, H., Deininger, K. and Feder, G. 1993. Power distortions, revolt and reform in agricultural land relations. Prepared for T. N. Srinivasan and J. Behrman (Eds), *Handbook of Development Economics, Vol. III.* Amsterdam: North-Holland. Forthcoming.

Binswanger, Hans and von Braun, Joachim. 1991. Technological Change and Commercialization in Agriculture. The Effect on the Poor. *The World Bank Research Observer,* 6(1):57–80.

Binswanger, H., Young, M., Bowos, A. and Mundlak, Y. 1987. On the determinants of cross-country aggregate agricultural supply. *Journal of Econometrics,* 36:111–131.

Blakemore, H. and Smith, C.T. 1975. *Latin America: Geographical Perspectives.* London: Methuen.

Gardner, Bruce. 1989. Recent Studies of Agricultural Trade Liberalization. In A. Maunder and A. Valdes (Eds), *Agriculture and Governments in an Interdependent World.* Proceedings of the Twentieth International Conference of Agriculture Economists held at Buenos Aires, Argentina. Dartmouth: University of Oxford.

Haggblade, S. and Hazell, P.B.R. 1989. Agricultural technology and farm–non-farm growth linkages. *Agricultural Economics,* 3:345–364. Amsterdam: Elsevier Science Publishers.

International Agricultural Trade Research Consortium. 1994. The Uruguay Round Agreement on Agriculture: An Evaluation. Commissioned paper No. 9.

Krueger, Anne O. 1992. A synthesis of the political economy in developing countries. *The Political Economy of Agricultural Pricing Policy,* Vol. 5. Baltimore: The Johns Hopkins Press.

Lipton, Michael. 1977. *Why Poor People Stay Poor.* London: Temple Smith.

Lipton, Michael and Ravaillion, Martin. 1994. Poverty and Policy. In T. N. Srinivasan and J. Behrman (Eds), *Handbook of Development Economics, Vol. III.* Amsterdam: North-Holland. Forthcoming.

McGuirk, Anya M. and Mundlak, Y. 1992. The transition of Punjab agriculture: a choice of techniques approach. *American Journal of Agricultural Economics,* 74: 133–43.

Mundlak, Y., Cavalho, D. and Domenech, R. 1988. Agriculture and Economic Growth, Argentina 1913–84. Washington, D.C.: International Food Policy Research Institute; Cordoba, Spain: IEERAL. Processed.

Organization for Economic Co-operation and Development (OECD). 1987. *National Policies and Agricultural Trade*. Paris: OECD.

Parikh, K.S., Fischer, G., Frohberg, K. and Gulbrandsen, O. (IIASA.) 1986. Towards Free Trade in Agriculture. Laxenburg: International Institute For Applied Systems Analysis.

Ravallion, Martin and Datt, Guarav. 1994. How Important to India's Poor is the Urban-Rural Composition of Growth? Washington, D.C.: World Bank Policy Research Department.

Roningen, V., and Dixit, P. 1988. Economic Implications of Agricultural

Policy Reform in Industrial Market Economics. *Paper presented at the International Agricultural Trade Research Consortium Symposium.* Annapolis, MD: 19-20 August.

Schiff, M. and Valdes, A. 1992. A Synthesis of the Economics in Developing Countries. *The Political Economy of Agricultural Pricing Policy,* Vol. 4. Baltimore: The Johns Hopkins Press.

Schultz, T.W. 1964. *Transforming Traditional Agriculture.* New Haven, Conn: Yale University Press.

Tyers, Rod and Anderson, Kym. 1988. Liberalizing OECD Agricultural Policies in the Uruguay Round: Effects on Trade and Welfare. *Journal of Agricultural Economics* 30:197–216.

3

Evolution of the agrarian structure in South Africa

Masiphula Mbongwa, Rogier van den Brink and Johan van Zyl[1]

Introduction

The dual structure of agriculture in South Africa and the comparatively low productivity of small African farms observed today is not the result of genuine economies of scale in the large farm sector, but of decades of government policy that was guided – in large measure – by the general political and economic philosophy of white domination known as apartheid. The results of these policies include distortions in land and labour markets, output and input markets, infrastructure, agricultural credit and services, and the creation of large-scale white farms. This outcome is not unique to South Africa, but is common to colonial and post-colonial societies in most of Africa and Latin America.

This chapter has two themes. The first is that African family farming was viable and successful in responding to the increased demand for agricultural products emanating from the mining centres in the latter half of the 19th century. During that period, African owner-operated or tenant farming proved to be as efficient as large-scale settler farming based on hired labour. African farmers adopted new agricultural technologies, entered new industries and outcompeted large-scale settler farming in some of the emerging agricultural markets. Moreover, the ineffectiveness of colonial government to intervene in agricultural markets on behalf of settlers, combined with the reluctance of manorial estate[2] owners – who relied on African tenant farmers for the operation of their farms – to support anti-African agricultural policies, resulted in

1 This chapter is a modified version of Chapter 3 of 'South African Agriculture: Structure, Performance and Options for the Future', Discussion Paper 6, Southern Africa Department, The World Bank, Washington, D.C. 1994. It builds on preparatory work done by Michael Roth, Helena Dolny and Keith Wiebe (*1992*).

2 A manorial estate is an area of land allocated temporarily or as a permanent ownership land holding to a manorial lord who has the right to tribute, taxes or rent in cash, kind or in corvée labour of the peasants residing on the estate. Manorial estates can be

a situation in which African farmers were able to accumulate agricultural capital, wealth and farming skills.

The second theme is that the formation of a stronger, richer and unified settler state in 1910 – the Union of South Africa – ushered in a policy environment which suppressed and isolated African farmers from mainstream agriculture in order to facilitate their transformation into rural and urban labourers. The leverage of manorial estate owners – who previously frustrated farming policies aimed at constraining African tenancy – was greatly eroded by the mining industry and junker estate[3] farmers. The process by which the transformation of African farmers to labourers was effected involved the progressive closure of African access to most markets (land sale and rental, agricultural capital, inputs and outputs), the exception being a racially-segmented labour market. Currently, African agriculture is associated with the economy of the former homelands, where it represents only a minor part of total income and in general fails to provide even the basic subsistence needs of the population.

Farming in 19th century South Africa

The agrarian economy of South Africa in the mid-19th century consisted of large-scale white farms with hired labour, manorial settler estates with indigenous tenant farmers, and free indigenous farming on black-owned land. There were two main geographic subsectors: *coastal* and *interior* farming. Coastal farming – made up of all three types of farms – produced horticulture, livestock and crops, and exported, wool, wine and fruit to Europe. Interior farming, mainly by indigenous farmers, raised livestock and crops for home consumption and marketed the surplus. The land tenure arrangements of these farms varied from communal land ownership to private or state land ownership (both with numerous forms of tenancy, most often quitrent). Trade within and between the coastal and interior areas was dominated by livestock, hides and ivory in exchange for guns, ammunition, textiles and transport equipment.

organized as haciendas or as landlord estates. A hacienda is a manorial estate in which part of the land is cultivated as the home farm of the owner and part as the family farms of serfs, usufructuary-right holders, or tenants in exchange of rent in kind, labour or cash. Landlord estates are cultivated entirely by tenants (*Binswanger et al, 1993:1*). South Africa's manorial estates were associated with settler landlords and land companies.

3 A junker estate is a large land holding which produces a diversified set of commodities operated under a single management with hired labour. Labourers receive a plot of land to use for a house and for garden farming (*Binswanger et al, 1993:1*).

The discovery of diamonds in the 1870s and gold in the 1880s in the interior changed the interior farming system and the South African economy and state completely. Large and rapidly growing urban and industrial population centres sprang up around mining areas, creating substantial markets for agricultural products. The subsequent supply response from African farmers confirmed for South Africa what has been observed elsewhere in the world with respect to the viability and efficiency of family farming over large-scale farming based on hired labour at low levels of technology.[4] Numerous studies confirm the efficiency of African family farming during the 19th century in response to the growing demand for agricultural products from the mining areas.[5]

With simple technology and plentiful land, labour was the most critical factor in the success of farming. However, the relative inefficiency of large settler farming implied low profitability, and the resulting difficulty of offering sufficient wages to attract indigenous labour away from their own farms. This led to labour shortages for the settler farmers in some regions of the country.

Between 1850 and 1870, African farmers supplied the major towns of the English colony of Natal with grain and exported the surplus to the Cape. In 1860, over 83 per cent of the nearly half million hectares of white-owned land was farmed by African tenants. Their accumulation of capital and wealth caused the Native Affairs Commission (1852–3) to comment that Africans were becoming wealthy, independent and difficult to govern.

Many settler farmers agreed with this view because they were unable to compete with African farmers who produced higher grain yields per hectare and cultivated more land than they did. The settlers argued that labour shortages kept them from competing effectively with the African farmers who had technology, equipment, farming skill and little or no need to work on settler farms. For instance, in the Orange Free State, competition from African farmers in Basotholand was seen as a menace. The burghers complained about the 'lower costs of the black farmers, who allegedly did not have to maintain the "civilized standard of living" that the whites had to maintain, and could therefore undercut

4 Family farming is based on family labour and includes sharecropping, cash rent and labour tenant farming, and independent farming. Large farming, on the other hand, is based on externally acquired labour which includes slave, indentured or wage labour.

5 Wilson, 1971; Trapido, 1975; Morris, 1976; Bundy, 1979; Beinart, 1982; Lacey, 1982; Keegan, 1986. This section relies heavily on Bundy (1979 and 1986).

the Free State farmers. There was much talk of "unfair competition" from "primitive peoples" with few consumption requirements or responsibilities' (*Keegan, 1986:210*).

African farming was widespread both within and outside settler manorial estates in the Transvaal and the Orange Free State. In 1870, the Transvaal set aside 240 000 hectares of treaty areas plus an additional 496 000 hectares as African locations. However, the big landlords and land companies continued to enter into various tenancy arrangements with African farmers and, by the end of the century, half of the Africans lived on privately-owned settler lands despite a wide range of anti-squatter laws.

Short of labour, the large settler farmers persuaded the colonial government to intervene on their behalf by: (a) limiting African competition in the market place; and, (b) setting up native reserves of tiny pieces of land to create an artificial land shortage in order to force African farmers to seek work on manorial farms. To this end, various measures were used including livestock, hut and poll taxes; road rents; location, vagrancy and pass laws; and confinement to the reserves. These measures, while successful in constraining African owner-operated farming, did little to reduce competition among manorial estates for African tenant farmers. Obviously, the estates which offered African farmers better tenancy conditions received more tenants. In response to the settlers' request to reduce such competition, the state then intervened in the land rental market and sought to reduce the number of rent-paying African tenant farmers on manorial estates.

Competition from black transporters of agricultural products was also deemed unfair by white transporters. After levying a road tax in 1896, the Volksraad of the Orange Free State used the threat of rinderpest to further proscribe Africans travelling with livestock (other than a white employer's) in any part of the state (*Keegan, 1986:211*). Hence, African opportunities for trade were also curtailed.[6]

When their problems seemed insoluble, some settler farmers in

6 This was aggravated by trading licensing policies which attempted to curtail itinerant trade in general in favour of non-African merchants with a fixed abode. The curious absence of rotating village markets in much of the previously settler-ruled economies of Eastern and Southern Africa seems to find its root in these restrictive business practices. Currently, the simple absence of a time and place where buyers and sellers can meet in rural areas is an almost too obvious constraint on rural development. However, markets are social institutions which need to be purposefully created and maintained – their emergence is not automatic or natural.

Queenstown (Cape Colony) proposed that African farming be abolished altogether. But a group of white merchants, whose well-being depended on marketing African produce, campaigned successfully to have this proposal rejected. They argued that settler communities elsewhere in the country which had adopted such measures ended up in a depressed economic condition because Africans were better farmers than the white settlers and represented a substantial share of purchasing power in the local economy. They warned that the forcible removal of African farmers from the land, and their replacement with a small number of settlers, would result in economic disaster for business and the white urban community in general.

Influenced by European settlers, the government restricted black African land rights very early and created reserves that were too small to support independent African agriculture. The Glen Grey Act of 1894 was a further measure introduced in this regard and restricted farm ownership in the reserves to one parcel of no more than slightly above three hectares. It also levied a labour tax on all men living in the reserves who did not own land; and banned the sale, rental or subdivision of land by introducing a perverted form of communal tenure. Many colonial powers have interfered with the flexibility and adaptability of communal tenure systems and with the democratic institutions that prevailed in some of them, by rigidly codifying the systems and by awarding hereditary rights to chiefs to manage land relations in the community and to allocate land to community members (*Noronha, 1985*). Chiefs were chosen based on their willingness to co-operate with the colonial powers. While chiefs often did not interfere with local adaptation of the systems and allowed practices inconsistent with the law, the laws codifying traditional tenure often restricted the rights of community members to rent out or sell land, even to other community members. The capacity to enforce such restrictions was particularly strong in South Africa, and such constraints, together with migration and the continuing lack of profitability of agricultural investment, may have contributed to the underutilization of land in the homelands.

The restrictions on black agriculture led to tenant farming (with tenants called squatters) being the main mode of production accessible to black Africans. In 1882, 55 per cent of the native population in Natal lived as tenants, 35 per cent on privately owned land and 20 per cent on Crown land (*Bundy, 1986*). Absentee landlords and large companies normally imposed cash rents, but the system of 'farming on the halves' in which the African sharecropper, owning oxen and plough, ceded half of the output to the landlord for the right to cultivate the land, was common with resident landlords.

Tenancy was even more pronounced in 1904 in the Transvaal where, of 900 000 black Africans, 14 per cent farmed their own land, 20 per cent lived on Crown land, and 49 per cent lived on European-owned land, leaving 11 per cent in the reserves and less than 6 per cent in wage employment (*Bundy, 1986*). Given these large numbers, early attempts to use legal action to reduce the number of 'idle squatters' – in 1866, 1871, 1873, 1881 and 1883 – were not enforceable. A strong interest by the mining sector in reducing wage levels, and the desire to obviate food imports, led to new laws in the 1890s (*Bundy, 1986*). By 1892, all tenants residing on white farms had to be registered. In 1895, the maximum allowable number of tenants per farm was restricted to five to spread the available supply of labour more evenly, and native commissions were set up to enforce this law. In 1899 a license fee for each tenant, payable by the landlord, but usually passed on to the tenant, was introduced.

The situation at the turn of the century for the African farmer

After the discovery of diamonds and gold, British imperial interests had compelling economic reasons to bring the Transvaal and the Orange Free State under control in a land federation. The British attempts to take control of the Boer states ultimately resulted in the Boer War beginning in 1899; it was a bitter and divisive conflict from which Britain emerged the victor in 1902.

During this period many Africans, especially in the Transvaal and the Orange Free State, bought land as individuals and in groups as land syndicates. Missionaries were often used as 'fronts' because buying land was difficult for Africans. Land buying was given a boost during the brief post-war period of British rule (1902–1910). Africans were allowed to convert their labour tenancies to sharecropping or fixed-rent (cash) tenancies, or to purchase land outright. No exact information is available regarding the amount of land bought during this period,[7] but some settlers speculated that Africans would succeed in buying back all that they had lost during the colonial wars (*Plaatje, 1987*).

By the 1890s, European farmers in South Africa were suffering from serious competition from African farmers who could supply food items

7 One estimate is that, by 1900, around 250 000 acres had been bought by black farmers in the Transvaal alone.

at lower prices. The productivity advantages of the African family farms over the large European farms are suggested by a wide range of anecdotal evidence from that period, including petitions by European farmers to restrict African competition to ensure their own survival. These advantages are also suggested by the fact that the smaller African farmers not only made a profit and provided revenue for local traders, but also won prizes in agricultural shows (*Binswanger & Deininger, 1993; Bundy, 1979*).

African farmers achieved their success not in an ideal competitive market or a supportive policy environment, but in a hostile society determined to undermine them. A series of levies and fees imposed on Africans between 1903 and 1905 forced them to pay higher income taxes than whites. Restrictive taxes and laws already passed in the 19th century were strengthened by the 1908 Natives Tax Act which imposed twice as much tax on sharecroppers and other fixed rent-paying squatters as on labour tenants in an effort to reduce the attractiveness of the former types of tenancy which were more profitable to the black tenants than the labour tenancy contracts. Measures to remove squatters by force were undertaken in 1909 and 1910. Tenants were also increasingly forced to sell off or remove their allegedly excess cattle, i.e. all cattle other than the cattle that the tenants used to plough the fields. Measures were taken to restrict Africans from the profitable rural transportation sector.

The state, however, did more than hinder black farming; white farmers received substantial support from the government in the form of subsidies, grants and other aid for fencing, dams, houses, veterinary and horticultural advice, as well as subsidized rail rates, special credit facilities and tax relief during the period 1890–1908. In 1908, the Transvaal Director of Agriculture declared that during the past 20 years more money had been spent per head on South Africa's white farming population than in any other country in the world (*de Kiewiet, 1942:260*).

Despite all of these efforts to restrict African farming either to the reserves or manorial estates and to subsidize white farmers, the black farmers continued to maintain a competitive edge in the market place. The chronic labour shortages on white farms were intensified by the emergence of the mining and manufacturing industries with their massive labour demands.

Agrarian development: 1910–1947

The Union of South Africa was formed as a dominion of the British Empire in 1910. At this time there were nearly six million people living in South Africa. More than two thirds of these were Africans, more than a quarter whites, about a tenth 'coloureds', and a very small number Asians. The South African Party, representing white landlord and mining interests, headed the first government. One of the goals of the new government was to ensure adequate supplies of labour to the mines. To achieve this objective, numerous restrictions were imposed on African farmers, thereby forcing them to serve as a much needed source of labour in other sectors, but chiefly in mining. Many of the manorial estate owners who had been supportive of African farmers in the past and had often prevented hostile government policies, were co-opted and became part of the new social order.

The Glen Grey Act of 1894 was followed by the Natives Land Act, passed in 1913 and confirmed in 1936, which limited the area where black Africans could establish new farming operations to the reserves (totalling 7,8 per cent of the country's area). The Natives Land Act also confirmed the Glen Grey Act provisions concerning communal tenure, i.e. maximum holding sizes and restrictions on land transactions. It also barred blacks from buying land from whites, and prohibited them from sharecropping and cash rentals. The main intention of the law, which was 'almost exclusively the basis of the country's future policy of apartheid' (*Wilson, 1971*), was to transform tenants into wage workers for the mines. The law was called 'a law made for the mining houses' (*Davenport, 1987*). The law was also intended to 'curb black farming practices at a time when white farming was beginning to pick up ... to check black sharecropping ... and to prevent the purchase of land by syndicates of blacks who ... were beginning to move ahead fast' (*Davenport, 1987*). The immediate effect of the law was to force African families, who were formerly independent farmers on sharecropped land, to accept wage labour and give up their equipment. The longer-term effect was to end African farming above the subsistence level and to degrade the reserves to 'dormitories' (*Hendricks, 1990*) for a cheap African labour force.

This effect can be illustrated by comparing black African production and consumption data for 1875 and 1925 in Victoria East, a typical Ciskei village in the reserves. Between 1875 and 1925, the population dependent on agriculture in this village more than doubled, but sales per family in 1925 were only a quarter of what they had been in 1875. Integration into the market for non-agricultural goods also decreased substantially and, by 1925, when agricultural income no longer covered food costs, most cash was spent on food and groceries, and earnings from wage labour had to be used to cover the deficit. By 1925, even an exceptionally good harvest could provide only about half of the population's nutritional requirements; the village, in fact, had become a structural food deficit area with funds for investment, as well as human capital, in serious shortage. In the reserves as a whole, the situation was the same. As early as 1918, agricultural production in the reserves amounted to only 45 per cent of the population's food subsistence requirements. Rapid migration enabled people in the reserves to hold the percentage of subsistence requirements met out of their own agricultural production almost constant until 1955; but then, when the state's relocation policy became fully effective, population density increased considerably, and the food requirements that could be met from reserve production declined to about 20 per cent (*Simkins, 1981*).

Table 3.1: African Population on Various Classes of Land, 1916

					EUROPEAN FARMS		TOTAL POPULATION	
	Reserve lands	Mission lands	Native-owned lands	Crown lands	Occupied by Africans	Occupied by Europeans	Rural	Urban
Cape	1149,4	24,3	39,3	12,5	7.6	240,4	1473,5	128,0
Natal	479,8	44,6	39,3	37,1	86,6	357,9	1044,1	37,9
Transvaal	283,2	24,0	40,4	71,5	232,1	408,6	1059,9	322,5
OFS	17,2	1,8	4,7	–	–	279,4	303,0	48,8
Union	1929,6	94,6	123,7	121,1	325,2	1286,3	3880,5	537,2

Source: Beaumont Commission Report, 1916.

The Land Acts

'Awakening on Friday morning, June 20, 1913, the South African native found himself not actually a slave, but a pariah in the land of his birth.'[8] On that date, the Natives Land Act No. 27 of 1913 drew a firm line between white and black landholding, prohibiting each from 'entering into any agreement or transaction for the purchase, hire or other acquisition … of any such land (in the area allotted to the other)

8 Sol Plaatje (1987), First Secretary General of the African National Congress.

or of any right thereto, interest therein, or servitude thereover'. The Land Act of 1913 segregated Africans and Europeans on a territorial basis by establishing codified native reserves, referred to as 'scheduled areas'. Independent black agriculture and cattle-raising could now only be undertaken in the native reserves. About 7,8 per cent of the country's farm land was in the 'schedule' for the reserves. This became the only area where African subsistence farming could legally be conducted. The Act was specifically designed to end both landlord and hacienda manor estate farming which relied on various forms of tenancy, and to establish in its place large land-holding companies (junker estates) which would operate under a single management with hired labour. Labourers would receive a plot of land for use only for house and garden farming.

The Act defined natives as members of an aboriginal race or tribe of Africa and prohibited them from renting, buying or otherwise acquiring land outside the reserves without the approval of the Governor-General. It also excluded freehold, mortgaging and land sale rights within the reserves and stipulated further that no person other than a native could acquire land rights within these reserves, except sites for trade or business, without state approval.

In 1916, after the Land Act was put into effect, about 50 per cent of rural Africans lived on the reserves, 2 per cent on mission lands, 3 per cent on native owned lands, 3 per cent on state lands, 8 per cent on landlord estates, and 33 per cent on haciendas. Of the total population, 12 per cent lived in urban areas (Table 3.1). Afrikaners were the white majority in rural areas, while the English were dominant in the towns and cities. Table 3.2 shows the ownership and occupation of land in 1916.

The Beaumont Commission was appointed under provisions of the Land Act to organize the reserves. The Commission reported that the scheduled land was only sufficient for approximately half of the native population and recommended that further land be released to the reserves to ensure territorial segregation, specifying the areas which should be added.

As Table 3.2 indicates, the reserves were limited to 7,8 per cent of the total land area. Outside the reserves, natives owned only 0,7 per cent of the land and lived on state and European-owned lands (another 3,6 per cent); thus the total land for native use was 12,1 per cent. Yet the white population strongly opposed the Beaumont Commission's recommendation for an increase of the reserves. Consequently, the situation remained unchanged until 1936 when the Native Trust and Land Act No.18 was passed, establishing the Native Land Trust. The Trust released the recommended 6 209 858 hectares (quota land) to the original 1913 scheduled reserve areas (non-quota land), increasing the size of the reserves to 13,7 per cent of the country (see Table 3.3 for the breakdown of this land).

Table 3.2: Land Areas by Land Tenure Systems in the Union of South Africa (hectares)

Native reserves	9 538 300
Mission reserves	460 000
Native-owned lands	856 100
Crown lands occupied	805 100
EOL: Occupied by Europeans	90 314 000
EOL: Occupied by Africans	3 550 900
Vacant Crown land, reserves and other	17 002 400
Total:	122 526 800

Note: EOL: European-owned land occupied either by (a) Europeans; or (b) Africans.
Source: (a) Report of the Beaumont Commission, 1916, pp. 3 and 4.
(b) 1990 South African Statistics, p. 26.2; and DBSA, South Africa: An Inter-Regional Profile, p. 34. Excludes Walvis Bay.

Table 3.3: Land Quotas Delimited by the 1936 Act

	Area (hectares)	Per cent
Transvaal	4 306 643	69,4
Cape Province	1 384 156	22,2
Natal	450 536	7,3
Orange Free State	68 523	1,1
Total:	6 209 858	100,1

Source: Bantu Trust and Land Act No. 18 of 1936.

Other measures undertaken to support the Land Acts included the Native Administration Act of 1927 which introduced yet another encroachment by the state on the indigenous tenure system. The

Governor-General (the State President today) was imposed on Africans as their 'traditional' chief with decisive powers on African land matters. African chiefs were reduced to salaried officers of the government.

The need for a cheap supply of labour also led to several new government acts. The Masters and Servants Acts of 1911 and 1932 tightened the grip of junker estate managers on farm workers. They prohibited breaking contracts, changing employers, or assigning family members to other employers. The Native Regulation Act of 1911 was modified and used to register with a labour bureau all African male and female workers over the age of 16 and seeking work with a labour bureau. Once registered as farm workers, they could not switch to industrial employment (*United Nations, 1976*). The Prison Act allowed estate farmers to serve as wardens on African prison farms and to have the benefit of cheap prison labour.

African farmers were not allowed to join state-sponsored marketing co-operatives or farmers' unions. Without membership in these organizations, it was difficult to secure credit, market output, or to obtain research and extension services. Moreover, outside these agricultural institutions, little independent activity was permitted by law.

The combination of all these measures began to erode the development of African farming and, by the 1920s, increasing population pressure caused African households in the reserves to spend 60 per cent of their income on food. Infant mortality increased and signs of malnutrition began to surface. However, landlord estates relying on African tenants managed to survive the Land Acts for some time. In 1925 they still accounted for 4,5 million acres with fixed-rent or sharecropping tenants. But, as these changed into junker estates, African tenant farming declined. Tenant shares became smaller and smaller, and labour tenancy contracts became longer and stricter (*Morris, 1976*). By gradually destroying tribal institutions and closing many income-earning opportunities – the exception being labour markets, the capital, wealth, farming skills and information base that African farmers had accumulated over generations began to wither away.

The planned transformation of the agrarian structure from haciendas to junker estates did not resolve the problem of low profitability or labour shortages for large-scale farming. African owner-operated farming was effectively eliminated, but landless farmers often chose to work in the mines, in new manufacturing industries or in towns, rather than on junker estates which, almost from the beginning, were notorious for their low wages and bad working conditions.

When the 1932 worldwide recession hit on top of these repressive labour measures, the flow of workers from the farms to urban areas became a deluge. Between 1936 and 1951, the largest single source of newly urbanized African people was the white rural sector. Five times as many Africans came from white farms as from the reserves. Ever-increasing amounts and kinds of state support were required to keep junker estate farming alive.

1910–1947: Protection of white agriculture

The white farming subsector experienced a completely different treatment from that of its African counterpart. The support system for white farmers, already long in place, was strengthened substantially to enable the effective occupation and use of the land set aside for white ownership only. The Land and Agricultural Bank of South Africa (the Land Bank) was established in 1912 to give loans to farmers who did not have access to commercial banks. The Co-operative Societies' Act of 1922 began the co-operative movement, and the Agricultural Marketing Act of 1937 created marketing schemes and control boards to administer it. Research and extension services as well as the necessary infrastructure were provided to ensure the success of white farming. Between 1910 and 1935, there were 87 Acts passed in the Union Parliament rendering permanent assistance to farmers, financed directly and indirectly by the revenues from the mining sector. Between 1910 and 1936, the State spent £112 000 000 from revenue and loan funds on agriculture. Between 1932 and 1936 alone, the operations of the Land Bank and the direct expenditure by the government on behalf of the white farmers amounted to £20 428 092. On export subsidies alone, a total of nearly £11 000 000 was spent between 1931 and 1937 (*de Kiewiet, 1942: 253–261*).

White land settlement

Whereas the Natives Land Act of 1913 restricted African access to land, the Land Settlement Act of 1912 standardized the acquisition, exchange and disposal of state lands for white settlement. The terms and conditions of the Act were updated in the Land Settlement Amendment Act

9 These amendments, among others, include the Land Settlement Acts Further Amendment Act No. 28 of 1920 and No. 21 of 1922; the Land Settlement Laws Further Amendment Act No. 26 of 1925; The Land Settlement (Amendment) Act No. 6 of 1928, No. 1 of 1931, No. 57 of 1934, No. 47 of 1935 and No. 45 of 1937; the Land Settlement Relief Act No. 25 of 1931; and the Land Settlement Act No. 21 of 1956.

of 1917 and in many subsequent amendments. These Acts established the procedures by which white settlers could apply for state and privately-owned lands.[9] The Minister of Lands, on the recommendation of the Land Board (a counterpart of the Land Trust), was empowered by parliament to allot state lands and to use public funds appropriated to buy privately-owned land for subdivision into suitable agricultural holdings for white farmers.

The size of the farms acquired ranged from large cattle ranches to smallholdings for market gardens. The Land Board prepared a report on the holdings; if the results were favourable, the land was surveyed, divided into farms, appraised and advertised. Listings of farms would be published from time to time in the Government Gazette and local newspapers, with particulars of the land offered, the governing Land Acts, the terms and conditions of the lease or purchase, and when and how to apply. Applicants could obtain farms by leasing them for five years (renewable) with the option to buy, or by transfer of land purchased especially for them.

- **Leased farms:** The applicant had to be of good character and at least 18-years-old, possess qualifications sufficient for utilizing the land, and intend to occupy, develop and work the holding. The land was initially leased for five years. No rents were charged in the first year; in the next two years, 2 per cent of the value of the land was charged and 3,5 per cent in the fourth and fifth years. The Minister could extend the lease for a further five-year period at a rental of not less than 4 per cent of the land value. The option to purchase the holding, by paying the purchase price over a period of 20 years, could be exercised at any time with the Minister's consent, subject to fulfilment of the conditions of the lease. Instalment payments amounted to about 7,33 per cent of the purchase price per year, which included interest at 4 per cent with the residual applied to principal (*1916 Year Book*).

  By 1941, the conditions for leasing had become more generous. No rent was payable the first two years; the other charges remained almost identical. The option to purchase was extended to 65 years although the purchaser had to pay instalments of at least £25 per year. Reduced rent and interest applied in certain semi-arid areas.
- **Purchased:** The government set a maximum price for the agricultural holdings of £1 500 which included survey fees, transfer costs, and two years of accrued interest. The government loan was to be no more than 80 per cent. The buyer was required to pay 20 per cent

down to purchase the land, with no further payment for the next two years; the repayment schedule for the next 18 years was in six-month instalments of 7,85 per cent of the remaining balance. Four per cent of this was interest, the rest principal (*1916 Year Book*).

By 1941, the government would loan up to £2 250 of the price, and the repayment period was increased to 65 years. By 1956, this amount was increased to £5 150 and the purchaser's down payment was reduced to 10 per cent.

Both mechanisms for obtaining land required personal occupation, mandated care of improvements and trees, and specified additional land improvements amounting to 10 per cent of the value of the holding within the first five years; they also prohibited sub-letting, mortgaging or transfer without permission. Leases were subject to cancellation in situations of non-compliance with provisions in the Acts. The 1956 Resettlement Act further strengthened residency requirements in the holdings, but permitted the Minister to grant exemptions for a temporary period on the recommendation of the Land Board.

Although the Land Settlement Act of 1912 was the standard bearer, the allotment of state land to whites for agricultural or pastoral purposes was also implemented under many earlier Acts. Table 3.4 provides an

Table 3.4: Allotment of Agricultural Holdings during 1916

	No. of holdings	Area (hectares)	Amount (£)	Rent (£)
Land Settlement Act, 1912	210	168 636	110 053	–
Crown Land Disposal Ordinance (Transvaal)	134	90 557	58 215	–
Crown Land Disposal Ordinance 1903 (Transvaal)	26	21 414	10 654	–
Act 15 of 1887 (Cape): Sales	13	4 356	993	–
Act 26 of 1891 (Cape): Leases	25	19 291	–	523
Act 26 of 1891 (Cape): Sales	1	7 621	395	–
Natal Proclamation	35	28 711	13 026	53
Irrigation Settlement Act 31 of 1909	22	120	3 353	–
Act 13 of 1908 (OFS): Leases	7	2 085	–	145
Total Land Alienated	473	322 791	196 689	721

Source: South Africa Official Year Book 1916.

overview of the various leases and purchases granted under these Acts in 1916 (*1916 Year Book*). Between 1910 and 1936, about 700 farmers per year were settled, supported by massive state subsidies.

In addition to help in acquiring land, loans were made to help white farmers obtain stock, seeds, equipment, and other items needed to develop their farms. Permanent improvements such as drilling operations could also be funded, and the cost added to the rent or purchase price of the land. These advances could not exceed £250 (£500 by 1941). Short- and medium-term loans were also available to lessees through the Land Bank for fencing, construction of dipping tanks, and other improvements. One result of this period of strong government support was the growth of the number of white farms from 81 432 in 1921 to a peak of 119 556 in 1952.

1948: Apartheid and the rural African household

In 1946, the African population living in the reserves had dropped from 50 per cent in 1916 to 40 per cent. The Social and Economic Planning Council reported that the quality of the reserve lands was deteriorating, and the 1948 Fagan Commission found that many Africans were becoming permanent inhabitants of the towns.

Campaigning on a platform of more rigid racial separation for blacks, and power and wealth for whites, especially Afrikaners, the National Party came to power in the 1948 election. The previous government, although adhering to segregation, had tried to address African land grievances in a conciliatory spirit; the new government abandoned all such pretences. The new 'apartheid' went further than previous segregation policies by also segregating African ethnic groups from one another and forcing them to live in separate tribal areas.

The objectives of the new policy were to: (a) facilitate white political control by dividing African natives along tribal and ethnic lines; (b) protect the white working class from African competition through various discriminatory laws; (c) further racial segregation so as to preserve the cultural identity of whites; and, (d) reduce the cost of development of industrial regions (*Cobbett, 1987*). Unstated objectives were to retain the black agrarian structure established by the Land Acts, to continue white land settlement, and to advance the white agrarian economy by establishing large commercial farms. These farms would differ from junker estates in that the main supply of labour would be migrant workers (i.e. African farmers from the homelands). These objectives were largely achieved. Under apartheid, the old agrarian economy dominated by junker estates was transformed into a new agrarian economy based on large commercial farms.[10]

Constitutional framework

The Native Authorities Act of 1951 and the Promotion of Bantu Self-Government Act No. 46 of 1959 artificially created eight national units out of the Pedi, Sotho, Tswana, Swazi, Tsonga, Venda, Xhosa and Zulu ethnic communities. The boundaries for these 'national units', not surprisingly, coincided with the reserve boundaries as defined by the Land Acts. The Transkei became the first self-governing homeland in 1963. Nine other homelands followed.

In 1970, the Bantu Homelands Citizenship Act was passed making every African in the Republic a citizen of some homeland. The Bantu Affairs Administration Act of 1971 transferred control over Africans, regardless of where they lived, from white local authorities to Bantu Affairs Administration Boards. The Bantu Laws (Amendment) Act of 1972 justified forced resettlements of African people and stated that 'a Bantu tribe, community, or individual could be removed from where they lived without any recourse to Parliament, even if there was some objection to the removal'. These Acts gave the government the right to banish to a homeland any Africans considered 'redundant' in urban or white areas, and then abrogated their constitutional right to belong to the Republic of South Africa.

The Commission for the Socio-Economic Development of the Bantu Areas within the Union of South Africa of 1954, known as the

Table 3.5: Target African areas (ha) designated by the Commission

Scheduled black areas (1913)	9 190 101
Released black areas (1936)	5 815 460
Balance of quota land outstanding (1955)	1 630 692
Land owned by blacks situated outside scheduled and released black areas (i.e. 'black spots') (1955)	161 593
Total:	16 797 846

10 A large ownership holding which produces several different commodities, operating under a single management with a high degree of mechanization, using a few long-term hired workers who may reside on the farm and seasonally hired workers who do not reside permanently on the farm (*Binswanger et al, 1993:1*).

Tomlinson Commission, placed the ultimate size of the homelands at about 17 million hectares (Table 3.5). This included: (a) the original scheduled land from the 1913 Land Act; (b) the land recommended to be added to the homelands at that time by the Beaumont Commission, but actually not released until 1936 (quota land); (c) other land acquired by the Trust and land owned by blacks prior to 1936; and, (d) 'black spots' or isolated small areas of land occupied by Africans but located outside the homelands, as well as land planned for the homelands.

The Land Acts had long attempted to consolidate the 'black spots' by eliminating some that were 'poorly situated', i.e. in white areas, and combining others to make them into homelands. The elimination of the spots, however, meant that the land area had to be subtracted from the amount available for the particular homeland.

The Tomlinson Commission of 1954 was concerned that the quality of the land in the reserves could not support the 500 000 African families living in these areas. The Commission proposed a drastic cut in the number of families in the homelands and a series of Betterment or Closer Settlement Schemes to stop soil degradation through land use planning, relocation of people and livestock, stock-culling, fencing, contour ploughing, water conservation and erosion control. It also urged the Trust to hasten the process of buying the outstanding quota land. By March 1967, however, the Trust had only managed to buy five million hectares of quota land; 1,3 million hectares were still outstanding. In 1968, the administration of the Native Land Trust was handed over to the Department of Bantu Affairs. This removed the acquisition of land and its allocation from the agenda of the 1936 Land Act and placed it within the agendas of the homeland and industrial development policies. This essentially implied the resettlement of African people in homelands, and a further allocation of land only if the homelands opted for independence. Table 3.6 describes in detail the land situation of the homelands in 1976.

Although the stated basis for the exchange of holdings was the quality of land, i.e. African farmers should be resettled on land equal to their former holding; in practice this was not the case. Much of the land released for the homelands, often pieces of non-contiguous scrubland, certainly did not meet any land quality standard.

The relocation of Africans to the homelands from white rural and urban areas, from African owned areas, and from one place to another within the homelands, was never voluntary. The Surplus People Project has estimated that 3,5 million people (predominantly Africans) were forcibly resettled between 1960 and 1980 for various reasons: eviction of

	Quota Land[1]				Non-quota Land[2]				
	(a) Trust vested	(b) Trust acquired	(c) Black acquired	Total	(d) Black owned prior to 31/8/36	(e) Black acquired after 31/8/36	(f) Trust vested	(g) Trust acquired	Total[3]
Transvaal:									
Lebowa	278 503	1 081 361	230 922	1 590 786	222 325	455	414 767	39 824	677 371
South Ndebele	560	69 143	4 902	74 606	530	–	–	–	530
Gazankulu	387 711	215 933	7 907	611 551	18 350	–	44 483	–	63 013
Swazi	151 589	114 001	4 028	269 558	5 167	–	27 336	2 088	34 591
Bophuthatswana	49 185	657 304	129 955	836 445	506 489	118 619	139 800	15 177	680 082
Venda	347 489	101 071	5 862	454 422	17 271	–	177 807	700	195 778
KwaZulu	3 255	51 060	–	54 315	–	–	13 843	–	13 843
Total:	1 218 232	2 289 873	383 576	3 891 683	770 312	19 074	818 036	57 789	1 665 211
Cape Province:									
Ciskei	53 966	115 462	8 867	172 295	32 282	1 661	729 662	11 969	775 575
Transkei	5 239	143 766	9 031	158 036	84 471	3 551	3 482 708	24 893	3 595 623
Bophuthatswana	74 833	495 523	11 986	582 343	25 552	4 433	1 386 481	194 911	1 611 377

Total:	134 038	754 751	23 884	912 674	142 035	9 645	5 598 851	231 7735	982 575
Natal:									
KwaZulu	48 022	355 372	17 847	421 240	138 945	1 854	2 604 016	7 209	2 752 023
Orange Free State:									
Bophuthatswana	–	53 972	1 993	55 965	26 903	723	20 725	59 878	
QwaQwa	–	5 417	–	5 417	–	–	42 827	42 827	
Total:	–	59 389	1 993	61 382	6 903	723	63 552	102 705	
Grand Total:	1 400 292	3 459 385	427 300	5 286 979	1 078 465	31 296	9 054 455	308 298	10 502 514

Source: Department of Bantu Administration and Development; Unpublished data.

Notes: 1. **Quota Land**

(a) *Trust vested land:* Government land in the release areas (Act No. 18 of 1936) which was vested in the Trust in terms of Section 6(1)(b).

(b) *Trust acquired land:* Land acquired by the Trust since 1936 outside the scheduled black areas (Act No. 27 of 1913).

(c) *Black acquired land:* Land acquired by blacks in 1936 outside the scheduled black areas, which is situated in released areas and adjoining them or adjoining scheduled black areas.

2. **Non-quota Land**

(d) *Black owned land prior to 31/8/36:* Land owned by blacks prior to the Bantu Trust and Land Act, No. 18 of 1936.

(e) *Black acquired land after 31/8/36:* Land acquired by blacks in scheduled black areas.

(f) *Trust vested land:* Government land in scheduled black areas which was vested in the Trust.

(g) *Trust acquired land:* acquired by the Trust in scheduled black areas.

3. Due to the rounding of two figures to the nearest hectare, the sum of the figures may differ slightly from the figures actually given in the 'total' column.

Table 3.7: Non-quota Land

	Land owned by blacks prior to 31/8/36 (ha)	Non-quota land acquired by blacks after 31/8/36 (ha)	State land vested in the Trust (ha)	Non-quota land acquired by the Trust (ha)	Total (ha)
Transvaal	748 754	19 074	818 042	74 338	1 660 208
Cape Province	141 148	9 824	5 514 404	264 389	5 929 765
Natal	135 757	1 877	2 598 657	13 468	2 749 759
Orange Free State	26 727	723	63 552	11 703	102 705
Total:	1 052 386	31 498	8 994 655	363 898	10 442 437

Source: Department of Development Aid, *1990 Annual Report*, p. 15.

black tenants and 'redundant' workers from white farms (32,1 per cent); intra-city removals due to the Group Areas Act[11] (23,7 per cent); homeland consolidation and clearing of black spots (19,1 per cent); urban relocation from white areas to homelands (19 per cent); removal of informal and unauthorised urban settlements (3,2 per cent); relocation due to development schemes (0,7 per cent); politically motivated removals (1,4 per cent); and other factors (0,9 per cent).

The situation from 1960 to democratization in 1994

South African whites voted narrowly in October 1960 to adopt a new constitution and make the country a republic, which was established on 31 May 1961. The Republic subsequently withdrew from the British Commonwealth. Racial strife increased, leading to international financial sanctions and to a reassessment of apartheid by moderate political leaders. Part of that reassessment was to decide what constituted a viable and sustainable basis on which to build a new agrarian structure.

The land owned by the Trust prior to the 1936 Land Act, and the amount of land owned and acquired by blacks and the SADT Fund from 1936 to 1990, are set out in Tables 3.7 and 3.8. A total of 10 442 437 hectares of non-quota land was made available under the 1913 Act. The

11 The Group Areas Act of 1959 designated all South African land for the exclusive use of one racial group or another.

Table 3.8: Quota Land from 31 December 1990

	State land vested in the Trust (ha)	Land acquired by the Trust (ha)	Land acquired by blacks (ha)	Total (ha)
Transvaal	1 249 050	3 213 274	414 134	4 876 458
Cape Province	195 252	1 323 260	28 869	1 547 381
Natal	66 617	492 135	20 587	579 339
Orange Free State	253	197 742	1 993	199 988
Total:	1 511 172	5 226 411	465 583	7 203 166

Source: Department of Development Aid, *1990 Annual Report*, p, 15.

area of quota land acquired by the Trust by the end of 1990 was 7 203 166 hectares: 72,5 per cent came from land acquired by the Trust; 21,0 per cent from state land, and 6,5 per cent from land purchased by blacks.

The present agrarian structure has been built up systematically since the turn of the century. By the end of the 1980s, the African family farming sector had all but been eliminated, and African peasants had been transformed into wage workers on large farms, in mines and in secondary industries. Nearly 90 per cent of the agricultural land was in white areas, supporting a total rural population of 5,3 million people, more than 90 per cent of whom were Africans. The remaining agricultural land was in the homelands and supported over 13 million people. Originally the homelands were justified as areas where Africans would do subsistence farming; today up to 80 per cent of household incomes in the former homelands come from migrant earnings and pensions.

In 1916, every African except those on State (Crown) and European owned and occupied lands, owned at least four hectares. By 1990 the individual holdings had dropped by 75 per cent, to one hectare per person. This happened mainly because of the phenomenal population growth in the reserves, whose area has remained relatively static over the period. Today, most former homelands are peripheral, overcrowded, poverty-stricken and lack a proper infrastructure, despite some family farmer support development programmes. Genuine developmental strategies were frustrated by the overwhelming problems of these areas. Although the proportion of state spending for agriculture in the homelands increased in the 1980s, only a small amount represented actual transfers to farmers. State spending for services for Africans was also lower in the homelands than in the white rural areas.

- **Land tenure in the homelands.** Communal tenure in the homelands is officially defined by the Proclamation R188 of 1969 as 'unsurveyed land' or 'permission to occupy'. Under this Proclamation, a male person holds rights to various land allotments for residential use, arable farming and grazing. Land access is usually by virtue of membership to a community, not through sale, lease or rent. Only men are entitled to inherit land rights. People do not legally own their residential and arable allotments. Rather, they are allowed the right of occupation and cultivation, subject to conditions stipulated by the homeland authorities (*de Wet, 1987*).

  According to Murray (*1989*), approximately 15 per cent of the land in the homelands is held on freehold or conditional (quitrent) title. But African freeholds rarely belong to a single entrepreneur who farms using either family, hired labourers or tenants. Most freehold purchases were undertaken by extended families or syndicates. These farms are heavily populated by descendants of the original purchasers and extended clientele. Today they are more peri-urban in nature than rural farmlands. Many freeholders have become landlords to residential tenants. Freeholds once represented a 'progressive ideal' of an area where Africans could lead a lifestyle without government interference. There has been a steady erosion of those ideals. Class differentiation and conflicts have emerged. In some freehold areas of Natal, landlords collect their rents under armed guard (*Cross, 1990:22*). In other areas, patron–client relationships are exploited to further political disputes. Violence in Natal's tenancy areas is exacerbated by 'politicized local patron–client followings controlled by warlords' (*ibid*). Pockets of black freehold areas have also been involved in the state's relocation policy. Property has been expropriated and the occupants resettled.
- **Betterment planning.** Bembridge (*1986*) estimated that about 70 per cent of South African black rural areas are officially under 'betterment planning'. Betterment planning is South Africa's attempt at villagization, i.e. planned village land use. Before 'betterment', people lived in clusters of homesteads, along hills or ridges, with their fields near rivers and streams. They grazed their cattle on the hills and in the forests, or further from home. With 'betterment', they changed to new fields and to new residential areas. The new land use system was

inflexible; people found themselves with smaller fields and gardens than before, and had to walk greater distances to fetch fuel, water and thatching grass (*de Wet, 1987*). This was accompanied by very unpopular stock-culling measures triggering peasant resistance to 'betterment' in the 1940s and 1950s throughout the homelands (*Cross, 1990*).

- **Migrant labour.** More than a century of contract-wage labour in the mining and commercial farming sectors, and more recently in the manufacturing and service sectors, has perpetuated migrant labour in South Africa. Until the mid-1980s, workers who left the rural areas to seek better opportunities in towns were prevented by law from taking their families with them. The removal of these influx control measures has meant, theoretically, that black workers have been permitted to sell their labour and settle where they please within the confines of the group areas. However, cyclical rural–urban migration has become a way of life in South Africa, where the majority of rural households are better viewed as members of dislocated urban communities (*Nattrass & May, 1986*).

Data provided by Halbach (*1988*) demonstrates the pervasiveness of migrant labour in the South African economy. In 1982, almost 60 per cent of the black workers in South Africa were non-resident, or working only temporarily in urban South Africa. Lacking important manufacturing, mining and agricultural industries, the homelands primarily became transfer and service economies – labour reserves

Table 3.9: Black Out-migration from the White Rural Areas, 1980–1985

	Net out-migration (persons)	Percentage drop in population 1980–85 (%)
Orange Free State	220 000	2,4
Eastern Cape	110 000	4,1
Natal	530 000	8,9
Eastern Transvaal	400 000	5,4
Northern Transvaal	210 000	7,5
Others	140 000	–
Total:	1 610 000	

Source: Simkins, C. (*1990*) derived from an analysis of South African statistics, 1986.

for the rest of the economy. Social discrepancies exist between households with and without migrant earnings, with and without land, and with and without pensions, which has led to a steady worsening of income distribution within the homelands (*Simkins, 1984*).

Demographic studies indicate that the rural areas outside the homelands (i.e. the white commercial farming areas) were the main source of out-migration in the 1980s.[12] Net out-migration from white rural areas from 1980 to 1985 was 1,6 million (Table 3.9), while net in-migration into the metropolitan areas for the same time period was only around 900 000. Metropolitan in-migration includes that from homelands and white commercial farming areas. Thus, the population of the homelands increased as a result of natural population growth as well as in-migration from white commercial farming areas. A significant portion of the out-migration from white rural areas resulted from apartheid policies. For example, 180 000 people scattered throughout the Orange Free State and classified as 'unemployed' were sent to Botshabelo – an artificially-created settlement with no economic base, now housing three quarters of a million people.

Conclusion

Two important lessons emerge from this chapter regarding the evolution of the contemporary agrarian structure in South Africa. The first is that some African farmers, either on their own land or as tenants on manorial European settler estates, were able to respond successfully to the needs of new and rapidly growing markets which arose around industrial mining and urban population centres in the latter half of the 19th century. Not only did commercial African farmers participate in certain product markets, some also effectively competed against large settler farming based on hired labour. This rise of African commercial farming took place under conditions of relative land abundance, weak and ineffective government interventions, and relatively undistorted markets. Under these conditions, some African farmers accumulated a significant stock of capital, wealth, farming skills and experience.

The second lesson is that the development and prospects of African farming and rural development were crippled by a long list of government policies including, but not restricted to, creating an artificial land

12 Simkins (*1990*) estimated the black rural population outside the homelands to be 3,8 million in 1983, of which 3,4 million resided on white farms.

shortage for African farmers; prescribing their participation in the sales and rental markets; confining their access to land to the reserves; excluding them from credit markets; blocking them from output markets; denying them access to marketing co-operatives and farmers' unions; refusing them extension services and access to public sector investments; subjecting them to rigid and authoritarian state and state-made communal land tenure systems; forcing them onto the labour market through extortionary taxes and levies; and forcibly resettling millions to densely populated reserves (homelands). The decline of African farming implied a gradual loss of agricultural capital, wealth, farming skills and experience. Farming ceased to be a window of entrepreneurial opportunity and managerial and technical development for Africans.

South Africa illustrates the devastating effects of continued and forceful depression of the profitability of small-scale agriculture. Numerous measures made the accumulation of physical or human capital in black smallholder agriculture unprofitable, and led to the virtual disappearance of independent black farmers soon after the Natives Land Act in 1912 which, as explained above, limited black farmers' access to land outside the reserves and defined the communal land right system in the peasant areas in excessively restrictive and counter-productive ways. Having lost their right to purchase land outside the reserves, black farmers were removed from their farms ('black spots') in the more fertile and accessible white farming areas. In subsequent periods, intervention shifted to the labour and product markets which supported European agriculture through artificially depressed wages of black workers, direct transfers, credit, marketing monopolies and output subsidies. Such policies remained in place long after they had been discredited elsewhere, with support to peasant farming in the homelands being confined to a rigid and paternalistic model of the small full-time farmer. These policies were complemented by stringent pass laws and by the Land Subdivision Act of 1970 (*Budlender & Latsky, 1991*). In contrast to other countries, the enforcement capacity of the South African state has been unrivalled.

References

Beaumont, W.H., Burger, S.W., Collins, W.R., Stanford, W.E. and Wessels, C.H. 1916. Report of the Natives Land Commission, Vol. 1. Cape Town: Cape Times Ltd, Government Printers.

Beinart, W. 1982. The Political Economy of Pondoland, 1860–1930.

Cambridge: Cambridge University Press.

Bembridge, T.J. 1986. An overview of agricultural and rural development in less developed areas of Southern Africa. *Development Southern Africa,* 3(1) : 20–36.

Binswanger, Hans and Deininger, K. 1993. South African Land Policy: The Legacy of History and Current Options. *World Development,* 21(9) : 1451–1476.

Binswanger, H., Deininger, K. and Feder, G. 1993. Power, distortions, revolt and reform in agricultural land relations. *World Bank Discussion Paper.* Washington, D.C.: Agriculture and Natural Resources Department, World Bank.

Boserup, E. 1965. *The Conditions of Agricultural Growth.* Chicago: Aldine Publishing Company.

Budlender, G. and Latsky. 1991. Unravelling rights to land and to agriculture activity in rural race zones. In De Klerk, M. (Ed), *A Harvest of Discontent: The Land Question in South Africa.* Cape Town: IDASA.

Bundy, C. 1979. *The Rise and Fall of the South African Peasantry.* Berkeley and Los Angeles: University of California Press.

Bundy, C. 1986. *Remaking the Past, New Perspectives in South African History.* Cape Town.

Christodoulou, N.T. and Vink, N. 1990. The potential for black smallholder farmers' participation in the South African agriculture economy. Mimeo. Johannesburg: Development Bank of Southern Africa.

Cobbett, M. 1987. The land question in South Africa: a preliminary assessment. *Paper presented at a Workshop on the South African Agrarian Question.* Johannesburg: University of the Witwatersrand: May 24–26.

Cross, C. 1990. Land tenure in black rural areas: social and political underpinnings. *Paper presented at the IDASA Workshop on Rural Land*, Houwhoek.

Davenport, T.R.H. 1987. Can sacred cows be culled? A historical review of land policy in South Africa with some questions about the future. *Development Southern Africa,* 4(3) : 88–400.

De Kiewit, C.W. 1942. *A History of South Africa – Social and Economic.* London: Oxford University Press.

De Wet, C. 1987. Land tenure and rural development: some issues relating to the Transkei/Ciskei region. *Development Southern Africa,* 4(3) : 459–478.

Halbach, A.J. 1988. The South African homeland policy and its consequences: an evaluation of separate development. *Development Southern Africa,* 5(4) : 508–526.

Hendricks, F.T. 1990. *The Pillars of Apartheid: Land Tenu– Rural Planning and the Chieftaincy.* Uppsala, Sweden: Almqvist & Wiksell.

Houghton, D.H. 1973. *The South African Economy,* third edition. Oxford: Oxford University Press.

Keegan, T. 1986. *Rural Transformations in Industrialising South Africa: The Southern Highveld to 1914*. Johannesburg: Ravan Press.

Keegan, T. 1986. Trade, accumulation and impoverishment. *Journal of*

Southern African Studies, 12(2) : 196–216.

Lacey, M. 1982. *Working for Boroko: The Origins of a Coercive Labour System in Southern Africa.* Johannesburg: Ravan Press.

Lipton, M. 1985. *Capitalism and Apartheid: South Africa, 1910-84.* Totowa, New Jersey: Rowman and Allanheld.

Migot-Adholla, S., Hazell, P., Blaret, B. and Place, F. 1991. Indigenous land right systems in sub-Saharan Africa: a constraint on productivity? *World Bank Economic Review,* 5(1) : 155–175.

Morris, M.L. 1976. The development of capitalism in South African agriculture: class struggle in the countryside. *Economy and Society,* 5(3) : 292–343.

Murray, M. 1984. The origins of agrarian capitalism in South Africa: a critique of the social history perspective. *Journal of Southern African Studies,* 15(4).

Natrass, N. and May, J. 1986. Migration and dependency: sources and levels of income in KwaZulu. *Development Southern Africa,* 3(4) : 583–599.

Noronha, R. 1985. A review of the literature on land tenure systems in sub-Saharan Africa. *Report No. ARU 43*, Agriculture and Rural Development Department. Washington, D.C.: World Bank.

Plaatje, Sol. 1987. *Native Life in South Africa.* London: Longman.

Roth, Michael, Dolny, Helena and Wiebe, Keith. 1992. Employment, efficiency and land markets in South Africa's agricultural sector: opportunities for land policy reform. Madison: Land Tenure Center, May.

Simkins, C. 1981. Agricultural production in the African reserves of South Africa, 1918–1969. *Journal of Southern African Studies,* 7 : 256–283.

Simkins, C. 1984. What has been happening to income distribution and poverty in the homelands? *Development Southern Africa,* 1(2) : 142–147.

Simkins, C. 1990. Financing rural development. *Development Southern Africa,* 7: 591–601, (Special Issue), October.

Trapido, S. 1978. Landlord and tenant in a colonial economy: the Transvaal, 1880–1910. *Journal of Southern Africa Studies,* 5(1) : 26–58.

United Nations. 1976. *Land Tenure Conditions in South Africa.* Rome: Center against Apartheid, Department of Political and Security Council Affairs.

Wilson, F. 1971. Farming, 1866–1966. In Wilson, M. and Thompson, L. (Eds), *Oxford History of South Africa,* Vol. II : 104–171.

4

South African land policy: the legacy of history and current options[1]

Hans Binswanger and Klaus Deininger

Introduction

South Africa has repealed most components of legal structures supporting apartheid, its former official policy of racial segregation. Several of the racially motivated land laws, such as the Native Lands Act, were abolished in the early 1990s. One of the most important questions in the debate about a future land policy, is whether the current land distribution provides a good basis for future growth of output and employment in the agricultural sector and for a gradual elimination of the inequalities in access to land, or whether land reform and resettlement are necessary to achieve these goals.

These issues depend critically on the nature of economies of scale in agriculture, the first topic to be reviewed in this chapter. Historically, economies of scale were rare – limited to a narrow selection of plantation crops – and the literature contains no single example of economies of scale arising for farm sizes exceeding what one family with a medium tractor could comfortably manage. Based on this finding, we examine two central questions: first, how and why have dualistic farm size structures, in which a highly mechanized commercial farm sector is juxtaposed with a subsistence-oriented smallholder sector, emerged over history; and how have they been sustained in a number of developing countries, including South Africa? Second, if large farms are not efficient, why do owners continue to find it profitable to accumulate large landholdings? An answer to this second question is central to the issue of whether a more equitable distribution of land can be accomplished through market transactions alone, without some type of land reform. Theoretical considerations suggest that even if markets are not distorted,

1 Reprinted (slightly adapted) from *World Development,* Vol. 21 no. 9, H.P. Binswanger and K. Deininger, South African land policy: The legacy of history and current options, pp. 1451–1476, 1993, with kind permission from Elsevier Sciences Ltd, The Boulevard, Langford Lane, Kidlington OX5 1GB, UK.

mortgage-based acquisition of land by the poor is impossible, and therefore land reform is the most promising way of achieving a rapid transition to a more efficient and equitable distribution of landholding. This proposition leads us to examine the success and failure of a variety of land reform programmes, as well as the costs accruing to countries with dualistic farm size structures when they have failed to undertake land reform or have implemented it only after protracted conflict.

Following this general review based on worldwide evidence, we then examine the congruence with, or divergence from the international experience of the specific history of South African agriculture (discussed in Chapter 3). We also consider the prospects and institutional requirements for land reform. We conclude the chapter with a strong case in favour of a rapid and large-scale land reform programme, and sketch out the mechanisms by which such a task could be accomplished using judicial and market-assisted processes for the transfer and resettlement of land.

The international experience

Policies to establish and sustain large-scale agricultural production in the face of competition by smallholders, which resulted in the formation of a highly dualistic agricultural sector, have not been limited to South Africa. In this section, we discuss the theoretical issues involved, portray the historical development of landlord estates and haciendas, examine the potential and cost of reform, and illustrate the different paths of reform in each of these systems as well as the implications of not undertaking such reform.

Economies of scale in production and marketing

Economies of scale in agriculture arise from the indivisibilities of inputs such as tractors and management. The indivisibility of such inputs leads to an initial segment of declining average costs in the production function until the lumpy input is fully utilized. Draft animals, threshers, tractors and combines can be fully utilized (i.e. reach their lowest cost of operation) only on farms of a certain minimum size. Management skills are another example of an indivisible input. The better the manager, the larger the optimal farm size. Rental markets for machine services and management can, in principle, circumvent the lumpiness of some but not all of these inputs[2], for example, in Europe since the 19th century,

2 Binswanger and Rosenzweig (*1986*) explain that due to moral hazard problems, rental markets for draft animals are unlikely to be feasible. Furthermore, rental markets are more difficult to organize for time-bound operations.

threshers that were too large for individual farms have been rented out. The rental market for combines in the US is very efficient, involving large-scale movement of the machines from south to north. Similarly, rental markets for specialist services or publicly financed extension, possibly organized by farmers themselves, can greatly reduce the threshold for such economies of scale.

Economies of scale arising from indivisibility of inputs are offset by agency costs (*Jensen & Meckling, 1976*) which result from the need to manage wage labour and enforce effort on large-scale operations. The lack of incentive of wage workers to work hard, and the ensuing need to supervise labour or to offer incentive contracts, has received considerable attention in industrial organization literature (*Stiglitz, 1987*). It is recognized to have profound implications for the organization of production, in particular the optimal size of the firm (*Calvo & Wellisz, 1978*). The potential losses from imperfect information are particularly large in agricultural production due to spatial dispersion of the production process and the need to adjust constantly to micro-variations of the natural environment.[3] It has therefore long been recognized that supervisory capacity is an important determinant of the mode of operation of large tracts of land (*Eswaran & Kotwal, 1985*). The fact that family members are residual claimants to profits and thus have higher incentives to provide effort than does hired labour, as well as the fact that they share in the risk and can be employed without incurring hiring or search costs, is one of the main reasons for the superiority of family over large-scale wage operations.[4] This lack of economies of scale in agriculture, and the resulting competitive superiority of farms operated predominantly with family labour, is confirmed by a large body of literature[5] on the 'negative farm size–productivity relationship' which draws on experience from both developing and developed countries. Much of the

3 Exceptions are plantation crops like bananas and sugarcane where, due to perishability of the raw produce, economies of scale associated with the capital-intensive processing/marketing stage can be captured by integrating the operation of the plant with those of the farm, which is operated by wage labour. An alternative way to capture these economies of scale is contract farming, as practised by smallholders in many areas of the world, including the sugar industries of Natal. For a more detailed discussion of this topic, see Binswanger and Elgin (*1988*).

4 With simple unmechanized technology, large landholders' supervisory capacity would soon become binding and lead them to rent out their land to independent family farmers on a cash-rental or share-rental basis (*Binswanger & Rosenzweig, 1986*).

5 See Barraclough and Collarte (*1973*) for six Latin American countries; Berry and Cline (*1979*) for studies incorporating Brazil, Colombia, the Philippines, Malaysia,

literature on the diseconomies of scale comes from regions where agricultural mechanization is incomplete and where technical change has been slow. But even in the developed world, studies of economies of scale suggest that these arise mainly from the indivisibility of the management input, combined with the high opportunity cost of managerial and family labour in a high wage economy. It can thus be taken for granted that agriculture in low wage environments is characterized by constant or even slightly decreasing returns to scale.

In contrast to production itself, related activities such as input supply, marketing and credit provision might be characterized by the existence of significant economies of scale. These economies arise from the cost of acquiring information about market opportunities and the creditworthiness of a borrower, the cost of the mortgage instruments for credit, and so on. With constant transaction costs, it is more profitable for banks and marketing firms to transact in large quantities and large loans than in many small ones and, for this reason among others, a formal credit market for very small farmers often does not exist. It has long been recognized that these economies of scale can be partly circumvented by well-designed marketing and credit co-operatives[6] and by careful group lending approaches. Moreover, the costs of assembling products from many sellers, or of providing inputs to many small farms, can be reduced by eliminating barriers to entry for small informal traders and by contract farming.

The emergence of large farms under low population density

Under low population density, individual peasants or workers always have the opportunity of leaving large farms to establish their own family farms in the bush. With unmechanized technology, no economies of scale, and a non-negative supervision cost, they would be able to produce output at a lower cost than large farms based on wage labour. This is supported by the fact that, in the absence of outside intervention, operational holdings all over the world and under a wide variety of land ownership systems have been cultivated by family farmers (*Grigg,*

and India; Kutcher and Scandizzo (*1981*) for the north-east of Brazil; Cornia (*1985*) for 15 countries in Africa, Asia and Latin America; Sen (*1981*) for the Indian Punjab and West Bengal; Bhalla and Roy (*1988*) for the whole of India disaggregated into 78 agroclimatic zones. A more comprehensive review of the literature on this subject is provided by Binswanger *et al* (*1993*).

6 Indeed, the use of service co-operatives instead of production collectives in the Soviet Union was proposed by Chayanov (*1991*) as early as the 1920s.

1992). Large farms under these conditions would therefore not be profitable and their continued economic success would require either restrictions on competition by family farms or special support to large farms.

The literature on tenancy (*Otsuka & Hayami, 1988; Otsuka et al, 1992*) has long emphasized that for landlords to attract tenants or workers, they must offer a contract that gives potential tenants at least the same utility they would obtain as workers elsewhere in the economy. This measure, called the reservation utility level, is generally equivalent to what a family could earn from its own independent farm.[7] However, in collaboration with the state, landlords can devise ways to depress the profits or the reservation utility of independent farmers, or to limit the choices open to them, and thereby obtain tenants or workers at lower cost. In a large number of countries, large-scale farms were in fact established under conditions of low population density in precisely this way. Five mechanisms were used:

- **Reduction of availability of land for peasant cultivation** by granting or auctioning rights to 'unoccupied' lands to members of the ruling class, and confining free peasant cultivation to infertile or remote areas with poor or no infrastructure and market access. Farm profits and utility were reduced by the higher labour requirements for producing a unit of output on poor land, by increased transport and marketing margins, and by increased prices for consumer goods imported to the region.
- **Differential taxation** of the free peasant population through tribute requirements, or by hut, head or poll taxes that had to be paid in cash, kind or labour services; but waiving these taxes for workers or tenants in manorial estates.[8] Many of these tax systems survived until after the Second World War.[9]

7 The reservation wage which must be offered by the landlord corresponds to the expected utility available in family farming, including the risk attributes of the corresponding income stream.

8 In Alto, Peru, Indians could evade both tribute and the dreaded labour service in the silver mines (the mita) by moving to European haciendas as *yanaconas,* and giving up their tribal identity. The only other way for them to escape was to move to highly inaccessible areas and confine their lives to subsistence cultivation without the benefit of trade with the outside world.

9 Tribute systems were used widely in Western Europe during the feudal period, as well as in conjunction with forced supply systems in East India, Java, Sumatra and Guatemala. They survived until the middle and late 19th century in Eastern Europe and Japan, and in Kenya, Mozambique and other parts of Africa until after World War II.

- **Further restriction of market access** by co-operative or monopoly marketing schemes that bought only from the farms of the rulers. For example, rights to extraction of labour and tribute, together with a monopoly on inputs and outputs, were combined under the prazo system in Mozambique. Similarly, in Kenya, the production of coffee by native African smallholders was prohibited outright until the 1950s; and European monopolies on sales of tobacco in Zimbabwe and Malawi were directly transferred to the large farm sector after independence. In Malawi, they survived until the beginning of the 1990s.
- Once large farms had been established by the rulers, **coercive interventions in the labour market** were sometimes used together with the above distortions to help retain workers or tenants on manorial estates. Pass and vagrancy laws, debt peonage and agrestic slavery were common examples of such coercive interventions.
- If such coercive interventions were no longer legal, the profits of peasant farms were often reduced by **confining agricultural public goods and services** (roads, extension, credit) to the farms of the rulers,[10] or by subsidizing these farms directly.

Binswanger *et al* (*1993*) provide 23 cases illustrating the long tradition and widespread use of these distortions to establish large farming operations under conditions of low population density. The earliest recorded incidence is in the Arthasastra in the 4th century BC, but their use has been remarkably consistent over time and across continents. Pure tribute extraction was the method favoured as long as the ruling class did not engage in production. When the rulers wanted to establish economically viable production, they always intervened in more than one market, and often adopted a combination of restrictions on land use with tax requirements. Groups with widely different cultural, religious and

10 In Zimbabwe, African maize cultivation which was the main source of domestic food supply during the initial mining boom, was actively supported by the so-called 'Master Farmer Programme' to educate African farmers in the late 1920s, when Europeans found it more profitable to grow tobacco and cotton. Collapse of the markets for those crops, and decreased maize prices during the great depression, led to the adoption of monopoly marketing and the dual price system, and also to the abandonment of the Master Farmer Programme. The responsible officials had to declare publicly that they never intended to 'teach the Natives to grow maize in competition with European producers' (*Phimister, 1988*). Following the revived interest of Europeans in tobacco during World War II and their neglect of maize cultivation, the Master Farmer Programme was reinitiated, indicating a renewed dependence on African supplies of maize.

ethnic backgrounds (the Ottomans, the Haas and Fulani in Africa, the Fujiwara in Japan, and every European colonial power) all imposed these same systems on peasants or conquered people when faced with similar material conditions. The emergence of these distortions, therefore, appears to have been determined primarily by material conditions of production rather than by culture.

The potential and cost of land reform

After World War II, uprisings by peasants in the context of independence struggles, external influences and the low productivity of large-scale agriculture, led to the implementation of land reform measures in many countries with dualistic landholding structures. Redistributive land reform can increase equity and also efficiency, as argued above, if there are no economies of scale or an inverse relationship between farm size and productivity levels.

The land market cannot be expected to lead to an efficiency-enhancing redistribution of land because poor family farmers who do not have much equity cannot acquire land even if they have access to mortgage credit.[11] This is the case because the price of land includes a collateral or risk premium over and above the capitalized value of agricultural profits, on account of the preferential access that land ownership provides to credit markets through its collateral value. Once poor farmers are provided with credit to buy the land, therefore, they can no longer pay for the mortgage out of farm profits without recourse to wage income. Distortions such as differential subsidies for large farms, income tax exemptions for agriculture, and use of land as a speculative asset in the presence of macroeconomic instability, all aggravate the problem by creating additional income streams that only benefit large landowners. These income streams are also capitalized into the value of land, and drive the land price above the capitalized value of farm profits, i.e. the productive value of the land.[12] Two major consequences emerge: first, a free land market alone will not be able to transfer land

11 For a more detailed discussion of this issue, see Binswanger and Elgin (*1988*) or Binswanger and Rosenzweig (*1986*). This counter-intuitive result arises because covariance of agricultural incomes across farms makes the establishment of rural credit and insurance markets very difficult, and land ownership provides one way to self-insure or improve access to the imperfect credit market.

12 See Just and Miranowski (*1988*) for an empirical discussion of these factors in the US; and see Brandao and Rezende (*1992*) for Brazil. In South Africa, land values have persistently exceeded the productive value of the land.

to smaller and poorer farmers, unless they are provided with grant financing in addition to, or instead of, mortgage financing. Such a grant element is required to provide the equity that the poor do not have. Second, elimination of all differential subsidies is a precondition for the long-term success of any redistribution of land.

Beneficiaries of land reform who lack equity cannot sustain viable farms or an improved standard of living (compared to their pre-reform situation) while repaying a land mortgage that covers the full value of their land.[13] Beneficiaries can play a supplementary role, and the requirement to provide some equity in the form of a downpayment can be a useful screening device, but a grant element is needed to provide equity if a land reform programme is really to redistribute land to the poor. Land reform without an explicit grant element, therefore, will either come to a grinding halt because of the extraordinary financial requirements, or will require *de facto* expropriation from large landowners who are compensated with bonds – an approach that is politically feasible only under extraordinary circumstances. In the first case, government funds earmarked for land purchases will soon become exhausted, possibly because of default by beneficiaries on loan repayments, and the programme will stop. Many ambitious land reform plans have ended because of the fiscal requirements associated with full compensation, for example in Venezuela (where, despite oil revenues, the area redistributed to beneficiaries remained very small), the Philippines and Brazil. In the second case, the bonds used to compensate landowners might have built-in features that erode their real value over time. So, although landowners receive the nominal value of their bonds, time erodes the real market value of the bonds received, and the government offers no compensation for this loss. Most landowners naturally oppose such thinly disguised confiscation, and only in special circumstances (Korea, Japan, Taiwan, Cuba), or under violence (Vietnam), is such a step politically feasible. A third case that would avoid these predicaments requires another form of financing land purchases. Foreign grants, internal tax revenues, or a combination of grants and taxes have been applied successfully in Kenya in order to provide the grant element

13 Organizations such as the Penny Foundation in Guatemala and other organizations in Ecuador have been able to buy land from owners and distribute it to small farmers with little apparent help from the government (*Forster, 1992*). These cases usually involve some grant element or subsidy of the services and credit provided to the smallholders; or they can be based on the purchase of land below market price on account of the liabilities of the former owner or of the workers, which are forgiven as part of the deal.

required for the poor to purchase land, and to transfer a significant amount of land to the poor while compensating landowners at market values.

A precondition for sustainable land reform is the elimination of all implicit and explicit distortions favouring large farms.[14] If governments introduce land reform into a distorted environment that differentially favours large ownership holdings, the capitalization of such subsidies and tax breaks into higher land values for large farmers would lead the recipients of redistributed land – the small farmers – to sell out to larger farmers, thereby defeating the purpose of land reform. The crucial importance of the macroeconomic and political environment, the impact of land reform policies on non-reformed sectors, and the possibility of land reform resulting in higher output, are illustrated by the case of Chile,[15] where expropriation and redistribution of almost 20 per cent of the total agricultural area between 1964 and 1970 was accompanied by

14 In Brazil, disproportional support for large farms at the frontier led to the emergence of an agricultural structure dominated by large farms (*Binswanger & Elgin, 1988*). It is argued that 'what happens in the settled agricultural areas of the country is probably more important in curbing resource destruction than what is done physically in the nature preserve itself' (T*hiesenhusen, 1991*), and that removal of differential subsidies is the first step towards a settlement pattern that is more compatible with ecological principles. The significance of institutional factors for the negative ecological consequences of frontier settlement, rather than technological or economic constraints, is also emphasized by Southgate (*1990*).

15 It has been argued (*Jarvis, 1989*) that – at least in the short run – the macro-economic environment in Chile was much more important than any single steps undertaken through land reform. This assertion is supported by the rates of growth for agricultural output during the four main periods of concern:

(a) A high rate of growth of agricultural output (3,5 per cent) prevailed during the 1964–1970 period, despite disruptions caused by the expropriation and redistribution to collectives (*asentamientos*) of 18 per cent of the country's total farmland. Jarvis argues convincingly, and in line with evidence from other countries (*deJanvry & Sadoulet, 1989)*, that this growth was due to (subsidized) investment by large landowners who hoped to avoid expropriation, as well as to considerable government investment in the reform sector.

(b) Expropriation of an additional 25 per cent of the farmland during the Allende years, 1970 to 1973, led to a decline of 18 per cent in agricultural value added. Microeconomic data for 1972 and 1973 (*Brown, 1989*) suggests that institutional factors (decreased incentives) associated with the land reform were of minor importance compared to such policy variables as price controls, inflation and civil disturbances. This is confirmed by a 27 per cent increase in agricultural production in 1974, which – with almost no changes in the institutional environment yet implemented – must have been due to higher agricultural prices, elimination of input shortages, increased availability of credit, and increased political stability.

considerable output increases, presumably due to high investment. By contrast, the break-up of reform collectives in Chile into family farms – in the presence of extremely unfavourable government policies – failed to increase output significantly. This type of reform only became fully effective after some of the debts incurred to pay for the land had been forgiven, and structural impediments to the success of small farmers had been eliminated. In general, once distortions are removed, the grant element required to bring about land reform becomes much more modest.

Different paths to land reform and their implications

Until recently, most large landowners operated their holdings either as haciendas or as landlord estates. In a hacienda, the owner operates the home farm with labour provided by tenants, squatters, or traditional holders of usufructuary rights. The latter cultivate produce for their subsistence (and sometimes for the market) on the plots allocated to them, using the time available to them after they have provided labour for the owner's home farm. In a landlord estate, the owner rents out all of the land and does not engage in own account farming. There is a basic difference between land reform as it applies to landlord estates and to haciendas.

Reform in **landlord estates**, which has been widespread in Asia, essentially addresses the equity concerns of society by transferring the

(c) Annual growth of agricultural output averaged only 1,3 per cent between 1975 and 1983, less than during the pre-1964 and the 1964–74 periods. While about one third of the area in the reform sector was restored to former owners by the post-Allende government, there were no provisions to assist the ex-*asentados*, who had to start operation on their individualized parcels with no capital, a huge debt on land, and responsibility for their share in the debt of the former asentamiento. Lack of capital (30 per cent of ex-*asentados* did not have any machinery, and 44 per cent had to harvest all their crops by hand) led to serious undercultivation in the decollectivized small farm sector. Together with decreased internal demand (due to a worsening income distribution), unfavourable price policies, and limited technical assistance to the small farm sector, this lack of capital finally forced between 40 and 50 per cent of the remaining beneficiaries of land reform to sell out by 1986 (*Jarvis, 1989*).

(d) Agricultural production during the 1984–86 period increased at 6 per cent a year, a rate of growth that had been surpassed only during the first four years of land reform. The growth can be attributed to the partial forgiveness of debts incurred for purchases of land, increased government assistance to small farmers, and a more favourable policy environment characterized by devaluation and increased internal prices.

entitlement to land rent without much change to the operational farm structure. Potential efficiency gains are associated with improved investment incentives and with increased security of tenure, as explained below. In Latin America and parts of Africa, where the farm structure has been based on **haciendas,** the threat of land reform legislation, together with the availability of subsidized credit, had led to widespread tenant evictions, thus reducing the resident work force (*Castillo & Lehman, 1983; de Janvry & Sadoulet, 1989*). The large owner-operated commercial farms that have emerged as a result of these policies have been difficult to subdivide since redistribution of that land requires major changes in the organization of production (such as resettlement) and because, in many cases, neither the infrastructure nor the investments in physical capital provide an appropriate basis for the initiation of smallholder cultivation. Unlike tenants on landlord estates or haciendas, the potential beneficiaries of redistribution of land from commercial farms, i.e. the resident labour force or the external proletariat, have little or no independent farming experience. Therefore, even if the issue of financing is resolved, the availability of appropriate technology and credit, as well as the competitiveness of input and output markets become crucial to the success of such reform.

Reform of landlord estates in market economies

Rapid transition from a landlord estate to a family farm in a market economy (Figure 4.1, number 7) has led to stable systems of production relations. Under tenancy on landlord estates, operational units were already being managed by the potential beneficiaries prior to reform. This situation did not require changes in the organization of production, since it transferred ownership from large landlords to tenants who already farmed the land and had access to the skills and implements necessary to cultivate their fields. Major policy decisions concerning the ceiling on landholdings, the amount and type of compensation for expropriated land, and the criteria for selecting beneficiaries and determining their financial obligations, however, have had considerable impact on the final results of such reforms. At the end of World War II, many landlord estates in Iran, East Punjab, Eastern India, China, Taiwan, Korea, Japan, Bolivia and Ethiopia were transferred to tenants and, in all of these countries, the new systems have proven successful. These reforms have led to increases in output and productivity and to systems of land ownership that are highly stable, so there are virtually no landlord estates left which could be subjected to land reform.

In line with theoretical expectations, productivity gains from this type of reform have been associated with improved investment incentives for former tenants, and with increased security of tenure. In general, such gains will be modest if tenants have to compensate landowners at near-market prices, if security of tenure was already high in the previous system, or if cash-rent contracts prevailed.[16] Empirical evidence[17] shows that reform of landlord estates has led to considerable increases in productivity, and that costs to the government of complementary investments (infrastructure, housing, draft animals, training in management skills) to support the newly formed farms were minimal, because the agrarian production system was already in place, or could be established at low cost by the beneficiaries themselves.

Reform of hacienda systems in market economies

By contrast with Asia and India, reform of the hacienda systems prevalent in most of Latin America, in Algeria and in southern and eastern Africa, has been a lengthy and difficult process and has usually led to the emergence of large owner-operated Junker[18] estates (Figure 4.1, number 10). These estates are large farms based on cultivation by the owners themselves, with the help of wage labour.[19]

16 Alternatively, the disincentive effects associated with share tenancy could be low, an interpretation suggested by Otsuka and Hayami (*1988*).

17 Land reforms in Japan and Taiwan were associated with higher investment, rapid adoption of already available technological innovations, and higher use of family labour (*Callison, 1983*). In Taiwan between 1953 and 1960, annual increases of inputs and outputs, by 11 per cent and 23 per cent respectively, resulted in increased factor productivity of 12 per cent per annum (*Koo, 1968, quoted in King, 1977*). In Japan, labour and land productivity increased by 5 per cent and 4 per cent per annum respectively between 1954 and 1968 (*King, 1977*). Land reform in Korea had a positive impact on agricultural growth, which increased at a rate of slightly below 4 per cent a year (*Dorner & Thiesenhusen, 1990*).

18 Junkers were members of the landed nobility in Prussia who were prompted by the threat of land reform to resume self-cultivation. We follow deJanvry (*1981*) and others in using the term generically.

19 This labour force is often organized in a hierarchy of workers, foremen, supervisors, permanent workers with a house and garden plots, and external workers hired on a seasonal or daily basis.

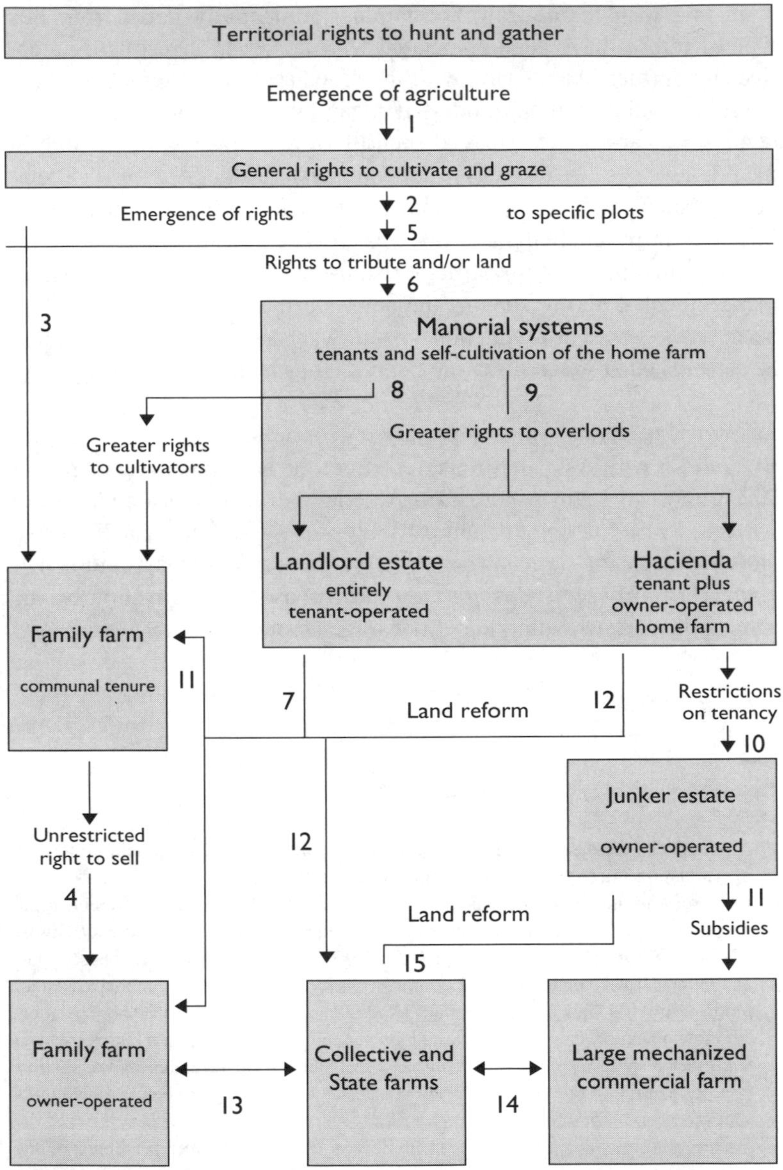

Source: Binswanger *et al*, 1993.

Figure 4.1: Evolution and Structure of Production Relations and Property Rights in Agriculture

If, as argued, the above wage-labour based operations are less efficient than tenant farming, then increases in owner cultivation at the cost of tenanted land would be associated with losses in efficiency. This relationship has two important implications. First, rational landowners would not establish Junker estates unless induced to do so by exogenous constraints such as the threat of land reform or legal restrictions on tenancy. For example, hacienda owners have often tried to reduce their exposure to expropriation through land reform by evicting tenants, the main potential beneficiaries of expropriation (*Castillo & Lehman, 1983*). Restrictions on tenancy, although often intended to protect tenants' rights, have had the same effect. Second, the lack of competitiveness of Junker estates compared with the smallholder sector has made them economically invariable. As soon as they were established, the owners of such estates began to lobby either for protection or for subsidies to modernize through rapid and large-scale mechanization. By substituting subsidized capital for labour, Junker estates were thereby transformed into mechanized large-scale commercial farms that were no longer dependent on large amounts of labour (Figure 4.1, number 11).[20]

It was the threat of land reform that led to the establishment of Junker estates in Prussia[21] in the early 19th century, in Ecuador in the 1950s (*de Janvry, 1981*), in Colombia from 1961 (*Zamosc, 1989*), and in Chile in the 1930s (*Loveman, 1976*). The emergence of Junker estates and large

20 A high degree of mechanization reduces the potential for reform of large commercial farms, primarily because there are not enough families living on such farms with the skills that would be necessary to work their own plots if the farms were subdivided. Furthermore, if capital is directly (e.g. through cheap credit) or indirectly (e.g. through overvalued exchange rates) subsidized, such highly mechanized farms often appear to be very efficient as long as these subsidies are not explicitly accounted for.

21 In Prussia, land reforms beginning under Frederick the Great transformed the rights of hereditary tenants on crown land (in 1750) and on the manorial Junker estates (in 1806) to freehold property rights. These reforms required tenants to give up between one third and one half of the land they possessed as hereditary usufruct to the Junkers as compensation for the latter's loss of labour and plough services. This step led to considerable increases in Junkers' home farms. The initial reform was confined to hereditary tenants and did not include tenants or holders of non-hereditary usufruct rights. Fearing further land reform, however, the Junkers evicted a large number of the remaining tenants and reverted to self-cultivation with hired labour. Junkers relied on wage labour from evicted tenants, from the increasing number of freeholders, and from migrant labour from more labour-abundant Poland. The threat of workers acquiring usufructuary rights that might eventually be converted into ownership rights, led to an ordinance in 1892 that prohibited Polish workers from staying on the farms during the winter months, thus effectively converting them into an external proletariat.

mechanized farms was aided by direct prohibitions on tenancy, construed to be exploitative; by ill-conceived attempts to increase tenants' welfare by imposing ceilings on rental rates and crop shares; by granting high tenure security to tenants after a few years, which caused owners to issue only short-term leases and shift tenants from plot to plot; and by labour laws. Whatever the motivation for these legal provisions, all of them induced owners of haciendas (and less frequently of landlord estates) to either shift to land uses with low labour requirements, such as ranching, or to resume self-cultivation and mechanize in order to circumvent the labour supervision constraint.

The emergence of Junker estates as a specific response to pending land reform, and to restrictions on tenancy, supports the view that these estates are inherently inefficient. Once established, they must find ways to reduce their labour costs or increase their revenues in order to compete successfully with family farms. After losing their rights to either rent or labour services from tenants, landowners often seek rents from the expanding urban and industrial sectors, a change described in detail by de Janvry (*1981*). One method of securing rent has been to ban foreign competitors from entering domestic agricultural markets, which forces consumers to subsidize Junker estates or commercial farms indirectly. Outstanding examples of this method are the formation of the German Zollverein at the end of the 19th century (*Gerschenkron, 1965*); the imposition of tariffs on beef imports in Chile in 1897 which were maintained even in the face of consumer riots in 1905 (*Kay & Silva, 1992*); and selective price support to products from large-scale farms in Kenya, Zimbabwe and South Africa (*Deininger & Binswanger, 1993*).

The other instrument for securing rents has been subsidization of mechanization which has led to the transition to mechanized commercial farms (Figure 4.1, number 11) in almost all existing Junker estate systems.[22] The huge sums involved[23] have been provided through direct subsidies for machines as in Kenya; or through cheap credit, as in South

22 deJanvry and Sadoulet (*1989*) argue that the threat of land reform and expropriation, and the large landowners' ability to lobby effectively (in coalition with the urban sector) for subsidies and one-sided provision of public goods, led these landowners to mechanize and make the transition from traditionally oriented haciendas to large mechanized commercial farms in Colombia (1961–68), Ecuador (1936–57), Peru (1964–69), Venezuela (1959–70), and Chile (after 1972). In Ecuador, two separate stages can be distinguished. Widespread eviction of tenants and emergence of large-scale entrepreneurs, i.e. the formation of Junker estates, until 1957, was followed by a period of increased emphasis on the family farm sector together with widespread

Africa, Zimbabwe, and virtually all of South America. Mechanization, which eliminates the need to rely on hired labour, has resulted in widespread labour/tenant evictions, even in countries where labour remains very cheap and mechanization is hardly optimal from a social point of view.

Reforms in socialist economies: collectivization and state farms

Landlord estate systems in the former Soviet Union, Vietnam and China, were initially converted into *family farms* (Figure 4.1, number 10), in much the same way as in reforms of the landlord estate systems in market economies. Collectives were created in 1929 in Russia, from 1957 in China, and from 1958 in North Vietnam, by conglomerating redistributed farmlands into single management units (Figure 4.1, number 13). In collective farms, land is owned and operated jointly, under a single management; by contrast, in systems of communal ownership, families operate their own plots, but their land can still be subject to periodical reassignment.

Where Junker estates or large commercial farms have been subject to reform (as in East Germany, Peru, Chile, Nicaragua, Peru, Mozambique and Algeria), land reform has resulted in the direct formation of *state farms* (Figure 4.1, number 11) rather than collectives. In most of these cases, workers continue as employees under a single management without a change in internal production relations (Figure 4.1, numbers 12 and 14). Over time, the organizational differences between collectives and state farms tend to disappear.

Collective farms established in socialist economies, like their capitalist counterparts, have developed into large, mechanized state farms that are subject to severe inefficiencies. As indicated by recent attempts to decollectivize in a number of socialist countries (China, Vietnam,

mechanization (1958–73). As a consequence, redistributive land reform would have been associated with at least a temporary loss of output and would have required provision of similar public goods to the small farm sector, making such an option very costly and in many cases politically undesirable.

23 Abercombie (1972) shows that, in addition to exemption from import tariffs, and other measures such as bulk imports and controlled prices to reduce the cost of machinery, real interest rates on mechanization loans in most of Latin America during the 1950s and early 1960s were actually negative; and that farmers in Chile, Argentina, Brazil, and Venezuela paid back only 50 to 80 per cent of the amount received as equipment loans. Subsidies and selective price supports to products from large farms were also employed in Kenya and South Africa after World War II.

Peru, Algeria), the transformation of large farms into collectives rather than family farms does not increase efficiency.

Aside from the ideological underpinnings of collective or state farms, the potential benefits of this type of organization were assumed to be associated with economies of scale; equitable distribution according to need rather than work; the ability to provide insurance; and the provision of public services such as health care, education and defence. Most of these benefits, however, have proved to be either elusive or not contingent on the maintenance of collective production. If economies of scale were present, rational action by self-interested farmers would result in the formation and maintenance of collective forms of production[24] without extra-economic incentives or coercion; payment only according to need rather than on the basis of work would reduce overall production (*Israelson, 1980*) and not be Pareto-efficient. Moreover, the potential of collectives to substitute for insurance markets is extremely limited[25] due to the covariance of risks in agriculture and, in most societies, cheaper mechanisms than collective production are available to ensure against both covariate and non-covariate risks. Finally, economies of scale in production-related activities such as input supply, marketing and credit can be utilized through co-operative approaches without collective production.

There are three fundamental disadvantages associated with collective production in agriculture:

- First, even if effort were perfectly observable, co-operative production is likely to be inefficient because remuneration for each individual depends on the productive performance of the collective as a whole. The problem is aggravated by the fact that monitoring in agricultural production is particularly difficult. Therefore, even if the

24 Putterman and diGiorgio (*1985*) show that, in theory, iterative voting on the amount of land to be farmed collectively, and the amount of product to be distributed according to work performance, will lead to a Pareto-optimal choice of these two parameters. While economies of scale would result in an interior solution, their absence would lead to the emergence of fully individualized farming.

25 Carter (*1987*) has shown that, in the absence of other possibilities to ensure against non-covariate risks, rational risk-averse individuals would choose some degree of co-operation in order to enjoy the associated insurance benefits, even if there were no economies of scale. While theoretically valid, empirical evidence (*Walker & Ryan, 1990*) indicates that covariate risks are of much higher importance and that social ties can provide a less costly means of ensuring against non-covariate risks.

need for work-based remuneration in collectives is realized, the implementation of such schemes may be costly or impossible. This is illustrated by the cases of China and Cuba where, after production had declined significantly under need-based remuneration schemes, sophisticated work-point schemes were developed. These schemes proved to be too cumbersome to implement and were soon replaced by wage schemes with lower informational requirements (*MacEwan, 1981; Ghai & Peek, 1988*). A further complication is that, in collective agriculture, individual plots may be available to cultivate for subsistence or for sale on parallel (or officially recognized) markets. If, because of depressed official prices, for example, the marginal product of work on collective lands is low, it would be rational to concentrate effort on the private plot. To counter this tendency, governments were forced to impose – with varying degrees of success – mandatory work requirements or delivery quotas for collective production.

- Second, successful collectives tend to degenerate into capitalist enterprises or wage-labour operated state farms by successively substituting cheaper wage labourers for more expensive members (*Ben Ner, 1984*). McGregor (*1977*) provides theoretical justification for, and empirical examples of, the tendency of co-operative enterprises to reduce membership in order to increase current consumption by members.[26] It is not surprising, therefore, that in many cases collective farms voted to redistribute land into family-sized farms once they were given the chance to do so.[27]

26 McGregor (*1977*) notes that such problems can be overcome by paying rents to all factors, including the capital and land contributed by members, which of course, would imply some form of individual ownership. In most empirical cases, members of co-operative ventures established as a result of land reform neither receive rights to land nor are allowed to withdraw part of their land from co-operative cultivation.

27 Quantitative evidence for the decline of the collective sector throughout Latin America is given by Ortega (*1990*). The absence of economies of scale in coastal rice and cotton farms, as well as in farms in the highlands of Peru, led reform beneficiaries to *de facto* subdivide reformed collectives by concentrating effort on their private plots. The ensuing need for increased use of wage labour, which had to be paid the minimum wage, further decreased the collective farms' expected profits and led to the gradual expansion of private plots and pressure for legal possibilities to subdivide and obtain individual land titles (*Kay, 1983*). In Zimbabwe, the attempt to establish collectives was soon abandoned in favour of a smallholder-oriented strategy (*Weiner, 1985*). In Ethiopia, competition for labour between individually owned and collective plots is reported by Pickett (*1988*). In the Dominican Republic, private parcels (of about

- Third, in collectives, decisions to invest, save and distribute profits are made jointly and, if there is no secondary market for equity, it is rational for members to under-invest (*Bonin, 1985*), leading to disappearance of the capital stock over time.

These difficulties have prompted even those countries that have adopted collective forms of production to abandon all or parts of the collective model in favour of more pragmatic but not necessarily more efficient solutions,[28] often in the form of state farms. If large mechanized collectives or state farms are to be transformed into family farms, the number of beneficiaries has to be increased above the workers currently residing on the farms, and competitive input and output markets, as well as facilities for smallholder credit, have to be re-established.[29]

The social cost of incomplete or long-delayed reform

The maintenance of a highly dualistic agricultural structure based on relatively inefficient haciendas or commercial farms involves not only

3 hectares) from the land reform in 1961 were collectivized in 1972, without success; and in 1985, the reality of farmers producing on individual plots was acknowledged, and co-operatives (*asentamientos asociativos*) based on individual production but collective marketing, credit and input purchases were officially recognized (*Meyer, 1991*). In Panama, land reform co-operatives are highly indebted; they use labour at much less than the profit maximizing level, and are characterized by very low land productivity (*Thiesenhusen, 1987*). In Honduras, co-operatives decreased their membership over time by about one fifth, which enabled the remaining members to attain incomes much higher than those outside the reform sector and created tensions between co-operative members and workers employed by them (*Barham & Childress, 1992*). In Algeria, production co-operatives were characterized by low productivity, membership desertion, resistance to admission of new members (*Pfeiffer, 1985*) and high mechanization, together with considerable underemployment of the workforce, who received an average income of only about two thirds of the legal minimum wage (*Trautman, 1985*). The same pattern of declining output and transformation into a 'collective Junker estate' has been observed in Mozambique, where collectives have used government-subsidized credit to mechanize and proletariatize the peasant workforce associated with the former haciendas (*Wuyts, 1985*).

28 In Nicaragua, collective organization of production on state farms was a political success but an economic failure, with only 4 out of 102 of the collectives making a profit (*Colburn, 1990*). Collectives in Cuba, after an initially dismal performance, increased work incentives by adopting a more effort-based system of remuneration and by forming more decentralized production brigades. Despite better access to credit and other subsidies than the voluntary small farmer co-operatives, however, the Cuban collectives still provided their members with an income that was an average of 23 per cent lower than the small co-operatives. The small co-operatives, however, faced serious

static losses of allocative efficiency,[30] but also dynamic losses associated with the reduced profitability of smallholder cultivation, and the ensuing lack of incentives to invest in physical and human capital in the smallholder sector. Furthermore, the literature tells us that rents associated with distortions are often dissipated in competition for these rents, and these rent-seeking costs have to be added to the static and dynamic efficiency losses. In many cases, duality in farm sizes has also been associated with protracted and violent struggles that have significantly reduced the performance of the agricultural sector or the economy as a whole. The losses from such conflicts are even more elusive to economic measurement, but their magnitude can be gauged from the length and intensity of such struggles. Moore (*1966*), Wolff (*1968*), Migdal (*1974*), Skocpol (*1982*), and Scott (*1976*), emphasize the important role of peasant unrest in many current and past incidents of violence. While peasants have rarely been protagonists of radical class struggle or initiated revolutionary movements, (although in some cases they have turned fascist to protect their land claims), the importance of peasant discontent is indicated by the fact that remote areas of limited agricultural potential – sometimes designated as communal areas, reserves or homelands in which free peasants have found refuge – have provided active and passive support for guerrilla fighters in many social conflicts.

Even in Brazil, where failure to reform has not resulted in massive peasant uprisings and civil war, the social costs of continued massive distortions in favour of large farms has been great. Between 1950 and 1980, while agricultural output in Brazil grew at a remarkable 4,5 per cent a year and land area expanded at 1,5 per cent a year, agricultural

problems of membership desertion once the possibility of private marketing in a 'small farmers' market' became available in 1980. (The threat of desertion was so serious that, after a number of temporary suspensions, this private farmers' market was permanently abolished in 1986 (*Ghai et al, 1988*).) In Ethiopia and Mozambique, collectively operated state farms received the lion's share of public sector investment (*Ghose, 1985; Wuyts, 1982*), but these farms were characterized by sharply declining yields and mounting deficits (*Griffin & Hay, 1985*).

29 For a more detailed discussion of decollectivization issues, see Binswanger *et al*, (*1993*).

30 Quantitative estimates of the size of this efficiency loss are scarce. Loveman (*1976*) provides a rough estimate for Chile, indicating that during the 1949–64 period, more than US$100 million could have been produced on the more than 40 per cent of land that was left uncultivated (or under natural pasture) by large landlords. This estimate, of course, includes the resource cost of producing such output.

employment only grew at 0,7 per cent a year. During this period, the large-scale farms evicted most of their internal tenants and workers, many of whom became insecure seasonal workers without farming skills, or migrated to urban slums. An alternative growth path could have provided rural employment and self-employment opportunities to a substantial number of these people, and could have gainfully absorbed a substantial share of the rapidly growing urban population.

In Mozambique, measures to extract peasants' labour have included forced cultivation and vagrancy laws (both instituted in 1930), and forced labour (instituted in 1949). As a result, from the beginning of the guerrilla struggle in 1961–62 until independence in 1975, Frelimo guerrillas were able to recruit and obtain support from peasants in inaccessible rural areas, who received credit and some marketing services in return for their support (*Isaacman, 1983*). These areas (about 25 per cent of the total land area of the country) became the 'liberated zones' in 1968. Attempts at land reform after independence created highly mechanized collective farms but did not address the problems of the freehold sector. Violence continues to this day.

In Zimbabwe, large-scale eviction of about 85 000 black African families from European land between 1945 and 1951 (*Mosley, 1983*) led to a general strike among blacks in 1948 (*Ranger, 1985*); then to guerrilla attacks by ZANU (Zimbabwe African National Union) in 1964 (*Scarritt, 1991*); and finally to substantial acceleration of guerrilla warfare from 1971 until independence in 1980. The guerrilla uprising adopted the strategy of a people's war by taking up the peasants' many grievances over unequal distribution of land and state interference with production. Guerrilla fighters used the Tribal Trust Areas as bases for frequent attacks on European farms which, as a result, were abandoned in great numbers towards the end of the war (*ibid*). Policy distortions in favour of large farms, some of which were acquired by black Africans, remain in place, however, in spite of evidence that these farms are less efficient than smallholdings (*Masters, 1991*). A substantial settlement programme has provided farms to blacks, but many of the farms may be too large, given the limited capitalization of the farm families. Land reform continues to be a major political issue.

In Guatemala, communal lands were de facto expropriated in 1879 by legislation that gave all proprietors three months to register land titles, after which the land was considered idle or abandoned. The legislation resulted in widespread allocation of land to large coffee growers, some of which was redistributed in the land reform effort of 1951–54.

Following a military coup in 1954, virtually all the land that had been redistributed was reinstated to the old owners, and farms expropriated from foreigners were allocated to nationals in average chunks of more than 3 000 hectares (*Brockett, 1984*). Since then, resistance to the land system has become more radical, and suppression has become harsher. Co-operative movements in the 1960s were suppressed, leading in 1972 to the formation of the Guerrilla Army of the Poor (EGP) which had its main base in the Indian highlands. This was followed by a wave of government-supported assassinations in 1976; the formation of the Committee for Peasant Union (CUC) in 1978; and government massacres of protesting peasants, the most spectacular of which took place in 1980 (*Davis, 1983*). Almost 40 years after the first attempt at land reform, peasants were still demonstrating for reform in 1988.

In El Salvador, smallholder land was appropriated in a similar way by an 1856 decree that declared all communal land not at least two-thirds planted with coffee to be under-utilized, and therefore the property of the state. Communal land tenure was then abolished in 1888, and sporadic revolts led the government to establish such measures as a 'security tax' on exports to finance rural police forces (*McClintock, 1985*). Rural unions were banned in 1907, and a National Guard was created in 1912. Areas where land pressure was particularly severe were the main centres of the revolt of 1932, in which between 10 000 and 20 000 peasants were killed (*Mason, 1986*). During, 1932–1979, the situation was characterized by continuing and escalating violence. Throughout the period, guerrillas promised rural credit and reform of the land marketing and wage systems, and thereby considerably increased their support in rural areas. Renewed eviction of tenants ensued during the 1960s in the haciendas of the cotton-growing lowlands, and between 1961 and 1970, house plots available to *colonos* decreased by 77 per cent, while the number of *colonos* fell from 55 000 to 17 000 (*McClintock, 1985*). In 1979, a coup by reform-minded officers led to land reform as an attempt to pre-empt a critical shift in popular support to the FMLN-FDR guerrilla forces. Large-scale eviction of tenants and narrow eligibility rules sharply limited the number of beneficiaries, however, and more than a decade of civil war ensued. The peace accord of 1992 mandates still more land reform.

Experience in Colombia also indicates that incomplete land reform measures are unlikely to create social stability. During the 1920s, conflicts over land between *colonos* and large-scale farmers at the frontier were only of a local character; but more co-ordinated tenant actions in the late 1920s led to the consideration of different kinds of reform legislation in the 1930s. Law 200 of 1936, however, vested rights to once-public lands with large landlords rather than providing plots for tenants,[31] and thereby formed the basis for immediate and future tenant evictions (*LeGrand, 1986*). A long period of *violencia* then ensued (*1940-65*), during which guerrillas recruited and received support from peasant groups. The effect of land reform legislation in 1961 and 1968 was limited to ex post regularization of previous land invasions, and had no effect on improving the distribution of operational holdings. In fact, the number of beneficiaries were significantly lower than the number of tenants who had previously been evicted, and the size of the holdings they received was smaller than what they had lost (*Zamosc, 1989*). Ideological commitments to collective farms as reform enterprises further reduced the success and economic welfare of reform beneficiaries. Peasant invasions of large farms intensified in the early 1970s, leading to a declared state of emergency after 1974. Regional peasant mobilizations, strikes and blockades flared up again in 1984, indicating that the conflict is not yet resolved.

In Peru, the effective exclusion of most highland Indians from the benefits of the agrarian reform of 1973, together with an acute crisis in subsistence agriculture, led the indigenous communities to provide substantial support to Shining Path, especially after the 1980 drought. As a result of Shining Path activity, more than half of the administrative areas in the country have become virtually inaccessible to government forces (*McClintock, 1984*), and public investment in these regions has become unfeasible, inducing large-scale migration to the cities. Partly because Shining Path remains active, and partly because of poor economic management during the 1980s, Peru is now suffering capital flight and sharp economic decline. Other countries with similar prolonged conflicts over land distribution include Chile, Nicaragua and Angola. The only country that had a highly dualistic farm size structure at the end of World War II, and that has escaped protracted violent struggle, has been Kenya for the reasons explained in the following section.

31 Squatters who had farmed land for five years 'in good faith', i.e. without knowing that it was private property, acquired titles to the land; but landlords who were able to provide titles to their land could, after paying compensation for improvements, legally evict squatters. Maximum sizes of land grants, which had been part of the proposed 1933 legislation, were not imposed.

Land reform in South Africa: lessons from Kenya and Zimbabwe

The emergence of the present agrarian structure in South Africa is discussed in Chapter 3. African agriculture was suppressed through a variety of measures. The events in South Africa were typical of many colonial territories in Africa where governments, influenced by European settlers, restricted black African land rights very early, and created reserves that were too small to support independent African agriculture. As a consequence, tenant farming (with tenants called squatters) was the main mode of production accessible to black Africans. In addition, white large-scale agriculture benefited from an extraordinary set of privileges and favourable policies. Given this background of a dualistic agricultural sector, this section briefly discusses land reforms in Kenya and Zimbabwe to demonstrate that land reform within such a dualistic structure is feasible. Key elements of a successful reform strategy are illuminated.

Potential for land reform

Given the extraordinary discrimination against black South African farmers, it is not surprising that agriculture in the areas allocated to them is generally poorly developed. Even in the few former homeland areas with high quality land, the returns an African farmer can expect to derive from his/her labour, and from investing in land improvements, machines or farming skills, is only a small fraction of the returns that would accrue from the same effort or investment on similar quality land in the white farming sector. It is not surprising, therefore, that few blacks have been motivated to put their talents and money into agriculture. But this, of course, was what the system aimed to achieve. Thus, the current low productivity of black South African agriculture cannot be used to predict the potential of smallholder farming. To observe that potential, one has to go to the few countries, such as Kenya, in which discrimination against agriculture in general, and against smallholders in particular, has been modest in the post-colonial period. Since Africa as a whole has had the most severe discrimination against smallholder agriculture by worldwide standards (*Lipton & Lipton, 1992*), such examples where smallholder farming could achieve its full potential are very few. The depressed profitability of smallholder farming has also limited the success of the few interventions in favour of the sector, therefore past experience cannot be a guide to what development programmes could achieve if they were carried out in undistorted environments.

In both Kenya and Zimbabwe, large-scale eviction of tenants after World War II led to violent eruptions of peasant protest (*Mosley, 1983; Ranger, 1985*). Commissions to investigate the land question in these countries, as well as in South Africa, in the 1950s, concluded that any lasting solution must at least remove constraints limiting the access of African farmers to output markets and to land outside reserves; provide secure titles to smallholders to allow them access to credit; and open up job opportunities outside the reserves to enable African farmers to establish economically viable holding sizes, in and outside the reserves.

In Kenya, where European producers never achieved the degree of control over agriculture that they did in South Africa, simply eliminating restrictions on access to output markets in response to the Swynnerton Plan of 1954 evoked considerable response from smallholders. This belied arguments that they would not be responsive to price incentives and lacked the ability to acquire the skills and establish the infrastructure needed to make smallholder farming a success. In terms of output, it is estimated that resettling blacks on former white farms in the course of the million acre scheme[32] led to production increases of between 15 and 90 per cent (*Ruthenberg, 1966*). Between 1955 and 1968, total coffee production increased more than threefold, with the percentage of coffee produced by black smallholders climbing from less than 10 per cent to more than 50 per cent of the total. Also, the total output of pyrethrum quadrupled and the smallholders' share jumped from about 10 per cent to almost 90 per cent. These examples confirm that smallholders are able to respond to economic opportunities. At the same time, maize production remained constant which alleviated widespread fears that production of food staples would drop if smallholders were allowed to diversify.

Productivity comparisons between small and large farmers, although not numerous, convey the same general picture. Comparisons of settlement farms and large farms in agro-ecologically similar locations indicate higher partial productivities of labour, capital and recurrent inputs on small settlement farms (*Nottidge & Goldsack, 1971*). In 1974, the Integrated Rural Survey (*Kenya Ministry of Agriculture, 1978*) showed that holdings of less than 0,5 hectares in the resettled

32 The million acre scheme is the main component of the resettlement programme executed in Kenya during the 1960s. It involved the purchase of about a million acres from white farms by the state, and the sale (at around two-thirds of market value, the difference being covered by a British grant) of this land to individual farmers in small parcels.

areas marketed more than six times the volume of produce of farms in the five to eight hectare class, in terms of value per hectare marketed. Economic evaluation of four settlements established under the million acre scheme indicated that all except one had positive net present values, even if family labour was valued at market wages. At accounting prices, all of the settlements had positive net present values; and the 'high density schemes' with smaller farm sizes, were found to be superior to the 'low-density schemes' which, with larger farm sizes, were supposed to form the basis for commercial 'yeoman' farmers (*Scott et al, 1976*).

In contrast to Kenya, the performance of settlements in Zimbabwe was far less successful.[33] Determinants for this unsatisfactory performance can be identified by comparing settlement programmes in both countries (i.e. Kenya 1964–70, Zimbabwe 1980–90).

Financing and organization of land purchases

Since government agencies in both countries had to purchase the land, select settlers and provide the necessary infrastructure, implementation required intensive use of skilled administrators. Land was given away free rather than for payment by the beneficiaries – as was the case in Kenya, a task which proved much more complicated in Zimbabwe where resettlement was a long and bureaucratically cumbersome procedure (*Munslow, 1985*). The need to finance land purchases, without recovering expenses from farmers, strained government resources and also hindered the proclaimed goal of greater fairness.[34] The selection of settlers became highly politicized, so neither the most needy nor the most able benefited from the programme.

Limits on subdivision

Limits on subdivision were a serious constraint in both countries. In Kenya, they prevented an extension of smallholder settlement to the remaining large farms which, in the meantime, had been acquired by politically influential blacks (*Leys, 1974*). In Zimbabwe, such limitations affected even government-directed resettlement programmes,[35]

33 See the data provided in Deininger and Binswanger (*1993*).

34 The need to compensate owners was part of the 'willing buyer–willing seller' agreement that was made part of the Zimbabwean constitution. Land purchases by individuals or groups of blacks were not precluded by this agreement; but the prohibition on subdividing large estates made it impossible for smallholders to purchase such farms.

35 Until recently, the government could only buy complete farms.

and made purchases of small tracts of land by individual farmers virtually impossible.

Land rights

Although theoretically equitable, granting settlers usufructuary rights rather than ownership or long-term transferable leases hindered their economic success in three ways: first, the restricted rights reduced incentives for long-term improvements in land; second, the absence of ownership rights prevented the emergence of a rental market, and precluded specialization and dynamic adjustment to life cycle phenomena or differentials in settlers' abilities; and third, without land titles, settlers could not obtain commercial credit; the credit they were able to get from the state was limited to the minimum required by foreign donors. Farmers used the donor-funded credit for short-term consumption, and the repayment rate was poor.

Non-agricultural activities

The explicit prohibition of off-farm employment in Zimbabwe, motivated by the adherence to the ideal of the full-time farmer, severely limited farmers' ability to accumulate working capital and to adjust to temporary shocks such as drought. In Kenya, off-farm employment has repeatedly proved crucial to the ability of farmers to repay loans and to fund agricultural investment, for example, establishing cash crops.

Farm sizes

Both in terms of area and number of settlers, the programme in Kenya was smaller than in Zimbabwe. While settlers in both countries lacked knowledge and capital, the large farm sizes in Zimbabwe (64,4 hectares as compared to 12,6 hectares in Kenya), together with the low level of government services, the lack of access to credit markets, and the prohibition of rental and off-farm employment, made it particularly difficult for reform beneficiaries in that country to establish viable farming operations and make productive use of their land.[36]

Provision of technology and services

In Kenya, extension and technical services were key elements of the settlement strategy.[37] The huge financial requirements associated with

36 An extreme example of this situation is that in some collective settlements, set up in high-potential regions to maintain economies of scale, farms larger than 1 000 hectares only had access to a single tractor (*Weiner, 1988*).

land purchases in Zimbabwe made it difficult to fund adequate technical services.

Output markets

Kenya and Zimbabwe have made little progress in dismantling state monopoly marketing systems that were established specifically to discriminate against smallholder agriculture. In Zimbabwe, the difficulties arising from state monopolies on outputs are compounded by similar monopolies on inputs. These monopolies confer structural advantages to the large farm sector, distort prices and also drain government funds. In 1986 alone, annual losses of the marketing parastatals amounted to Z$820 million, or more than 20 per cent of total government spending. Parastatals for marketing and input supply have failed to achieve the goals set for them. For example, only 50 per cent of settlement farmers use fertilizers, and the lack of inputs appropriate for smallholder cultivation is widely quoted as one of the reasons for the failure of smallholder agriculture to develop under resettlement.

Bureaucratic control

Many reform processes, in particular in Zimbabwe, have been hampered by an excessively administrative and paternalistic approach in which a land reform agency selects and purchases farms, selects beneficiaries, designs the farming schemes, and redistributes the farms to beneficiaries. These land reform agencies are often expected to provide infrastructure, production support, marketing services and social services, which creates confused objectives and operating guidelines. As a consequence, land reform takes a very long time,[38] as it did in Mexico where 60 years were required to distribute half the national farm land. A high degree of decentralization, as in Malaysia and Kenya, and the use of market-assisted mechanisms, as in Kenya and Guatemala, can greatly reduce these problems.[39]

37 Settlements were often ridiculed because of the comparatively lavish provision of these services compared to other regions.

38 In South America, it typically takes land reform agencies about five years to purchase a farm and an incredible 20 years to turn over land title to a beneficiary. Bribes which are often used to speed up this process reduce the scope of the land reform to benefit the poor.

39 For a more detailed review of the experience with resettlement, see Kinsey and Binswanger (*1993*).

Implications for resettlement and land reform

While experience in Kenya and Zimbabwe indicates that land reform is possible under conditions resembling those in South Africa, it also illustrates the importance of institutional arrangements that can help the reforms to succeed. We summarize these types of institutional arrangements below.

Resettlement and its organization

Land reform involving mechanized commercial farming systems requires that such farms be subdivided and resettled because the efficiency of family farms is associated with the use of family labour. Too few families reside on highly mechanized farms to work these farms efficiently if the land is redistributed exclusively to the residential labour force. Especially if the beneficiaries are poor, as they are likely to be, they do not have the access to the credit market and credit subsidies that would allow them to use either mechanization or hired workers.

Size of settlement units

The desire to establish viable farming units has often led authorities to provide parcels that are too large in comparison to the settlers' capital and skill endowments. Together with other restrictions, this practice has resulted in continued underutilization of land, negative effects on equity, and failure to reap the potential benefits from resettlement.[40] Allotment of smaller parcels would enable more people to benefit from resettlement. In contrast to the highly administrative and paternalistic approaches used by land reform agencies in the past, market-based approaches combined with lump sum transfers to the poor which would allow them to purchase land, would lead to higher flexibility and thus increase efficiency as well as equity.

Elimination of subsidies and legal restrictions

Tax preferences for agriculture, differential subsidies for large farms, and macroeconomic instability leading to the use of land as a speculative asset, are now capitalized in the land value and prevent the emergence of a competitive land market. These tax preferences, therefore,

40 Leo (*1978*) shows that, in the case of Kenya, settler success was not related to the size of the initial landholding.

have to be dismantled prior to land reform. If tax preferences and other differential farm subsidies are not eliminated, market prices of land are likely to be higher than the capitalized value of agricultural profits, which makes the use of market-based approaches to effect redistribution of land unfeasible. The groups (classes) of farmers favoured by such subsidies would be able to pay higher prices for land. Therefore, if such distortions are not dismantled, market processes might lead to reconcentration of landholdings even after a redistributive reform. Similarly, existing legal restrictions on subdividing large estates, and on land sales and rentals, will have to be abandoned if land reform is to lead to the emergence of a productive smallholder agriculture.

Beneficiary selection

Selecting settlers based on their commitment to farming, as evidenced by their willingness to pay part of the cost of land, is most likely to result in a resettlement programme that increases efficiency as well as equity. Provisions for beneficiaries to purchase (with credit) relatively small plots of land, and to add more land (through rental or purchase) if they become successful, are likely to be conducive to this goal. Unsuccessful settlers, likewise, should be allowed to rent out or sell their land and take on off-farm employment.

Technology and extension

The establishment of profitable agricultural enterprises is contingent on the availability of appropriate technology and extension. While farmers' participation in the design of institutions to provide such services is essential, government transfers will be necessary, at least in the initial phase.

Input and output markets

The small farmer option is only viable if there are competitive input and output markets that can serve them. Otherwise, the land and entrepreneurial rents from agriculture will be captured by monopolistic input suppliers and output marketers. In the same way, if sufficient credit to overcome temporary shocks is unavailable, periods of bad weather might lead to distress sales by new landowners who do not possess other assets. The creation of competitive input and output marketing systems, and of a viable financial system for small farms, has to be addressed prior to, or at the same time as, the break-up of large farms into family farms.

Outline of a plan for action

Based on international experience, South Africa seems to have two options: rapid and massive redistribution of land to black and 'coloured' groups, which would involve substantial resettlement from the former homelands onto land now in the commercial sector; or decades of likely peasant insurrection, possibly civil war, combined with capital flight and economic decline. At this time, peasant unrest is still unorganized and confined to intra-racial conflicts and sporadic attempts at land invasion. If the hopes raised by the repeal of most of the racially motivated land laws and restrictions on mobility are not confirmed by rapid restitution of land in the former 'black spots', and by additional assignment of land from the commercial sector, land invasions are likely to increase. At first, such invasions will be unco-ordinated and sporadic, and will concentrate in KwaZulu-Natal and other areas close to the former homelands. While Afrikaner farmer groups will be able to defend areas that are not contiguous to the former homelands, and are arming themselves to do so, they will be powerless against murders committed by current and former workers and tenants. Over time, some political group will probably organize a well-disciplined peasant movement capable of co-ordinated insurrection and terrorism, using the rapidly increasing stock of arms in rural areas. Limited appeasement schemes, at that point, will only be able to postpone a co-ordinated peasant insurrection for some period, the length of which is difficult to predict. But failure to act decisively and rapidly is likely to bring it about sooner or later.

Could South Africa avoid this fate by concentrated rural development in the former homelands? Such development is obviously necessary whether or not land reform is undertaken; but even an aggressive programme to develop human resources, roads, water supply and agriculture in the former homelands, will not be able to generate a sufficiently large number of farm and non-farm jobs to avert a massive unemployment crisis. Converting commercial sector farms to small- or medium-sized part-time or full-time farms is one of the cheapest and fastest ways to generate productive employment, both farm and non-farm, on the massive scale that is required. If a non-disruptive process to achieve this transition can be found, it will also lead to more intensive land use and higher production. Both the reallocation of land and increased employment will reduce the extraordinary current tensions.

Given the complexity of existing land relations and the widely diverging aspirations of different social and political groups, a new land law should be anchored in the constitution (*see also* the discussion in

Chapter 8). It should include provisions for secure private ownership, for more flexible forms of communal tenure than are currently allowed, and for government ownership of ecological reserves and perhaps national forests. Private ownership should be secure and not subject to effective use. Communal ownership could continue to be an option in the former homelands, and could also be used by land reform beneficiaries on subdivided farms. It should be based on the minimum common denominator of all communal rights systems: the constraint on sales to outsiders imposed by the community on its members. Just as company law ensures shareholder democracy, community law should specify minimum democratic rules for decision-making. Communities would have to fulfil these in order to enjoy the protection of the law, guaranteeing their land rights and their legal status.

Redress for violations of democratic rules for minority groups in these communities should be available through an arbitrage process, perhaps including traditional tribal institutions, with ultimate recourse to the judiciary. The law should leave to local communities the right to determine their internal land rules. Internal arrangements could include collectives in which plots are not tradable, co-operatives with inalienable individual rights, and condominiums with largely unrestricted individual rights to rent and sell land to other members of the community. Allowing communities to amend their own bylaws would enable them to adapt to local circumstances and to changing external conditions over time. By voting to eliminate the ban on sales to outsiders, a community would join the private property regime for the plots it has allocated to individual families. A clear legal basis for a land reform programme also needs to be created. Relevant provisions should include cut-off dates for claims, principles governing compensation, arrangements for financing the reform programme, and so on. Experience suggests that lack of legal clarity leads to delays and breakdowns in implementation.

A judicial approach could be used to settle *claims to specific plots* by groups evicted from their land during the apartheid regime's black spot removal. (See Chapter 15 for a full discussion on the land claims procedure.) The *general claim* for restitution arising from past land policies and from the systematic destruction of smallholder farming cannot be handled by judicial means; but it could be handled by a market-assisted land reform, as in Kenya, where the poor were given grants to help them buy land from the commercial sector. The *government would not buy or expropriate any land,* but land would be transferred from willing sellers to groups of eligible beneficiaries. The beneficiaries would receive a partial grant to buy land, perhaps in the form of a land purchase voucher. They would combine this grant with their own equity and a loan from the Land Bank.

The beneficiary groups would be free to choose between communal tenure (in a great variety of forms) or private ownership. By having the freedom to choose their farms, internal management schemes and subdivisions, they could select locations and farming systems most appropriate to the capital and skill endowments of their members. The farming systems could range from suburban residential communities with small kitchen and market gardens to part-time, full-time or commercial farming. Beneficiaries would be assisted with additional transfers or vouchers for feasibility studies, land subdivision and recording, subsistence for the first year, and agricultural extension for subsequent years. The land vouchers should be targeted to the poor, who would be selected through some form of means testing. Without a means test, the scheme would likely benefit the middle class, bureaucrats and tribal chiefs.

Financing for the programme would have to come from external and internal sources, some of which might include an external donor consortium, the uncollectable debt claims of the commercial farm sector, or an income or value added tax levied on the past beneficiaries of apartheid.

Injecting purchasing power into the land market could cause the price of land to rise, which would make the programme excessively expensive. The potential rise in land prices could be held in check by establishing a level playing field for all farm sizes; i.e. by eliminating all distortions that favour the commercial sector in the tax code, the credit system and the marketing system, as well as in access to services and technology. Levelling the playing field would prevent land prices from rising much above the productive value of the land, and would prevent distortions from forcing a reaggregation. In addition, loans from the Land Bank for land purchases should be limited to the productive value of land, an approach that is already largely in place.

Achieving agreement at the constitutional stage requires that current tensions over land policy, and the mutual distrust of the negotiating parties be rapidly diffused. One way to diffuse current tensions over land policy might be to implement the following short-run measures: (a) abolish the land subdivision act; (b) recognize that labour tenants and workers who have resided on farms for a long time have a claim to some land and housing on a retroactive basis (i.e. whether or not they were recently evicted); (c) establish a moratorium on the distribution of government (trust) land; and, (d) establish a moratorium on land invasions.

Substantive and rapid market-assisted land reform and resettlement is the greatest if not the only hope for peaceful development in South Africa. It is also the greatest hope for the rapid growth of productive employment and self-employment opportunities. The international community has a great interest in helping South Africa to finance such a programme for economic and humanitarian reasons; and the short-term measures suggested here would contribute to mobilizing international support.

References

Abercombie, K.C. 1972. Agricultural mechanization and employment in Latin America. *International Labour Review,* 106(1) : 11–45, July.

Barham, B.L. and Childress, M. 1992. Membership desertion as an adjustment process on Honduran agrarian reform enterprises. *Economic Development and Cultural Change.* Vol. 40(3) : 587–613, April.

Barraclough, S. and Collarte, J.C. 1973. *Agrarian Structure in Latin America: A Resumé of the CIDA Land Tenure Studies of Argentina, Brazil, Chile, Colombia, Ecuador, Guatemala and Peru.* Lexington Books.

Ben-Ner, A. 1984. On the stability of the co-operative type of organization. *Journal of Comparative Economics,* 8(3) : 247–260, September.

Berry, R.A. and Cline, W.R. 1979. *Agrarian Structure and Productivity in Developing Countries.* Baltimore: Johns Hopkins University Press.

Bhalla, S.S. and Roy, P. 1988. Mis-specification in farm productivity analysis: the role of land quality. *Oxford Economic Papers 40.* Oxford: Clarendon Press, Oxford University.

Binswanger, H.P. and Rosenzweig, M.R. 1986. Behavioural and material determinants of production relations in agriculture. *Journal of Development Studies,* 22(3) : 503–539, April.

Binswanger, H.P. and McIntire, J. 1987. Behavioural and material determinants of production relations in land abundant tropical agriculture. *Economic Development and Cultural Change,* 36(1) : 73–99, October.

Binswanger, H.P. and Elgin, M. 1988. What are the prospects for land reform? In A. Maunder and A. Valdez (Eds), *Agriculture and Governments in an Interdependent World.* Proceedings of the Twentieth International Conference of Agricultural Economists. Buenos Aires: International Association of Agricultural Economists.

Binswanger, H.P., Deininger, K. and Feder, G. 1993. Power, distortions, revolt and reform in agricultural land relations. In T.N. Srinivasan and J. Behrman (Eds), *Handbook of Development Economics, Vol. III.*

Bonin, J.P. 1985. Labour management and capital maintenance: investment decisions in the socialist labour-managed firm. In D.C. Jones and J. Svejnar

(Eds), *Advances in the Economic Analysis of Participatory and Labour Managed Firms, Vol. I.* Greenwich, CT: JAI Press.

Bonin, J.P. and Putterman, L. 1986. Economics of Co-operation and the Labour-managed Economy. *Fundamentals of Pure and Applied Economics 14.* Providence: Department of Economics, Brown University.

Boserup, E. 1965. *The Conditions of Agricultural Growth.* Chicago: Aldine Publishing Company.

Brandao, A.S.P. and deRezende, G.C. 1992. Credit subsidies, inflation and the land market in Brazil: a theoretical and empirical analysis. Mimeo.

Brockett, C.D. 1984. Malnutrition, public policy, and agrarian change in Guatemala. *Journal of Inter-American Studies and World Affairs,* 26(4) : 477–497, November.

Brown, M.R. 1989. Radical reformism in Chile: 1964–1973. In W. Thiesenhusen (Ed), *Searching for Agrarian Reform in Latin America.* Boston: Unwin Hyman.

Callison, C.S. 1983. *Land to the Tiller in the Mekong Delta: Economic, Social and Political Land Reform in Four Villages of South Vietnam.* Lanham, MD: University Press of America.

Calvo, G.A. and Wellisz, S. 1978. Supervision, loss of control and the optimum size of the farm. *Journal of Political Economy,* 86(5) : 943–952, October.

Carter, M.R. 1987. Risk sharing and incentives in the decollectivization of agriculture. *Oxford Economic Papers 39.* Oxford: Clarendon Press.

Castillo, L. and Lehman, D. 1983. Agrarian reform and structural change in Chile, 1965–79. In A.K. Ghose, *Agrarian Reform in Contemporary Developing Countries.* New York: St. Martins Press.

Chayanov, A.V. 1991. *The Theory of Peasant Co-operatives.* Columbus: Ohio State University Press.

Colburn, F.D. 1990. *Managing the Commanding Heights: Nicaragua's State Enterprises.* Berkeley: University of California Press.

Cornia, G.A. 1985. Farm size, land yields and the agricultural production function: an analysis of fifteen developing countries. *World Development,* 13(4): 513–534, April.

Davenport, T.R.H. 1987. Can sacred cows be culled? A historical review of land policy in South Africa with some questions about the future. *Development Southern Africa,* 4(3) : 388–400, August.

Davis, S.H. 1983. State violence and agrarian crisis in Guatemala. In M. Diskin (Ed), *Trouble in Our Backyard: Central America and the United States in the Eighties.* New York: Pantheon.

Deininger, K. and Binswanger, H.P. Rent seeking and the development of large scale agriculture in Kenya, South Africa and Zimbabwe. *Economic Development and Cultural Change.* Forthcoming.

De Janvry, A. 1981. *The Agrarian Question and Reformism in Latin America.* Baltimore, M.D.: John Hopkins University Press .

De Janvry, A. and Sadoulet, E. 1989. A study in resistance to institutional change: the lost game of Latin American land reform. *World Development,* 17 : 1397–1407.

Dorner, P. and Thiesenhusen, W.C. 1990. Selected land reforms in East and Southeast Asia: their origins and impacts. *Asian Pacific Economic Literature,* 4(1) : 69–95, March.

Eswaran, M. and Kotwal, A. 1985. A theory of contractual structure in agriculture. *American Economic Review,* 75(3) : 352–367, June.

Forster, N.R. 1992. Protecting fragile lands: new reasons to tackle old problems. *World Development,* 20(4) : 571–585, April.

Gerschenkron, A. 1965. Agrarian Politics and Industrialization in Russia 1861–1917. In *Cambridge Economic History of Europe, Vol. 6.* Cambridge, UK: Cambridge University Press.

Ghai, D., Kay, C. and Peek, P. 1988. *Labour and Development in Rural Cuba.* Geneva and New York: International Labour Office and Macmillan Press.

Ghose, A.K. 1985. Transforming feudal agriculture: agrarian change in Ethiopia since 1974. *Journal of Development Studies,* 22(1) : 127–149, October.

Griffin, K. and Hay, R. Problems of agricultural development in socialist Ethiopia: an overview and suggested strategy. *Journal of Peasant Studies,* 13(1): 37–66, October.

Grigg, D. 1992. *The Transformation of Agriculture in the West.* Oxford and Cambridge: Basil Blackwell, Oxford and Cambridge.

Hendricks, F.T. 1990. *The Pillars of Apartheid: Land Tenure, Rural Planning and the Chieftaincy.* Uppsala, Sweden: Almqvist & Wiksell.

Isaacman, A. and Isaacman, B. 1983. *Mozambique: From Colonialism to Revolution, 1900–1982.* Boulder, CO: Westview Press.

Israelson, L.D. 1980. Collectives, communes and incentives. *Journal of Comparative Economics,* 4(2) : 99–124, June.

Jarvis, L.S. 1989. The unravelling of Chile's agrarian reform, 1973–1986. In W. Thiesenhusen (Ed), *Searching for Agrarian Reform in Latin America.* Boston: Unwin Hyman.

Jensen, M.C. and Meckling, W.H. 1976. Theory of the firm: managerial behaviour, agency costs, and ownership structure. *Journal of Financial Economics,* 3(4) : 305–600, October.

Just, R.E. and Miranowski, R.E. 1988. US land prices: trends and determinants. In Maunder and Valdez (Eds), *Agriculture and Governments in an Interdependent World.* Proceedings of the Twentieth International Conference of Agricultural Economists. Buenos Aires: International Association of Agricultural Economists.

Kay, C. 1983. The agrarian reform in Peru: an assessment. In A.K. Ghose (Ed), *Agrarian Reform in Contemporary Developing Countries.* London: St. Martin's Press.

Kay, C. and Silva, P. (Eds). 1992. *Development and Social Change in the Chilean Countryside: From the Pre-Land Reform Period to the Democratic Transition.* Amsterdam: CEDLA.

Kenya Ministry of Agriculture. 1978. *Large Farm Sector Study,* 4 volumes. Nairobi: Ministry of Agriculture.

King, R. 1977. *Land Reform: A World Survey.* London: G. Bell and Sons.

Kinsey, B.H. and Binswanger, H.P. 1993. Characteristics and performance of resettlement programmes: a review. *World Development,* 21(9).

Koo, A.Y.C. 1968. *Land Reform and Economic Development: A Case Study of Taiwan.* New York: Praeger.

Kutcher, G.P. and Scandizzo, P.L. 1981. *The Agricultural Economy of Northeast Brazil.* Washington, D.C.: The World Bank.

LeGrand, C. 1968. *Frontier Expansion and Peasant Protest in Colombia, 1850–1936.* Albuquerque: University of New Mexico Press.

Leo, C. 1978. The failure of the 'progressive farmer' in Kenya's million-acre settlement scheme. *Journal of Modern African Studies,* 16(4): 619–678, December.

Leys, C. 1974. *Underdevelopment in Kenya: The Political Economy of Neo-Colonialism 1964–71.* Berkeley: University of California Press.

Lipton, M. 1985. *Capitalism and Apartheid: South Africa, 1910–84.* Totowa, NJ: Rowman and Allanheld.

Lipton, M. and Lipton, M. 1992. Some policy and institutional factors affecting the impact of agriculture on the poor: lessons for South Africa from experience elsewhere. *Paper presented at Experience with Agricultural Policy: Lesson for South Africa.* Mbabane, Swaziland: The World Bank/United Nations Development Programme, November 2–4.

Loveman, B. 1976. *Struggle in the Countryside: Politics and Rural Labour in Chile, 1919–1973.* Bloomington: Indiana University Press.

Masters, W.A. 1991. Comparative advantage and government policy in Zimbabwean agriculture. *PhD dissertation.* Stanford: Stanford University.

Mason T.D. 1986. Land reform and the breakdown of clientelist politics in El Salvador. *Comparative Political Studies,* 18 : 487–516.

McClintock, C. 1981. *Peasant Co-operatives and Political Change in Peru.* Princeton: Princeton University Press.

McClintock, C. 1984. Why peasants rebel: the case of Peru's Sender Luminoso. *World Politics,* 37(1) : 48–84, October.

McClintock, M. 1985. *The American Connection Vol 1: State Terror and Popular Resistance in El Salvador.* London: Zed Books.

McGregor A. 1977. Rent extraction and the survival of the agricultural production co-operative. *American Journal of Agricultural Economics,* 59(3): 478–488, August.

Meyer, C.A. 1990. A hierarchy model of associative farming. *Journal of Development Economics,* 34(1/2) : 371–383, November.

Migdal, J.S. 1974. *Peasants, Politics and Revolution: Pressure Toward Political and Social Change in the Third World.* Princeton: Princeton University Press.

Migot-Adholla, S., Hazell, P., Blarel, B. and Place, F. 1991. Indigenous land right systems in sub-Saharan Africa: a constraint on productivity? *World Bank Economic Review,* 5(1): 155–175, January.

Moore, B. 1966. *Social Origins of Dictatorship and Democracy: Lord and Peasant in the Making of the Modern World.* Boston: Beacon Press.

Mosley, P. 1983. *The Settler Economies: Studies in the Economic History of Kenya and Southern Rhodesia 1900–1963.* Cambridge: Cambridge University Press.

Munslow, B. 1985. Prospects for the socialist transition of agriculture in Zimbabwe. *World Development,* 13(1): 41–58.

Noronha, R. 1985. A review of the literature on land tenure systems in sub-Saharan Africa. *Report no. ARU 43*, Agriculture and Rural Development Department. Washington, D.C.: World Bank.

Nottidge, C.P.R. and Goldsack, J.R. 1971. *The Million Acre Settlement Scheme, 1962–1966.* Nairobi: Kenya Department of Settlement.

Ortega, E. 1990. De la reforma agraria a las empresas asociativas. *Revista de la CEPAL*/Naciones Unidas, Comision Economica para America Latina y el Caribe, Vol. 40, April: 105–122.

Otsuka, K. and Hayami, Y. 1988. Theories of share tenancy: a critical survey. *Economic Development and Cultural Change,* 37(1): 31–68, October.

Otsuka, K., Chuma, H. and Hayami, Y. Land and labour contracts in agrarian economies: theories and facts. *Journal of Economic Literature*, 30(4) : 1965–2018, December.

Pfeiffer, K. 1985. *Agrarian Reform under State Capitalism in Algeria.* Boulder: Westview Press.

Phimister, I. 1988. *An Economic and Social History of Zimbabwe 1890–1948: Capital Accumulation and Class Struggle.* London, New York: Longman.

Pickett, L.E. 1988. *Organizing Development through Participation: Co-operative Organization and Services for Land Settlement.* A study prepared for the International Labour Office. London: Croom Helm.

Putterman, L. and DiGiorgio, M. 1985. Choice and efficiency in a model of democratic semi-collective agriculture. *Oxford Economic Papers 37.* Oxford: Oxford University Press.

Ranger, T. 1985. *Peasant Consciousness and Guerrilla War in Zimbabwe: A Comparative Study.* London: James Currey.

Ruthenberg, H. 1966. *African Agricultural Production Development Policy in Kenya, 1952–1965.* Berlin: Springer.

Scaritt, J.R. 1991. Zimbabwe: revolutionary violence resulting in reform. In J.A. Goldstone, T.R. Gurr and F. Moshiri, *Revolutions of the Late Twentieth Century.* Boulder, CO: Westview Press.

Scott, C. 1976. *The Moral Economy of the Peasant.* New Haven: Yale University Press.

Scott, M.F.G., MacArthur, J.D. and Newbery, D.M.G. 1976. *Project Appraisal in Practice: The Little Mirrlees Method Applied in Kenya.* London: Heinemann.

Sen, A.K. 1981. Market failure and control of labour power: towards an explanation of 'structure' and change in Indian agriculture, Parts 1 and 2. *Cambridge Journal of Economics,* 5(3 & 4): 201–228 and 327–350, September and December.

Skockpol, T. 1979. *States and Social Revolutions: A Comparative Historical Analysis of France, Russia and China.* Cambridge: Cambridge University Press.

Skockpol, T. 1982. What makes peasants revolutionary? *Comparative Politics,* 14: 351–375.

Southgate, D. 1990. The causes of land degradation along 'spontaneously' expanding agricultural frontiers in the Third World. *Land Economics,* 66(1): 93–101, February.

Stiglitz, J.E. 1987. The causes and consequences of the dependence of quality on price. *Journal of economic literature,* 35 : 1–48.

Thiesenhusen, W.C. 1987. Incomes on some agrarian reform asentamientos in Panama. *Economic Development and Cultural Change,* 35(4): 809–831, July.

Thiesenhusen, W.C. 1991. Implications of the rural land tenure system for the environmental debate: three scenarios. *Journal of Developing Areas,* 26: 1–24.

Trautmann, W. 1984. Rural development in Algeria: the system of state-directed co-operatives. *Quarterly Journal of International Agriculture,* 24(3): 258–267, July–September.

Walker, T.S. and Ryan, J.G. *Village and Household Economies in India's Semi-arid Tropics.* Baltimore: John Hopkins University Press.

Weiner, D. 1988. Land and agricultural development. In C. Stoneman (Ed), *Zimbabwe's Prospects: Issues of Race, Class, State and Capital in Southern Africa.* London: Macmillan.

Wolf, E. 1968. *Peasant Wars of the Twentieth Century.* New York: Harper and Row.

Wuyts, M. 1985. Money, planning, and rural transformation in Mozambique. *Journal of Development Studies,* 22(1): 180–207, October.

Zamosc, L. 1984. The agrarian question and the peasant movement in Colombia: struggles of the National Peasant Association 1967–1981. Cambridge: Cambridge University Press.

Zamosc, L. 1989. *Peasant Struggles and Agrarian Reform.* Meadville, PA: Allegheny College.

Learning from Experience

PART II

5

Characteristics and performance of settlement programmes: a review[1]

Bill Kinsey and Hans Binswanger[2]

Introduction

This chapter examines the characteristics and performance of settlement and resettlement programmes. We were prompted to undertake this review because of our interest in possible land reform in South Africa. The key question is whether land belonging to about 50 000 large-scale commercial farms, and covering about 87 per cent of the arable surface of the country, should be subject to land reform for the benefit of black and 'coloured' populations. Like other countries with dualistic farm size structures, the commercial farm sector is highly mechanized and, as a consequence, only relatively few workers (or labour tenants) remain on the large farms. Redistributing the farms to these workers alone will create holdings that have too few family workers to operate efficiently, or to maintain the current high degree of mechanization, without hiring casual workers. The beneficiaries, however, are unlikely to be sufficiently wealthy or to have access to subsidized credit to enable them to hire the necessary workers or buy the required machines. In order to maintain or increase the productivity of the distributed farms, therefore, additional beneficiary families will have to be resettled onto them. This approach is also necessary to extend the benefits of land reform to rural populations that no longer reside on commercial farms. Land reforms involving settlement are much more difficult than reforms of landlord estate systems; in most of Asia under these estates, reform involved transferring ownership rights to tenants who were already in possession

1 Reprinted from *World Development*, Vol. 21 no. 9, Characteristics and performance of resettlement programmes: a review, pp. 1477–1494, 1993, with kind permission from Elsevier Sciences Ltd, The Boulevard, Langford Lane, Kidlington OX5 1GB, UK.

2 The authors gratefully acknowledge the many helpful comments of the participants at a workshop in Swaziland in November 1992 where an earlier version of this paper was presented. We would also like to express our appreciation for the generously detailed comments on the workshop version of the paper provided by S. M. Jinya and E. A. Attwood.

of the land, had implements and draft animals, and knew the local climate, soils and farming techniques. Only if the settlement component of the land reform programme is properly designed can land reform involving large-scale commercial farms succeed.

The scope of our review, therefore, is the worldwide settlement experience, whether or not in the context of land reform, and regardless of the administrative nature of the programmes. The programmes we have analysed include planned settlements executed by governments; private settlement by companies or social or religious groups; and regularization of, and support for, spontaneous settlements or land invasions.

Categories of analysis

Resettlement with agricultural production as one of its core objectives has been undertaken on every continent, creating a large literature.[3] As noted by Hulme (*1988*), much of the literature is fragmentary and idiosyncratic. There is no generally accepted methodological approach or theoretical basis for the analysis of resettlement schemes. Hulme characterizes three broad categories of analysis: (a) conventional evaluations, the majority, which are based on empirical approaches to scheme or policy performance; (b) the social consequences approach, as practised mainly by sociologists and anthropologists, which is concerned with the impact of schemes on individuals, families and communities; and, (c) radical and political approaches, which are derived from theories of the role of the state in development.

We have examined a wide range of literature in all three of the categories and rather than impose our own ideas about the success or failure of settlements from a distance, we have accepted the judgement of the researchers, government and agency officials who were involved. Because the bulk of the literature comes from the first of Hulme's categories, assessments of performance are weighted heavily by the disciplines in which researchers have been trained – principally agriculture, economics, geography and public administration. Moreover, evaluations in this category tend to present static descriptions and to gloss over

3 The extensive literature on the experience with planned settlement in Africa alone contains numerous examples of programmes judged to have been unsuccessful in many, if not most, dimensions. See, for example, Lewis, 1954; Hilton, 1959; Christodoulou, 1965; de Wilde *et al,* 1967; Moris, 1968; Chambers, 1969; Takes, 1975; World Bank, 1978; Hansen and Oliver-Smith, 1982; Reining, 1982; Kinsey, 1982 and 1983; Leo, 1984; McMillan, 1987; and Cook and Mukendi, 1992.

processes and relationships. Unfortunately, this deficiency cannot be remedied by balancing conventional evaluations with those drawn from the social consequences literature, since there is too little of the latter to serve as an effective counterbalance. Finally, because the radical and political approaches are generally unconcerned about the success or failure of the specific characteristics in which we are interested, this approach is least represented in this review.

In addition to the individual studies, we have drawn on reviews by other researchers, on institutional reviews such as those of World Bank-assisted programmes, and on some government documents. We also undertook a more detailed review of the characteristics of seven such programmes (see Table 5.1 below).

Table 5.1: Characteristics of Resettlement Programmes

Burkina Faso	River valley resettlement following *onchocerciasis* clearance.
Ethiopia	Imperial and post-revolution resettlement.
Guatemala	Resettlement based on a market-directed alternative to land reform.
Indonesia	Transmigration resettlement (both official and spontaneous).
Kenya	Irrigated resettlement (pre- and post-independence); dryland resettlement (pre- and post-independence); spontaneous resettlement.
Malaysia	Official resettlement based on perennial crops and creation of employment.
Zimbabwe	Irrigated resettlement (pre- and post-independence); dryland resettlement (post-independence); spontaneous resettlement.

Methodological issues

We are seeking to evaluate individual characteristics of resettlement rather than pass definitive judgements on the success of the complete package of activities in resettlement programmes. We distinguish two broad groups of interrelated variables.

The first group contains the *implementation characteristics* of resettlement such as: scale (overall size, average holding size and number of settler families); organization (public, private, spontaneous or mixed); sources of funding for land acquisition, infrastructure and credit; settler selection rules and practice; land allocation rules and practice; land rights (freehold/leasehold/tenancy, pledging, subdivision and inheritance); restrictions (crops grown, markets, employment and others); provision of infrastructure and speed of construction; access to and modality of supply of production services (credit, extension, tillage, irrigation, storage, marketing and transport); access to and modality of supply of social services (health care, education, housing, water and sanitation); quality of the resource base (soils and water); cost per beneficiary family; grant elements (in land cost, infrastructure, credit, and social services); cost recovery rules (for land, infrastructure, social services, and credit); costs and timing of cost recovery; the speed of formation, organization, autonomy, and the participation of settler groups.

We want to evaluate the impact of the implementation characteristics on the second group of variables, the *outcomes* of resettlement. This group of variables includes yields and production levels; family income levels; asset accumulation and savings; consumption levels; poverty alleviation; responsive capacity; diversification of production patterns; land trading and reaggregation and/or subdivision of holdings; environmental impacts; stability of settlement; and sustainability of farming systems.

Few, if any, settlements will be successful in all outcome variables; and very few studies contain details on more than one or two outcome variables. But since we are interested in the specific characteristics associated with success, and not in the package as a whole, even judgements of success in some dimensions will be illuminating.

Economic investigations often suffer from a severe lack of reliable data, and few evaluations cover more than one point in time. There is the additional difficulty of choosing an appropriate time frame for studying resettlement. In the short- or medium-term, agricultural settlement projects that remain on a subsistence level will not generate many stimuli to economic development. Yet, in the longer term, even projects of this nature can produce beneficial development effects by concentrating population in ways that allow growth opportunities to be realized in other economic spheres. Resettlement projects judged to have been successful in one or more dimensions appear to have passed through a series of three to five year stages of evolution, during which settlers adjusted to their changed circumstances in ways that affected the performance and impact of the settlement project (*Chambers, 1969; Colson, 1971; Moran, 1979; Scudder, 1973, 1975 and 1984; Scudder & Colson, 1979 and 1981; Kebschull, 1986*).[4] Most of the evaluations we have examined, however, are set in static cross-sectional frameworks.

Characteristics and performance

The land tenure system defines rights and obligations with respect to the acquisition and use of land in agricultural settlements, and the land settlement projects incorporate a wide array of tenure systems, each of which reflects local customs, differing legal systems, and specific programme objectives.

Land tenure, allocation rules and practice

It is clear on the basis of the review that no settlement programme can succeed without a clear tenure policy. There are four major issues concerning tenure: (a) whether to allow individual holdings of arable land or insist on collective methods of farming; (b) whether to grant permanent ownership rights or only usufruct rights; (c) whether to allow sale and rental of land or constrain land transactions; and, (d) if land sales to outsiders are unrestricted, whether or not to issue title.

As in the socialist world, where collectives are being abandoned, the record shows that with few exceptions, collective forms of land use in developing countries – where a farm is managed as a single large-scale unit and members receive labour points and profit shares – are associated with failure.[5] World Bank-assisted irrigation projects involving collective farming and tenure in Mexico, for example, performed poorly relative to those where beneficiaries were given individual plots. The review also found that where beneficiaries were organized as collective farms, not only in Mexico, but also in Peru, Honduras and Nicaragua, they often subdivided the holdings informally. In addition, in

4 In utilizing the concept of stages in the development of resettled areas, Nelson (*1973*) maintains that the stages are generally of longer duration than most analyses. He describes three stages in the development of successful settlements:

- The pioneer stage (5–10 years), during which settlers move into an area and establish basic life-support systems such as water, food, shelter and access to consumption goods, markets and credit.
- The consolidation stage (5–10 years), during which community organizations, educational and health services, transport improvements and permanent housing are established, and better use is made of landholdings and credit.
- The growth stage (duration unspecified), during which farms are capitalized, permanent agricultural systems are established, related agro-industries are developed, credit becomes generally available, and supply and marketing systems are firmly grounded.

5 See, for example, the case of Zimbabwe as recounted by Akwabi-Ameyaw (*1990*). A similar pattern of failure occurred on the communal plots in Tanzania's *ujamaa* villages (*Oberai, 1988b*).

Zimbabwe, collective co-operative settlements – Model B schemes[6] – managed, but have poor group cohesion (*Akwabi-Ameyaw, 1990*). In a single year, more than 20 per cent of the country's collectives disbanded, and groups willing to take over the facilities have been hard to find (*Zimbabwe Government, 1991*). The one collective co-operative in Zimbabwe judged to be successful has in fact parcelled out its land and operates as a quasi-Model A scheme, a smallholder model with nucleated settlements (*Mumbengegwi, 1984*). Likewise, in Kenya where resettlement schemes have taken a variety of forms, there was at one time a pattern of private individuals and companies purchasing large farms that subsequently failed. These farms were then subdivided, resulting in a better match between farm sizes and household resources and managerial skills (*Deininger & Binswanger, 1992*).

In a number of resettlement programmes, individual tenure on arable land has been combined with common ownership and management of pasture land and forest resources, as in the Mexican *ejido* system and Zimbabwe's Model A schemes. The literature on common property management in settlement projects, however, is scanty.[7]

The review shows no clear advantage of ownership over secure usufructuary rights or long leases, so long as land rights are unequivocally spelled out. During an initial period, spontaneous settler communities in Bolivia often provide only for usufructuary rights, which can be reassigned when unsuccessful settlers leave. These usufruct rights only harden to ownership rights over time, as under the *ejido* system in Mexico, where they now have been converted into full ownership rights. The Mexican irrigation systems, for example, were settled mainly by *ejidatarios* who, until recently, only enjoyed permanent and inheritable rights of usufruct (*Heath, 1992*). Their five to ten hectare plots in Northwest Mexico were the original home of the Green Revolution and have been associated with highly modern farming ever since. A positive example for leases is the experience of the Family Farms resettlement schemes in Zambia, where 14-year inheritable leases were granted to

6 In Zimbabwe, there are four distinct resettlement models: Model A comprises smallholder farming based on individual allocations of arable land, communal grazing land and nucleated villages; Model B comprises collective co-operatives; Model C is satellite production and a core estate; and Model D is an experiment to explore options for livestock-based resettlement.

7 Most of the literature on this theme that we reviewed dealt with the issues of resettling pastoralists and managing common grazing lands. See, for example, Oxby (*1984*) and Adams and Devitt (*1992*).

settlers very quickly (*Bomford, 1973*). China is another case where the use of leases has been positive.

Technical efficiency and economic viability are best achieved by a programme design that promotes individual landholding through the issue of freehold titles or other long-term usufructuary agreement. Despite the obvious association between security of tenure and success of settlement, there is an almost universal tendency to withhold title to ownership, even where full private ownership is the intended legal framework. Formal documentation of usufruct rights is even more seldom achieved. Withholding title is often justified by the settlers' indebtedness to the land reform or settlement agency, even long after it is clear that the debt will never be recovered or after inflation has reduced the debt to nominal amounts. But withholding title or clear rights only perpetuates administrative approaches. Such was the case in Paraguay, where state land was allocated but titles were withheld for two decades, and in Burkina Faso, where the government remained unclear about its land ownership policy almost 20 years after settlement began (*Murphy & Sprey, 1980*). In the case of the Dandakaranya project in India, the failure to confer land ownership rights for as long as two decades after land allotment led to unrest and a mass exodus of settlers (*Bose, 1983*).

Associated with poorly defined land rights and with the withholding of title or documentation of usufruct rights, are poor access to credit and a lack of investment – as in the cases of Burkina Faso, Costa Rica, Zimbabwe, Kenya, and Indonesia – and an exacerbation of intergenerational problems. The experience of Kenya and Thailand shows that granting titles can be managed, even where the agricultural sector is characterized by very large numbers of smallholdings (*Jones & Richter, 1982; Haugeraud, 1983; Scholz, 1986; Green, 1987; and Manshard & Morgan, 1988*).

Constraints to selling land in settlements are widespread. Where communities form spontaneous settlements, as the Mennonites did in Brazil or Bolivia, and Indian groups and farmers' unions did in Bolivia, sales to outsiders are usually prohibited or sharply constrained, which makes such settlements, in effect, communal tenure systems. In these cases, land title provides little advantage for accessing official credit schemes since alienability is restricted. On the other hand, sales and rental to members of the community are usually allowed.

In official settlement schemes as well, there is a uniform tendency to outlaw sales and rental, both to outsiders as well as among the beneficiaries of the schemes. These restrictions often reflect the desire to retain tight administrative control over settlers – a position often justified by the argument that without such conditions, settlers would either sell or pledge the land and become landless, or they would subdivide it among their heirs and make the holdings uneconomic. Prohibitions on the rental or sale of land, however, are almost uniformly damaging since they lead to conflict and result in land being left idle. In Costa Rica, when land was abandoned by plotholders, no one else could use it. In Zimbabwe, settlers are not only expected to forego their rights to land in their area of origin, but are also prohibited from selling or leasing their plots; they enjoy no intergenerational security of tenure on settlement land. The early emphasis on welfare objectives in Zimbabwe's resettlement programme resulted in significant numbers of destitute settlers who could not use all the land allotted to them and who, unless they engaged in illicit land sales or leasing, became victims of 'land-poaching' (*Zimbabwe Government, 1991*). The review shows that restrictions on rental and sales can rarely be enforced and that, if they are partially enforced, they tend to create a climate of insecurity of tenure leading to efficiency losses and idle land.

The evidence clearly shows that static tenurial arrangements, if rigidly enforced, lead ultimately to stagnation or, if the administering authority lacks the power or will to enforce tenure regulations, to unrest and upheaval. Programmes need to be designed deliberately, therefore, as experiments that incorporate a range of possible approaches. One way to do this might be to configure the initial allocation so that not all the land is allocated, but that some is held in reserve to permit later flexibility. Another way would be to allow rent, sale, transfer and subdivision. Although learning-by-doing is a principle accepted by many resettlement agencies, a number of these agencies nevertheless pursue rigid programmes.

Scale of programme

There is little we can say about whether or not scale is associated with success. The literature gives equally positive ratings to programmes as small as the Family Farms schemes in Zambia and as massive as the programmes of the Malaysian Federal Land Development Authority (FELDA). Other large-scale programmes, as in Ethiopia or the case of villagization in Tanzania, have largely been judged failures. More significant than physical scale appears to be the time scale for implementation, flexibility in the programme, and the quality of staff involved with implementation. Rapid, overly ambitious implementation is always associated with failure in at least some dimensions, and frequently involves costly and socially disruptive resettling of populations that have already been dislocated once.

Size of holdings

The size of allocated holdings is an important factor in attracting settlers and determining agricultural incomes; the more appropriate the size, the more successful the result. As a recent review of the literature on economies of scale has suggested (*Deininger & Binswanger, 1992*), the area that a farm family can operate efficiently depends on the amount of family labour, the farmer's managerial skills, the machinery and capital stock that the farmer owns, and the farmer's access to credit markets. An approach based on these considerations was taken in the Family Farm settlements in southern Zambia, where holding sizes and the ratio between arable and grazing land were determined in response to each settler family's labour, livestock and initial capital endowment.

Few settlement programmes, however, explicitly allocate land according to the factors discussed above. Instead, two principles are frequently used to determine holding size: (a) farms must not be larger than settlers can cultivate (often using only family resources); and, (b) farms must be large enough to provide settlers with an income higher than they had previously, but not so high that tensions develop between favoured settlers and existing farmers. The review found that these principles are often inflexibly employed; for example, extreme rigidities in land allocation have characterized the programmes in Zimbabwe and Indonesia (*Kinsey, 1983; Uhlig, 1984; Kebshull, 1986; Fasbender & Erbe, 1988*). By contrast, FELDA in Malaysia has modified the size of the standard holdings allocated to its settlers several times over two decades in response to changing market prices and income possibilities (*Ogawa & Chan, 1985; Bahrin, 1988*).

In light of the evidence on economies of scale and the skills and support needed for successful settlements, it is to be expected that excessive allocations of land to farmers with little family labour, poor farming skills and poor access to credit would lead to poor results. In Sudan's Gezira Scheme, for example, many settlers became absentee landlords because the land they were allocated far exceeded the labour capacity of their families (*Sabry, 1970*). Additional evidence of the benefits of allocating relatively small holdings comes from the official low- and high-density settlement programmes in Kenya. While the low-density schemes were on better land and required greater levels of capital assets and farming skills than the high-density schemes, they returned lower levels of social profit. Moreover, in Kenya, early studies of the economic performance of resettlement schemes also demonstrated that smallholders, working more labour-intensively, could make more profitable use of land previously cultivated by large farms (*von Haugwitz, 1972*). One estimate put the increase in production arising from the shift from large, low-density farm units to small, high-density units at 15 to 90 per cent (*Deininger & Binswanger, 1992*). Similarly, in the higher potential areas of Zimbabwe, yields, production and incomes from the five hectare plots of arable land outstripped planners' conservative targets for Model A schemes. The Model B schemes, on the other hand, with an average size of more than 9,8 hectares of mostly prime land, have been economic and social disasters (*Zimbabwe Government, 1991*).

There are instances, however, when the rule about economies of scale does not seem to apply. In Indonesia's resettlement schemes, for example, the long-standing practice was to allocate two hectares of land based on an irrigated rice paddy farming system (*Hardjono, 1977*) but, under rainfed farming, this amount of land was clearly insufficient. Similarly, in Zimbabwe's lower-potential areas, where the standard holding has been unable to generate sufficient output, government reports show that growing numbers of settlers have been extending their arable areas illegally (*Zimbabwe Government, 1991*). Much the same response pattern was found in Indonesia (*Fasbender & Erbe,1988*).

The characteristic most critically associated with the success of settlements appears to be *flexibility with respect to farm size*. There is evidence, for example, that larger holdings are more privately remunerative (Kenya and Malaysia), while smaller holdings are socially more profitable (Kenya) and can produce crops at lower cost (Bolivia). The point is that flexibility in allocating appropriately sized holdings is positively associated with the use of purchased inputs, lower levels of social conflict, greater income-earning possibilities, growth of mutual-help efforts, and fewer administrative problems. Nevertheless, many rigid administrative systems do not respond to errors in farm size, but persist in allocating holdings of a size that has proven ineffective.

Type of scheme organization and administration

Management and administration are also important in determining the degree of success of a settlement programme, and the review revealed a wide variety of patterns. A broad typology of styles of organization and administration includes the following:

Administered official settlement

The classic example of this approach is Mexico's land reform in which land was administered and settled over a 50-year period, with the process controlled by an enormous land reform bureaucracy that withheld title from settlers for decades.[8] Other examples include programmes in Indonesia, Zimbabwe, Ethiopia, Guatemala and Burkina Faso, and irrigated resettlement in Kenya. Typically, the administering authority is involved in every aspect of resettlement: from land acquisition to settler selection, infrastructure and irrigation development, extension, and post-production activities.

Perhaps the most dominant and damaging characteristic in almost all the administered programmes examined is excessive paternalism on the part of the administering authority and associated personnel. The adverse consequences of paternalism are worse when it is centrally administered than when administration is local. The costs of centrally administered paternalism are seen most vividly in the case of Ethiopia, but the evidence strongly suggests that central organizations perform rather poorly everywhere. Support for this claim is also found in Indonesia, where resettlement performance improves when responsibility for administration is handed from the central agency to the provincial administration at the end of a project's first five years (*Oey & Astika, 1978*).

Paternalism aside, however, there have been important differences in administrative style among resettlement projects, many of which have developed in response to the challenge of managing what are inherently multi-sectoral projects. In Zimbabwe, for example, the Ministry of Agriculture does the planning and technical preparation work, including land surveys and setting out plots. The Department of Rural Development of the Ministry of Local Government selects the settlers; moves them onto the plots and chairs the co-ordinating committee that secures infrastructure investments, social services and agricultural extension from the ministries in charge. It also controls the development of rural service centres. In an effort to strengthen implementation capacity and avoid time-consuming attempts at co-ordination, Zimbabwe's major resettlement agency is currently seeking to expand its authority and, in the face of resistance from other agencies with settlement-related roles, to play a more central role in all resettlement programmes (*Zimbabwe*

8 See Ireson (*1987*) and Heath (*1992*) for analyses of the consequences of Mexico's long land reform programme.

Government, 1991). Land reform agencies in Central America, as well, have attempted to become centralized multi-sectoral development agencies for their clientele, but with very poor results. A centralized administrative approach also exists in Indonesia but there, as noted above, the management is handed over to local authorities after the fifth year of each settlement project. In Malaysia, a centralized settlement institution has handled execution at the local level very effectively through the extensive involvement and collaboration of local government institutions, although the development of even more localized institutions suited to settler needs has been discouraged (*Bahrin, 1988*). Kenya largely managed to avoid the problems of centralised administration by maximizing the involvement of existing local institutions and using a short-term task force approach toward each project, with temporary staff seconded from other institutions.[9]

In a comprehensive analysis of 24 land settlement schemes in Latin America, Nelson (*1973*) recognized five different categories: directed settlements, semi-directed settlements, spontaneous settlements, private land settlement schemes, and foreign colonization. His ranking of the 24 schemes shows that directed settlement projects have the poorest performance record: 'Few spheres of economic development have a history of, or a reputation for, failure (that can) match that of government-sponsored colonization … the evidence is irrefutable' (*Nelson, 1973: 265*).

While many examples from the rest of the world confirm the experience in Latin America, the cases of Kenya and Indonesia show that the pitfalls of government-sponsored settlement can be reduced by strong involvement of local governments. The finite horizon of the Kenyan task force approach, and of the Indonesian handover to local governments, has prevented paternalism from developing in these cases, as it did in the FELDA schemes in Malaysia or in Zimbabwe's open-ended commitments.

Paternalism is usually based on good intentions – the wish to see settlers do well and to shield them from excessive hardships in the early stages of the resettlement process. But these good intentions are almost always linked to a profound failure to appreciate that continued control and coddling of settlers in the later stages of the resettlement process is more damaging than inadequate support in the earliest stages.

9 C. P. R. Nottidge, personal communication.

Paternalism is also strongly linked to institutional self-interest, to a misconception of the processes that make resettlement effective, and to the assumption that resettlement programmes can be successful only if administrative support is maintained.

Paternalism expresses itself in excessive constraints on settlers' decisions regarding production techniques and crop choices, land rental and sale, labour market participation, investment and marketing. Several of these constraints are discussed in other sections of the chapter. Some constraints arise when settlement agencies pursue a full-time entrepreneurial farmer model and seek to create small replicas of commercial farms or plantations. In Zimbabwe, for example, the concern with settlers' commitment to full-time farming went so far that, for more than a decade, settlers were prohibited from participation in the non-farm labour market.[10] Commitment to this kind of model, however, ignores the fact that small farmers all over the world, including in the developed countries, combine farm and non-farm sources of income within the same household. Given the severe credit constraints faced by small farmers, their reliance on several uncorrelated income sources must be considered a wise portfolio diversification.

An example of paternalism being motivated by concern is the belief that settler-controlled purchases of inputs or sales of outputs would result in less favourable prices, due to the settlers' lack of experience and information. This concern, however, could also reflect a wish to control the harvest in order to ensure credit repayment or at least know the precise credit repayment capacity of each settler. For example, in the programme of the *Fundación del Centavo* (FUNDACEN, or the Penny Foundation) in Guatemala, settlers are denied the knowledge and experience that would equip them to manage their own input purchases and commodity marketing (*Dunn, 1992*). FUNDACEN, a private foundation, has a committed field staff, but staff members tend to design paternalistic approaches irrespective of their sectoral affiliation.

10 This prohibition has recently been relaxed in response to twin crises. First, most settlers in the early years of the programme received institutional credit but subsequently defaulted under the strain of the successive droughts of 1982, 1983 and 1984. They have thus been ineligible for further loans from the agricultural credit agency. Second, the severe drought in the 1991/92 season meant that many settler households were unable even to feed themselves and had to liquidate assets in order to purchase food. Condoning off-farm work to allow farmers to acquire operating capital, rebuild assets and procure food, was therefore a prudent reaction to a situation for which a remedy was beyond the scope of the state's resources.

Technical staff often fail to understand that they are participating in a new and different system. They fail to recognize that settlers are to be encouraged in their new role as owners or rights-holders, and treated as independent decision-makers who must learn to operate within the constraints imposed by nature, their skills, their assets, and their access to markets. Instead, technicians often behave as if they were foremen or overseers for their institutions. In Zimbabwe, some resettlement officers have been accused of dictatorial behaviour, while technical staff in Guatemala behave as if settlers could be 'fired' for small or imagined infractions (*Dunn, 1992*).

Unauthorized or spontaneous settlement

The global importance of spontaneous settlement was made clear in a World Bank study (*1978*), which showed that almost 75 per cent of new rural land settlement in the late 1970s was spontaneous. This type of settlement is far more flexible in the possible arrangements for land acquisition than administered settlements. Spontaneous settlers often negotiate for land with existing farmers, as in Indonesia, where land can be purchased, rented or acquired under traditional arrangements. Other spontaneous settlements are community-based, as in the Eastern Lowlands of Bolivia, where Mennonites purchase lands and then parcel out plots to their members, and where highland Indians establish settlement outposts for some of their members in the unoccupied lowland forests (*Solem et al, 1985*). Some communities even relocate in their entirety, as did the Tangwena people in eastern Zimbabwe.

The pre-existing, self-selected settler groups that engage in spontaneous settlement tend to have strong cohesion and resilience but, by and large, they have been prohibited from participating in official programmes. In Bolivia, for example, the official programmes were changed to admit such groups only after administered settlement failed. In Brazil, where the Mennonites carried out a resettlement programme entirely on their own, their projects, as well as many private schemes, have generally been far more successful than public ones.

There have been numerous administrative responses to spontaneous settlers, or squatters, as they are often called. One response, as in Costa Rica and Honduras, where hundreds of farms were taken over by syndicates of settlers, has been to incorporate these groups into the land reform programme, expropriate and (partly) compensate the former landowners, and prohibit settlers from selling or renting their five to ten hectare parcels. At the other extreme are overtly hostile responses, with both colonial and post-independence governments burning and bulldozing settlers into at least a temporary withdrawal. A more moderate response has occurred in Zimbabwe, where spontaneous settlers have taken over pasture land and stream banks in existing resettlement schemes and undertaken pre-emptive settlement on new properties acquired for resettlement. Here political and social sensibilities have been paramount and, only after a decade of tolerance, have the official settlement agencies and the beneficiaries of official settlements become more hostile toward the squatters.

Despite the high costs associated with many official programmes, the literature contains no comparative, *ex-post facto*, cost-benefit analysis of spontaneous settlement. This lacuna partly reflects the fact that few public resources are used by spontaneous settlers but, even in the absence of any analysis, it can be assumed that private individuals or communities continue to expend resources because they are reaping positive returns. Very little work has been done, as well, on the self-management capabilities of spontaneous private sector settlements. Brazil, Paraguay and Bolivia seem to have had positive experiences with this settlement (*Solem et al, 1985; Ozorio de Almeida, 1992*), but more systematic enquiry is needed into the conditions required for community-based management of settlements.

Assisted spontaneous settlement

Another alternative to administered settlement is to allow land to be invaded and then to regularize the occupiers.[11] Although spontaneous settlers at times seem to choose self-managed schemes in order to avoid defects in administered schemes, in other cases they model themselves after official settlements, sometimes even surpassing official settlements in importance and success, as in Indonesia (*Hardjono, 1986; Scudder, 1986*) and Zambia. In Kenya, as well, it has been argued that large-scale spontaneous settlement was a direct consequence of the official programmes (*Mbithi & Barnes, 1975; Leo, 1984; Mwagiru, 1986*). These experiences have led to programmes that provide selected infrastructure, land regularization and agricultural services to spontaneous settlers. Such a strategy was pursued with apparent success in Bolivia after the directed settlement approach failed, but it was abandoned by external funding agencies and was unable to get domestic budget support.

11 If there is no reconciliation, and negotiations become stuck in stalemate, this may well happen in South Africa.

Assistance before and after settlement can come from either private or public organizations, or both. In the case of the Family Farms settlements in Zambia, for example, former squatters were helped to become productive settlers by a non-governmental organization that co-ordinated inputs from the government, and expertise and resources from a commercial bank (*Bomford, 1973; Kinsey, 1979*).

Combining settlement with contract farming

Some of the classic plantation crops lend themselves very well to resettlement schemes that incorporate contract farming. Because of the need to closely co-ordinate the production and processing of sugar, for example, that crop lends itself either to contract farming, as in some of the irrigated resettlement in Zimbabwe's lowveld, or to Model C type schemes – large-scale farming on core estates with nearby settlers having the chance to supplement their income through seasonal wage labour (*Kelly, 1972; World Bank, 1984; Stanning, 1984; Rukuni, 1988*). Contract farming has also worked well for settlements that produce tea in Kenya and palm oil in Malaysia; and settlement agencies have operated successful contract schemes in Sudan (the Gezira scheme) and in Zimbabwe. FUNDACEN also successfully uses contract farming models for vegetable production in its Guatemala settlements (*Dunn, 1992*). On the other hand, the weak performance of Zimbabwe's Model C tobacco-growing schemes suggests that contract farming alone cannot circumvent other problems (*Zimbabwe Government, 1991*). Moreover, although dairying shows potential as a productive activity for resettlement, and has been highly successful in the Nandi medium-scale resettlement areas north-west of Nairobi in Kenya (*Metson, 1979*), it too has been plagued with problems when implemented in the form of a Model C scheme in Zimbabwe.

Modalities of land acquisition

Where land for settlement is owned by governments, as in the case of Bolivia, or can be acquired under customary law, as in Indonesia, land acquisition presents no major problem. Elsewhere, the government must either purchase land at market prices or expropriate it as part of a land reform programme, with varying degrees of compensation. Alternatively, land can be purchased by communities such as the Mennonites, by non-profit organizations such as FUNDACEN, or by private settlement companies. There is no evidence that either the mode of acquisition of land, or the price paid for it, has a strong impact on the ultimate performance of settlements.

The level of compensation that buyers pay for land, however, strongly influences the scope of the settlement programmes since the budgets for such programmes are often sharply constrained. As discussed in Binswanger, Deininger and Feder (*1992*), only a fraction of the land costs are likely to be recovered from poor settlers if compensation is at the market price. Limited budget allocations have confined land reform and settlement programmes in Costa Rica, Honduras, Guatemala, and Brazil to very small areas. British budget resources were used to finance land purchases in Kenya; and in Zimbabwe, the British government's share in resettlement programmes was 50 per cent of all costs including land, which allowed for greater scope of the programme, if not necessarily for greater speed of implementation (*Leo, 1984; Cusworth & Walker, 1988*).

Settler selection rules and practice

Settlement is a difficult process in which the settlers must adapt to new environments and, at the same time, create productive enterprises. Failure is common and dropout rates can be high. To deal with the consequences of dropout requires greater flexibility in land allocation, as discussed in the previous section. Better settler selection can reduce later dropout, even though it cannot eliminate it.

Issues of settler selection arise in relation to directed, spontaneous and private resettlement. Of course, in the case of spontaneous or private settlement, it is the settlers or the communities that do the selecting themselves. Relying on such self-selection is likely to lead new communities which are more socially representative and cohesive than those created by applying a narrow set of criteria. Since official resettlement programmes have received the most attention in the available literature, however, it is this experience that we draw upon here.

A point-scoring system has been adopted by FELDA in Malaysia and is under consideration in Zimbabwe. In this system, the specific traits needed for resettlement communities to be successful are clearly spelled out and defined objectively. The traits most commonly taken into consideration in selection procedures are described in Table 5.2.

Although we found no consistent evidence to support favouring settlers who have capital, we did find fairly consistent evidence that choosing settlers mainly on the basis of equity leads to problems. It seems to be a fact of life that agricultural settlement schemes, except in special circumstances, do not make good welfare programmes. While the landless may be able to make good use of the new resources provided by settlement, the elderly, sick and disabled, and those with an inadequate access to labour, usually cannot. A successful programme, therefore, must maintain a balance between a settler's suitability and need.

In the point-scoring scheme used in Malaysia, the basic criteria have remained largely unchanged over time, but the weights attached to different criteria have changed frequently (*Bahrin, 1988*). Criteria have been reweighted, for example, to favour those with particular skills where these skills were in short supply in the resettlement programmes, or to give preference to younger candidates where unemployment among school-leavers had become a problem.

One other point should be made about settler selection. Where formal selection procedures are employed, it is important that they include a briefing for would-be settlers. In every case that we examined, problems arose where settlers went into schemes with unrealistic expecta-

Table 5.2: Selection Criteria for Resettlement

Selection criteria	Performance outcome
Age	Experience shows that those aged under about 45 are generally the most successful.
Education	There is strong evidence that better-educated settlers are more successful than those less well-educated.
Family labour force	Agricultural and economic performance has a strong positive correlation with the number of family members able to work.
Marital status	Married settlers almost invariably outperform those not married.
Farming experience	A background that includes farming experience and skills is strongly predictive of good performance.
Capital assets	There is no consistent evidence in favour of selecting settlers who already have capital or assets. Settlers who have gone into schemes well-equipped have fared no better than those who have had little. In sound projects, capable settlers who begin with little seem able to accumulate capital and to acquire assets remarkably quickly.
Nationality	No clear evidence.
No prior criminal record	No clear evidence.

tions about what resettlement involved or how long it would take for their incomes to reach certain levels.[12] Those who are most disillusioned are usually the first to leave.

Restrictions imposed on settlers

In addition to restrictions imposed on land dealings, settlers are almost always subjected to rules and regulations concerning what crops they may grow, how they may grow them, when and where they may sell their produce, for whom they may and may not work, where and how they may build their houses, and so on.

We found no single case where rules limiting purchasing of inputs and marketing of outputs had a positive impact. In Guatemala, for example, scheme administrators have denied settlers the opportunity to develop skills and acquire knowledge that is vital to their success as independent operators (*Dunn, 1992*). Except for those growing horticultural products, settlers are insulated from anything that looks like a market.

Restrictions on crops that may be grown are so frequently ignored, in whole or in part, that it is surprising to see them reappear in one new project after another. Except where the restrictions have a sound market or agronomic foundation, such as the need to meet certain rotational requirements, or the restrictions on cropping patterns in Malaysia where settlers were required to grow the tree crops that FELDA found had the best market potential, settlers tend to grow the crops that meet the requirements of their families and that find a ready local market. The experience of the AVV (*Autorité des Amanagements des Vallées des Volta*) in Burkina Faso shows this very clearly (*Murphy & Sprey, 1980; McMillan, Painter & Scudder, 1990*). Where administrators have managed to enforce restrictions, as in Guatemala and parts of Indonesia, the result has been resentment, malnutrition and lower levels of income. Where settlers have been free to alter cropping patterns, as in Zimbabwe and the spontaneous settlements in Indonesia, there is evidence of diversification and steady income growth.

Provision of infrastructure and speed of construction

Depending upon specific circumstances, certain types of infrastructure – roads and water supplies, for instance – have to be provided through a

12 For a good example, see the discussion of the Indonesian transmigration programme in Kebschull (*1986*) and Fasbender and Erbe (*1988*).

government agency. The evidence clearly shows that safe water supplies should be present when settlers move in and that roads should follow without much delay, the latter because poor roads result in restricted access to a wide range of services, as in Guatemala (*Dunn, 1992*), and also in higher costs and lower farmgate prices, as in Indonesia (*Hardjono, 1977*). New roads have been particularly important to frontier settlements and to settlements based on the reallocation of land from former estate or large-scale farms, where the few original roads served only the central estate or homestead.

Despite the need for an infrastructure, however, there is no merit to the idea that everything must be provided by the state before the settlers arrive. In Zimbabwe's Model A schemes, settlers have arrived a few months before the rains and faced the daunting tasks of clearing, preparing and planting the land while, at the same time, building houses and other structures. They have also moulded bricks and contributed labour for schools and other public buildings. In Zambia's Family Farms schemes, for example, settlers built schools and raised funds to pay for teachers before the Ministry of Education provided them.

Excessively rapid government-directed implementation has a more negative effect on resettlement than does slower implementation by settlers.[13] The one exception has been Malaysia, where the involvement of settlers in all stages of implementation – clearing, planting and building – was slowing the pace of resettlement to an unacceptable degree. FELDA's response in that case was to use contractors for much of the work in the development phase, but to stipulate that half the contractors' casual employees had to come from the settlement scheme (*Bahrin, 1988*). That experience clearly shows the possibility of using contractors to carry out time-sensitive operations. Similar forms of contracting – for tractor-hire or building services, for example – can often be organized by the community itself, with settlers who have the resources contracting out their labour or their assets.

Access to and modality of supply of services

Extension

All the evidence points clearly to a positive association between extension and agricultural success. In the earliest stages of Zimbabwe's

13 See Wakjira (*1977*) for an extreme example of an over-zealous pace of implementation in Ethiopia.

intensive resettlement programme, extension coverage was virtually universal, and helped settlers to make a major shift in production technology (*Kinsey, 1987*). Coverage in Zimbabwe has subsequently returned to presettlement levels, however, which probably reflects a reduction in settler demand. Universal coverage also exists in Burkina Faso, even though the anticipated intensification of cropping has not materialized (*McMillan, 1987*). In Guatemala, where extension has been top-down and synonymous with management, it has helped settlers to adopt new technologies but has been less helpful with adaptations to cropping systems (*Dunn, 1992*). Settlers in Indonesia have received little or no extension service, and have learned only from their neighbours about local cropping systems, pests and diseases (*Fasbender & Erbe, 1988*). The outcome there has been long delays in devising suitable cropping models in areas where the assumed conditions did not exist to support the standard cropping models. In Ethiopia, also, extension services have been almost non-existent and certainly ineffectual (*Chole & Mulat, 1988*).

Tillage services

While there is evidence that tillage services can help settlers to cope with a specific seasonal bottleneck or overcome a conflict in timing of activities, there is no evidence that these services have to be administratively provided. Assuming a hire-service sector exists, settlers could contract for tillage services, as in Kenya and Thailand; coupons could be provided to enable them to do so without directly involving a resettlement agency or the government. Groups could also purchase tractors and then hire them out to neighbours. There should be no administrative barriers to any possible options.

Transport

Precisely the same arguments apply to transport services as to tillage services.

Marketing

In general, the kind of marketing system to facilitate success is related, first, to the crops that are grown and, second, to the destination. Marketing systems fall into three broad patterns. The first involves central processing, often with credit recovery tied to marketing. Two crops in this category are oil palm and sugarcane. The second pattern involves long-distance marketing of fruits and vegetables. In Guatemala, settlers growing horticultural products for the long distance market have operated successfully with a contract farming system (*Dunn, 1992*). The same has been true of small-scale growers in Kenya (*Schapiro & Wainaina, 1989*). Marketing co-operatives can also work in these circumstances, although intensive co-operative development support must be anticipated for some time. Finally, in the case of grains and subsistence crops, and of fruit and vegetables for local markets, no marketing problems are likely to emerge.

Credit

Credit can be considered under two categories – for land purchase and for other purposes.

Where credit has been made available for land purchase with no grant element included, efforts to recover the full cost of the land have invariably failed, leading to large-scale default and withholding of title. The evidence clearly shows, therefore, that there must be a substantial grant component in resettlement programmes to make-up for the missing equity of many poor beneficiaries. Alternatively, own-equity can play a part in the purchase of land by settlers who are not-so-poor, and credit can then be used to top up their resources. Assistance to the poor and those more asset-rich can be successfully combined.

Credit for other purposes also has a poor recovery record. In Costa Rica, all recovery has been poor. In Guatemala, FUNDACEN itself recovers loans, but it is too early to judge overall patterns. In Indonesia, where the settlement programme itself administers credit, recovery has also been very poor, at least during the first five year period (*Otten, 1986; World Bank, 1988*). The Malaysian experience, however, has generally been positive, mainly because all credit recovery can easily be linked to marketing (*Bahrin, 1988*), and because credit has been used to support tree crops with long maturity periods allowing for relatively long payback periods. In Zimbabwe, after a short trial period, the pre-existing credit system has been used, but defaults caused by drought have kept many settlers outside the formal credit system for almost a decade (*Zimbabwe Government, 1992*).

The evidence indicates that alternatives to institutional and other credit need to be considered in resettlement programmes. In Indonesia, for example, it was only the fact that many spontaneous settlers were able to work as wage labourers that enabled them to accumulate the assets to begin farming on their own behalf (*Hardjono, 1986*). In other cases, similar arrangements will be critical of the accumulation of assets or funding of seasonal inputs. The relaxation of Zimbabwe's strict prohibition against off-farm employment, for example, has finally enabled settlers who have fallen out of the institutional credit net to meet their own needs for funding.[14]

Settlers must have surpluses on a continuing basis for recovery to be possible. If commercially viable crops are grown so that credit is not burdensome, recovery is not a problem. Recovery of credit through output marketing has been the most successful approach.

Irrigation

There are two possibilities for resettlement based on irrigation. The first is to subdivide existing irrigation schemes, but we found no cases in which this has happened. The second is to create new irrigated resettlement schemes, an approach that is unlikely to be cost-effective in South Africa.

Of all the highly successful examples of irrigation investment for land reform beneficiaries, none has involved the subdivision of existing irrigation schemes or resettlement over long distances. The experience with irrigation in Mexico, Zimbabwe, Kenya, Indonesia and Ethiopia clearly shows that it is better to create irrigation schemes before parcelling out plots of land. Allocating land on the assumption that irrigation will be put in later nearly always creates subsistence holdings when – as happened in Ethiopia and Indonesia – the necessary pumps and equipment are not provided, or the area is later judged unsuitable for irrigation (*Oey & Astika, 1978*). Experience also shows that irrigation in resettlement schemes demands even more effective extension services than does rainfed farming (*Clayton, 1978; Stanning, 1984; Rukuni, 1988; Wakjira, 1977*). Finally, the review found that irrigated projects, even more than dryland projects, tend to withhold secure rights to land and to maintain settlers on short-term, renewable tenancies, particularly in state-sponsored irrigation in Africa.

Social services

The approaches to the provision of social services in settlement programmes are varied and performance is mixed.

14 The prohibition might also have been relaxed because it was difficult or impossible to enforce. Adams (*1991*), for example, reports that in one resettlement scheme, 48 per cent of women engaged in off-plot employment at a time when the prohibition was still in effect.

In Tanzania, the provision services such as clean water, primary education and health care were the primary justification for the massive villagization programme. In this case, the improvements had significant positive benefits, albeit at great economic and social cost (*Oberai, 1988b*). In cases such as Kenya, Ethiopia and the Brazilian Amazon, however, there is evidence that settlement increased the need for services such as health care because the resettlement process itself heightened exposure to diseases, raised the incidence of malnutrition and increased the rate of child mortality (*Clayton, 1978; Wood & Schmink, 1979; Goldwater, Colchester & Luling, 1986; Woldemeskel, 1989; Dejene, 1990*). The failure of new settlement to provide safe water supplies is also strongly associated with increased health risks.

In the case of education, many opportunities exist for settlers to participate in providing facilities and services. In many cases, notably Zambia and Zimbabwe, settlers have been willing to help to provide their own social overheads.

We found no clear-cut instances where housing can be justified as a social overhead. Settlers in most cases appear more than willing to build housing for themselves, although grants for materials are often needed. If grants are provided, they are most likely to make a strong impact at the beginning of a project.

Resource base

Resettlement projects on good soils and with adequate moisture to grow crops have often been very successful. Where soil and climate have been marginal, as in certain parts of Indonesia, Ethiopia, Zimbabwe and the Brazilian Amazon, outcomes have been poor. In such cases, especially when settlements are remote, only very strong communities are successful. For example, the Mennonites, a cohesive, motivated community who possess a long-term perspective and the ability to tolerate isolation and the slow build-up to reasonable levels of living, have successfully colonized the semi-arid Chaco of Paraguay. There is also evidence from Zimbabwe that irrigation in resource-poor areas is likely to amount to little more than a costly but futile attempt to circumvent the limitations of the environment (*Kelly, 1972; World Bank, 1984*).

Cost per beneficiary family

If all costs – public and private – are factored in, then the full costs of settlement are likely to be far higher than many studies have estimated. In spontaneous settlement, particularly, an enormous amount of work is done by the beneficiaries themselves. The capitalization of settler effort into land improvements, housing and infrastructure should be valued appropriately, encouraged and taken into account in settlement programmes. If a large grant element is included in the allocation of land, on the other hand, the public cost will be high. The critical, and largely unresearched, issue here appears to be that of the minimum public investment needed to stimulate a strong private investment response on the part of the beneficiaries.

The studies examined provide no direct evidence that higher public costs per beneficiary family are significantly associated with performance outcomes. This conclusion is supported by the World Bank's review of settlement (*1978*), which found that complex and costly projects are no more likely to succeed than more modest ones, and that project success is not closely correlated with government outlays per beneficiary, even though a threshold level of physical infrastructure and social services is needed. This conclusion was contradicted by the more-limited review conducted by Gosling and Abdullah (*1979*), in which the authors stated that land settlement requires heavy levels of investment and that more investment leads to greater chances of success. Their conclusion, however, did not appear to be based on actual cost comparisons.

Grant elements in settlement programmes

The extent of grants varies widely. In Zimbabwe or Ethiopia, settlers made no contribution to either the land costs or the development costs. In Indonesia, Malaysia, Kenya, Zambia and Guatemala these costs are treated as credit. In Indonesia they are not recovered, so they amount to a de facto grant.

For given settlement budget allocations, larger grants reduce the number of beneficiaries. Excessive grant elements also foster dependency and paternalism. We have seen, however, that where land is acquired at market prices, a grant element is necessary to provide poor beneficiaries with the necessary equity to engage in risky agricultural production and to stand a chance of recovering the credit provided to them. Some grant elements may be essential, such as for initial subsistence, technical assistance, and perhaps building materials. The proper mix of grants and credit, therefore, is a compromise among several competing objectives.

Cost recovery

Recovery of land loans has been poor, except in Guatemala, Kenya and Malaysia, although in Malaysia, the time taken for full recovery has been in excess of 20 years. The pattern for recovery of seasonal production credit is almost universally poor, except where recovery of loans is linked to output marketing. An exception is Zambia, where settlers pay ground rent, have to open an account with a commercial bank in order to obtain loans, and are required to deposit one-third of the amount of any loan before it is disbursed. Recovery is on a voluntary repayment basis, while the penalty for non-repayment is loss of the deposit and of further credit facilities (*Bomford, 1973*).

Settler groups

A great deal of evidence points to the conclusion that the emergence of strong, self-reliant groups among settler communities is a vital ingredient in ensuring the sustainability of settlement programmes. We know from experience in Bolivia and Brazil that settlements carried out by pre-existing groups have been more successful than when done without such groups. In some cases, cohesive groups have been discriminated against. In Malaysia, for example, settlers who proposed group-based institutions to meet their own needs were regarded as disruptive by FELDA staff (*Bahrin, 1988*). In Zimbabwe, the only way a group could choose to be resettled as a group was to agree to the collective co-operative model of settlement, rather than the more popular pattern of individual holdings grouped around nucleated villages (*Zimbabwe Government, 1981*). By contrast, where groups have been artificially created – as with the farmers' associations in Ethiopia – outcomes have been almost universally negative (*Dunning, 1970; Chole & Mulat, 1988*).

Available literature reveals very little about the process of group formation in settlements where groups did not exist before the settlement was formed. In Kenya, studies were carried out almost exclusively by economists rather than other social scientists, and even these studies ended before the settlement process had been completed.[15] We therefore do not know much about whether organizations are forming during the resettlement process itself, how they form, or how they contribute to the success of settlement efforts. It would be useful for researchers to revisit some of the settlement sites in Kenya, and also to visit settlements in other areas, to investigate these issues.

15 See, for example, the later studies by Clough (*1968*) and von Haugwitz (*1972*), which are largely based on the farm management survey work done in the mid-1960s.

Conclusion

A number of links between core characteristics of resettlement and specific outcomes appear to be fairly well established. An understanding of these links are useful for avoiding costly mistakes in the initial design and establishment of settlements. We concur with Cook and Mukendy (*1992*), however, that the available literature offers few clear guidelines for specific additional actions, or their sequencing, which would assist settlements in their long-term development. The main lessons from this review are that:

- Directed schemes are designed on the premise that 100 per cent of settlers will or should be successful. Such outcomes never materialize. Instead, schemes should be designed with the flexibility to anticipate that some settlers will leave because of failure, or for other reasons. Therefore constraints on sale, leasing and other forms of land reallocation should not be imposed since they create idle land and other inefficiencies.
- Most programmes have erred on the side of excessive administration. Too much weight has been given to maintaining centralized models of settlement rather than to supporting spontaneous settlement, greater local and community participation and responsibility, and shared implementation based on contracting. If resettlement is not decentralized, it cannot be flexible. Settlement agencies tend to adopt, implicitly or explicitly, infinite horizons for implementation, so that resettlement is never 'finished' and areas are returned to local administrations only after a very long period, if ever.
- The assumption, in many cases, has been that every element in resettlement programmes must be provided to the settlers, and usually before they arrive. Very little research has been done on the sequence of activities, however, or on the minimum level of public sector investment needed to generate sufficient private and community investment response. The elements of a minimum public sector package appear to include: safe water, roads, relatively good land, extension, and subsistence allowances. There is little evidence, however, about the need for the provision of education and housing.
- The capacity of settlers to generate capital has often been seriously underestimated.
- Settlements will be more successful if farm sizes are adjusted to agricultural skills, experience, the family labour force, and the capital available to the settler families. Higher educational levels of settlers

will further enhance outcomes.

- Land rights must be clearly defined as ownerships or long-term leases, and settlers should be allowed to sell or rent out their land to other settlers.
- Some grant finance is required to provide poor settlers with the equity necessary to engage in risky own account farming and to repay the remaining credit grants for initial subsistence; technical assistance is also required.
- Paternalistic constraints on crop choice, technology, marketing or labour market participation are either not enforceable or have adverse impacts on settlement success.
- Settlements based on collective co-operatives do not work. Such programmes have broken down everywhere they have been tried.

References

Adams, Jennifer M. 1991. Female wage labour in rural Zimbabwe. *World Development,* 19(2/3) : 163–77, February/March.

Adams, M. and Devitt, P. 1992. Grappling with land reform in pastoral Namibia. *Pastoral Development Network Paper 32a.* London: Overseas Development Institute.

Akwabi-Ameyaw, Kofi. 1990. The political economy of agricultural resettlement and rural development in Zimbabwe: the performance of family farms and producer co-operatives. *Human Organisation,* 49(4) : 320–38.

Bahrin, Tunku Shamsul. 1988. Land settlement in Malaysia: a case study of the Federal Land Development Authority projects. In A. S. Oberai (Ed), *Land Settlement Policies and Population Redistribution in Developing Countries: Achievements, Problems and Prospects.* New York: Praeger.

Binswanger, Hans P., Deininger, Klaus and Feder, Gershon. Power, distortions and reform in agricultural land markets. In T.N. Srinivasan and J. Behrman (Eds), *Handbook of Development Economics, Vol. III.* Forthcoming.

Bomford, L. 1973. An evaluation of the Family Farms resettlement schemes, Zambia. MPhil dissertation. Reading, UK: University of Reading.

Bose, A. 1983. Migration in India: trends and policies. In A. S. Oberai (Ed), *State Policies and Internal Migration: Studies in Planned and Market Economies.* London: Croom Helm; New York: St. Martin's Press.

Chambers, R. 1969. *Settlement Schemes in Tropical Africa: A Study of Organizations and Development.* London: Routledge & Kegan Paul.

Chole, Eishetu and Mulat, Teshome. Land settlement in Ethiopia: a review of developments. In A. S. Oberai (Ed), *Land Settlement Policies and Population Redistribution in Developing Countries: Achievements, Problems and Prospects.* New York: Praeger.

Christodoulou, D. 1965. Land settlement: some oft-neglected issues.

Monthly Bulletin of Agricultural Economics and Statistics, 14(10) : 1–6.

Clayton, Eric. 1978. A comparative study of settlement schemes in Kenya. *Agrarian Development Unit, Occasional Paper No. 3.* Ashford, Kent, UK: Wye College, University of London.

Clough, R.H. 1968. An appraisal of African settlement schemes in the Kenya highlands. PhD dissertation. Ithaca: Cornell University.

Colson, Elizabeth. 1971. *The Social Consequences of Resettlement.* Manchester: Manchester University Press.

Cook, Cynthia C. and Mukendi, Aleki. 1992. Involuntary resettlement in bank-financed projects: lessons from sub-Saharan Africa. *AFTEN Working Paper No. 3.* Washington, D.C.: Environment Division, Africa Technical Department, The World Bank.

Cusworth, John and Walker, Judy. 1988. Land resettlement in Zimbabwe. *Evaluation Report EV 434.* London: Overseas Development Administration.

Deininger, Klaus and Binswanger, Hans P. Are large farms more efficient than small ones? Government intervention, large-scale agriculture, and resettlement in Kenya, South Africa and Zimbabwe. *World Bank Policy Working Paper.* Washington, D.C.: The World Bank. Forthcoming.

Dejene, Alemneh. 1990. *Environment, Famine and Politics in Ethiopia: A View from the Village.* Boulder: Lynne Reiner Publishers.

de Wilde, J. C., Mcloughlin, P., Guinard, A., Scudder, T. and Mabouche, R. 1967. *Experiences with Agricultural Development in Tropical Africa: The Case Studies.* Baltimore: The Johns Hopkins University Press.

Dunn, Elizabeth G. 1992. The FUNDACEN experience: factors for success and failure in a Guatemalan land purchase-sale program. *Land Tenure Centre, Research Paper No. 107.* Madison, WI: Land Tenure Centre, University of Wisconsin-Madison.

Dunning, Harrison C. 1970. Land reform in Ethiopia: a case study in non-development. *UCLA Law Review,* 18(2) : 271–307.

Fasbender, Karl and Erbe, Susanne. 1988. *Towards a New Home: Indonesia's Managed Mass Migration–Transmigration between Poverty, Economics and Ecology.* Hamburg: Verlag Weltarchiv GMBH for the Hamburg Institute of Economic Research.

Goldwater, Mike, Colchester, Marcus and Luling, Virginia. 1986. *Ethiopia's Bitter Medicine, Settling for Disaster: An Evaluation of the Ethiopian Government's Resettlement Programme.* London: Survival International.

Gosling, L. A. P. and Abdullah, H. 1979. Rural population redistribution. In L. A. P. Gosling and L. Y. C. Lim (Eds), *Population Redistribution: Patterns, Policies and Prospects.* New York: United Nations Fund for Population Activities.

Green, Joy K. 1987. Evaluating the impact of consolidation of holdings, individualization of tenure, and registration of title: lessons from Kenya. *Land Tenure Centre Paper 129.* Madison, WI: Land Tenure Centre, University of Wisconsin-Madison.

Habitat. 1986. *Spontaneous Settlement Formation in Rural Regions, Volume Two: Case Studies.* Nairobi: United Nations Centre for Human Settlements [Habitat].

Hansen, A. and Oliver-Smith, A. (Eds). 1982. *Involuntary Migration and Resettlement: The Problems and Responses of Dislocated Peoples.* Boulder: Westview Press.

Hardjono, Joan M. 1977. *Transmigration in Indonesia.* Kuala Lumpur: Oxford University Press.

Hardjono, Joan M. 1986. Spontaneous rural settlement in Indonesia. In *Spontaneous Settlement Formation in Rural Regions, Volume Two: Case Studies.* Nairobi: United Nations Centre for Human Settlements [Habitat].

Haugerud, Angelique. 1983. Consequences of land tenure reform among smallholders in the Kenya highlands. *Rural Africana,* 15(16) : 65–89, Winter/ Spring.

Heath, J. R. 1992. Evaluating the impact of Mexico's land reform on agricultural productivity. *World Development,* 20(5) : 695–711, May.

Hilton, T. 1959. Land planning and resettlement in Northern Ghana. *Geography,* 44: 277–40.

Hulme, David. 1988. Land settlement schemes and rural development: a review article. *Sociolgia Ruralis,* 28(1) : 42–61.

Ireson, W. R. 1987. Landholding, agricultural modernization and income concentration: a Mexican example. *Economic Development and Cultural Change,* 35(2) : 351–66, January.

Jones. G. W. and Richter, H. V. (Eds). 1982. Population resettlement programmes in South-east Asia. *Development Studies Centre Monograph No. 30.* Canberra: The Australian National University.

Kebschull, Dietrich. 1986. *Transmigration in Indonesia: An Empirical Analysis of Motivation, Expectations and Experiences.* Hamburg: Verlag Weltarchiv GMBH for the Hamburg Institute of Economic Research.

Kelly, S. P. A. 1972. African and European settler and estate agriculture in the Rhodesian Lowveld. *PhD dissertation.* Salisbury: University of Rhodesia.

Kinsey, B. H. 1979. Agricultural technology, staple food crop production and rural development in Zambia. *Monographs in Development Studies No. 6.* Norwich, UK: University of East Anglia.

Kinsey, B. H. 1982. Forever gained: resettlement and land policy in the context of national development in Zimbabwe. *Africa,* 52(3) : 92–113.

Kinsey, B. H. 1983. Emerging policy issues in Zimbabwe's land resettlement programmes. *Development Policy Review,* 1(2) : 163–96, November.

Kinsey, B. H. 1986. The socioeconomics of nutrition under stressful conditions: a study of resettlement and drought in Zimbabwe. *Occasional Paper Series.* Harare: University of Zimbabwe, Centre for Applied Social Sciences.

Kinsey, B. H. 1987. *Agricultural extension in intensive resettlement schemes: a case study within the framework of the National Extension Framework Study.* Harare: Ministry of Lands, Resettlement and Rural Development, Department of Agricultural Technical and Extension Services.

Leo, Christopher. 1984. *Land and Class in Kenya.* Toronto: University of Toronto Press.

Manshard, Walther and Morgan, William B. (Eds). 1988. *Agricultural Expansion and Pioneer Settlements in the Humid Tropics.* Tokyo: United Nations University Press.

Mbithi, Philip M. and Barnes, Carolyn. 1995. *The Spontaneous Settlement Problem in Kenya.* Nairobi: East African Literature Bureau.

McMillan, Della E. 1987. The social impacts of planned settlement in Burkina Faso. In Michael H. Glantz (Ed), *Drought and Hunger in Africa: Denying Famine a Future.* Cambridge, UK: Cambridge University Press.

McMillan, Della E., Painter, Thomas and Scudder, Thayer. 1990. Land settlement review: settlement experiences and development strategies in the onchocerciasis controlled area of West Africa – Final Report. *IDA Working Paper Number 68.* Binghamton, New York: Institute for Development Anthropology.

Metson, Joan. 1979. Mixed dairying and maize farming in the Nandi high-potential areas of Kenya. *PhD dissertation.* Norwich, UK: University of East Anglia.

Moran, E. 1979. Criteria for choosing successful homesteaders in Brazil. In G. Dalton (Ed), *Research on Economic Anthropology, Vol. 1.* Greenwich, Connecticut: JAI Press Inc.

Moris, J. 1968. The evaluation of settlement schemes performance: a sociological appraisal. In R. Apthorpe (Ed), *Land Settlement and Rural Development in Eastern Africa.* Kampala: Nkanga Editions.

Mumbengegwi, C. 1984. Agricultural producer co-operatives and agrarian transformation in Zimbabwe: policy, strategy and implementation. *Zimbabwe Journal of Economics,* 1:47–75, July.

Mupawose, R.M. and Chengu, E.T. 1982. Determination of land policy in Zimbabwe. *Paper presented at the Workshop on Land Policy and Agricultural Production.* Gaberone, Botswana: UN University, February 14–20.

Murphy, Josette and Sprey, Leendert H. 1980. *The Volta Valley Authority: Socio-economic Evaluation of a Resettlement Project in Upper Volta.* West Lafayette, Indiana: Purdue University, Department of Agricultural Economics.

Mwagiru, Wanjiku. 1986. Planning for settlement in rural regions: the case of spontaneous settlements in Kenya. In *Spontaneous Settlement Formation in Rural Regions, Volume Two: Case Studies.* Nairobi: United Nations Centre for Human Settlements [Habitat].

Nelson, M. 1973. *The Development of Tropical Lands.* Baltimore: John Hopkins University Press.

Nottidge, G.P.R. 1982. *Personal communication: the Workshop on Land Policy and Agricultural Production,* Gaberone, Botswana: UN University, February 14–20.

Oberai, A.S. (Ed). 1988. *Land Settlement Policies and Population*

Distribution in Developing Countries: Achievements, Problems and Prospects. New York: Praeger.

Oberai, A. S. 1988. An overview of settlement policies and programmes. In A. S. Oberai (Ed), *Land Settlement Policies and Population Distribution in Developing Countries: Achievements, Problems and Prospects.* New York: Praeger.

Oey, Mayling and Astika, Ketut Sudhana. 1978. A report from the Institute for Economic and Social Research Faculty of Economics of the University of Indonesia in collaboration with the Research and Development Board, Department of Manpower, Transmigration and Co-operatives. Mimeo, Jakarta: Institute for Economic and Social Research, Faculty of Economics, University of Indonesia.

Ogawa, Naohiro and Chan, Paul T. H. 1985. Land settlement programmes and issues on the second generation: the case of Malaysia. In *Urbanization and Migration in ASIAN Development.* Tokyo: National Institute for Research Advancement.

Otten, Mariel. 1986. Transmigration: Myths and Realities – Indonesian Resettlement Policy, 1965–1985. *IWGIA Document No. 57.* Copenhagen: International Work Group for International Affairs.

Oxby, Claire. 1984. Settlement schemes for herders in the sub-humid tropics of west Africa: issues of land rights and ethnicity. *Development Policy Review,* 2(2) : 217–33, November.

Ozorio de Almeida and Luiza, Anna. 1992. *The Colonization of the Amazon.* Austin: University of Texas Press.

Reining, C. 1982. Resettlement in the Zande development scheme. In A. Hansen and A. Oliver-Smith (Eds), *Involuntary Migration and Resettlement: The Problems and Responses of Dislocated Peoples.* Boulder: Westview Press.

Rukuni, Mandivamba. 1988. The evolution of smallholder irrigation policy in Zimbabwe: 1982–1986. *Irrigation and Drainage Systems,* 2 : 199–210.

Sabry, O. A. 1970. Starting settlements in Africa. In *Land Reform, Land Settlements and Co-operatives No. 1.* Rome: Food and Agriculture Organisation.

Schapiro, Morton Owen and Wainaina, Stephen. 1989. Kenya: a case study of the production and export of horticultural commodities. In *Successful Development in Africa: Case Studies of Projects, Programs and Policies.* Washington, D.C.: Economic Development Institute, The World Bank.

Scholz, Ulrich. 1986. Spontaneous rural settlements and deforestation in south-east Asia: examples from Thailand and Indonesia. In *Spontaneous Settlement Formation in Rural Regions, Volume Two: Case Studies.* Nairobi: United Nations Centre for Human Settlements [Habitat].

Scudder, T. 1973. Summary: resettlement. In W. C. Ackermann, G. F. White and E. B. Worthington (Eds), *Man-made Lakes: Their Problems and Environmental Effects.* Washington, D.C.: American Geophysical Union.

Scudder, T. 1975. Resettlement. In N. F. Stanley and M. P. Alpers (Eds), *Man-made Lakes and Human Health.* London: Academic Press.

Scudder, T. 1984. The development potential of new lands settlement in the tropics and subtropics: a global state of the art evaluation with specific emphasis on policy implications. *AID Program Evaluation Discussion Paper No. 21.* Washington, D.C.: United States Agency for International Development.

Scudder, T. 1986. Interrelationships between government-sponsored and spontaneous settlement. In *Spontaneous Settlement Formation in Rural Regions, Volume Two: Case Studies.* Nairobi: United Nations Centre for Human Settlements [Habitat].

Scudder, T. and Colson, Elizabeth. 1979. Long-term research in Gwembe Valley, Zambia. In G. M. Foster *et al* (Eds), *Long-term Field Research in Social Anthropology.* New York: Academic Press.

Scudder, T. and Colson, Elizabeth. 1981. From welfare to development: a conceptual framework for the analysis of dislocated people. In A. Hansen and A. Oliver-Smith (Eds), *Involuntary Migration and Resettlement: The Problems and Responses of Dislocated Peoples.* Boulder: Westview Press.

Selassie, Hadas Haile, Ali, Ato Ahmed and Kebede, Hanna. 1979. *Evaluation of the Settlement Authority and Settlement Projects.* Addis Ababa: National Land Settlement Authority.

Solem, R. R., Greene, R. J., Hess, D. W., Ward, Carol B. and Taylor, P. L. 1985. Bolivia: integrated rural development in a colonization setting. *AID Project Impact Evaluation Report No. 57.* Washington, D.C.: United States Agency for International Development.

Stanning, J. L. 1984. Sabi River Basin irrigation projects: facts and issues relating to land tenure and agricultural productivity. MS. Madison, WI: Land Tenure Centre, University of Wisconsin-Madison.

Takes, C. 1975. *Land Settlement and Resettlement Projects: Some Guidelines for their Planning and Implementation.* Wageningen, Netherlands: International Institute for Land Reclamation and Improvement.

Uhlig, Harald. (Ed). 1984. Spontaneous and planned settlement in South-east Asia. *Giessener Geographische Schriften, Vol. 58.* Hamburg: Institute of Asian Affairs.

von Haugwitz, Hans-Wilhelm. 1972. Some experiences with smallholder settlement in Kenya. *Afrika-Studien No. 72.* Munich: Weltforum Verlag for ILO-Institut für Wirtschaftforschung.

Wakjira, Fekadu. 1977. Recent institutional innovations and structural transformation of rural economies in Ethiopia. *Paper presented at the International Seminar on Agrarian Reform, Institutional Innovation and Rural Development: Major Issues in Perspective.* Madison, Wisconsin: Land Tenure Centre, University of Wisconsin-Madison.

Woldemeskel, Getachew. 1989. The consequences of resettlement in Ethiopia. *African Affairs,* 88(352) : 359–74, July.

Wood, C. H. and Schmink, M. 1979. Blaming the victim: small farmer production in an Amazon colonization project. *Studies in Third World Societies,* 7: 77–93.

World Bank. 1978. *Agricultural Land Settlement.* Washington, D.C.: The World Bank.

World Bank. 1984. *Zimbabwe Irrigation Sub-sector Review.* Washington, D.C.: The World Bank.

World Bank. 1988. Indonesia: The transmigration programme in perspective. *World Bank country study.* Washington, D.C.: The World Bank.

Zimbabwe Government. 1981. *Resettlement Programme: Policies and Procedures.* Harare: Ministry of Lands, Resettlement and Rural Development.

Zimbabwe Government. 1991. Resettlement progress report as of June 1991. Harare: Ministry of Local Government, Rural and Urban Development, Department of Rural Development.

Zimbabwe Government. 1992. Second report of settler households in normal intensive Model A resettlement schemes – Main report. Harare: Ministry of Lands, Agriculture and Rural Resettlement, Monitoring and Evaluation Section, Planning and Research Unit.

6

The political implications of alternative models of land reform and compensation

Hans Binswanger

Introduction

This chapter focuses on the administrative, cost and political economy implications of alternative land reform processes and modes of compensation. The main alternatives considered are land reform based on:

- a willing buyer – willing seller model using grants to enable beneficiaries to enter the market;
- expropriation with compensation at market prices;
- expropriation with compensation at a price just below the market price.

The interim constitution of South Africa recognizes existing land rights and explicitly does not allow confiscation of land from current owners. Confiscation, i.e. the seizing of land without compensation, is therefore not discussed. All elements of the propertied classes usually unite to oppose confiscation. Indeed, history demonstrates clearly that confiscation is only a possibility during revolutionary political change associated with considerable violence.

Expropriation of land

It is clear that, when considering judicial restitution of plots of land lost by black or 'coloured' communities in the process of black spot removal, expropriation of the new owners must be an option, and only

1 This chapter grew out of discussions in South Africa at the African National Congress (ANC) Constitutional Committee on Land Reform in Sambonani, Eastern Transvaal in May 1993, where alternative modes of land reform and of compensation of previous owners were discussed. At that time, it already seemed clear that the interim constitution of South Africa (already under preparation) would not allow confiscation of land from current owners. Confiscation, i.e. the seizing of land without compensation, was therefore not discussed.

the level of compensation of the new owners of the land must be considered. But restitution claims will inevitably be a small fraction of any land reform programme in South Africa which intends to transfer a large proportion of land to victims of apartheid.

For the much larger programme of redistribution, the case for expropriation with compensation at prices below the market price rests on three considerations:

1. An argument that it is not just to compensate beneficiaries of apartheid for benefits they derived from the land and agricultural policies associated with the regime. (I will not address this issue in this note.)
2. Expropriation is needed because it is unlikely that the market will make sufficient land available for redistribution. (Other chapters in this book suggest that this is not likely to be the case, and expropriation may only be needed as a last resort.)
3. Expropriation will allow land for the programme to be acquired at a lower cost than a market-assisted land reform, especially when it is associated with just compensation at below the market price. Arguments against expropriation are the following:
   (a) The market price is a policy variable, and eliminating privileges and policies which favour the large farm sector will bring it down to lower programme cost, an issue analysed in other chapters.
   (b) Expropriation is likely to be a lengthy and protracted process, especially when just compensation at below market prices is contemplated, and expropriation judgements can be appealed by the higher level courts. Not only does this increase the administrative costs, but also delays benefits from accruing to beneficiaries. Even relatively small discount rates will translate delays of several years into huge costs for beneficiaries.
   (c) Reliance on the willing buyer – willing seller model will greatly reduce the political opposition to land reform and may increase its political sustainability over the decade or so required to implement any sizeable programme.

The remainder of this chapter deals with the cost issue and the political issue.

The cost and political aspects of land reform

There are two fundamental political hurdles to land reform which must be overcome. The first is to design an approach which can win a sufficiently broad coalition to make it acceptable to the constitutional assembly and a subsequent parliament, and enacted in the respective constitutional and legal provisions, such as a restitution act, revisions to the land law, etc. The second is to design the land reform programme so that a sustained coalition emerges which, year after year, will insist on sufficient appropriation in the budget process to carry out the land reform and settlement programme for the many years required to implement it.

The advocates of compensation at less than market prices correctly point out that, even when a land reform law specifies the use of an ear-marked tax or a wealth levy for financing the programme, there is no guarantee that these funds will be applied to purchase farms for the poor. Moreover, the political coalition to eliminate policies and privileges in favour of the large-scale commercial sector may break down, leading to higher land prices as well as the diversion of fiscal resources from land purchases to support of the commercial farm sector, making it impossible to achieve a sufficient level of counterpart funding to trigger foreign loan and grant funds for land reform.

Advocates of expropriation argue that legal means must be found to reduce the cost of land acquisition in order to ensure the sustainability of the long-term programme. These include provisions which allow the expropriation of idle land, under-utilized land and vacant land; the use of land ceilings; compensation formulas which take account of how the land was acquired originally, or which pay only market value for land improvements, but not the raw land. Many land reform laws, of course, have tried all these approaches, and experience shows that they lead to enormous complexities. Idle land, for example, can always be converted to forests or pasture. Ceilings need to take account of the agro-climatic potential and land quality, and also lead to evasion through sub-division of farms or sale to close relatives. Land improvements and farm buildings have no market value independent of the land, so only a replacement value can be estimated. Circumstances of land acquisition will be hard to codify and operationalize.

Moreover, expropriation at below market prices requires that the state purchase the land rather than the beneficiaries. While not inevitable, this is likely to lead to the emergence of a land reform or settlement agency whose personnel will eventually engage in rent-seeking behaviour of its own (for example, in Mexico, Colombia, Honduras and Costa Rica).

Positions of interest groups

Table 6.1 lists the likely position of different interest groups to expropriation at below market prices versus market prices at the stage of legislative approval, and at the stage of annual appropriations. The following conclusions emerge:

- For **urban tax payers** or the urban poor the modalities of expropriation do not matter much. Urban taxpayers will oppose any cost of land reform at either stage. The urban poor or their representatives may support land reform legislation because they may hope that property reform will spread to other assets. But they would oppose annual appropriations, because these may cut into their social spending.
- **Lawyers and other agricultural experts** would gain from the complicated regulations associated with expropriation at below market prices, because they would generate a stream of demand for their expert services. They could not be counted on to strongly support annual appropriations, however.
- **Mining, banking and industrial interests** would oppose any expenditure and approval of a programme because of its budget costs, irrespective of the mode of compensation. But they would be more strongly opposed to legislative approval of expropriation with compensation below market price, for fear of a spread of such expropriation to urban assets.
- Major differences appear for **commercial farming interests**. They can be thought of as consisting of farmers in trouble who would like to leave farming and take their net worth along with them, and those who are viable and committed to farming in the long run. Both groups would sharply oppose models based on just compensation below market prices, at both the legislative and the annual appropriation stage.

  Compensation at market prices or market-assisted reform, on the other hand, splits the commercial farmers into those who would like to sell, and those who are committed farmers, since the former would be able to realize their intentions. Moreover, at the annual appropriation stage, the sellers will become vocal supporters of larger appropriations because a lack of appropriation reduces their chance of selling and possibly depresses the market price of land.

  Even efficient and committed farmers could be persuaded to be neutral to such a programme, although they are more likely to oppose both approval and appropriations for fear of diversion of funds from programmes which favour them, or because they would dislike having black and 'coloured' neighbours, many of whom would be poor, at least initially.

- Major differences also arise among the **beneficiaries**. For a given set of judicial rules governing claims to restitution, (e.g. the cut-off dates which govern eligibility for restitution, evidence rules, court procedures), potential beneficiaries of judicial claims are likely to be indifferent to the amount of compensation the state would have to pay the current owners of any plot of land to be restituted. Their real interest is irrelevant, since the state will be compelled to pay whatever amounts the courts decide.

  For urban and rural beneficiaries of a general land reform, the situation is different. Apart from their sense of justice, they are likely to oppose compensation at market prices for justified fear that annual appropriations would not be forthcoming. Once a market-price based programme is adopted, however, their position is likely to change since annual appropriation will be the critical constraint to their obtaining the land.

Under a market-assisted scheme, where annual appropriations will determine the amount of grant resources, and where groups of beneficiaries buy the land, programme implementation mechanisms could be designed to sharply draw the attention of beneficiary groups and seller farmers to the appropriate process. If a land board were to determine eligibility of groups (based on proof of existence of membership, member deposits of their equity contribution to the land price, and democratic decision-making and bylaws), the eligibility could be declared independently of availability of funds, as was done in the Mexican land reform process. The speed with which eligible communities would get access to land would then depend on the level of annual appropriations, rather than on the speed of expropriation proceedings for specific farms they are waiting to obtain.

With all legal impediments to the land transactions eliminated, the piling up of eligible groups which have not received their grant would mobilize the group leaders and the farmers, with whom they have explored potential sales, into vocal political pressure groups likely to press elected representatives to act.

In the alternative scenario of below market price compensation, the critical constraint to acquiring a farm for redistribution to eligible

Table 6.1: Political Positions of Different Groups

	Below Market Prices		At Market Prices	
	Legislative approval	**Annual appropriations**	**Legislative approval**	**Annual appropriations**
Commercial farmers sellers committed farmers	opposed opposed	opposed opposed	in favour or neutral opposed or neutral	clearly in favour neutral or opposed
Beneficiaries *(rural and urban)* of judicial claims of general programme	indifferent because state will have to pay in favour	indifferent because state has to pay, in favour or indifferent because not critical to land acquisition, diverted	indifferent because state has to pay opposed	in favour clearly in favour very focused
Mining, banking, industrial interests	opposed for fear of extension of principle	opposed, for fear of extension of principle, or diversion of funds	indifferent or opposed because of cost	opposed because of cost
Lawyers and agricultural experts	in favour because of more work	indifferent	opposed	indifferent
Urban taxpayers	opposed	opposed	opposed	opposed
Urban poor (non-beneficiaries)	in favour in hope of extension or indifferent	opposed	in favour or indifferent	opposed

beneficiary communities is the administrative and/or judicial decision that farmland is idle, uncultivated, or beyond the land ceiling, etc. While annual appropriations are important, even to pay the lower compensation, the complexities of the administrative and legal procedures will divert the political energies of the beneficiaries and their leaders from the annual appropriation to the myriad of separate expropriation cases and their complexities, which must be solved even prior to the budget becoming an issue.

The resulting paralysis and stalemate may explain the slow progress of the Mexican land reform in spite of powerful peasant mobilization for seventy years, and in spite of eligibility being handled as discussed above.

Issues for land reform

Advocates of a land reform programme therefore face the following issues:

- Because annual appropriations cannot possibly be guaranteed, enlarging the scope of the judicial restitution process is a risk-averse strategy at the legislative stage, since clear rules for judicial restitution will eventually compel the state to pay. The risk in this strategy is that the legal proceedings may drag on over many years, which can be minimized by careful attention to the design of the rules and court procedures. Indeed, the clarity of the rules appears to be more important than the mode of compensation.
- For general compensatory land reform, the issue is more complex. If the beneficiaries and assorted legal and agricultural experts are powerfully organized, they may be able to push a land law based primarily on expropriation through the constitutional or legislative assembly, even if they propose compensation at below market prices. If they are not powerfully organized, they will have to accept compensation at market price in order to appease the mining, banking and industrial interests and split the commercial farmer groups.
- Accepting compensation at market prices may cost some pride and sense of justice, but substantially improves the sustainability of annual appropriations and counterpart funding. An incentive-compatible coalition may become possible between the beneficiaries and the potential sellers, and opposition of other interest groups will be confined to the competition for government funds, rather than to fighting over the principles and desirability of the reform.

Conclusion

Market-assisted land reform and expropriation with compensation at market prices seems to have several advantages: (a) a more poorly organized coalition of beneficiaries may be able to win approval at the legislative stage; (b) the annual budget process for funding the grants can rely on a broader and more focused coalition of supporters; and, (c) market prices can be influenced by policies which eliminate the privileges of the large-scale sector.

Executing a land reform through grants or vouchers to beneficiary groups who buy from willing sellers leads, of course, to compensation at market prices. It obviates the need for a land reform/settlement agency and chances for bureaucratic rent-seeking are therefore less. Beneficiaries are free to choose the land in the market and to do with it what they wish, rather than having to follow the guidelines of an agency. The cost and delays of expropriation proceedings are avoided. It appears, therefore, that expropriation – even at market prices – should only be used as a last resort.

7

Rural development and poverty reduction

Hans Binswanger

Persistent rural poverty

Five broad paths of agricultural and rural development emerge from an analysis of the extent to which countries have implemented misguided policies and to which they have been affected by developed countries' protectionist policies.[1] Only one of these groups of countries – the group which either did not use or which abandoned outdated and discredited policies early, including countries such as Taiwan, Indonesia, Malaysia, Thailand and China – has had substantial declines in rural poverty. In all other countries today there are many more rural poor than there were 45 years ago. In many countries the number of rural poor has doubled or tripled. There is little comfort in the fact that the proportion of the population living in rural areas and the proportion of the rural population in poverty has declined in the group of countries which taxed modestly and provided for public investment. Even in this group of countries, including India, Mexico, Kenya and the Philippines, the gains have been modest and, in the other three groups, major disasters have taken place.

The bankruptcy of the urban strategy

The time is long overdue for declaring the urban development strategy to rural development and poverty reduction intellectually bankrupt. In most of the developing world, the rural areas today contain many more poor people (in many countries two to three times as many) than they did in 1950 when this theory first became fashionable.

In many countries where rural poor populations have shifted to urban slums, urban poverty is seen today as a more severe problem than rural poverty. This ignores the fact that it is the premature shedding of labour from the commercial farm sectors and the failure to make alternative arrangements for the rural labour forces in expanding and modernizing

1 These five patterns of agricultural and rural development are discussed and elaborated on in Chapter 2.

smallholder sectors, which has been a root cause of the urban poverty problem.

Often advocates of an urban migration and development strategy suggest that it is highly unlikely that the urban poor will find rural livelihoods attractive. This ignores several factors. Many of the urban poor have grown up in rural areas and maintain links to them, which they activate in times of hardship in urban areas. In many countries, urban people who become unemployed during periods of recession return temporarily or permanently to rural areas if they have land rights or can work on farms of relatives or acquaintances. And rural populations continue to grow, despite migration, in all but the most advanced developing countries, providing a new supply of farm workers and operators, if only there are opportunities for them. The neglect of a small farmer development strategy has left undeveloped many opportunities for small farmers which commercial farmers would have found unattractive since their development would have required heavy labour input. Finally, just because countries have proceeded on an erroneous path for so long does not imply that a drastic course correction would not be beneficial.

The evolution of World Bank policy

Ever since the World Bank made rural poverty reduction a priority in the early 1970s, it has advocated smallholder development as the main strategic element under the enlightened leadership of an eminent South African, Montague Yudelman. By and large, the Bank followed its own prescription in lending through integrated rural development programmes and in subsector programmes for irrigation, research and extension.

But the smallholder strategy had far from universal support, even among Bank staff. It was undermined, especially in Bank support for agricultural credit. While these projects consistently tried to limit subsidies and direct more credit to the smallholder sectors, good intentions affirmed in loan documents and loan covenants were often undermined in practice. This failure to prevent mis-targetting of credit has led to a sharp reduction of agricultural credit operations in the Bank's portfolio, and partly explains the decline in agricultural lending by the Bank.

Since the early to mid 1970s, Bank policy was also supportive of land reform. Several attempts were made to translate this commitment into lending programmes, e.g. in the Philippines, Zimbabwe and Brazil, but the political climate did not permit this. The Bank's commitment to land reform has been reaffirmed in a recent agricultural strategy paper. The demise of the cold war, the fiscal unsustainability of large-scale commercial farm sectors, and new approaches to land reform (see below) may finally open opportunities in this area.

The failure of integrated rural development

The opposition to land reform in many countries, such as Colombia, made integrated rural development directed at smallholders the next best smallholder strategy. Many programmes and projects were supported by the Bank. These consisted of an integrated package of support to smallholder agricultural development for a specific area or region. Some of these projects were called area development projects.

The projects typically consisted of synergistic interventions in agricultural extension, research (if little technology was available), marketing, input supply, credit, rural roads, water supply, electricity infrastructure and small-scale irrigation. Sometimes the projects included social infrastructure such as primary schools and health centres.

These interventions were planned by technicians from the particular countries, with the assistance of Bank teams. Methods for beneficiary consultations were developed and applied. Execution of the components was generally delegated to government organizations, often highly centralized ministries or parastatals. Co-ordination proved difficult, and the key remedy was to introduce project management units, staffed by Bank-selected professionals, which maintained authority over the disbursement of funds and the supervision of procurement and implementation efforts in the project area.

These projects typically proved unsuccessful. Diagnosis of the causes of failure differ, but the following elements are generally agreed upon:

- **Adverse policy environment**. It quickly became apparent that these projects, when pursued in an adverse policy environment for agriculture as a whole or for the small-scale sector, amounted to pushing on a string, and could not succeed. Reform of the policy environment was seen as a prior condition for success. The greater success rate of integrated rural development projects in Asia compared to Latin America and Africa supports this diagnosis.
- **Lack of government commitment**. Often governments did not provide the counterpart funding required for implementation of the programmes, for the entire programme or vital components thereof, despite assurances given in negotiations.

- **Lack of appropriate technology**. This proved important in unirrigated areas, especially of Africa, where there was no history of past commitment to agricultural research, or where colonial research efforts had decayed. An early remedy was to include project-specific research components, most of which failed; in addition, these undermined the national agricultural research systems by robbing them of talented researchers.
- **Lack of beneficiary participation**. The programmes were often designed with a top-down approach in which beneficiaries were not given any authority for decision-making or programme execution. Even if they were consulted in advance, they could not be sure that their preferences were given adequate weight. Most often, therefore, they chose the only decision-making option they had: voting with their feet.
- **The complexity or co-ordination problem**. It is ironic that complexity should have become the Achilles' heel of rural development. After all, building rural roads, small-scale infrastructure or providing agricultural extension must be dramatically simpler tasks than the construction of large-scale irrigation infrastructure or ports, where the Bank did not encounter a co-ordination problem. The co-ordination problem emerged as a consequence of delegating sub-programme execution to government bureaucracies or parastatals which were typically highly centralized and had their own objectives. Many of them were out of touch with beneficiaries who could much more easily have co-ordinated the relatively simple tasks at the local level. There the issues are often quite simple, and information is readily available to local decision-makers. Indeed, one might classify integrated rural development as the last bastion of central planning, swept away by reality like all other central planning schemes.

The failure of integrated rural development has left experts interested in rural poverty reduction inside and outside the Bank in disarray. The Bank has retreated from the ambitious agenda of the 1970s into the support of subsector-specific programmes or projects, each dealing with a specific component of rural development, such as agricultural extension, small-scale irrigation, rural roads, primary education or health care. This means that support for rural poverty reduction has become highly selective within the Bank's programme, even fragmented, since nowhere has it been possible to support the full array of interventions which are required for successful rural poverty reduction.

The worst consequence of the failure, however, has been the inability to assist countries with advice on policies and programmes which would enable them to implement rural development programmes and reduce rural poverty successfully. Policy advice rightfully concentrates on eliminating direct and indirect distortions, supporting infrastructure and social investment in rural areas and, for the poor, implementing land reform, reducing interventions through parastatals, strengthening agricultural research and extension, etc. How to implement the investment and support strategies which are recommended in the rural areas of entire countries is left unanswered. By withdrawing from rural development, the Bank has left the complexity and other implementation problems in the hands of the governments of the particular countries. They have not disappeared just because the Bank has withdrawn from them.

Administrative and fiscal decentralization

Countries, other donors and multilateral lenders, as well as the World Bank, of course, have tried to find alternatives to integrated rural development. Most of these initiatives deal with administrative and/or fiscal decentralization, and the greater involvement of beneficiaries, which are discussed below.

The first approach has been privatization of infrastructure and service delivery to the private sector, especially of marketing functions. This relieves the government of a fiscal burden and often improves the delivery of services once the private sector has taken over the functions. The private sector can provide most production and infrastructure services to large-scale entrepreneurial sectors, often at a lower cost than government. But partial or full government finance is required for poverty-reducing rural development efforts based on a small farmer strategy.

Another approach is channelling resources for specific small-scale productive or social projects to beneficiary groups, either directly or through inter-mediation with non-government organizations. This approach has flourished in countries where bureaucratic or political institutions have been severely discredited including, for example, Zambia or South Africa. Governments, bilateral donors and multilateral lenders have increasingly resorted to this method. Social funds delegate planning and execution to beneficiary groups or their NGO agents, but they leave ultimate approval and disbursement authority with central project units and the social fund administrators.

A more radical evolution of rural development programmes has taken place in Mexico and Colombia and, recently, on a pilot basis in Brazil, where the programmes have evolved into matching grant mechanisms for rural municipalities or districts or for poor beneficiary groups, without necessarily losing their multi-sectoral approach. Matching grants will be further discussed below. Within these programmes, genuine decision-making power over project funds is delegated to municipalities and/or beneficiary groups through such mechanisms as municipal funds. Within certain budget limits, the municipalities are empowered to choose from a menu of poverty-reducing community projects. Project selection takes place according to rules which increase the transparency of decision-making to the ultimate beneficiaries and assist in proper targeting to the poorer groups. Tens of thousands of small-scale community projects have been executed in this way.

Many countries have recently gone much further with administrative and fiscal decentralization of rural development. Administrative decentralization can take place either through deconcentration of administrative powers in government bureaucracies, and/or through delegation or devolution of rural development functions to lower level governments and/or communities. Fiscal decentralization involves the assignment of revenue sources to lower level governments, and/or the transfer of resources to such governments through unrestricted revenue sharing and/or through restricted or matching grants.

An extremely successful fiscal and administrative decentralization effort occurred in China in the late 1970s, along with the elimination of collective farms and the reduction of farm taxation. In China, most revenues are collected by local entities and shared with higher level governments. The extraordinary rural development performance which followed has already been commented on. However, the central government has found itself in great fiscal difficulties and is currently reforming the tax and revenue sharing system to resolve the fiscal imbalances which have emerged.

Much administrative deconcentration and delegation of functions has occurred in Mexico under the Solidaridad and other public sector reform programmes, and in Indonesia. These have been associated with greater revenue sharing for poorer regions, and the development of sophisticated matching grant mechanisms to lower jurisdictions and community groups. Enormous improvements have occurred in productive and social infrastructure, water supply and sanitation, as a consequence of these decisions.

Similar successes have been achieved in the state of West Bengal, India, where fiscal and administrative decentralization were associated with political decentralization in the form of a revival of the elected governments at village, block and district levels. This approach is now being generalized to the whole nation by the ratification of a constitutional amendment mandating the same elected government structure and fiscal commissions for all states of the Union.

Colombia has gradually and fairly carefully transferred additional fiscal resources to municipalities, much of them earmarked for health and education. At the same time, it has reformed and strengthened central government matching grant systems such as the Integrated Rural Development programme (DRI) and a social fund programme. These changes are still being implemented.

But decentralization has also had its failures (*Crook & Manor, 1994; Meenakshisundaram, 1994*). In Ghana and the Ivory Coast, elected new local governments were created but the enthusiasm they engendered dissipated rapidly because they were starved of resources. In Karnataka, India, a reform effort similar to that in West Bengal was reversed for political reasons.

In Brazil, the 1988 constitution transferred a total of 68 per cent of government fiscal resources to states and municipalities through general revenue sharing without assigning corresponding responsibilities, thus only deepening the fiscal crisis of the central government. Nor did it reform the matching or conditional grant systems of central ministries, missing a major opportunity to influence the spending of lower level jurisdictions in socially productive ways. Some lessons emerge from these experiences:

- Decentralization is not a simple panacea or a recipe. It matters how it is put together. The many different elements – fiscal, administrative and political – must be consistent with each other to avoid fiscal imbalances, failure or backlash.
- Decentralization is politically difficult since most bureaucrats and central politicians tend to oppose the implied loss of power.
- It appears to work better if deconcentration precedes or accompanies delegation by placing professional staff into local offices.
- It cannot work if elected governments are not given adequate fiscal powers or transfers from higher level governments.

These lessons do not imply that centralization works better. There is plenty of evidence that it does not, namely, the failed integrated rural development strategy. But it does imply that it must be done deliberately, consistently and carefully.

Empowering the poor

Another lesson emerging from these experiences is that consulting the poor is not enough to empower them for their own development, even with the most genuine intentions. Nor is administrative and fiscal decentralization sufficient. In Karnataka, for example, the decentralization effort improved the match of development expenditures with local preferences, accelerated implementation without increasing costs, and made local government employees (such as teachers) more assiduous in their attendance. It also reduced the amount, if not the frequency, of corruption by shifting it from the state level to lower levels. But it neither improved nor reduced the effectiveness of targeting development programmes to the poor (*Crook & Manor, 1994*). Additional steps will be required.

The first is the earmarking of conditional or matching grant resources for poverty alleviating projects or programmes and the delegation of their execution to poor communities, where technically feasible. This is done in social fund or municipal fund programmes discussed earlier. The second is to strengthen the political representation of poor and disadvantaged groups in local political bodies, as has been done by reserving seats for women and scheduled and backward castes in the constitutional reform of the Panchayat Raj system in India.

Where such constitutional change is not feasible, as in Mexico or Brazil, the rules of earmarked matching fund systems can be designed to ensure greater representation. In Mexico, all decisions about fund allocations must be taken in open assemblies at the municipal level, and a proportion of the funds must be allocated to outlying settlements, which usually are poorer than the municipal headquarters. In Brazil, a special municipal council has been created for the allocation and administration of the funds, which ensures adequate statutory representation of poor rural communities in these non-elected bodies.

Accountability to the poor can also be improved by additional rules which encourage openness and transparency, such as representation of small farmers, women and rural workers on boards of research stations, supervisory committees of extension systems, or on land or labour committees which deal with rural land and labour issues.

Market-assisted land reform

Even where countries attempted land reform, they often entrusted it to a centralized land ministry or to a parastatal, a Land Reform Institute. The government acquired land through expropriation, with or without compensation, depending on the historical development of legal provisions. Much of the land was acquired after it had already been invaded and the arrangement amounted to regularization of a *fait accompli*. These arrangements were costly and slow.

Where land was acquired or regularized by purchase, a bilateral bargaining game ensued between each of the sellers and the government, where each could threaten to use provisions of the law to improve its bargaining position. Economic theory does not suggest any reason to expect such a process to lead to low acquisition prices if the sellers are wealthy and the government is under political pressure from the peasants. Where land was expropriated with no or minimal compensation, each case led to protracted legal battles, frittering away the energies of the peasants and the land reform agency alike into numerous legal battles, many of which were lost.

Disillusion with the slow pace often led to loss of political momentum and budget erosions for land acquisition. Using these processes, it took Mexico some 60 years after the end of the revolution to achieve the task of redistribution. And Mexico has been the most successful of all Latin American countries in transferring land in a peaceful manner.

Market-assisted land reform avoids the bilateral bargaining game and leads to competition on both sides between the buyer groups and the sellers. It avoids years of delays associated with disputes about compensation levels. It privatizes and thereby decentralizes the essential processes. The process should not be left unsupervised, however. District land committees, reporting to regional and national committees, are a promising policy. And parties will ultimately have to be provided with recourse to the courts. A decentralized land or agrarian court system, to which disputes that cannot be resolved through arbitration or by the land committees can be appealed, could ensure this recourse.

The conditions for such a process to work are well known: beneficiaries must receive partial grants to enable them to buy land without starting out with impossible debt equity ratios. A decentralized structure, capable of assisting with the provision of infrastructure and social and agricultural services, is needed. In addition, policies and programmes must not reward large-scale farming through privileges.

Employment generation

Additional rural employment generation can be achieved by insisting that rural infrastructure be constructed using labour intensive techniques. Many countries have gradually improved their ability to ensure this through their contracting procedures and by other means.

Employment generation programmes are also a useful tool, especially during periods of macroeconomic recession, sharp agricultural price declines or drought. Zambia and other countries have shown how such programmes can be implemented in a decentralized way, with much involvement of NGOs and community groups.

Employment generation may also be needed even where there are few agricultural or non-agricultural development opportunities but where, nonetheless, an immobile labour force is unemployed and poor. They may be immobile because of legal or economic limitations on migration or, in the case of poor women, because they also have child-rearing responsibilities.

Summary

This chapter discussed the failure of integrated rural development, and indicated how various countries, other donors and multilateral lenders, as well as the World Bank, have tried to find alternatives to integrated rural development. Most of these initiatives deal with administrative and/or fiscal decentralization and greater involvement of beneficiaries.

These new initiatives are based on a broad professional consensus on rural development and poverty reduction that includes the following key elements:

- Where land is unequally distributed, such a strategy requires a substantial prior or concurrent effort at land reform. Constraining land rental and insisting on expropriation without compensation has a perverse impact on the rural poor. Centralized ministries or parastatal bureaucracies are not good at implementing land reforms.
- Rural areas require substantial investments in economic and social infrastructure, in health, education and farm support programmes. Concentrating these investments into urban areas is not less costly and misses an important opportunity.
- Successful and cost-effective implementation of such development programmes requires the mobilization of the skills, talents and labour of the rural population through decentralized administrative, fiscal and political systems conducive to their genuine participation, and through private sector involvement.
- A special effort is required in the design of decentralized mechanisms so that the poor can participate effectively in decision-making, execution and accountability. Otherwise, rural elites of any colour will appropriate most of the benefits of the rural development programmes.

References

Crook, Richard and Manor, James. 1994. Enhancing participation and institutional performance: democratic decentralization in South Asia and West Africa. *A report to ESCOR, the Overseas Development Administration, on Phase Two of a Two Phase Research Project.*

Meenakshisundaram, S.S. 1991. *Decentralization in Developing Countries.* New Delhi: Concept Publishing Company.

Agricultural Policies, Law and Land Reform

PART III

8

Bedevilling agrarian reform: the impact of past, present and future legal frameworks

Heinz Klug[1]

Introduction

Agrarian reform involves a vast range of policy options and legal mechanisms. However, when we discuss access to land, there are primarily six ways for individuals and communities to obtain land: (a) colonization – the settling of new or unsettled lands; (b) inheritance; (c) purchase in a land market; (d) allocation through a policy of redistribution by the state; (e) the restoration or restitution of previously dispossessed properties; and, (f) land invasions – the settling of land owned by others.

Of these alternatives, the first two – colonization and inheritance – are practically irrelevant in situations where there are no longer vast areas of terra nullius and where those who need access to land have no chance of 'inheriting the earth'. It is the interaction of the remaining four alternatives – invasions, restitution, redistribution and the market – which have consequences pertinent to any conceptualization of a necessary legal framework for land reform. Most important, it is necessary to remain aware of the ways in which the structuring of any single one of these alternatives will impact on the functioning of the others. For example, the land market is only open to those with the existing capacity to meet the

1 This chapter is drawn in large part from the results of a study completed in October 1992 for a World Bank project on a Rural Restructuring Programme for South Africa. The original report was the product of the submissions of a team of contributors without whom it would not have been possible. I wish to acknowledge the contributions of the following individuals: Timothy Bruinders, Nomagcisa Cawe, Matthew Chaskalson, Catherine Cross, Cawe Mahlati, Charles Dlamini, Simon Forster, Steve Goldblatt, Johan Latsky, Tuli Madonsela, Mariam Mayet, Kgmotso Moroka, Gerald Nongauza, Mzamo Nxumalo, Kate O'Regan, Kobus Pienaar, Mpueleng Pooe, Theunis Roux, Peter Rutch, Henk Smith, Thoahlane Thoahlane and Rudolph Willemse. Although the final document and this chapter attempt to integrate their contributions, its different emphases do not necessarily reflect the opinions of all contributors. I therefore accept sole responsibility for all errors.

required purchase price, while the vast majority of land seekers will have to turn to the alternatives of restitution, allocation and, if all else fails, land invasions. These remaining three processes will also impact on one another. If the land claims process is tied up in legal challenges then more people will look to the process of land allocation; however, if both the processes of restitution and allocation are bogged down in political and bureaucratic conflicts, then the insecure and dangerous option of land invasions will present itself as a viable alternative.

In these circumstances of great inequality in land holdings, and a combination of historical exclusion and continuing economic incapacity to enter the market, the task is to imagine a legal framework that will ensure long-term security for all land holders, while addressing the legacy of apartheid and colonial land dispossession and exclusion. To this end, a legal framework must aim at establishing a working balance between the three alternatives to which the vast majority of land seekers will turn. This balance should work to ensure that, while the judicial process of restitution which will be slow but potentially empowering continues, a parallel process of land allocation exists which will address the needs of those who require access to land primarily for economic reasons. If there is no process of land redistribution, then those in need of land will demand that their claim to land be incorporated into the restitution process. Similarly, if the restitutionary process becomes bogged down in legal challenges, or if compensation is offered instead of land, then more and more claimants will turn to the political process, demanding a more effective policy of land acquisition and allocation. Finally, if the processes of restitution and allocation do not provide access to land, then communities and individuals will have no option but to resort to land invasions. Unfortunately, the history of land reform around the world demonstrates that land invasions, which governments then normalize through legal processes of expropriation and allocation, have been the most common and effective processes of land reform. Given the potential of violent conflict inherent in the process of land invasions, a legal framework must aim to reduce the likelihood of this alternative.

Legal frameworks: hindering and facilitating reform

Legal frameworks for land holding and land use are never neutral. The impact of any particular legal regime will be either to hinder or facilitate particular practices and policy options. While apartheid laws prohibiting black ownership and use of 83 per cent of the land in South

Africa had an obvious impact on the creation of inequality in land holdings, many other aspects of South African land law contributed to the system's illegitimacy among the majority of South Africans. Consideration of future options requires us to understand the basic elements which undermined the effectiveness of the inherited system.

A legacy of illegitimacy: the inherited framework

The framework of South African land law that has been inherited by a post-apartheid South Africa is characterized by a number of elements which undermine its legitimacy and have profound consequences for the establishment of a functional system of land law. These elements include: a hierarchy of land tenures in which freehold title is privileged; the fragmentation of land law in different parts of the country; the lack of an adequate system for recording all land rights; the prevalence of bureaucratic discretion over the land rights of the majority of land holders and even over the disposition of land claims under the Advisory Commission on Land Allocation; a racial and gender stratification of inheritance in land rights; and, the resorting to informal tenure systems by those excluded from the privileged, official system of land rights.

Uncertainty and the hierarchy of land tenures

The South African system of land ownership and deeds registration has functioned effectively for a very small percentage of the South African population. The major shortcoming of the registration system has been the privileging of freehold tenure over other forms of tenure which have not been incorporated into the system. A major challenge facing the registration system is the effective incorporation of these rights into the system.[2] The costs of providing security of tenure through formal processes of survey, titling and registration have been widely debated. However, the reduction of the levels of accuracy required for survey is unlikely to bring about reduced costs because, although the technology is already available to surveyors,[3] professional control of the process is preventing a fall in expenses.

2 In particular, the rights in Schedule 2 to the Upgrading of Land Tenure Rights Act 113 of 1991, and the occupation and site permits referred to in the conversion of Certain Rights into Leasehold or Ownership Act 81 of 1988.

3 The average surveying fee in four IDT capital subsidy upgrading projects represents a mere 2,6 per cent of the total costs of the projects. Section 7 of the Land Survey Act, 1927 provides that regulations may be made prescribing the fees which a land surveyor shall charge for the survey of land. Regulation 67 of the Land Survey Regulations

The fragmentation of land law

Apart from the approximately 130 statutes affecting land in South Africa, it is important to recognize that land law has been completely fragmented through the implementation of apartheid policies. In order to determine the legal framework affecting land in any of the former self-governing bantustans or the four former so-called 'independent' TBVC territories, four different sources of law need to be considered: (a) laws made by the South African Parliament on matters referred to in the schedule with regard to the Self-Governing Territories Constitution Act 21 of 1971 – as the laws stood on the date the particular territory achieved self-governing status without reference to any subsequent amendments made to that law by the South African Parliament; (b) amendments made by a self-governing territory's Legislative Assembly to the laws contained in the schedule; (c) new laws made by the self-governing territory's Legislative Assembly (which might replace South African laws or be completely new); and, (d) laws made by the South African Parliament on matters not referred to in the schedule.

Despite this complex array of legal systems and statutes, it is possible today to identify three primary forms of land tenure in the rural areas of South Africa: communal tenure on 'scheduled land' (defined in terms of Act 27 of 1913); freehold tenure; and rights under a PTO (permission to occupy) which apply mainly to 'released land' set aside for acquisition by the South African Development Trust in terms of Act 18 of 1936.

The existence of a limited number of tenure forms, however, disguises the fact that an already fragmented system of land legislation has been further splintered since 1991 by the introduction of 'reform' laws which, on the one hand, give the State President extraordinary administrative powers to regulate land on an ad hoc basis and, on the other hand, provide a system of individual tenure 'upgrading' which, since it is to be implemented by individuals and groups, will destroy any existing geographic coherence of the prevailing tenure forms. An important aspect of this process was the transfer of Trust Land by the South African government into the possession of the self-governing territories and its transformation into state property, thus destroying the rights of

(made under GN R1814 of 1962) provides that the tariff for services shall be in accordance with the tariff prescribed in Annexure A to the Regulations. Provision is made for charges to be agreed upon at a higher rate. Although the contrary is not stated, it is understood that surveys are also undertaken at rates lower than the tariff.

existing PTO holders. Before agrarian reform is initiated, therefore, it is essential to conduct extensive research into the specific legislation and tenure rights applicable in each region of the country. This task will be complicated by the lack of documentation on applicable legislation and regulations in some self-governing territories, and even the disappearance of land records in some areas.

Lack of adequate recording

The bulk of land in South Africa is held in freehold title; however, the land rights of the great majority of land holders are not registered or recorded adequately. This privileging of freehold tenure over other forms of tenure and ownership was perpetuated in the Upgrading of Land Tenure Rights Act of 1991 which purports to allow all holders of 'lesser' forms of tenure to upgrade their rights with the aim of achieving a universalized system of freehold tenure. It is apparent, however, that the state has placed a considerable burden on institutional resources with regard to the upgrading of land titles. The institutions involved include the Ministry of Land Affairs (in relation to various statutes concerned with the upgrading of title), and state resources adhering to that Department, including the various offices of the Surveyor-General[4] and the regional offices of the Registrar of Deeds.[5] In addition, considerable resources will have to be committed by the Provincial Administrations for the upgrading of land tenure. Legislation such as the Provision of Certain Land for Settlement Act, 1993[6] draws on institutional resources, (in this case the Provincial Administrations), in order to promote new rural settlements, including agricultural holdings. One of the dangers is that institutional resources will be deflected solely towards privatization with little recognition of the recording needs of other forms of tenure, including forms of collective tenure and indigenous rights.

The institutional requirements of certainty

The key institutional challenge in order to ensure future certainty of title in rural areas is the rationalization and simplification both of the legislation and the institutions involved in the administration of land tenure rights in the rural areas. Simply adopting a process of privatization in

4 In Bloemfontein, Cape Town, Pretoria and Pietermaritzburg.

5 In Cape Town, Port Elizabeth, Vryburg, Pretoria, Johannesburg, Pietermaritzburg, Kimberly and King William's Town.

6 Act No. 126 of 1993.

the context of the upgrading of registrable titles has major implications for the institutional resources that would be required on the part of the state in the future. The tendency in the inherited statutes towards deprofessionalization, especially of the registration function, is likely to significantly increase the load on the institutional resources brought to bear on land registration by the state. The technical process is difficult enough. In political terms, the key dangers lie in continued fragmentation of jurisdictions dealing with land matters in various parts of the country. From the point of view of creating institutional capacity and certainty of title, it is essential to unravel and rationalize the legacy of apartheid legislation and institutions on a uniform basis.

ACLA, state largess and disempowerment

In response to widespread dissatisfaction with the failure of the 1991 White Paper on Land Reform and the accompanying legislation[7] to address land restoration,[8] an Advisory Commission on Land Allocation was constituted in terms of Chapter VI of the Abolition of Racially Based Land Measures Act.[9] The Commission (ACLA) had limited powers of recommendation only. In particular, its objects were to make recommendations to the State President regarding: (a) the identification of land belonging to the state or any state institution acquired for the purpose of promoting the objects of racially based land legislation which had not been developed or allocated for a specific purpose; (b) the identification of rural land with a view to its acquisition by the state for the purposes of agricultural settlement; (c) the planning and development, within the financial means of the state, of this land; and, (d) the allocation of this land.

Although welcomed, there were nonetheless severe reservations amongst communities claiming land and their legal representatives as to whether ACLA had any real power to address the question of land restoration on the scale required. A number of facts were criticized: the fact that only certain State land (and no privately held land) could be

7 Five land reform bills were introduced to Parliament during March 1991: the Abolition of Racially Based Land Measures Bill, the Upgrading of Land Tenure Rights Bill, the Less Formal Township Establishment Bill, the Residential Environment Bill and the Rural Development Bill. Of these, only the first three were eventually enacted.

8 The White Paper simply dismissed the question of restitutionary land allocation as being too complex and divisive to deal with. *See* White Paper on Land Reform, 1991, A2.11 para (f).

9 Act 108 of 1991.

identified in terms of the Act; the unwieldiness of concepts like 'land … which has not yet been developed or allocated for a specific purpose'; the absence of specific guidelines detailing how ACLA would receive and hear claims, the method proposed for the selection of the Commission's chairperson and members,[10] and the fact that no obligation was placed on the State President either to act on any of the recommendations made to him in terms of the Act or even to give reasons for failing to do so.

ACLA began receiving applications in October 1991. It delivered two annual reports by 1993, the contents of which tended to confirm many of the misgivings. It seems as though individuals, rather than rural communities, made most effective use of ACLA.[11] Although a clear attempt was made to appoint a Commission 'representative of the community as a whole'[12] in so far as the racial composition of that body was concerned, the members of the Commission were not representative of either the gender composition of the country or the full ideological spectrum with regard to questions of land restoration.[13] Both of the Commission's chairmen were former justices of the Supreme Court of South Africa, and both followed narrowly legalistic interpretations of the Commission's terms of reference. Most importantly, very few of the Commission's recommendations to the State President were acted upon.[14]

Given these shortcomings, the fast changing political circumstances led to quick changes in ACLA's functioning. The major part of the Abolition of Racially Based Land Measures Amendment Act of 1991 was taken up with amendments to ACLA's terms of reference. There were two main amendments. The first was reflected in the removal of the word 'advisory' from the Commission's name. This changed the status of the Commission from that of a purely recommendatory body to

10 Presidential appointment (in terms of subsec 90(1) of the Act).

11 *See,* Advisory Commission on Land Allocation, Annual Report: 1 March 1992 – 28 February 1993, p. 3.

12 As required by subpara 90(1)(c) of the Act.

13 The Commission's members at the time of the tabling of its Second Annual Report were: Justice S.W. McCreath (Chairman), Prof. N.J.J. Olivier (Deputy Chairman), Mr N.J. Kotze, Dr D.C. Krogh, Prof. H. Ngubane, Bishop T.W. Ntogana and Dr R.E. van der Ross.

14 Only three of ACLA's recommendations to the State President were acted upon during the first two years of its existence. The Commission's major achievement was to facilitate the partial settlement of the claim of the Mfengu people to land in the Tsitsikamma region.

one which enjoyed final decision-making powers in certain circumstances. The second concerned the extension of the Commission's powers to land previously outside its jurisdiction.

Section 88B defined six categories of land to which the revised Chapter VI would apply henceforth. These categories were:

1. State land and land of a development body acquired to promote the objects of any racially-based land law which was not alienated or allocated in terms of any other law for a specific purpose, or which was not developed or utilized for public purposes.
2. State land and land of a development body which was acquired to promote the objects of the Community Development Act of 1966, and which was not alienated or allocated in terms of any other law for a specific purpose, or which was not developed or utilized for public purposes.
3. Other state land which had not been developed or utilized by the state for public purposes.
4. Land of a local authority which was acquired to promote the objects of any racially-based land law or the Community Development Act and which was declared by the Minister,[15] with the concurrence of the local authority concerned, to be land relevant to this process.
5. Any state land or land which may be acquired by the state and which may be developed for agricultural or residential purposes.
6. Land referred to the Commission by the Minister in order to investigate the disposal, acquisition or development of that land for a purpose mentioned in the referral, and to promote the objects of this law.

The full importance of this categorization only emerges if one analyses the powers and functions of the Commission in relation to each category. Nevertheless, a few preliminary observations can be made. The first category of land closely resembled the kind of land which fell under ACLA's original jurisdiction, with two important exceptions: the substitution of land belonging to a 'development body' (rather than 'any state institution'); and the qualification that all land in this category not be 'developed or utilized for public purposes'. 'Development body' was defined in s 88A(ii) of the amended Act as meaning any one of the Housing Development Boards established for each of the racial

15 The Minister of Regional and Land Affairs (s 88A(vii)).

categories represented in the tricameral parliament, or as a development corporation established in terms of s 5(1)(a) of the Promotion of the Economic Development of National States Act 46 of 1968 or similar institution. Although the inclusion of land belonging to these development bodies may appear to have been a significant extension of the Commission's terms of reference, this is mitigated by the omission from the first category, land belonging to state institutions generally, and also by the effect of provisions requiring a declaration from the Minister as a prerequisite for consideration of such land by the Commission. This arrangement ensured that, in so far as land belonging to a development body was concerned, the claims process would be tightly controlled by bureaucratic discretion rather than demand-driven, a theme which is in fact characteristic of the ACLA process as a whole.

The inclusion of the second category of land in the Commission's terms of reference removed any uncertainty there may have been surrounding the application of the ACLA process to claims based on removals undertaken in terms of the Group Areas Act.[16] However, this provision also made it clear that only land still vested in the Community Development Board which also satisfied the other requirements of this section would be available for restoration. Once again, the distinction between alienated and unalienated Community Development Board land was quite arbitrary from the claimant's point of view. A claimant's right to apply for restitution of Group Areas land, if the restoration process was to make any sense at all, should not have depended on whether or not land expropriated in terms of that Act happened to have been alienated or developed by the Community Development Board. It may well be that the degree of development makes restoration of the actual land claimed impractical, but this should not constrain the right of an individual who suffered prejudice by virtue of the provisions of the Group Areas Act to at least obtain alternative relief.[17] That the Legislature was aware of the potential hardship of this distinction was made clear by the later attempt in the legislation to introduce a partial moratorium on the sale of Community Development Board land. The

16 Quite a few Group Areas claims were brought in terms of the Commission's former terms of reference, with varying degrees of success. See, for example, the Advisory Commission on Land Allocation's Annual Report: 1 March 1992–28 February 1993 at 3.5 (iv).

17 In cases where inadequate compensation was paid to persons removed from their homes in terms of the Group Areas Act, this might mean additional 'topping-up' compensation.

relevant section provided that first and second category land could only be sold by the State or development body concerned with the concurrence of the Commission, whilst another section of the Act provided that, for as long as the Commission was carrying out an investigation with regard to first category land, it had the power to direct the responsible public authority not to utilize or develop it. Of course, for two reasons, these two safeguards provided only a partial solution to the dilemma posed by the distinction between unalienated, unallocated, unutilized and undeveloped Community Development Board land, on the one hand, and land which has not been alienated, allocated, utilized or developed, on the other. First, to be at all effective, the Commission would need to have the capacity to conduct a thorough investigation into the likelihood that land, which the State or development body concerned intended to sell, would become subject to a land claim. Judging by ACLA's record, unless any future land claim commission's structures are greatly expanded, it is unlikely to have the capacity to fulfil this kind of 'audit' function, particularly if the terms of reference lead to an increased workload with regard to demand-driven claims.[18]

The inclusion of fifth category land within the Commission's terms of reference appeared to be a statement of intent that, where the land actually claimed could not be awarded for one or other reason then, in some cases, the Commission might have recommended that alternative state land be purchased in order to accommodate such claims. However, the Commission's power, in certain circumstances, to make orders binding on the state was subject to so many qualifications as to be almost meaningless. Section 88B(2), for example, provided that the Minister may give effect to orders or recommendations made by the Commission, *inter alia*, by regulating the manner in which land is to be awarded to successful claimants. The Commission's orders, in other words, would not have been directly binding on the Deeds Registrar concerned, but would have been subject to ministerial intervention. Unless the Commission's order or recommendation dealt specifically with the form of tenure to be awarded to the successful claimants concerned, the

18 Even if the Commission could have been effective in ensuring that no pieces of state land to which there are potential claims are in future alienated, some additional form of control over the sale of all state land is needed given the importance of such land for land redistribution generally. The policing of such a comprehensive moratorium is arguably not within a Land Commission's intended objects (since it is strictly speaking a land restitution body only), but the point remains that the sale of state land was actively promoted by government policies in the late 1980s and early 1990s.

Minister would have been in a position to promote the institution of private property regimes on restored land, in keeping with the general privileging of freehold tenure in other portions of the 1991 land law reforms.[19] A further restriction on the binding nature of the Commission's orders stems from the fact that, with regard to third category land, its order could only have been made 'with the concurrence of the Minister or Administrator concerned'. Finally, the proviso stating that the Commission's orders could not bind the state until the payment of compensation to existing rights holders was finalized, would have allowed the state to delay the land claims process by arguing that insufficient funds were available to pay the required compensation.

On the whole, even after the attempts to improve the process, the Commission on Land Allocation did not, as the former Deputy Minister of Regional and Land Affairs claimed when introducing the amending legislation to the tricameral parliament, 'provide an enabling framework to deal with cases of restitution and obviate the need for a land claims court'.[20] The omission of privately owned land from the Commission's terms of reference (unless referred to by the Minister in terms of s 88B); the tight control maintained over claims to land belonging to state institutions; the absence of any detailed procedural rules governing the Commission's functioning; the Act's preoccupation with individual ownership in respect of group claims; and, the lack of any real bite in the Commission's power to make binding orders, all meant that the land claims process envisaged in the Act did not address the legitimate demands or empower those who had been forcibly removed from their land in terms of apartheid laws and policies.

A racial and gender stratification of land rights

Laws relating to succession were structured on a basis which differentiated by race. Although the Administration of Estates Act of 1965 is of general application to the estate of any deceased person, leaving the property or a document purporting to be a will within South Africa, the administration of the estates of 'Black persons' was partially governed by section 23 of the Black Administration Act, 1927. The Administration of Estates Act, in these cases, applied only to the property of a 'Black person' which was capable of being devised by will. Section 23

19 The language of s 88B(2) certainly seemed to indicate that land which was successfully claimed by a group of persons would be subdivided rather than communally owned.

20 Debates of Parliament (18 June 1993) col. 1161.

of the Black Administration Act, 1927 provided that all movable property belonging to a 'Black person', and allotted by him or accruing under 'black law or custom' to any woman with whom he lived in a customary union, or to any 'house', shall upon his death devolve and be administered under 'black law and custom.'

The system of customary transmission of land

Succession in African customary law has as its *raison d'être* the perpetuation of the family head's name, and the creation and maintenance of a permanent family fund. Heads of households are granted access to land, and the whole family is expected to be catered for within that land grant. Children gain access to their father's land until it becomes too small. An individual has the right to part with his rights to cultivate the land. He may transfer his right over the whole or part of the land to his children, relatives or friends. The individual may further enter into sharecropping arrangements with others.

The customary law of succession is that of primogeniture. Although male heirs are eligible for allocation of land, women do not have an independent right to land since the family is the basic unit of the communal system and membership of the community is the basis of entitlement. Women, therefore, are not allocated land nor are they eligible to succeed to any property. However, the percentage of female-led households in South Africa is growing in the rural and urban areas, under both customary law and civil law. Thus women living in rural areas under communal tenure systems and 'customary law' are acutely disadvantaged.

A positive development was a process of negotiation begun in some rural communities which were anticipating land reform. These negotiations were aimed at granting women access to rights to land and to sit in village councils as and when the communities resettled on old traditional land – which had been taken from them – or on land newly allocated to them.

Informal tenures

In addition to the range of officially recognized, if neglected, forms of tenure, Catherine Cross has demonstrated through her research that there exists a whole range of practices that amount to a system of informal tenure arrangements. While these exist primarily in the peri-urban 'squatter' settlements or as forms of unrecognized tenancies, Cross shows how these dynamic informal tenures are based fundamentally on indigenous notions of land holding and social norms.

It has long been accepted that the overall direction of change in African and other indigenous tenure systems has been from an earlier, predominantly social version to a modernizing, individual form considered to be more compatible with a world economy. It has also been generally noted that over-rights allowing other parties to block transfer of land, tend to loosen as the world economic system penetrates the countryside. Recent work[21] has stressed the contested nature of change in land rights, emphasized the role of local power processes, and examined the possibility that forms of individual tenure recognizable as 'private' may not be the necessary outcome of the encounter between indigenous African tenure and the cash economy. It is also worth pointing out that the newer informal individual tenures are often easier and faster to transact than the earlier, more classical versions of indigenous tenure.

Informal tenures, then, represent the sum at any given time of how social tenures are managed by users on the basis of the prevailing political economy together with their perceptions of what is legitimate, as opposed to what is prescribed in the official tenures. On this basis, they can be expected to reproduce themselves in relation to any programme of reconstruction or land redistribution. It then becomes essential to provide in advance for the probable dynamics of informal practice for self-defined and self-organized community groups settling on redistributed land.

Informal processes of transmission

Overall, the classical indigenous tenures represent social principles of landholding which provide household and individual social rights with social oversight. This social oversight is provided mainly by other right holders and at the neighbourhood level. In addition, tribal officials are involved in the process to varying degrees, depending on the local situation. These processes, although participatory, are relatively slow and are not open to women. Classical indigenous tenures allow secure and reliable exchange of the residential land rights that are seen as important socially, but are less strong for production rights. In addition, since grazing rights represent opportunities for capital accumulation, savings and disaster insurance, these tenures are relatively rigid with regard to ensuring grazing rights for all cattle owned by community members.

By comparison, peri-urban informal systems of property rights are faster and more individualized tenures, and support less interventionary rights from other concerned parties. Control over entry of outsiders

21 *See* Sara Berry (*1984, 1988*).

through kinship group and neighbourhood institutions also declines as these structures weaken. With the tenure system changing, there seems to be greater scope for abuse in informal transactions, but overall security appears to remain satisfactory. However, the rapid course of change is often viewed with unease and dismay by residents, who link loss of control over settlement and the entry of outsiders with the spread of violence.

As regards arable land, the main difference appears to be a higher rate of conversion to residential use, resulting in a decline in the size of arable fields. However, security of temporary arable transfers may also be less. In addition, loss of grazing land to residential use accompanies a decline in the investment value of livestock as density increases and housing stock comes to represent an alternate form of savings. Housing stock is unequivocally individual property, and can usually be disposed of without reference to right holders outside the household.

The historical and contemporary records indicate that tenancy is an extremely flexible and adaptable tenure, but also one which is subject to internal tensions and often becomes exploitative when landlord control is strong. In common with other informal tenures, it responds to outside market conditions. At the same time, social entitlements can also be extremely powerful and, under some circumstances, can subvert the economic logic of production on privately held land, leaving landlords with little control over land use. While landlords and tenants are sometimes reported to assert mutual respect and regard, in many ways their fundamental interests are opposed.

The history of the small pockets of black-owned private land in South Africa suggests that the informal development of tenancy was nearly inevitable on individual holdings, though the extent to which tenancy will impact on redistributed land will depend partly on the amount of land that is redistributed and its location. Given South Africa's history of dispossession and removals, the extreme densities now found on African private land are not surprising. However, private land has become significantly more crowded than communal land under social tenures in the former bantustans, and the difference here appears to relate to the difference in the informal settlement systems involved.

In the early stages of redistribution, pressure from prospective tenants is likely to be strong. There is no clarity from existing examples as to whether or not tenant systems would be economically efficient under the range of possible conditions of location, support and policy environment. However, it is very likely that some forms of production tenancies will be viable. At the same time, the relation of subsistence and residential tenancies to redistribution is a matter of policy which has not yet been resolved. It appears that these informal land use regimes might be unwelcome to a production-oriented reform plan.

What seems clear is that tenancy needs to be acknowledged and provided for in legislation so that expectations on both sides can be clarified, and both parties can have access to mediation and dispute settlement as necessary. In particular, there may be a need to allow for the variety of different tenancy contracts which have existed in South African history, and for the new ones that may develop under an agrarian reform programme.

The transitional framework

The adoption by Parliament of an interim constitution on December 22, 1993 – by a vote of 237 to 45 – brought into 'legal existence' the constitutional agreement reached after months of negotiations at the World Trade Centre outside Johannesburg. The implementation of the 1993 Constitution in April 1994 put in place a new constitutional framework establishing the boundaries within which a new legal framework of land law and agrarian reform policy will have to be developed. While this is only an interim constitution, many of its features – including a justifiable bill of rights and a basically federal structure of government – will be entrenched in the final constitution through the framework of constitutional principles included in Schedule 4 of the constitution.

The new constitutional framework

The constitutional parameters of agrarian reform may be traced by considering the impact of a number of specific elements of the 1993 Constitution. Although these elements – including the property clause of the bill of rights; the structure of government; the recognition of traditional authorities; the provisions for advancing gender equality; and the recognition of indigenous law, are not the only parts of the 1993 Constitution which will impact on agrarian reform, they will establish the boundaries of future land law. The most important part of the 1993 Constitution affecting land rights – the clauses providing for the restitution of land rights and the establishment of a land claims commission and court – will be discussed in Chapter 15.

The property clause

Although substantial state interference with property rights is a fact of 20th century life, it is also true that, given the nature of the democratic transition in South Africa, a constitutionally protected property right is inevitable. The extent of interference will therefore be policed by judicial interpretation. The tendency of the courts in similar situations is to grant strong protection to owners – often trumping all other provisions in the constitution. This tendency can lead to conflict between the court and a democratically elected legislature and executive. The response of the ruling party in India, for example, was constant constitutional amendment and conflict with the court, with severe implications for the upholding of the rule of law and the constitution itself.

The incorporation of a right to property in section 28 of the 1993 Constitution – despite concerns that a property right would merely constitutionalize the historic injustice upon which property rights in South Africa rest – is characterized by a serious tension. On the one hand, it seems the state's power of eminent domain is restricted to acts of expropriation in furtherance of a public purpose only while, on the other hand, the compensation clause provides a completely different vision of the constitution balancing contending claims to property.

Although South African common law jurisprudence has historically taken a broad view of the state's power to expropriate, the adoption of the language of public purpose in a constitutional context raises the possibility that more restrictive notions of public purpose adopted in other constitutional settings may be incorporated into South African law, with the effect of frustrating government efforts to reallocate resources such as land, even when previous owners are offered full compensation. However, it has been argued with reference to the interpretation of the public use requirement of the Fifth Amendment of the United States Constitution, that public purpose may be interpreted to require that the 'taking produce(s) a public benefit or advantage,'[22] which in effect is no different to the public interest analysis advocated by critics of the present clause's public purpose formulation.[23]

This interpretation would reconcile the tension within the clause created by the obvious incongruence between a stunted interpretation of public purpose and an expansive compensation provision which explicitly incorporates the particular circumstances surrounding property relations in South Africa. The court is required to determine what compensation would be just and equitable by considering a range of factors,

22 *See* A. Cachalia *et al, Fundamental Rights in the New Constitution* (1994), p. 95.

23 *See* Marais, 'Snatching defeat from the jaws of victory', *Work in Progress 95* (February/March 1994), p. 26.

including the use to which the property has been put; the history of the property's acquisition; the market value of the property; the value of the investments in it by those affected; and finally, the interests of those affected. By implicitly balancing market value against specific historical factors concerning the acquisition and use of the property, this test moves in the opposite direction to a narrow definition of the public purpose formulation of the right to expropriate, in that it allows the court to balance contending interests.

Although the interpretation of these clauses is likely to be greatly contested, it can only be hoped that South Africa's new Constitutional Court will take a more nuanced and contextualized view of property than that taken by the Namibian High Court in *Cultura 2000 and Another v Government of The Republic of Namibia and Others*.[24] In that case, despite the obvious political and possibly racial motivation behind the gift of state property to Cultura 2000 prior to Namibian independence, the court merely asserted the Namibian Constitution's property clause and suggested that had 'the present Government wanted to acquire the farm Regenstein in the public interest for the State, it could certainly have expropriated it in terms of art 16(2) of the Namibian Constitution and have paid "just compensation".'[25]

Structure of government

The 1993 Constitution adopts 'strong regionalism' as the form of government structure to frame the geographic division of power in the new South Africa. Although the protagonists of a federal solution for South Africa advocated a national government of limited powers, the 1993 Constitution reverses the traditional federal division of legislative powers by allocating enumerated powers to the provinces. This allocation of regional powers according to a set of criteria incorporated into the constitution and its principles, was rejected by the IFP on the grounds that the constitution failed to guarantee the autonomy of the provinces. Despite the ANC's protestations that the provincial powers guaranteed by the constitution could not be withdrawn, the IFP argued that the allocated powers were only concurrent powers, and that the national legislature could supersede local legislation through the establishment of a

24 1993 (2) SA 12 (NmHC).

25 Id, p. 25 H-I. See the partially successful appeal against this decision by the Namibian Supreme Court in *Government of the Republic of Namibia v Cultura 2000* 1994 (1) SA 407 (NmSC).

national legislative framework covering any subject matter. This tension between provincial autonomy and the ANC's assertion of the need to establish national frameworks guaranteeing minimum standards and certain basic equalities, led to an amendment to the 1993 Constitution before the constitution even came into force. According to the Constitution of the Republic of South Africa Amendment Act 2 of 1994, the provinces are granted pre-emptive powers in the enumerated areas of provincial legislative authority. Schedule 6 lists the following areas in which regional legislation is to prevail, except under specified circumstances, as: agriculture; gambling; cultural affairs; education at all levels except tertiary; environment; health; housing; language policy; local government; nature conservation; police; state media; public transport; regional planning and development; road traffic regulation; roads; tourism; trade and industrial promotion; traditional authorities; urban and rural development and welfare services. Difficulty arises, in this context, in distinguishing the exact limits of a region's 'pre-emptive' powers and the extent to which the national legislature is able to pass general laws, under the rubric of concurrent powers, affecting broad areas of governance.

Furthermore, although the provinces have the power to assign executive control over these matters to the national government when they lack the administrative resources to implement particular laws, the constitution provides that the provinces have executive authority over all matters in which the region has legislative authority, as well as matters assigned to the provinces in terms of the transitional clauses of the constitution, or delegated to the provinces by national legislation. The net effect of these provisions will be continued tension between provincial governments and the national government over the extent of regional autonomy and the exact definition of their relative powers. The Constitutional Court's role in this potential struggle between regional autonomy and the establishment of minimum national standards or frameworks, will be further complicated by the constitutional provision entitling each province to an equitable share of nationally collected revenue, a determination which is to be made according to both enumerated constitutional criteria and the recommendations of the constitutionally established Financial and Fiscal Commission.

These constitutional developments make it clear that any legal framework for the creation and implementation of agrarian reform will have to contend with different regional emphases, not only with respect to the implementation of a reform programme, but also in relation to aspects of the legal framework. Although there is no specific mention of land law – we can assume that property issues, as such, will be accounted for by provisions in the national bill of rights – the granting of pre-emptive powers in the interim constitution will directly impact upon the design and implementation of an agrarian reform programme.

(a) Regional government

The 1993 Constitution explicitly recognizes and constitutionalizes nine provinces, thus guaranteeing the existence of a regional level of government. The Constitution of the Republic of South Africa Amendment Act 2 of 1994 fundamentally changed the original allocation of powers between the national and regional legislatures. While the original 1993 Constitution granted provincial legislatures concurrent powers to make laws with respect to a list of subjects defined in Schedule 6 of the constitution, the amendment shifted the allocation of powers so that any law passed by a provincial legislature in terms of the constitution's Schedule 6 will prevail over an Act of Parliament, except where certain defined national priorities[26] are involved. The national priorities which trigger a reassertion of national legislative authority reflect the standards employed internationally to delimit national and regional powers. The acceptance as a matter of national priority of any 'matter that cannot be regulated effectively by provincial legislation'[27] introduces the principle of subsidiarity which is employed in the European Union to determine the level of government to which a particular matter should be assigned. On the other hand, Section 126(3)(d), which empowers the national government to determine matters necessary for national economic unity and the promotion of inter-provincial commerce, seems to point in the direction of the overwhelming power of the commerce clause of the United States Constitution.

The autonomy of the provinces is further advanced by the constitutionally determined distribution of resources. Section 155(1) requires that a province 'shall be entitled to an equitable share of revenue collected nationally to enable it to provide services and to exercise and perform its powers and functions'. Despite this provision, an objection was raised to the original 1993 Constitution on the grounds that it only permitted a province to raise taxes locally if authorized to do so by an Act of Parliament. Although the amended constitution likewise limits a

26 1993 Constitution section 126(3)(a)-(e).
27 1993 Constitution section 126(3)(a).

province's taxing powers, the provinces are now empowered to exercise exclusive competence in taxing casinos, gambling, wagering, lotteries and betting. Since lotteries and other forms of indirect taxation are likely to become significant sources of government income, this concession is not merely cosmetic.

The likely impact of 'regionalism' on agrarian reform will be that separate initiatives will have to be launched in each region in order to contend with the region's exclusive legislative competencies in the areas of planning and development, town planning; traditional authorities and indigenous law; delivery of water, electricity and other essential services. In addition, the regions will share competencies with the national government, which will have limited concurrent powers, in a range of functional areas of significance to rural restructuring, including local government; agriculture; fish and game preservation; the environment; public works; and regional and local policing. Further, even when the national government is competent to pass laws in terms of concurrent jurisdiction, implementation will be in the hands of the regional government. Executive power relating to all these areas of exclusive and concurrent legislative competency will lie with the regional executives, with major consequences for the implementation of agrarian reform.

(b) Local/District government

Autonomous local government is guaranteed in terms of section 174 of the 1993 Constitution. This guarantee carries with it a constitutional obligation on the part of a local government to 'make provision for access by all persons residing within its area of jurisdiction to water, sanitation, transportation facilities, electricity, primary health services, education, housing and security within a safe and healthy environment.'[28] Although this provision recognizes the socio-economic rights demanded by the ANC's constituency, it is immediately constrained by the proviso that 'such services and amenities can be rendered in a sustainable manner and are financially and physically practicable'.[29] This attempt to place obligations on government to provide basic services in relation to the government's capacity to deliver these services bears resemblance to the framework for the advancement of socio-economic rights established under the International Covenant on Economic, Social and Cultural Rights of 1966.

28 1993 Constitution, section 175(3).
29 1993 Constitution, section 175(3).

Although s 179(1) of the 1993 Constitution requires local governments to be democratically elected, the electoral system for local government established by this section has the effect of establishing a consociational system of local government. Based on a combination of proportional and ward representation, with wards distributed according to old apartheid boundaries rather than in proportion to the number of voters in any particular area, the electoral system ensures that the realities of apartheid geography will provide a veto power over budget allocations at local government level.[30] This will perpetuate a degree of racial representation in local government until either the form of representation is changed or communities become effectively integrated.

Nothing in the constitution would seem to preclude the establishment of district level councils for the administration of land allocation and rural development. As far as district land committees are concerned, these may be established in terms of national land legislation and may be responsible for the allocation of land and the mediation of local land disputes. Although accountable to national government, they could be made up of local representatives of a predetermined list of constituencies, with national government retaining oversight responsibility.

However, due to the constitutional allocation of exclusive powers of regional development to the regions, district development councils – whether democratically accountable or not at the local level – cannot be constitutionally autonomous from regional government with respect to regional development priorities. Regional governments, however, may be encouraged to constitutionally enshrine district level administrations with relative autonomy into their own regional constitutions. Alternatively, the regions may implement their mandate constitutionally by devolving power to district level structures. However, this will have to be consistent with the local government powers secured in the national constitution, and in accordance with any powers constitutionally exercised by the national government. In this regard, a national framework for local government may provide a basis for district level administration of land redistribution and development.

(c) Traditional authorities

Another consociational element in the 1993 Constitution is the provision for traditional authorities at all levels of the constitutional structure.

30 1993 Constitution, section 176(a).

At the local level a traditional leader is entitled, ex-officio, to be a member of an elected local government within whose jurisdiction his community is located.[31] Traditional authorities at the provincial level will either nominate or elect a House of Traditional Leaders which, according to section 183(2)(a), 'shall be entitled to advise and make proposals to the provincial legislature or government on matters relating to traditional authorities, indigenous law and the traditions and customs of traditional communities within the province'. This form of representation is repeated at the national level with the establishment of a 20 person Council of Traditional Leaders elected by an electoral college constituted by the regionally based Houses of Traditional Leaders.

Apart from their right to advise government, traditional authorities have a right at both the regional and national level to delay legislation which has a bearing on traditional authorities, indigenous law and the traditions and customs of traditional communities for up to sixty days.[32] Although it may have been thought that the recognition of traditional authorities was merely a transitional feature, the Constitution of the Republic of South Africa Second Amendment Act 3 of 1994 makes it clear that traditional authorities are to enjoy an important place in the new constitutional order since specific recognition is given to the Zulu monarch in the regional constitution of KwaZulu-Natal. Furthermore, section 2 of Act 3 of 1994 amends Schedule 4 of the 1993 Constitution to ensure that any 'provision in a provincial constitution relating to the institution, role, authority and status of a traditional monarch shall be recognized and protected in the Constitution'. Given the traditional role played by traditional leaders in land matters, they are likely to continue to play a pivotal role in any land reform process.

Gender provisions

All women in South Africa, but African women in particular, have been legally discriminated against with regard to access to title in land, inheritance of such title, and their ability to bequeath such access. The 1993 Constitution formally recognizes gender equality in the interim bill of rights, and the interim constitution includes specific provisions for the establishment of a Commission on Gender Equality 'to advise and to make recommendations to Parliament or any other legislature with regard to any laws or proposed legislation which affect gender equality

31 1993 Constitution, section 182.
32 1993 Constitution, section 184.

and the status of women'.[33] The promulgation of a new property rights regime will thus provide an opportune time to address past gender discrimination in land rights.

Customary law and its practices are now subject to constitutionally protected rights, transforming the legal status of African women from that of minors under the perpetual tutelage of male relatives, to that of formal equality. This change will have far-reaching effects in rural society, and considerably increase the number of persons eligible for the allocation of land.

Despite this development, it must be recognized that a key institution determining women's social rights is 'the family'. Therefore, our understanding of what constitutes a family or a household in constructing land reform policies should not bypass female-headed households. Furthermore, a successful implementation of land reform should ensure that those directly affected – rural women – have their views solicited in constructing a land reform process since they, after all, will be the key to its successful implementation.

It should be noted that, as concerns the restitutionary process, there is a grave danger that women will once again be dispossessed. Because, as a class of persons, the capacity of African women to contract was restricted historically, title to most of the disputed areas was usually vested in the names of males – chiefs, missionaries or deed holders – and thus female beneficiaries of previously titled properties are not easily ascertainable. Even in instances where they can be ascertained, the patrilineal nature of succession in most communities poses difficulties for women.

For these reasons, the gender impact of landlessness, and the racial imbalances created by apartheid, cannot be redressed through a claims process alone. As far as land claims are concerned, the only way to address the consequent gender imbalance is to enable all people who can trace their descent to a removal, to bring a claim as a class and to subdivide the award. In addition to the specific legal constraints imposed by the subordinate legal status of women within customary law, the institutionalization of certain 'cultural' traditions has a number of consequences for women including: (a) lack of representation in decision-making processes; (b) a tendency on the part of actors, including policy makers, planners, architects and project managers, to target the 'community' as a unit of analysis, but not to address the issue as a

33 1993 Constitution, section 119(3).

specific issue of women's exclusion; and, (c) the exclusion of women from participation when the 'community' initiates and engages in negotiations with local authorities or development agencies.

Although women, who comprise over 80 per cent of the rural adult population, are by customary practice debarred from participation in decision-making structures such as the Imbizo or Kgotla,[34] women in rural areas hold the key to development. Their emancipation through a policy aimed at facilitating their participation in local government structures would empower them and ensure local democracy. Because they carry the disproportionate burden of securing the livelihood of their families in rural areas, women have a direct interest in rural development to ensure access to water, electrification, paved roads and all the infrastructural services necessary for the survival of a rural economy. The logic of this argument is that local democratic structures should be installed where rural populations can elect and dismiss their representatives. Consequently, the institution of chieftainship should be relieved of its administrative duties and control over local government. In addition, state-aided schemes or those administered by foreign development agencies, should require a percentage set-off calculated to reflect demographic representation or, in other words, rural development programmes should have a strong affirmative action component designed to address gender inequality.

Indigenous law

There are a number of difficulties which will accompany the constitutional recognition of 'indigenous law'.[35] These focus primarily on the question of what exactly is meant by indigenous law in the sense of 'tribal' or 'customary' or 'indigenous' land tenure. With colonial conquest, the nature, focus and content of indigenous law changed. What remained were those norms, mores and practices of decision that the colonial governments permitted or delegated to Africans, either traditional leaders or, if they were not sufficiently compliant, new appointees. This customary law, which is largely the customary law that we encounter now, was radically different in many ways from the precolonial custom that preceded it.

When confronted with this history and the consequent difficulty of

34 *See* Lydia Kompe and Janet Small, *Demanding a Place under the Kgotla Tree: Rural Women's Access to Land and Power,* TRAC 1993.

35 Indigenous law is recognized in section 181 of the 1993 Constitution.

determining the exact content of a 'legitimate' indigenous rule, it is tempting to argue that it is impossible to recognize 'tribal tenure' constitutionally. However, millions of South Africans continue to identify with and hold land under systems of communal tenure and, as a historical legacy of African culture, its constitutional protection may well be asserted in terms of the cultural rights of particular communities, or as a part of the constitutional protection of indigenous law. It is therefore essential to any agrarian reform programme that the recognition of communal 'tribal' tenure be incorporated into the constitutional framework and its position *vis-à-vis* other forms of tenure, both private and common, be clearly determined.

(a) 'Renovating' indigenous tenure

The notion of 'renovating' indigenous tenure involves the resolution of existing tenure problems through the adoption of relatively modest changes in tenure rules, reorganization of land administration machinery (sometimes altering its legal basis and legitimacy), and the creation of new, supportive linkages with national and regional institutions. A significant element of community control over land – a 'communal' element – is retained. This approach seeks to adjust the tenure system to changes in the economic and social environment in which it operates. One example is in English-speaking West Africa, particularly in Nigeria and Ghana, where the courts developed a common law notion of 'family land' out of a variety of tribal lineage-ownership systems. Here the courts permitted transactions in family land with the consent of all those interested. This required a clear definition of the 'family', those persons whose participation was required for an effective transaction.

Another interesting experience is that of Botswana. A system of Tribal Land Boards was created shortly after independence, shifting powers over land allocation from chiefs to boards composed of indirectly elected and ex-officio members. Ex-officio members were appointed from local representatives of relevant ministries. Chiefs remained as members, sometimes as chairmen, although later they tended to drop out in pique over their diminished powers. Although ownership of land remains vested in the tribe, and chiefs retain an adjudicatory role in conflicts over land, the Land Board administers the land in trust for the tribe.[36]

The major attraction of this model is its promise of relatively cost-effective reform with a minimum of social dislocation. It begins with the

36 *See* Riddell and Dickerman 1986; Machacha, 1981.

assumption that tenure change is necessary and desirable, but seeks financial and social economies in tenure change by building on existing institutional arrangements to the extent that this is practical.

(b) A framework of allocation

There is no conclusive or empirical evidence that suggests that a general sweeping away of customary land law and its replacement by a statutory regime has resulted in a marked increase in efficient allocation and use of land. Instead, the slipping away of customary land law has often resulted in a marked increase in the development of open access systems, conflict and social costs – increased landlessness and urban drift – which are thrown onto the state and society at large. Privatization can also cause great inequity when those with knowledge or access to knowledge of a new legal regime and finance, can rapidly acquire land at the expense of those who suddenly find that they no longer have legally recognized rights to the land.

In South Africa, a number of community tenure models already exist and operate successfully in managing land access, production means and environmental demands. Communities of communal land have developed an informal land market as their own approach to property rights.[37] Cases exist where such institutions have managed and successfully supervised tenure arrangements on a modified common property basis. A study in the early 1990s undertaken by the Transvaal Rural Action Committee (TRAC) of different forms of tenure amongst Transvaal rural communities established that, in rural communities such as those at Mathopestad, Mogopa, Braklaagte and Bakubung, the communal system of acquisition and transmission of rights to land has, in general terms, been retained or will be retained.[38] When re-establishing themselves, these communities see communal land control over land allocation as being one of the mainstays of the new system. Access to land by women is one of the issues debated by these communities.

The task of framing modest changes in substantive rules of tenure to meet the specific new needs of farmers is challenging. The possibilities will vary from one circumstance to another, but the fundamental challenge is how to create an adequate institutional framework for such change. Traditional local land administration institutions may or may not be able to meet new needs; in some cases, they may not even be able

37 Catherine Cross, 1988 and 1992.

38 TRAC: Botho Sechabeng (*1993*).

to deal adequately with their traditional tasks under changing circumstances. For example, land allocation becomes increasingly difficult for the allocating institution as pressure on land increases and disputes over land rights multiply. Government policy may have undermined the legal authority or economic bases for authority of the traditional institution.

Land legislation

While the new democratic legislature has produced legislation implementing the constitution's mandate of restitution, the much larger task of developing comprehensive land legislation remains to be undertaken. Some of the major elements of such a process will include: (a) the creation of a single national legislative framework for land rights and administration; (b) provisions enabling the state to acquire land for redistribution; (c) new rules to regulate the tension between ownership rights and illegal use of land based on necessity; and, (d) the creation of a viable system of land administration at the district level which may be coupled with the recognition of land rights.

Nationalization of land law

There is an urgent need to adopt at least a single national legislative framework in order to prevent a complete fragmentation of land rights. This 'nationalization' of land law,[39] however, will have to take into account the recognition of different forms of tenure and also the degree of local control over land use and development granted in the constitution to the different levels of government. At the regional level, it will be necessary to determine the exact level of administration over land affairs and rural development since it will, according to the 1993 Constitution, be the regions that have the power of implementation over region development and agriculture. However, land administration may include a national component, and it will be necessary to determine exactly how this level of land administration will interact with regional or district based structures.

Land acquisition

It has been argued that much of the existing land hunger could be addressed by redistributing the vast tracts of land presently held by the state, particularly that land which was held in terms of the former

39 *See* A. Sachs, Rights to the land. In A. Sachs. 1990. *Protecting Human Rights in a New South Africa*, p. 34.

Development Trust and Land Act of 1936.[40] But this land is either already occupied by black communities, providing no additional land for allocation, or is land held for environmental reasons. Furthermore, any restoration or land restitution process will require mechanisms whereby the state can intervene to make land available to successful claimants who have either been granted a right to alternate land, or where present occupiers are refusing to vacate land granted in adjudication to a particular claimant.

Land acquisition by the state may therefore play an important role in securing an increase in black land holdings beyond the 13 per cent stipulated in the repealed racially based land acts. Besides the purchase of land, there are several other criteria and mechanisms which could be used to facilitate a less costly and more speedy process. Expropriation is the most controversial means of land acquisition, yet a land reform policy based in part on a process of expropriation remains the most effective way to ensure a relatively timely and thorough process of redistribution. The adoption of a Land Acquisition Act is essential and must address the three central issues of a programme of expropriation and redistribution. It must identify: (a) the land to be taken in the reform; (b) what compensation is due to the current owners; and, (c) who shall be the beneficiaries.

Expropriation in terms of a Land Acquisition Act must be conducted on the basis of established criteria which, when applied, will adequately address these issues. Once a land reform process is initiated on the basis of the needs of a particular community or individuals for land, there are several appropriate mechanisms for obtaining the required land. It is in this instance that the state agency with responsibility for land reform will have to apply criteria established by a Land Acquisition Act. Farms could be subject to expropriation according to a number of criteria, including their size, under-utilization, abandonment, unauthorized subdivision, corporate ownership, failure to comply with labour and criminal laws, as well as location.

The excess-size provision of the Frei Agrarian Reform Law[41] in Chile 'established that rural properties larger than 80 basic irrigated hectares were subject to expropriation regardless of their productivity'.[42]

40 Formally the Native Trust and Land Act No. 18 of 1936.

41 Law 16.640 of 1967. *See* Thome, Law, conflict, and change: Frei's Law and Allende's Agrarian Reform. In W. C. Thiesenhusen. (Ed). 1989. *Searching for Agrarian Reform in Latin America*, p. 188.

42 Id. at 194.

Although it may be possible to establish similar criteria according to the size of a viable farm for each of the different agricultural zones in South Africa, the uneven and interspersed distribution of agricultural potential makes such an approach problematic in the South African context. An alternative approach would be to limit individual ownership to the holding of one viable farm. However, this raises the problem of defining the extent of a viable farm – by title deed, operational viability, etc.

Under-utilization often proves to be a difficult criteria to assert because it is subject to a difficult factual enquiry about the capacity and correct usage of a particular area of land. In Chile, expropriations on the basis of low productivity 'were not only subject to judicial review but also required an elaborate verification process involving technical, economic and social criteria for establishing whether the law applied'.[43] Although the burden of proof rested with the landowner to show that the land was in fact productively used, government lawyers had to be prepared to rebut the landowner's evidence. Zimbabwe provides an instructive history of legal mechanisms aimed at encouraging land utilization, later used in the process of land reform.[44] Adoption of under-utilization as a criteria for expropriation in South Africa may have the effect of increasing production among farmers fearing compulsory acquisition of their land, but it will also lead to complex litigation over the definition of under-utilization and its application to particular areas and types of agricultural production.

Although it should not be difficult to create criteria for identifying derelict or abandoned land, for example, land in an urban context for

43 Id. at 196

44 As early as 1925, an attempt was made to impose a tax on 'unoccupied land' as a means of putting pressure on owners to put the land to beneficial use. The Rhodesian government adopted an Integrated Plan for Rural Development in 1978 which identified five categories of under-utilized land. However, the 1979 Muzorewa Constitution defined land subject to compulsory taking due to under-utilization as 'a piece of land registered as a separate entity in the Deeds Registry'. This blunted the effectiveness of the power of expropriation because, even if a farmer used only a relatively small portion of the land for agricultural production, the government could not expropriate any part of it.

The Muzorewa Constitution was even more restrictive in its definition of under-utilization than the Rhodesian government's 1978 Integrated Plan for Rural Development. Instead of the five categories of under-utilized land defined in the Integrated Plan, the 1979 Constitution required that the owner had failed to put the farm to 'substantial use', that is, had not engaged in substantial agricultural production for five years. This limited the government's power of compulsory acquisition of those farms that had been practically abandoned.

which rates have remained unpaid for a period of years, the process of expropriation in these cases will require a period of notification and public advertising to ensure that the owners receive due process, including adequate notice and a fair chance to respond.

Unauthorized land-use or subdivision criteria do not seem to refer to the Subdivision of Agricultural Land Act No. 70 of 1970, but rather could be designed to prevent the circumvention of the excess-size provisions through the division of large landholdings among family members or straw persons.

In legislating against company ownership of agricultural land, it may be argued that companies, with different financial and investment bases to the family farmer, constitute unfair competition to the family farm and therefore should not be allowed to own arable land. In Sweden, company and foreign ownership of arable land has been prohibited since 1906. However, exceptions to the prohibition on institutional land holdings would have to be made to allow for producer co-operatives and other collective enterprises.

Inclusion of criteria based on criminal convictions arising out of failure to comply with labour laws or abuse of farm workers as bases for expropriation, will have a profound effect on the implementation of labour laws on the farms and the abatement of assaults and other illegal behaviour inflicted upon farm workers and other farm inhabitants. Under the present property laws, landowners can expel people from the land, demolish their homes, prevent people from entering, crossing or remaining on 'their' land. The result is that control over land is both control over a productive resource and over the lives of people. However, the difficulty of securing criminal convictions, and the question of the unintended collective punishment of the dependants and heirs in the case of forfeiture, may make such an option highly problematic.

The Lancaster House Constitution, however, gave the government the power to take any land that was under-utilized, even a portion of a farm, but failed to define what was meant by under-utilized. In attempting to interpret the Lancaster House Constitution's reference to under-utilized land, the Zimbabwean government will have to adopt criteria based on the production capacity of the land. For this purpose, it may rely on criteria including such variables as slope, size of soil fractions and their relative frequency in the topsoil, wetness, permeability of the upper topsoil, erosion, surface characteristics, and ecological zone, used by government agriculturalists to classify land in Zimbabwe into eight land capability classes. *See,* Seidman, Land Reform Legislation in Zimbabwe. (Unpublished paper.)

Finally, there is the criteria of location which will enable the government to obtain land for the implementation of agricultural programmes involving the creation of specific development areas without being thwarted by an unwilling seller. Application of this criteria will need to be closely scrutinized to ensure that it is not used for individual gain or with improper political motivation.

Squatting, trespass and nuisance

In deciding future legislation, it must be recognized that the squatting and trespass acts were part and parcel of the system of enforcing apartheid. Therefore, they should be removed from the statute book. In reformulating appropriate legislation, alternative ways of protecting property rights have to be found. These have to take into account the rights both of landowners and of the homeless and landless. For example, the right of the landowner to seek a court order for the eviction of illegal squatters should be subject to statutorily entrenched procedural safeguards. Homeless communities should be duly notified of any application to evict them or demolish their structures so that they are given sufficient time to prepare argument against it or make alternative arrangements.

The burden of providing suitable alternatives and services should rest with the state, whether land is privately owned or not. This responsibility would only be enforceable if it was to form part of legislation dealing with or providing for informal settlements or less formal townships. However, if the state is unable to find suitable alternatives, or fails to do this in a reasonable time period, then landowners could have a claim against the state for the continued illegal occupation of their land. Despite decriminalization, landowners would retain their civil remedies and, if the state is politically unable to evict or enforce a valid court order, landowners could claim an implicit 'taking' and require compensation from the state.

Disputes between squatters, landowners and the state could possibly be taken to a special land jurisdiction. This could be added to the functions of the Land Court which is to be established to adjudicate competing land claims. The Land Court, anyway, will have jurisdiction to decide conflicts that arise out of the claims of labour tenants and communities to reside on land where they are presently considered to be squatters. Although the courts, at one point, seemed prepared to engage, through the common law of nuisance, in struggles between landowners and the state over the settlement of black communities,[45] more recent case law has veered away from such intervention. The courts have indeed argued that if an administrator is able to prove that the infringement of private rights will result, no matter how he exercises his power, then the administrator may rely on the fact that the intrinsic nature of the acts authorized by legislation are such that their execution necessarily and inevitably involves disturbing common law rights.[46] Whether this line of argument will survive scrutiny by the Constitutional Court in its interpretation of the Property clause of the 1993 Constitution remains to be seen.

District level land committees and the recognition of land rights

One possibility, which will have the added benefit of enhancing certainty, will be for district level land administration to be coupled with a registration of land rights which, although initiated at the district level, would be duplicated into a nationally controlled system to provide a back-up to the land court and mediation processes. These three aspects – registration, adjudication and mediation – could be allocated constitutionally to the national level, providing a structure in which a constitutional separation of powers would secure local control and national oversight of land holding, but leave the regions free to determine development priorities.

A future framework: the Constitutional Assembly

The establishment of a Constitutional Assembly under the 1993 Constitution for the purpose of producing a final post-apartheid constitution, means that many of the issues covered by the interim constitution will be reopened. Although the constitutional principles contained in Schedule 4 of the 1993 Constitution will ensure that the protection of property and division of legislative powers between different levels of government remain features of the final constitution, the exact form these will take remains to be resolved. Because of the centrality of constitutionalized property rights to the design of any agrarian reform process, it is necessary to discuss alternative ways in which the Constitutional Assembly may approach the issue of property rights and land in the final constitution.

45 East London Western Districts Farmers' Association and others v Minister of Education and Development Aid and others, 1989 (2) SA 63 (AD).

46 Diepsloot Residents and Landowners Association and others vs Administrator, Transvaal and others, 1993 (3) SA 49 (TPD).

What kind of property clause?

If a property clause is to be adopted, it must be restricted in its scope and have a clear compensation clause specifying the relevant factors the court must consider in a compensation determination. The property clause should make a clear distinction between the regulation of property and when the state uses its power to acquire an individual's property. This distinction is vital to set clear boundaries for the protection of property. On the one hand, while the deprivation of property through regulation – the use of police power – is subject to procedural and other forms of review, it should not create an obligation for the state to compensate. On the other hand, the acquisition or use of an individual's property will give rise to a claim for adequate compensation as a taking or exercise of the power of eminent domain.

Another important distinction to be included in the provision is whether the state may exercise its powers of eminent domain in the public interest, or whether it may only take property, even when paying compensation, in furtherance of a public purpose. Although South African case law, up to now, has interpreted public purpose so widely as to collapse this distinction, in other jurisdictions this has, on occasion, been an important hindrance to any form of redistribution.

Compensation is a separate but specific aspect of a property provision in the constitution. While most jurisdictions have tended to use market value, regardless of the specific description of the standard of compensation in the constitution, it would be best to adopt an adequate compensation formula explicitly specifying what factors must be considered by the court in reaching its determination. These factors include the use to which the property is being put; the history of its acquisition; its market value; the value of the owner's investment in it; the interests of those affected; and the degree of past state investment in the form of subsidies and aid. The issue of delayed or immediate compensation and its specific form must also be specified, for example, cash or government bonds or, as in Zimbabwe, a convertible form.

Land clause

Even if we accept the need for a property clause in the constitution, it is vitally important for the potential success of any agrarian reform programme to give land a separate constitutional status. This may be justified in two principle ways: first, in terms of the role the law has played historically in the dispossession and loss of land and the need to deal with this legacy; second, in terms of the fact that land is a specific, limited resource tied to issues of nationhood and cultural identity in the South African context.

Taking precedence over the general property clause, a specific land clause in the constitution must provide a basis for restitution. This must include the specific restoration of land through a land claims mechanism and also through a general claim of restitution arising out of a historic dispossession and denial of land rights. This general claim provides a basis for a redistributory or allocatory land reform, whether through state initiated market reforms or state programmes of appropriation and redistribution. Together, these provisions aim to insulate land reform effectively from constitutional attack and to allow for the adoption of a demand-led affirmative action based land reform.

More specifically, the land clause must recognize different forms of property and tenure relations, and provide for the establishment of specific legal regimes for each, including private property; indigenous (tribal) property; and common (community) property. The land clause may also provide for the recognition and registration of a range of interests in land outside of ownership; a land ceiling, for individuals, or with respect to different types of tenure; gender equality within land tenure relations; and, a specific land jurisdiction, together with or separate from the land claims process.

A general constitutional claim to restitution

A more effective approach to this wider form of historical land claim may be through the inclusion of a general claim to restitution in the land clause of the constitution based on historic dispossession. Although providing no individual or direct remedy, a general claim to restitution would provide a constitutional basis for a redistributory land reform process similar to the affirmative action clause in relation to the guarantee of equality and equal treatment. This could serve to insulate a land reform or rural restructuring programme from effective constitutional challenge.

An interesting example of a general restitutionary clause is subsection 54 of the Constitution of Papua New Guinea which exempts from the general property clause any law that provides for the recognition of claimed title to land where: (a) there is a genuine dispute as to whether it was acquired validly or at all from customary owners; and, (b) if the land was acquired compulsorily, the acquisition would comply with the present constitution's protection from unjust deprivation of property.

Creating space for common property regimes

It is important to distinguish between tribal tenure and other common property regimes, both in constitutional recognition and in the provision of a specific legal regime for each. For example, in Mexico, the constitution recognizes *ejidos* (tribal tenure) and communal tenure, where communal tenure involves a voluntary framework of collective ownership or control over common property.

Group and individual rights

Recognition of communal and collective forms of tenure requires the adoption of a democratic framework within which the participants in these forms of tenure are empowered to constitute their own rules. The important point with regard to the construction of community constitutions and bylaws, is adherence to the flexibility of the underlying informal systems while providing a democratic institutional framework for new communities to hold and deal in land rights. The interests of women and of tenants are likely to be central with regard to equity.

It can be expected that the economic prospects of the beneficiaries in land reform communities will change significantly during the first years of the redistribution initiative. In addition, the class origins of the beneficiaries may also shift substantially. In order to keep community constitutions viable, provision has to be made for major changes to be enacted. However, changes in land transfer practices which develop will be relatively subtle in their onset, and will be held back by over-regulation.

Any effort to establish specific legal rules regulating land transfer in detail will be ill-conceived. Rather, two principles can be suggested. First, certification of land ownership in a common property framework should be provided by community-managed structures through clear procedures which recognize the rights of women to the land and, at the community's discretion, those of heirs. Second, classes of transfer can be given recognition in a broad framework, and dispute resolution procedures laid down which allow for rights of appeal within the community as well as outside it, and which provide ample access to mediation. No detailed attempt should be made to legislate which transfers are accepted or forbidden, or how they should be conducted. However, the group as a whole has the right of approval over any transfers involving outsiders.

In this type of enabling framework, informal practice can be expected to develop contract forms which will meet the needs of producers as they arise. The alternative of attempting to control procedures tightly is likely to result in informal practice ignoring or evading the bylaws and becoming an independent factor in a situation where law becomes irrelevant. At the same time, the problems of environmental management in relation to resources under pressure are serious. Research from other parts of Africa suggests that, even in the framework of newly established communities, it may be difficult for community procedures under common property to curtail prevailing expectations of general access to common resources. It is suggested that environmental bylaws be written in support of outside legislative provisions that offer incentives for compliance with stocking rates and water management provisions as laid down. These incentives may include access to additional grants or to service delivery, or incremental reduction in loan obligations, and might later extend to tax concessions. Alternatively, offenders may be disqualified from access to concessionary loans for housing or commercial expansion.

It is important that the benefits offered be made accessible to the whole community, to support the bylaws and counterbalance the social expectations supporting use practices that are legitimate, but benefit only a section of the membership. The alternative could be the reinstitution of minimal management regimes. These are viable in the long term, but are not highly productive and do not eliminate environmental degradation. Lastly, it is vital that communities should fully accept and support the bylaws they adopt. The use of general clauses which all beneficiaries of an agrarian reform programme are required to accept, is debatable in this light, and incentives should be considered instead. To the extent that communities are pressed to adopt provisions they do not want, the likelihood of evasion and of informal practice supervening is significantly increased.

Conclusion: implementing land laws and facilitating agrarian reform

The diversity of experience is considerable, but some themes appear fairly consistently. The experience with redistributive land reform in Asia and Latin America is being repeated in Africa. In Kenya, Zambia and Zimbabwe, redistribution has involved repurchasing large European farms operated as integrated units, and allocating that land to African settlers on what might be called a repurchase/subdivision/resettlement model. All these reforms involved substantial compensation for the land acquired. As in Latin America, it has been difficult to carry out reform on a scale which fundamentally alters the structure of landholding. This has been due to a number of factors, including constitutional constraints, shortage of funds for land purchase, and shortage of funds, trained staff, etc, for resettlement.

Co-operative and state farms have often failed in production terms. The creation of such operations causes major demands for managerial ability and capital and, because both are scarce, the states pursuing this model, in fact, have only been able to affect even smaller areas of land than those pursuing the repurchase/subdivision/resettlement model. However, co-operatives may be seen as a transitional mechanism, an institution to which title can be transferred quickly for a large farm, without the delays involved in subdivision. This may provide a structure for political organization to defend the land acquired from counter-reforms, and for the creation of internal tenure rules, subdividing land rights among its members and providing them with adequate security of title.

It is also important to consider what factors may impact on the implementation of agrarian reform. These include: (a) the accessibility of the legal system – how expensive is it for participants to engage the legal process, and does the system favour the demands of certain groups or classes over others?; (b) the prevailing legal culture – what are the dominant values that guide the legal system and process? Is it possible to ensure that a future legal process be both compatible with, and responsive to, the goals and practices of social reform in the context of rural restructuring?; (c) what values or interests guide the actions of the actors involved in the legal process (lawyers, judges, bureaucrats, legislators) or even those active in the agrarian reform process (bankers, agronomists, agricultural economists, etc)?; and, (d) what groups are affected by or interested in the reform process (including landowners, beneficiaries, as well as their political and social allies who may include foreign officials – USAID, World Bank, IMF, etc)? To what extent are these groups interested and capable of influencing the legal process, or even the wider rural policy environment and legislative process?

All these factors will impact on the chances of implementing agrarian reform. However, there are other material considerations which may hinder implementation, including the sheer lack of resources. For example, the agrarian reform agency may be completely starved of funds as in Bolivia; or be frustrated by the problem of corruption or informal extra-legal practices in the bureaucracy. Even well-meaning public officials may become guilty of imposing a top-down process which has little relevance to the actual conditions and needs of the intended beneficiaries, as was the case with the Miskito Indians in Nicaragua.

Finally, the whole process of implementation will hinge on the participation by the intended beneficiaries in the formulation of specific policies and programmes. However, the experience of officially sanctioned or organized participation rarely goes beyond token representation in advisory bodies, often with dire consequences for the future of agrarian reform. The most hopeful path lies in the construction of a facilitative environment in which participants and beneficiaries may organize effectively and ensure that their voices are heard in the development and implementation of land reform policies.

9

The contemporary agricultural policy environment: undoing the legacy of the past

Johann Kirsten and Johan van Zyl

Introduction

South African agriculture has had a long history of ever increasing governmental intervention, reaching a zenith in approximately 1980 with a hoard of laws, ordinances, statutes and regulations affecting all aspects of agriculture, including prices of and/or access to and/or use of natural resources, finance, capital, labour, local markets, foreign markets, foreign exchange, etc. Political and economic power had become highly concentrated (*Kassier & Groenewald, 1992*).

The evolution of the agricultural sector during the twentieth century can be divided into three distinct periods of structural change. These have been identified by a number of authors (*Vink & Kassier, 1991; Van Zyl & Van Rooyen, 1991; Brand et al , 1992*). The first phase consisted of the initial steps aimed at the territorial segregation of white and black farmers. This is discussed in Chapter 3. The second phase of structural change started just before World War II and lasted until the early 1980s. In the homeland sector, the period was characterized by increased pressure on food production within a policy environment aimed exclusively at large-scale development projects under expatriate management. This period saw the commercialization of white farming through the adoption of modern mechanical and biological technology, resulting in consistent growth in output within a policy environment heavily favouring increased production by large-scale owner operated farms. This policy environment led to the substitution of capital for labour. In 1970, the agricultural sector was the major employer in South Africa, employing 30,6 per cent of the economically active population (*Fényes & Van Rooyen, 1985*). Up to 1970, the total number of farm employees increased, but the impact of credit, labour and tax policies favouring capital substitution and mechanization led to considerable shedding of labour from agriculture, as indicated by the annual decrease of 2,67 per cent in the number of farm employees in the period from 1970 to 1980 (Table 9.1).[1]

Table 9.1: Annual Growth Rates in the Number of Farm Employees and Real Gross Capital Formation in Commercial Agriculture, 1950 to 1980

Period	Growth (%)	
	Total number of farm employees	*Real gross capital formation*
1950–1960	2,08	3,21
1960–1970	4,38	5,34
1970–1980	–2,67	5,09

Source: Adapted from Van Zyl *et al* (1987).

The policy environment of racial discrimination and price distortions, however, could not be sustained, and the pressures on agriculture for reversal of these policies began to mount during the 1980s, introducing the third period of structural change. This chapter details this period, starting in the early 1980s and continuing into the 1990s. The period is characterized by a reversal of the policies of the previous two periods, consisting of removal of the racial barriers between black and white agriculture, and increased liberalization and democratization of the agricultural sector. However, this process is not yet complete.

An understanding of the policy changes during this last period, and of the resulting policy environment, is necessary in order to contextualize the land reform debate in South Africa properly. This chapter shows how the changes in the agricultural policy environment, in effect, have prepared the ground for a market-assisted land reform process. This is mainly as a result of the removal of most of the policy distortions favouring the white, mainly large-scale, farmers during this period of structural change.

1 Two structural trends in South African agriculture are evident. The period 1950–1960 witnessed a large expansion in cultivated farm area, probably because tractors replaced draught oxen. Larger areas could be managed and more labour was required. Demands on labour for harvesting were heavy, especially as a result of the increase in yields throughout the 1960s and 1970s as a result of improved biotechnology. The introduction of the combine harvester during the 1970s alleviated this problem but, in return, caused large shedding of labour from the agricultural sector.

In chronological order, this chapter addresses the following aspects: it provides some background information on agricultural production trends during the period under review; it discusses the policy environment, including policy goals and instruments, during the 1980s; it deals with the changes in agricultural policy since the mid-1980s and discusses each of the major policy shifts during this period; the final section of this chapter deliberates upon the major effects of these policy changes on agricultural support, productivity and agricultural debt. The impact on the debt burden especially, is of particular importance for the approach to land reform outlined in Chapter 16.

Background

During the 1980s, South Africa witnessed a number of political changes and some political and economic instability. The new constitution in 1983 gave birth to the tricameral parliamentary system and the concepts of 'own and general affairs'. The violent uprisings during 1985/86 led to a state of emergency and the intensification of economic sanctions in the mid-1980s. As an important industry in the national economy, agriculture was also affected by numerous changes during the 1980s. In this context, the purpose of this section is to review the major production trends within South African agriculture since the early 1980s, with emphasis on staples.

The 1980s began with bumper harvests for maize in 1980 (1980/81) with a total harvest of 14,6 million tonnes. During the same period, a bumper crop for groundnuts was realized. The good harvests of the early 1980s, however, were followed by a period of drought between 1982 and 1984 resulting in widespread crop failures across the country. Between 1980 and 1990, bumper harvests for sorghum (1986), sunflower seed (1989), dry beans (1989), soybeans (1990) and sugarcane (1984) occurred. The field crop sector was again hit by periods of drought in 1988 and 1991/92. The total harvest for maize, sunflower seed, groundnuts, dry beans and sugarcane all showed growth rates of between 0,84 per cent and 17,03 per cent for the period 1980–1990, with maize showing the lowest growth rate and soybeans the highest.

Table 9.2 provides an analysis of the production and consumption of the most important agricultural commodities produced in South Africa during the period 1985 to 1993. This is done in order to establish total production, surplus production for the export market and the degree of self-sufficiency.

In spite of the periodic droughts experienced during the 1980s, South

African agriculture still succeeded in producing surpluses. This is confirmed by the self-sufficiency index (SSI) which indicates that South Africa is self-sufficient in all the important staples. Crop production can therefore drop (in total) before South Africa becomes a net importer of these products on a regular basis. Some individual commodities in this group, however, are imported on a net basis (e.g. oilseeds).

Table 9.2 also indicates that in horticultural production, particularly fruit, South Africa is not only self-sufficient but, to a large degree,

Table 9.2: Average Production and Consumption of Selected Agricultural Commodities in South Africa, 1985–1993

Commodity	Imports	Exports	Production	Consumption		SSI***
				Total*	Human**	
	(1 000 ton)					
Wheat	368	370	2 242	2 400	1 865	100,4
Maize (white & yellow)	515	2 106	8 019	7 012	2 839	114,4
Potatoes	4	11	1 161	1 142	942	101,7
Vegetables	5	27	1 776	1 755	1 580	101,2
Sugar	41	892	1 956	1 107	1 174	176,7
Beef	72	23	618	666	660	92,8
Mutton, goat's meat & lamb	17	0	176	193	191	91,2
Pork	2	2	117	117	116	100,0
Chicken	7	2	656	661	654	99,2
Eggs	0	3	199	196	186	101,5
Deciduous & sub-tropical fruit	0	511	1 484	974	876	152,3
Fresh milk	0	0	2 435	2 435	1 118	100,0
Dairy products	35	58	2 344	2 321	2 321	101,0
Sunflower seed oil	54	1	121	175	159	69,1
Citrus fruits (fresh & processed)	0	435	802	369	366	217,3

* Available for use = Opening stock + Productions – Closing stock + Imports – Exports

** Net human consumption = Available for use – Other uses – Losses, and further adjusted for extraction rate

*** SSI (self-sufficiency index) = Total production ÷ Total consumption x 100

Source: Adapted from the Annual Balance Sheets of the Directorate of Agricultural Economic Trends of the Department of Agriculture.

dependent on the export market. The situation in respect of horticultural products, therefore, is even more favourable than that of crop production. In contrast to crop and horticultural products, red meat has a self-sufficiency index of lower than 100. This implies that South Africa did not produce enough red meat during the years 1985 to 1993 to meet domestic requirements. These shortages were supplemented by imports from, among other countries, Namibia, Botswana and some European countries.

Red meat, coffee, rice, vegetables, animal fats and vegetable oils are the most important food products imported. The total gross value of agricultural production in South Africa was almost R15 000 million in 1987, whereas that of food imports amounted to about R1 200 million. Food exports in the corresponding period amounted to about R2 400 million (*Van Zyl & Van Rooyen, 1991*).

The area grown to crops fluctuated throughout the decade (see Table 9.3). The decline since 1986/87 in the area under maize is particularly noticeable. This is largely the result of the change in the price policy of the maize industry and, to a lesser extent, the land conversion scheme introduced to take land out of maize production and unfavourable climatic conditions. The considerable drop in the acreage under wheat production during the 1992/93 season is also related to unfavourable weather conditions; in particular, as a result of the 1991/92 drought.

The changes in area cultivated under crops such as maize, sorghum, groundnuts and sugarcane have been rather mixed. All showed declining rates ranging between –0,61 per cent and –10,78 per cent per annum between 1980 and 1990. The area under cultivation of sunflower seed, dry beans and soybeans increased by 5,24 per cent, 2,3 per cent and 11,57 per cent per annum respectively. Although the area under cultivation for maize, groundnuts and sugar cane showed a declining trend during the period 1980–1990, production of these commodities grew steadily during the same period. This may have been the result of the developments during the period 1980–1990 when research improved the yielding capacity of these crops. A combination of research and production shifted away from the marginally productive areas (farmers in drought-prone and less fertile areas reduced cultivation of, say, maize); and farmers became more intensive in their agronomic practices.

Agricultural policy during the 1980s

Agricultural policy in South Africa during the 1980s, to a large extent, was determined by the 1983 constitutional dispensation and the

Table 9.3: Area Grown under Selected Field Crops, 1983–1993 (1 000 ha)

Crop	1983/84	1984/85	1985/86	1986/87	1987/88
Maize	4 028	3 913	4 054	4 029	3 657
Wheat	1 809	1 919	1 951	1 926	1 729
Sorghum	283	315	307	314	265
Dry beans	54	47	56	54	60
Sugarcane	412	407	411	401	388
Tobacco	32	34	31	26	25
Potatoes	67	66	57	57	65
Crop	1988/89	1989/90	1990/91	1991/92	1992/93
Maize	3 778	3 475	3 026	3 452	3 623
Wheat	1 985	1 830	1 550	1 433	743
Sorghum	182	138	118	135	170
Dry beans	64	69	78	54	45
Sugarcane	380	375	373	379	386
Tobacco	25	25	22	24	24
Potatoes	72	63	66	59	55

Source: Abstract of Agricultural Statistics (RSA, 1994).

continuation of a dualistic agricultural policy and industry. Policy with regard to 'white' commercial agriculture was outlined by the White Paper on Agricultural Policy tabled in 1984. According to the White Paper, the objective of agricultural policy is to direct the development of agriculture in such a way that factors of production are utilized optimally with respect to economic, political and social development and stability, while also contributing towards the promotion of an economically sound farming community. This was to be achieved through pursuing production, marketing and other goals:

Production goals

- Striving towards optimum use of natural agricultural resources.
- The preservation of agricultural land. The government's objective would be to ensure that the potentially productive land was maintained as agricultural land and would retain any other land identified as agricultural land for agricultural purposes.
- Pursuit of a high number of well trained and financially sound owner occupant farmers.
- The optimum use of labour.

Marketing goals

- The pursuit of orderly marketing, duly considering the principles of the free market system. Since the government was advocating a free market system, the control boards needed to be applied with great circumspection to ensure that state involvement did not distort production, marketing and price structures.
- Maintenance of specific quality and hygiene standards of South African agricultural products.

General goals

- Self-sufficiency in food.
- Optimum participation in international trade of agricultural products.
- Maximization of agriculture's contribution to regional development, promotion of development in Southern Africa and the rest of Africa.

Several acts were passed aimed at the affirmation of these goals, most notably the Soil Conservation Act to ensure optimum use of agricultural resources, which came into effect on 1 June 1984. This Act also introduced the following schemes:

- Soil Conservation Scheme;
- Flood Relief Scheme;
- Bush Combat Scheme;
- Weed Scheme.

In terms of the Agricultural Resources Act (Act 43 of 1983), some of the important regulations aimed at the conservation of natural resources by maintaining the productive capacity of the soil were:

- No cultivator may plough or cultivate virgin soil without written permission. Permission should be sought from the local extension office at least three months before the planned cultivation.
- Any soil user should not allow excessive soil losses through water erosion on cultivated soil; this should be prevented by suitable conservation works, a crop rotation system, strip cultivation or by leaving sufficient crop residues. Any soil user that allows excessive wind erosion could be forced to protect it, i.e. erect wind breakers.
- Irrigated soils should be protected from water logging and becoming salinated through the necessary drainage works.
- Swamps or sponge areas may not be cultivated or drained, without written permission.
- Drainage water from a water course may not be re-routed to another course. A soil user should not erect any obstruction that will disrupt the natural pattern of the water course.
- No one should damage his/her natural grazing by over-stocking or mismanagement. A soil user exceeding his/her official grazing capacity will forfeit all claims for financial aid in the form of subsidies for soil conservation works and drought aid.

Food self-sufficiency

One of the major aims of agricultural policy in South Africa during the 1980s was 'self-sufficiency in respect of food, fibre and beverages and the supply of raw materials to local industries at reasonable prices' (*RSA, 1984*). The White Paper on Agricultural Policy (*RSA, 1984: 8–9*) motivates this policy aim as follows: *'For any country, the provision of sufficient food for its people is a vital priority and for this reason it is regarded as one of the primary objectives of agricultural policy. Adequate provision in this basic need of man not only promotes, but is also an essential prerequisite for an acceptable economic, political and social order and for stability.'*

In order to achieve this aim, the South African agricultural bureaucracy was geared in a biased manner to support the white commercial farmer, especially in livestock and field crops. Farmers were protected from foreign competition, received various forms of subsidies, often received producer prices at a premium relative to world prices, and had access to the latest and most productive mechanical and biological technology through an impressive research and extension network. Through these measures, South Africa maintained its position as a surplus agricultural producer and achieved the aim of self-sufficiency in the majority of commodities as shown earlier in Table 9.2.

Although these favourable circumstances encouraged farmers to produce and thereby contributed positively to the aim of self-sufficiency, they also encouraged some environmentally and economically unsound and unsustainable farming practices. These measures, for example, made the cultivation of maize so profitable that large stretches of marginal land in South Africa were planted to maize (*Brand et al, 1992*).[2]

The policy of food self-sufficiency should be seen in the context of the government's political agenda at the time. This policy was followed by many countries in the world, especially in the post World War II period. The South African policy was based, to some extent, on the world experience during the 1960s and 1970s. Surplus agricultural production was also seen as a way to earn foreign exchange in a world plagued by 'Malthusian views' of chronic food shortages. This policy was also necessary in order for South Africa not to rely for its basic foodstuffs on an increasingly antagonistic and hostile world. With the threat of sanctions becoming a reality in the 1970s and 1980s, the policy of food self-sufficiency fitted well into the total strategy to build the apartheid-based 'fortress of South Africa'.

The policy of self-sufficiency benefited producers considerably, at the expense of consumers. The strong agricultural lobby at that time, through parliamentarian representation and indirect interest in agriculture, ensured that agriculture received beneficial treatment for some time. It can therefore be said that many producers benefited largely from the agricultural policy. The policy, however, was at the cost of the consumers and also a total welfare loss to the country as a whole (*Van Zyl, 1989*). Because it encouraged unsound farming practices (*Brand et al, 1992*), it can be argued that the policy of self-sufficiency contributed to the present detrimental position of white commercial agriculture.

Apart from the problems faced by producers, the policy was also to the detriment of the consumers and rural producers. Food prices kept on rising and, despite the export of surpluses, more than 2 million people went hungry in South Africa every day.

The fact that overall food production has kept up with the population increase (and will in all probability still do so in the next two decades), does not indicate anything about the nutrition status of the population. The Committee for the Development of a Food and Nutrition Strategy

2 See Chapter 10 for details.

for Southern Africa (*1990*) made an effort to identify the nutritional deficient. It estimated that, in 1989, there were around 11,8 million people in South Africa and 4,5 million in the former 'independent territories' with an income lower than the minimum subsistence level (MSL); thus there were a total number of 16,3 million in South Africa as a whole, of whom 15,3 million or 93,5 per cent were blacks. According to Simkins (*1991*), there is substantial poverty among rural 'coloureds' and all black people: 33 per cent of urban blacks, 54 per cent of former homeland urban blacks, 58 per cent of rural 'coloureds', 72 per cent of rural blacks in 'white' areas, and 84 per cent of former homeland rural blacks live under the poverty line. This implies that 47 per cent of black people live under the poverty line.

However, it was indicated that the nutritional needy should be selected according to anthropometric rather than income criteria. Table 9.4 shows the estimated number of nutritional needy per population and target group in South Africa, determined according to anthropometric criteria.

Estimates according to these norms show that there are 2,3 million people in South Africa who can be considered for nutritional assistance, as against the 16,3 million according to income criteria. About 2 million or 86,7 per cent of the 2,3 million people are blacks. Table 9.4 also shows that 829 000 (35,9 per cent) are children of six months to five years, 1,3 million (55,8 per cent) are children of six to twelve years and 192 000 (8,3 per cent) are pregnant and lactating women.

Agricultural subsidies

One of the major instruments to achieve the goals of the White Paper of 1984, apart from the Agricultural Marketing Act, was agricultural credit. Agricultural policy during the 1980s was characterized by the large sums of government subsidies to farmers, usually in the form of drought aid and other disaster payments (see pages 209–210).

Apart from the subsidies to individual farmers, the government also paid industry subsidies to the wheat, maize and dairy industries, amongst others. The industry subsidy to the wheat industry was paid to keep consumer prices of wheat and wheat products (flour, bread) as low as possible. The payment to the maize industry was in terms of the government's subsidization of the Maize Board's handling and storage costs, in order to keep selling prices of maize as low as possible. The extent of the subsidies to the wheat and maize industry is shown in Table 9.5. Apart from the subsidization of the Maize Board's handling

Table 9.4: Number of Nutritional Needy in South Africa According to Anthropometric Criteria, 1989

	White	'Coloureds'	Indians	Blacks	Total
Children six months to five years:					
Urban	15 874	52 214	15 323	236 419	319 830
Rural	1 617	33 108	2 366	472 517	509 608
Total	17 491	85 322	77 689	708 936	829 438
Children 6 to 12 years	20 318	123 467	24 530	1 123 095	1 291 410
Pregnant and lactating women	2 061	16 492	1 260	171 938	191 801
Total	39 870	225 281	43 479	2 004 019	2 313 649

Source: Committee for the Development of a Food and Nutrition Strategy for Southern Africa (1990).

Table 9.5: Government Subsidies to the Wheat and Maize Industries (1980–1993)

Year	**Maize** (R mil)	**Wheat** (R mil)
1980	44,7	116,4
1981	59,5	162,1
1982	82,9	181,9
1983	69,9	193,4
1984	132,4	276,6
1985	215,0	194,3
1986	250,0	180,5
1987	151,0	147,0
1988	359,0	147,4
1989	79,9	132,0
1990	76,0	105,9
1991	100,0	60,0
1992	100,0	–
1993	–	–

Source: Abstract of Agricultural Statistics (RSA, 1994).

costs, the government was also responsible, from year to year, for payment of the Maize Board's export losses because the producer price was fixed by the Minister of Agriculture. The R200 million subsidy paid to the maize industry in 1991 and 1992 was used to repay the Maize Board's outstanding loan.

Changes in agricultural policy

Since the mid-1980s, agricultural policy was characterized by increasing deregulation and market liberalization. Vink (*1993*) argues that the deregulation of the agricultural sector started outside agriculture in the late 1970s when the financial sector was extensively liberalized following the publication of the De Kock Commission report. The immediate effect on agriculture came from changes in the external value of the currency and in the interest cost of farm borrowing. The decline in the value of the Rand resulted in farm input prices, which have a relatively large import component, rising faster than farm output prices. Changes to the reserve requirements of the banking sector made it impossible for the Land Bank to continue subsidizing farmers' interest rates. The use of interest rate policy by the Reserve Bank led to a rise in interest rates to very high levels which resulted in interest becoming the single largest cost of production in agriculture. These changes led to the increasing exposure of farmers to market-related interest and exchange rates.

Other changes in the broader political economy which led to changes in agricultural policy were the lifting of controls over the movement of labour in South Africa in the mid-1980s, and the considerable microeconomic deregulation leading to increased activity in the informal sector (*Vink, 1993*).

Within this climate of macroeconomic change, a number of shifts in agricultural policy took place during the 1980s (*Brand et al, 1992; Vink, 1993*):

- Budgetary allocations supporting white farmers declined by some 50 per cent since 1987; (*see also Vink & Kassier, 1991 & LAPC, 1993*).
- The real producer prices of important commodities such as maize and wheat declined by more than 25 per cent in real terms since 1984 and 1986 respectively.
- An extensive deregulation of controlled marketing in terms of the Marketing Act.

- Liberalization of price controls in large parts of the farm sector, again mainly in terms of the Marketing Act. This includes the change in price setting in the grain industries from a cost-plus basis to market-based systems leading to substantial declines in real farm output prices. Further examples include the eventual abolition of price control of dairy products, and later of flour, meal and bread and the termination of consumer price subsidies on maize meal and bread.
- Changing tax treatment for agriculture, for example, has seen the writing off of capital purchases extended from one to three years, thereby reducing the implicit subsidy.
- There has been a shift away from settlement schemes and large-scale projects as the major instruments of agricultural development in the developing areas (the former homelands), in favour of an approach based on the provision of farmer support services such as infrastructure, extension services and research, and access to credit and markets.
- The scrapping of the Land Acts and related legislation that enforced the racially-based segregation of access to land. This was the most visible of the policy changes in agriculture following the important political events of February 1990.
- Certain elements of labour legislation were made applicable to farm labour and the farm sector has now become part of the mainstream of industrial relations in South Africa. The Basic Conditions of Employment Act was made applicable to farm workers in May 1993.
- There was a reduction in the institutional confusion by the amalgamation of all the 'own' affairs and 'general' affairs departments of agriculture and through the dismantling of the Department of Development Aid.
- The removal of quantitative protection and the introduction of tariffs for farm commodities, mainly as a result of the pressures arising from the Uruguay Round of the GATT and the signing of the new GATT deal in April 1994.

Budgetary allocations to agriculture

During the 1980s, expenditure on agriculture, forestry and fishing increased in nominal terms from R833 million in 1982/83 to R2 240 million by 1990/91. When expenditure was deflated by the consumer price index, real expenditure rose between 1982/83 and 1984/85, but fell back for the rest of the decade (*LAPC, 1993*). Figures on budget expenditure provided by the Central Statistical Service indicate that white farmers' share of the agriculture budget was declining in the latter part of the 1980s. Between 1988/89 and 1990/91, white agriculture's share of the budget dropped from 72 per cent to 61 per cent. Conversely, over the same period, the former homelands received a greater proportion. Auditors' reports and expenditure estimates of the government indicate a similar trend. These figures show a steady fall in white agriculture's share of total expenditure from 79 per cent of the budget in 1985/86 to 52 per cent in 1990/91.

The decline in the 'white' budget vote, to a large extent, is attributed to the decline in industry subsidies. Government subsidies to the food industry amounted to an estimated R236 million in 1980. These subsidies were suspended gradually during the 1980s, up to the end of the decade, and eventually led to an increase in food prices (see Table 9.6). The most important of these subsidies was the bread subsidy which contributed 69 per cent to the total government subsidy (*BTT, 1992*).

The trend in agricultural financing (subsidies and loans to farmers), the largest single component of the budget, is less straightforward. Between 1985/86 and 1990/91 there was a steady fall in real expenditure on agricultural financing. As a proportion of the total budget, however, there is no marked trend – it fluctuated between a low of 27 per cent (1987/88) and a high of 47 per cent (1988/89).

During the early 1980s several financing schemes (subsidies and loans) of the Agricultural Credit Board were suspended. These were:

Table 9.6: Phasing out of Government Subsidies in the Food Chain

Item	1981 Subsidy (R)	Date of suspension
Bread	R162,75 million	March 1991
Maize marketing margin	R 59,4 million	March 1991
Butter	R 3,7 million	1983/84
Crop insurance	R 3,69 million	1987/88
Fertilizer	R 11,0 million	1987/88
Total	R240,54 million	

Source: Abstract of Agricultural Statistics (RSA, 1994).

- soil conservation works;
- eradication of invading weeds;
- the establishment and or management of private plantations or the financing of debts incurred for that purpose;
- housing for non-white farm workers;
- buying of private farm land.

A change in the composition of governmental support to agriculture was also noticed and is discussed on pages 210–213.

Decline in real producer prices

Real producer prices in many of the major commodities such as maize, wheat, red meat, oilseeds have shown a marked decline. Farmers also experienced a cost-price squeeze as a result of the prices of farm requisites rising faster than producer prices in nominal terms, as indicated in Table 9.7.

Reform of the agricultural marketing system

Agricultural marketing policy in South Africa, to a large extent, is, determined by the Agricultural Marketing Act (Act 59 of 1968 as amended). The Act enables the Minister of Agriculture to proclaim a marketing scheme to control the marketing of a particular commodity. The Act provides for the promulgation of subordinate legislation called

Table 9.7: Annual Increase in Producer Prices vs Increase in Prices of Agricultural Inputs (1980–1991)

Product	Producer price (% increase p.a.)	Prices of inputs (% increase p.a.)
Summer grains	9,7	12,4
Winter grains	9,0	9,8
Dairy products	11,2	11,3
Poultry	11,9	11,9
Red meat	11,1	12,2
Vegetables	10,1	10,1
Fruit	13,5	13,3
Average	10,6	12,0

Source: Abstract of Agricultural Statistics (RSA, 1994).

schemes. A total of 23 marketing schemes were established under the Marketing Act. With these schemes it was possible to (a) transform the agricultural output and input marketing system; (b) determine commodity prices, the level and stability of food prices; and, (c) to create certain concentrations of monopoly power, especially in agricultural processing industries.

A number of political economic pressures have led to a more market-related approach in the marketing of agricultural commodities in South Africa since the early 1980s. There was a reduction in the use of price control on a number of commodities. There were also shifts to more market-based pricing systems, away from the old cost-plus pricing procedure. Other pressures came from within the system, with many farmers becoming increasingly unhappy with aspects of the controlled marketing of many agricultural products. There was also a realization of the poor performance of the agricultural sector in aggregate, as measured by the very slow rate of productivity growth (*Thirtle et al, 1993*). In addition to the sectoral reforms, the economic environment for agriculture has been profoundly affected by changes to macroeconomic policy, most notably the tightening of monetary policy through increases in interest rates and exchange rate depreciation.

The trend of market liberalization was further enhanced by the pressures emerging from the GATT negotiations for the abolishment of quantitative import controls and the introduction of tariffs on all agricultural commodities. The replacement of quantitative controls on external trade by tariffs is intended to reduce the distortions created by quantitative administrative controls, to create a more commercial environment in the planning of imports, to reduce the role of government in the allocation of licenses, to limit the use of quantitative controls, and to increase the extent of competition. A general policy of tariffication has been in operation since 1985, but this has only begun to be applied to agricultural commodities since 1992. During 1994, tariffs were established for poultry, tobacco, vegetable oil, oilcake and red meat, and an overall strategy has been developed for submission to GATT.

The appointment by the Minister of Agriculture of the Committee of Inquiry into the Marketing Act (CIMA) in June 1992, was probably the main event which triggered the process of market deregulation since the beginning of 1993. Since the release of the CIMA report in January 1993, a total of eight marketing schemes and marketing boards were abolished, while the one channel pool scheme of the Wool Board was abolished, although the Wool Board remained intact to perform product

Table 9.8: A Summary of Recent Reforms to Marketing Schemes under the Marketing Act and other Legislation

Scheme/Product	Year of establishment	Recommendation by CIMA (1993)	Recent reforms (including those before 1993)
Single Channel Fixed Price			
Maize	1938	Change necessary	Shift to pool-type pricing (1987); prohibition on erection of grain silos repealed; grain sorghum established as surplus removal scheme (1986); scrapping of control measures on buckwheat under consideration; scrapping of price control on maize meal; change to buyer of last resort (April, 1995); one channel marketing system abolished.
Winter cereals	1938	Change necessary	Abolition of restrictive registration of millers and confectioners; elimination of bread subsidy (1990); price control on flour, meal and bread, and fixing of millers' margins scrapped (1991); simplification of grading system for wheat (1991).
Single Channel Pool			
Oilseed	1952	Change necessary	Abolition of import control measures on oilcake & fishmeal; groundnuts under surplus removal scheme.
Leaf tobacco	1939	Statutory power unnecessary	Discontinuation of single channel marketing system under the Co-operatives Act. Export subsidies suspended.
Deciduous fruit	1939	Moratorium on statutory powers	No change.
Citrus fruit	1939	Voluntary organization	Domestic market control abolished (1990).
Bananas	1957	—	Abolished in 1993.
Lucerne seed	1952	Statutory powers unnecessary	Switch to surplus removal scheme rejected (1990); Board permitted private imports and exports (1992).

Scheme/Product	Year of establishment	Recommendation by CIMA (1993)	Recent reforms (including those before 1993)
Wool	1972	Statutory powers unnecessary	Single channel pool scheme discontinued; Wool Board voluntary organization providing market information etc.
Dried fruit	1938	Statutory powers unnecessary	No change.
Chicory	1939	No intervention	Abolished in 1993.
Rooibos tea	1954	Statutory powers unnecessary	Abolished in 1993.
Mohair	1965	Voluntary organization	Abolished on 31 January 1994.
Dairy	1956	—	Consumer price control on fresh milk abolished (1983); price control on butter and cheese abolished (1985); price stabilization activities ended following court ruling ending levy income (1992); Dairy Board and dairy marketing scheme abolished (31 Dec 1993); Milk Board (Fresh Milk – voluntary organization) established 1 Jan 1994.
Surplus Removal Schemes (or Price Support Schemes)			
Red meat	1945	Change necessary	Abolition of restrictions on movement from uncontrolled to controlled areas (1992); abolition of restrictive registration of producers, abattoir agents, butchers, dealers, processors and importers.
Eggs	1953	Statutory powers unnecessary	Abolition of production and pricing control in 1993; abolition of Egg Board in 1994.
Potatoes	1951	Statutory powers unnecessary	Abolished in 1993.
Dry beans	1955	Statutory powers unnecessary	Abolished in 1993.

Scheme/Product	Year of establishment	Recommendation by CIMA (1993)	Recent reforms (including those before 1993)
Sorghum	1957	Statutory powers unnecessary	No change.
Supervisory and Price Regulation			
Canning fruit	1963	Statutory powers unnecessary	No change.
Cotton	1974	—	No change.
Control in terms of Promotion			
Karakul pelts	1968	—	Karakul scheme and board abolished *circa* 1985.
Control in terms of Other Legislation			
Sugarcane	1936#		Reform of cane quota system (1990).
Wine	1918		Abolition of production quota system (1992).
Ostriches and ostrich products	1958 * 1988 **	Statutory single channel control to be repealed	Abolition of single channel marketing system (1993).
Lucerne hay	1958	—	Abolition of single channel marketing system (1993). The last government notice allowing a co-operative to implement single channel marketing was withdrawn in 1993 (Oranje Co-operative).

The Sugar Act of 1936 established control measures in the sugar industry. The Act makes provision for a Sugar Agreement, established in 1943, to oversee the industry.

* Only ostrich products.

** Ostriches and ostrich products.

development, advertising and other services. The impact of these events on the reform and deregulation of South Africa's agricultural marketing system is evident from Table 9.8.

Liberalization of price controls in the food sector

The 1980s saw extensive deregulation of the agricultural marketing sector as discussed above. One of the important aspects of this deregulation was the liberalization of price control on a wide range of products. Examples of this price liberalization are contained in Table 9.9.

In their 1992 discussion document, the Board on Tariffs and Trade argue that abolishing price control was directly responsible for sharp price increases in all the products listed in Table 9.9.

Table 9.9: Abolition of Price Control in the Food Industry

Product	Level	Year abolished
Milk	Retail	1983
	Wholesale	1983
	Producer	1987
Cheese	Retail	1985
	Wholesale	1986
Butter	Retail	1985
	Wholesale	1988
Wheat flour	Retail and wholesale	1991
Bread	Retail and wholesale	1991

Source: BTT (1992).

Change in tax policy

In the past, the agricultural sector benefited from favourable fiscal policy. Lamont (*1990*) estimates that income tax concessions to farmers amounted to 70 per cent of their theoretical tax bill in 1981–84. Before the changes in fiscal policy in the 1980s, farmers could depreciate an entire asset for tax purposes within the first year of purchase.

Resources were not optimally deployed because capital formation occurred at the expense of a relatively cheap labour resource. Such tax concessions tend to result in over-investment in good years but lead to cash-flow problems in bad years (*LAPC, 1993*).

During the second half of the 1980s, tax concessions were reduced. Assets had to be depreciated over three years at rates of 50 per cent, 30 per cent and 20 per cent per annum respectively. Although this amounts to a significant reduction in tax concessions, depreciation provisions for agriculture are certainly more generous than for other sectors.

Agricultural and rural development policy

As a result of the dualistic nature of South African agriculture, different policies applied to white commercial agriculture and to black small-scale farmers in the former homelands. Policies with regard to homeland agriculture largely focused on a variety of agricultural development efforts. Three clearly defined approaches to agricultural development can be identified, i.e. betterment planning since 1936 to the late 1970s; centrally managed project farming and farmer settlement projects during the 1970s and 1980s; and the more broad-based farmer support programmes since the late 1980s (*cf. Ellis-Jones, 1987; Christodoulou & Vink, 1990; Van Rooyen et al, 1987; Van Rooyen, 1993; Bromberger & Antonie, 1993*). This dualism was further highlighted by the separation between agricultural and rural issues. In fact, rural issues were scattered over a wide range of departments, i.e. Water Affairs, Transport, Energy, Forestry, Agriculture, and Regional and Land Affairs.

Increasing emphasis was placed on large-scale centrally managed estate project farming during the 1970s (*Christodoulou & Vink, 1990*), particularly in the case of industrial crops 'where large units were desirable' (*Van Wyk, 1970:66*). The project farming approach obtained a further boost with the establishment in 1973 of an agricultural division in the Bantu Investment Corporation. According to Bromberger and Antonie (*1993*), Christodoulou and Vink (*1990*) and Christodoulou *et al* (*1993*), it appears that substantial financial losses were the norm with these schemes and the distribution of benefits was very limited in relation to total need and to aggregate resources available for development. Although higher levels of resource use and production and the creation of wage employment were promoted through modern farming enterprises managed by parastatal companies and consultants, little was done to promote a class of self-employed farmers or improved farming methods for smallholders outside these schemes. Schemes were later adjusted to settle selected persons as 'project farmers' operating under paternalistic control (*Van Rooyen, 1993*). Occupiers of plots were strictly selected; they had to farm according to direction and under supervision, and they were dismissed from their plots if they were unsuccessful (*Van Wyk,*

1970:66). This approach, commonly known as the farmer settlement approach, focused on large schemes but concentrated on settling selected labourers as project farmers operating under strict control. Participation by so-called farmers was accommodated by using farmer committees to assist the project manager. These farmers, however, were nothing more than paid wage labourers with virtually no control over their production activities. As such, a drive towards self-reliant farm businesspeople still did not materialize (*Christodoulou et al, 1993*). This approach was dominant in the late 1970s and early 1980s (*Christodoulou & Vink, 1990*).

With time, disillusionment developed about these projects. They were expensive, often loss-incurring, and rarely involved spill-overs or linkages with the surrounding communities. They were often viewed as 'islands of prosperity amidst an ocean of poverty' (*Bromberger & Antonie, 1993*). These models of development were thus viewed with increasing scepticism in terms of their undesirable impact on investment and operational costs, entrepreneurial establishment, fiscal affordability, upliftment of adjacent communities, project sustainability, and overall rural development (*Christodoulou et al, 1993; Van Rooyen et al, 1993*).

Acknowledging the limitations of agricultural development projects, an alternative approach to agricultural development was designed. The Farmer Support Programme (FSP) was introduced in 1986, trying to achieve a shift away from investment in projects to a programme which could provide access to support services for a large number of smallholders and rural households in a broad-based manner. An important motivation for this programme was the promotion of equitable access to support services, resources and opportunities.

Some effects of the changing farm policy

General

The discussion above shows that there have been many reversals of past policies since the early 1980s on macroeconomic as well as sectoral levels. These policy reversals include the scrapping of major racially-defined land laws; the reduction of subsidies on agricultural interest rates and public financial assistance; the decline of administered producer prices and the deregulation of the marketing environment. These changes in farm policy had significant effects on the agricultural sector, while it is important to note that different farming regions have experienced different circumstances. Aggregated data show that the sector is becoming more flexible in some parts of the country. This is highlighted by the improved aggregate debt service ratio along with financial difficulties for some groups of farmers; the increasing land use intensity in high potential regions and 'over-cropping' in more marginal regions; the aggregate decline in farm size; shifts in the cropping pattern; and the relative absence of yield effects.

The effects of these changes in farm policy can be traced through variables such as:

- the financial position of farmers;
- land use patterns;
- farm size and ecological considerations.

Much has been made of the increase in total farm debt in the period since 1980, and also about the fact that debt repayment has become the biggest input cost item for commercial farmers. The ability of farmers to service their debt, however, has improved in the period since the mid-1980s. This data refers to total farm debt; the preceding discussion makes it obvious that the size of debt and the ability to service debt will differ between regions. Examples include the successful use of credit to gear production by farmers in high-potential regions, especially where crops are produced for export; the more extensive production systems being followed by maize farmers in the Highveld, that is, by using fewer production inputs; and the higher rates of sequestration of farming enterprises in the lower-potential regions. Many of these changes are reflected in changing land use patterns.

The changing land use patterns in commercial farming have manifested themselves differently in the different regions of the country. They are related to the policy changes discussed earlier through changes in relative product prices and factor costs, the cash flow position of farmers, shifts in tax incidence, and so forth. A theoretical analysis of the effects of the changes in farm policy over the past decade would lead to the conclusion that such a decline in average farm size was indeed possible. However, this would be the aggregate effect of a number of more specific micro-level and regional changes. Those policy effects, which would lead to downward pressure on farm size, include the following (*Brand et al, 1992*):

- A higher incidence of part-time farming and of land rentals resulting from the need to find other sources of capital and to use less capital.

- More intensive farming in high-potential areas as farmers exploit growing local and foreign markets.
- Attempts to manage risk through mixed farming systems, that is, by more intensive management in the high-potential areas.
- The development of urban agriculture which, by definition, is suited to small-scale farming.
- Distress-selling of parcels of land in areas which have become vulnerable to the deregulation of controlled markets.
- The introduction of elements of labour legislation of farming which could result in innovations in the means of access to land, including farmer settlement, sharecropping, and sectional title arrangements.

On the other hand, there are a number of factors which could put an upward pressure on average farm size, including:

- The declining use of production inputs such as fertilizer and agrochemicals, leading to more extensive farming.
- The switching from crop production to livestock ranching in the more marginal cropping areas, including planted pasture.
- The switching to lower yielding but more drought resistant crop cultivars.
- The expansion of the corporate farming sector.

Agriculture is a prime user of natural resources. Although it supplies food and fibre, foreign exchange, and employment opportunities to the South African economy, a high price has been paid in terms of the degradation of natural ecosystems. The imbalances created by biotic simplification (monoculture), lack of managerial expertise and agricultural policies, are evident in many parts of the country. Recent studies by the Department of Agriculture have shown that at least 9 million ha of arable land and 21 million ha of grazing land in the 'white' farming areas are at present subject to some or other form of wind or water erosion. Of this, some 11 million ha or 13 per cent of the total agricultural land in these farming areas, have been damaged by mild or severe erosion. The erosion of topsoil is unacceptably high and much of the irrigation land has become degraded through salination, while natural grazing land is seriously overstocked.

Changes in domestic support to South African agriculture

Helm and Van Zyl (*1994*) calculated the total support received by the South African agricultural sector during the period 1988/89 to 1993/94 in terms of the Producer Subsidy Equivalent (PSE). The composition and changes thereof were also analysed. In order to determine the total domestic support to farmers in South Africa, the PSE was calculated on a sector wide basis and not on a product-specific basis. Certain policy measures, however, had to be calculated per product and then only could they be brought into the sector wide PSE. When formulating the PSE, there are two components that must be taken into account:

- The first component is the income transfers to producers as a result of agricultural policy. These transfers are calculated by means of a comparison between an internal market price and an external world price (*Van Heerden, 1992*). It is this component, the market price support, which has to be calculated on a product-specific basis.
- The second component is to bring into calculation the transfers from government sources. These transfers from either direct or indirect budgetary payments are calculated from government financial accounts. The calculations are done on a sector wide basis with the advantage that no proportionate allocation is necessary. According to Van Heerden (*1992*), the accuracy of these estimates depends on a reasonably accurate knowledge of the budgetary cost of these measures, which means not only information on budgeted funds, but also on the revenue foregone by governments (tax concessions) or costs not fully recovered (interest subsidies).

Table 9.10 shows the evolution of assistance to agricultural producers (as measured by the PSE) associated with South African support measures for the period 1988/89 to 1993/94, while Figure 9.1 depicts the evolution graphically.

The total PSE was at its lowest during 1988/89 with market price support accounting for only 11 per cent of total assistance, the remainder being financed by taxpayers. Of all the agricultural products, producer prices of only sugar, rye, chicory, eggs, beef, sheep and dairy products were higher than the representative world prices. The increase in the total PSE in 1989/90 was due to the higher production volume which led to a slight decrease in the percentage PSE from 11,70 per cent to 11,56 per cent in that year. Market price support accounted for about 31 per cent of total assistance in 1989/90. The reduction in the indirect income support component was mainly due to the fact that the production input subsidy paid to farmers was substantially reduced and then entirely eliminated the following year. In 1990/91, the total PSE

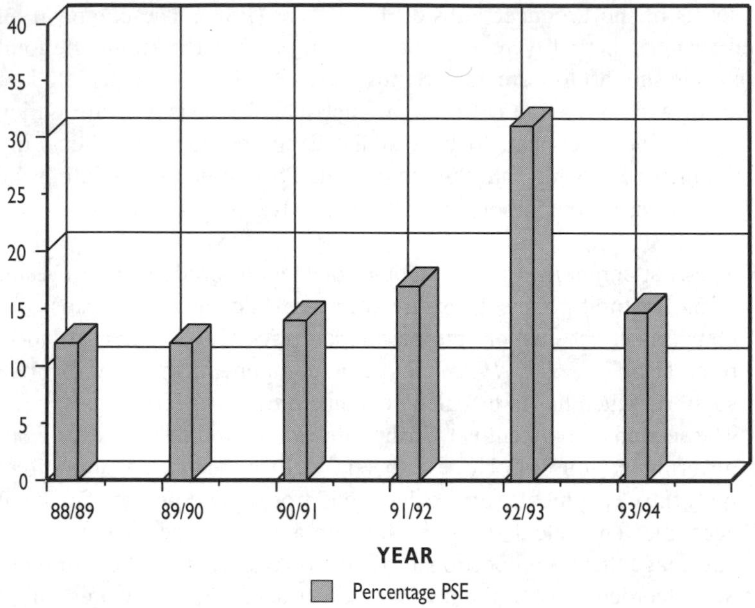

Figure 9.1: Total Domestic Support (1988/89–1993/94)

again increased as a result of substantially higher producer prices being paid for certain products, together with a decline in world prices. Market price support accounted for about 46 per cent of total assistance in 1990/91. With regard to direct and indirect income support, the amount of support involved remained the same, to a large extent, in comparison with the previous year. The percentage PSE increased to 13,69 per cent. Both the increases in producer prices and/or the decrease in world prices of agricultural products were once again the main reason for the higher market price support, together with the subsequent increase in the total PSE in 1991/92. Market price support accounted for about 60 per cent of total assistance and was 37 per cent higher than the previous year.

Van Heerden and Van Zyl (*1992*) calculated the aggregate measure of support for maize, which also indicates the changes in domestic support to maize producers since 1987 when much of the market liberalization commenced. These calculations are summarized in Table 9.11, indicating the low levels of support (direct or indirect) that maize farmers have received since the 1988/89 marketing year.

Table 9.10: The Calculation of Total Domestic Support (PSE)

Description	Unit	1988/89	1989/90	1990/91	1991/92	1992/93	1993/94
(a) Value of production: Products with MPS	Rand	11 321 897	13 454 158	13 784 297	15 736 341	12 872 328	16 467 791
b) Value of production: Products without MPS	Rand	5 231 386	5 965 538	6 910 111	7 497 910	11 193 516	11 860 609
(c) Direct payments	Rand	113 549	115 621	119 871	91 674	89 075	79 803
(d) ADJUSTED VALUE OF PRODUCTION (a+b+c)	Rand	16 668 832	19 535 317	20 814 279	23 325 925	24 154 919	28 408 203
Policy transfers to agriculture:							
(e) Market price support	Rand	216 819	701 428	1 308 831	2 321 722	2 448 684	2 119 873
(f) Direct income support	Rand	367 977	335 768	332 025	250 019	2 616 106	386 477
(g) Indirect income support	Rand	942 692	774 528	703 863	819 426	1 278 611	1 048 097
(h) General services	Rand	422 001	446 259	503 761	512 940	1 155 325	564 305
(i) TOTAL PSA (e+f+g+h)	Rand	1 949 489	2 257 983	2 848 480	3 904 107	7 498 726	4 118 752
PERCENTAGE PSE (i/d)	%	11,70	11,56	13,69	16,74	31,04	14,50

Table 9.11: The Aggregate Measure of Support (AMS) for Maize in South Africa, 1985 to 1991

Description	1985/86	1986/87	1987/88	1988/89	1989/90	1990/91
Total AMS (R)	459 749 902	950 276 162	1 010 668 482	–99 773 264	–2 714 886	–45 710 926
Unit AMS (R/t)	R55,42	R114,20	R137,44	–R14,22	–R0,23	–R5,25
Percentage AMS	17,61%	37,86%	40,99%	–4,41%	–0,07%	–1,45

Source: Van Heerden, (1992).

Effects on productivity in South African agriculture

The change in agricultural policy as discussed above also had some effect on the total factor productivity (the ratio of aggregate output to an aggregate of all inputs combined) of South African agriculture.

The results of TFP calculations by Thirtle *et al* (*1993*) show that between 1947 and 1991, the output index has grown by nearly 350 per cent, at a rate of 3 per cent per annum. The index of inputs has more than doubled, growing at 1,8 per cent per annum; this aggregate hides the fact that inputs grew at over 2,5 per cent per annum until 1979 and, since then, have been falling at 0,9 per cent per annum. This fall in inputs explains the recent growth in the TFP index. Over the full period, TFP grew rather slowly, at 1,3 per cent per annum, but there has been fairly rapid growth in total factor productivity of 2,88 per cent per annum since 1981.

These TFP results are useful in explaining the effects of agricultural policy. The growth rate is greater than would be expected on the basis of Liebenberg and Groenewald's (*1990*) preliminary study of productivity in grain production. The increasing rate of growth over the period is in accordance with Van Zyl and Groenewald's (*1988*) perception that farmers' profits came under increasing pressure as inflation gathered pace. The rapid growth of productivity since 1983 is in agreement with the regional econometric study by Van Schalkwyk and Groenewald (*1992*), which found evidence of substantial growth in some regions, since 1981. The growth in productivity can be explained by the increasing competitive pressures within the industry as a result of the policy reversals and removal of price distortions caused by credit, tax and macro policies.

In a further study on TFP growth, and growth in net farm income, Van Zyl *et al* (*1993*) calculated that total factor productivity grew 4,63 per cent annually since 1983, and countered the decline of –3,11 per cent in terms of trade during the same period; a growth of 6,24 per cent in real NFI resulted (Table 9.12). The growth in productivity since 1983 can be ascribed to a gain in capacity utilization which, amongst other factors, can be attributed to a longer replacement period of tractors.

Table 9.12: Average annual growth rates in real net farm income for the period, 1973–91 (%)

PERIOD	NFI	TFP	TERMS OF TRADE
1973–91	–1,06	1,48	–2,63
1973–83	–8,14	0,27	–3,27
1983–91	6,24	4,63	–3,11

Note: NFI = Net Farm Income
TFP = Total Factor Productivity
Terms of trade = Output prices / input prices

The analysis by Van Zyl *et al* (*1993*) shows that the agricultural sector experienced a steady decline in its financial performance since 1973, with the largest downswing in 1983 when a recovery phase started. The decline is attributable to the cost-price squeeze which obviously exerts considerable pressure on income. The negative trend, however, was countered by an annual growth in productivity of 4,63 per cent since 1983. However, agricultural policy, especially issues such as import substitution, import protection, price policies of marketing boards and general macroeconomics resulting in high inflation, should be addressed in order to rectify the unfavourable terms of trade of the agriculture sector if a sustained growth in profit is wanted.

The financial position of farmers

Declining farm profitability as a result of the reversal of distortionary policies (and adverse weather conditions) caused severe cash-flow problems in agriculture (*Van Zyl & Van Rooyen, 1991*). Liquidity problems have affected the financial standing of commercial agriculture in three ways: (a) debt loads increased; (b) loan arrears mounted; and, (c) sequestrations increased. The total debt of farmers has increased substantially since the mid-1970s as is shown in Table 9.13.

The decline in farm profitability also seems to have caused a substitution of short-term for long-term debt from 1970 until the mid-1980s. The ratio of short-term to total debt increased from 28,2 per cent in 1970 to 54,6 per cent in 1985, and peaked in 1991 at 57 per cent (*World Bank, 1994*). From Table 9.13 it can also be calculated that the share of total farm debt at commercial banks and co-operatives increased from 20 per cent and 8 per cent respectively in 1970 to 30 per cent and 25 per cent respectively in 1991, again indicating the switch to short-term debt.

The high growth rates of farm debt per annum for the period 1980–1985 (see Table 9.14), to a large extent, can be attributed to the drought and general economic conditions. Interest rates, drought, volume of field crop production, real GNP and the ratio of input to output prices have been shown to have had a relatively large impact on the real debt burden in the period from 1970 to 1985 (*Van Zyl et al, 1987b*).

While the total debt was expanding in the period 1980–1990, farmers' gross income was declining at a rate of –0,65 per cent in real terms; but net incomes showed a growth rate of 1,55 per cent in the period 1980–1990 and a rate of 2,38 per cent in the period 1985–1990. Yet, in both these periods, gross incomes were declining. The growth in net incomes can be attributed to the declining growth rate of farm expenditure and improved productivity, as discussed earlier.

In the mid-1980s, the South African Agricultural Union carried out a national survey on the financial situation of all farmers. The survey revealed that 49 per cent of farmers were financially sound at the end of 1983, but the percentage in this category was expected to fall below 39 per cent at the end of 1984. While the financial position of farmers older than 50 years is generally sound, 38 per cent of farmers aged between 25 and 35 years were in a critical financial position. This proportion increased to well over 50 per cent by the end of 1984.

Many of these farmers have left the industry, but the majority have been kept on their farms during the 1980s and even the early 1990s through government intervention in the form of 'cheap' credit and debt relief to insolvent or near-insolvent farmers, provided through the Financial Assistance Schemes of the Department of Agriculture, managed by the Agricultural Credit Board. In 1993, around 17 000 farmers still benefited from such assistance. If it is argued that these farmers are

Table 9.13: Total Nominal Farming Debt, 1970–1992 (R million)

Year	Land Bank	Commercial banks	Agricultural co-operatives	Department of Agriculture	Private persons	Other financial institutions	Other debt	Total debt
1970	295,8	281,8	113,4	136,2	242,9	286,9	45,0	1402,0
1971	345,8	272,4	127,9	154,1	231,7	279,2	49,6	1460,7
1972	360,0	275,2	145,8	159,5	234,5	291,0	50,2	1516,2
1973	368,3	373,4	158,8	154,6	290,0	317,5	62,1	1724,7
1974	389,8	384,0	197,8	148,6	292,0	310,8	62,5	1785,5
1975	429,1	454,5	247,4	151,2	327,3	324,1	70,0	2003,6
1976	478,8	485,4	344,3	149,5	369,4	393,2	77,0	2297,6
1977	507,8	560,3	462,2	150,1	414,7	426,2	85,7	2607,0
1978	537,0	620,2	556,3	156,0	452,2	453,4	95,0	2870,1
1979	592,4	690,9	654,3	164,3	485,6	523,2	107,6	3218,3
1980	675,6	801,5	866,9	180,0	579,9	612,3	122,4	3838,6
1981	855,9	1054,6	1129,7	201,8	601,3	833,2	162,2	4838,7
1982	988,5	1599,5	1367,6	247,3	634,0	774,6	174,0	5785,7
1983	1330,5	2253,8	1780,2	308,7	670,0	880,7	185,0	7408,5
1984	1923,0	2968,8	2233,7	443,3	720,0	999,5	207,0	9495,3
1985	2338,4	3315,3	2754,0	549,2	792,0	1128,2	240,5	11117,6
1986	2648,6	3436,6	3080,7	684,3	890,2	1420,3	251,4	12412,1
1987	2807,5	3355,2	3224,1	789,4	940,4	1500,0	263,7	12880,3
1988	2923,5	3477,7	3411,7	920,7	924,8	1295,0	407,5	13360,9
1989	3149,1	4650,0	3586,9	971,7	986,6	1160,0	405,5	14909,8
1990	3441,1	4949,6	3780,3	1013,1	1209,4	905,0	675,6	15974,1
1991	3512,5	5116,3	4300,8	1167,8	1301,9	800,0	727,6	16926,9
1992	3711,0	5181,8	3900,6	1348,6	1395,9	878,6	780,2	17196,7

Source: Abstract of Agricultural Statistics (RSA, 1994).

Table 9.14: Annual Growth Rates of Debt from Selected Sources

Category	1980–1990	1985–1990
Land Bank	2,98%	12,7%
Agricultural co-operatives	1,0%	10,29%
Department of Agriculture	5,49%	10,49%
Private persons	–6,48%	–6,48%

Source: Abstract of Agricultural Statistics (RSA, 1994).

also the most inefficient, it can be said that the policy of blanket debt relief and subsidies only adds to the financial unsustainability of the sector and the entrenchment of inefficiencies.

The preferential financial assistance for which farmers can qualify includes the following schemes (*Rimmer, 1993*):

- **Debt consolidation.** Farmers who did not qualify for commercial credit could consolidate their debt at a nominal interest rate of 8 per cent (1991) with a loan period of 15–20 years. Annual disbursements for consolidating debt increased rapidly and, in 1991, the accumulated total was R522,5 million.
- **Means of crop production.** Crop production credit is provided to farmers who, as a result of adverse farming conditions, cannot acquire short-term credit elsewhere. The nominal interest rate for these loans in 1991 was 8 per cent, compared to a commercial rate of 21 per cent. The accumulated total of loans provided to farmers under this scheme amounted to R833,6 million.
- **Interest subsidy schemes.** This is a special drought assistance scheme for producers of winter grain and summer crops. The terms in 1991 were 5,5 per cent for 5 to 10 years. The accumulated total from 1981 to 1991 stood at R1 077,8 million.

Apart from these schemes, financial assistance to farmers during the 1980s also took the form of the following programmes which provided loans and subsidies (*World Bank, 1994*):

- Disaster and drought relief: loans for R8,1 million and subsidies of R540,6 million;
- Flood assistance: loans for R10,9 million and subsidies of R271,3 million;
- Conversion of marginal lands: subsidies of R133,7 million;
- Purchase of agricultural lands: loans for R104,6 million;
- Emergency drought schemes: subsidies of R36,6 million;
- Interest subsidies on Land Bank loans in designated areas: R46,8 million;
- Allocations of state land: loans for R41,0 million;
- Safeguarding of residents: subsidies of R36,6 million;
- Water works: loans for R25,6 million;
- Soil conservation works: loans for R18,1 million;
- Water quota subsidies: R15,9 million;

- Purchase of livestock: loans for R10,4 million;
- Production cost subsidies: R6,4 million;
- Improvements: loans for R3,8 million;
- Purchase of implements and vehicles: loans for R1,5 million;
- Sinking of boreholes: loans for R0,4 million.

To summarize, the following should be noted. During the 1980s, the state granted financial assistance in one form or another to some 27 000 farmers. Direct financial assistance to these farmers over the decade amounted to R1 728,1 million, while subsidies totalled R2 353,6 million.

The declining profitability in many parts of the agricultural sector would have produced substantial declines in farm incomes had it not been for the state aid and financial assistance outlined above. But, in spite of the generous financial assistance offered to insolvent farmers, loan arrears on interest owed increased as the farm financial crisis worsened. It also did not succeed in countering the structural decline of farm profitability since the early 1980s, and the debt burden worsened increasingly. The increased importance of short-term debt as described earlier was a major sign of the worsening debt crisis in farming. An important component of the short-term credit (mainly at co-operatives) fell under a carry-over scheme for farm debt which was guaranteed by the government. This programme, initially introduced after the 1982/83 drought became a permanent feature, escalated as a result of the 1991/92 drought when the guarantee required by the government rose from an initial R800 million in 1983 to R2,4 billion in 1992.

The drought relief package announced by the Government in 1992 consisted of a R2,4 billion debt relief (the guarantee referred to above) plus an additional R1 billion drought relief amounting to a total of R3,4 billion. This constituted a substantial recapitalization of the least efficient sub-sectors of the agricultural sector, namely the livestock and grain producers in the summer and winter rainfall areas.

It is clearly evident from this discussion that the approach of blanket debt relief has been very costly, and has entrenched inefficiency and inequality in the commercial farming sector.

Conclusion

The evolution of the South African agricultural sector during the twentieth century witnessed three major trends. First, a period of territorial and racial segregation at the beginning of the century. Second, a period of modernization of white agriculture through the adoption of modern mechanical and biological technology, resulting in consistent growth in output within a policy environment heavily favouring increased production by large-scale owner operated farms. This policy environment led to the substitution of capital for labour, and resulted in a decline in the number of farm employees up to the early 1980s. The third trend was the reversal of the distortionary policies of the 1960s and 1970s brought about by external as well as internal pressures on the agricultural sector. This chapter paid particular attention to the changes in agricultural policy during the 1980s and analysed the changes in detail; it also showed the effects of the changes on farming debt and total factor productivity growth.

The undoing of past legislation and policies started first through the process of removing distortionary policies such as subsidies and tax concessions (all mainly as a result of budgetary cuts). Secondly, the process of market liberalization and deregulation undid most of the protective and distortionary effects of the Agricultural Marketing Act. Thirdly, the removal of a range of racially based laws since 1986 (the scrapping of the Pass Laws) and, in particular, since 1990 (the abolition of the Land Acts and Group Areas Act); the election of 1994 and the post-April 1994 dispensation, reversed virtually all the racially-based policies introduced earlier this century.

The policy changes in the commercial agricultural sector exerted considerable financial pressure on the sector, leading to increased bankruptcies during the mid-1980s and improved productivity to counter declining profitability in certain sub-sectors. The latter led to an improved financial situation for certain farmers, but still the majority depended on financial aid to remain in the industry. The costliness and unsustainability of continued financial assistance was highlighted in this chapter. The suspension of this policy, therefore, could be an important factor in contributing to a successful land reform programme, ensuring that only the efficient farmers remain in a newly restructured agricultural sector.

Despite the legislative and policy changes, it is still true to say that not much has changed in terms of improved equality and improved living conditions for the rural poor. The unequal ownership of land and the major effect of the past policies persists and, unless this is changed, it is unlikely that the conditions of the poor, of whom many reside in rural areas, will improve. The past policies have been removed, but it will now take a committed effort, including a well-planned and well-structured land reform programme, to rid the agricultural sector of the effects of the legacy of the past.

References

Board on Tariffs and Trade (BTT). 1992. Preliminary report on an investigation into the price mechanism in the food chain with recommendations for its improvement. *Unpublished discussion document.* Pretoria: BTT.

Brand, S., Christodoulou N., Van Rooyen, C.J. and Vink, N. 1992. Agriculture and redistribution. In R. Schrire (Eds), *Wealth or Poverty? Critical Choices for South Africa.* Cape Town: Oxford University Press.

Bromberger, N. and Antonie, F. 1993. Black small farmers in the homelands. In Lipton, M. and Simkins, C. (Eds), *State and Market in Post-apartheid South Africa.* Johannesburg: Witwatersrand University Press.

Christodoulou, N.T. and Vink, N. 1990. The potential for black smallholder farmers' participation in the South African agriculture economy. *Paper presented at a conference on 'Land reform and Agricultural development'.* United Kingdom: Newick Park Initiative, October.

Christodoulou, N.T., Sibisi, M.L. and Van Rooyen, C.J. 1993. Shaping the impact of the small farmer support programmes (FSPs) in South Africa. *Unpublished Mimeo.* Halfway House: DBSA.

Committee for the Development of a Food and Nutrition Strategy for Southern Africa. 1990. *Report of the Committee for the Development of a Food and Nutrition Strategy for Southern Africa.* Pretoria: Department of Agriculture.

Ellis-Jones, J. 1987. Guidelines for the role of the public sector in promoting agricultural development, with particular reference to Transkei. *Development Southern Africa,* 4(3) : 538–542.

Fényes, T.I. and Van Rooyen, J. 1985. South African agriculture and migrant labour. In P. Martin (Ed), *Migrant Labour in Agriculture: An International Comparison.* California: Giannini Foundation, University of California, pp. 177–191.

Helm, W. and Van Zyl, J. 1994. Domestic agricultural support in South Africa from 1988/89 to 1993/94 : A calculation. *Paper presented at AEASA conference*. Pretoria: 19–20 September.

Kassier, W.E. and Groenewald, J.A. 1992. Agriculture: An overview. In R. Schrire (Ed), *Wealth or Poverty? Critical Choices for South Africa.* Cape Town: Oxford University Press.

Lamont, M.P. 1990. International tax reform : implications for concessionary tax provisions in agriculture with special reference to South Africa. *Unpublished PhD Dissertation*. Stellenbosch: University of Stellenbosch.

LAPC 1993. A review of the South African agricultural budget. *Unpublished LAPC research document.* Johannesburg: August.

Liebenberg, G.F. and Groenewald, J.A. 1990. Die RSA-Landbouruilvoet. *Agrekon,* 29(3): 178–184.

Rimmer, M. 1993. Debt relief and the South African drought relief programme: an overview. *Unpublished working paper*. Johannesburg: Land and Agricultural Policy Centre.

RSA 1984. *White paper on Agricultural Policy.* Pretoria: Government Printer.

RSA 1994. *Abstract of Agricultural Statistics.* Pretoria: Directorate of Agricultural Information.

Thirtle, C., Sartorius von Bach, H.J. and Van Zyl, J. 1993. Total factor productivity in South African agriculture, 1947–1991. *Development Southern Africa,* 10 : 301–318.

Van Heerden, W.R. 1992. An economic analysis of an aggregate measure of support for maize in South Africa. *Unpublished MSc thesis*. Pretoria: University of Pretoria.

Van Heerden, W.R. and Van Zyl, J. 1992. Measuring agricultural support: An economic analysis of the AMS for maize. *Unpublished Mimeo.* Pretoria: University of Pretoria.

Van Rooyen, C.J. 1993. An overview of DBSA's (small) farmer support programme (FSP) 1987–1993. *Paper presented at the Evaluation of the FSP workshop*. Halfway House: Development Bank of Southern Africa, 29 and 30 April.

Van Rooyen, C.J., Vink, N. and Christodoulou, N.T. 1987. Access to the agricultural market for small farmers in South Africa: The farmer support programme. *Development Southern Africa,* 4(2): 207–223.

Van Rooyen, J., Vink, N. and Malatsi, M. 1993. Viewpoint: Agricultural change, the farm sector and the land issue in South Africa. *Development Southern Africa,* 10(1) : 127–130.

Van Schalkwyk, H.D. and Van Zyl, J. 1994. Is South African land overvalued: common misconceptions? *Paper presented at the 32nd AEASA conference.* Pretoria: University of Pretoria, 19–20 September.

Van Wyk, J.J. 1970. Agricultural development in South African bantu areas. *Agrekon,* 9(1) : 64–67.

Van Zyl, J. 1989. Interrelationships in maize markets in Southern Africa II: Welfare aspects of the farmer support programme. *Development Southern Africa,* 6(3) : 360–348.

Van Zyl, J, Fényes T.I. and Vink, N. 1987a. Labour related structural trends in South African maize production. *Agricultural Economics,* 1(3) : 241–258.

Van Zyl, J. and Groenewald, J.A. 1988. Effects of protection on South African commercial agriculture. *Journal of Agricultural Economics,* 39(3): 387–401.

Van Zyl, J., Van der Vyver, A. and Groenewald, J.A. 1987b. The influence of drought and general economic effects on agriculture. *Agrekon,* 27(2) : 1–9.

Van Zyl, J. and Van Rooyen, J. 1991. Agricultural production in South Africa. In M. de Klerk (Ed), *A Harvest of Discontent: The Land Question in South Africa*. Cape Town: IDASA.

Van Zyl, J., Van Schalkwyk, H.D. and Thirtle, C. 1993. Entrepreneurship and the bottom line: how much of agriculture's profits is due to changes in price, how much to productivity? *Agrekon,* 32(4) : 223–229.

Vink, N. 1993. Entrepreneurs and the political economy of reform in South African agriculture. *Agrekon,* 32(4) : 153–166.

Vink, N. and Kassier, W.E. 1991. Agricultural policy and the South African State. In M. de Klerk (Ed), *A Harvest of Discontent: The Land Question in South Africa.* Cape Town: IDASA.

World Bank. 1994. *South African Agriculture: Structure, Performance and Options for the Future.* Washington, D.C.: World Bank Southern Africa Department.

10

Natural resource management issues in rural South Africa[1]

Johan van Zyl, Craig McKenzie and Johann Kirsten

Introduction

The majority of South Africa's poor reside in rural areas, mainly in the former homeland territories. With employment opportunities in the formal sectors of the economy limited, many of the poor depend heavily on the natural resource base for their basic needs, such as food, energy, water and housing. Their livelihood is closely tied to the well-being of the resource base. Considering the over-utilized resources in the former homeland areas, mainly due to overpopulation, the question could be asked: Why do people overuse resources available to them when their survival is at stake? Pinstrup-Andersen and Pandya-Lorch (*1994*) argue that desperate hunger leads to desperate strategies for survival. At that point, conservation of natural resources for people's own future welfare or the welfare of their children is less important to them, particularly if they cannot be sure that their children will in fact benefit from such conservation.

Heath and Binswanger (*1994*) show that the closing off of access to land in Colombia has led to serious inefficiencies, inequity and negative environmental effects. A series of legal measures introduced in Colombia since 1936, and in particular Law 30 of 1988, closed off all possible avenues for tenants, small farmers and the poor to acquire land. All that was left for the poor was to occupy marginal, and often ecologically unstable, land. In many areas of Colombia, the Andean slopes are being denuded of vegetative and soil cover, and the resultant loss of moisture retention is having an adverse effect on stream flow. This, in turn, reduces the availability of water for agriculture, both for poor farmers on the slopes and richer farmers located in the valleys. In the

1 This chapter has benefited tremendously from the presentations and discussions at the 'LAPC Workshop on the Environment', Mariston Hotel, Johannesburg, 30–31 March 1994.

reserves, there is an acute problem of holding fragmentation. Due to the deprivation of land, farmers carve up their resources, while some move into marginal lands and higher and higher up hillsides.

In addition to the lack of access to land and resources, it is generally also true that the poor do not own resources or reap the benefits of conservation, and thus have few incentives to conserve soil, protect groundwater or preserve trees. Furthermore, many of the poor only have access to resources with open access which, due to the existence of externalities, are vulnerable to exploitation.

The combination of poverty, population increase, land constraints and lack of appropriate agricultural technology usually results in environmental degradation. It is for this reason that Pinstrup-Andersen and Pandya-Lorch (*1994*) argue strongly that poverty should be confronted to prevent the poverty-degradation cycle from being perpetuated. Poverty poses the most serious environmental threat in low-income developing countries and also the less developed areas of South Africa. The millions of people who live near the subsistence minimum will exploit natural resources to survive. The eradication of poverty, perhaps through improved access to land, could be one of the important ways to prevent further environmental degradation.

Against this background, this chapter discusses natural resource management issues in South Africa. It addresses the policy background, followed by discussions of land resources, water resources, legal issues and institutional aspects. It closes by providing a synthesis of all the issues involved, followed by some conclusions.

Background

Current natural resource issues in the rural economy of South Africa need to be examined in the context of the political and economic policies that have been in force – particularly with respect to agriculture, forestry, water resources and conservation – during the past century, as well as the unique opportunity of structural reforms in rural areas brought about by the achievement of a democratic society in 1994.

Although agriculture in South Africa is widely regarded as a successful sector – primarily because of the economy's self-sufficiency with regard to most agricultural commodities, specifically staples, this apparent success is largely based on a wide range of political and economic policy distortions that have favoured large-scale, white owned farms. This apparent efficiency was not accompanied by the other political and social goals of equity and sustainability, specifically at the political, economic, financial and environmental levels.

These same distortions have contributed significantly to the degradation of the natural resource endowment in the rural sector. In particular, three factors stand out as having caused much of the environmental degradation in rural South Africa (*McKenzie, Van Zyl & Kirsten, 1994*): (a) the poverty of large numbers of the rural population, especially those in the former homelands; (b) the impact of the prevailing structure of incentives faced by large-scale farmers with respect to natural resource use; and, (c) the property rights regime that allowed virtually unrestricted use of natural resources by favoured groups in society.

These factors were largely brought about and exacerbated by two sets of policies, namely (a) the racially motivated political policies of apartheid; and, (b) inward-oriented import replacement economic policies aimed at obtaining self-sufficiency at the national level, which distorted both society as a whole and the economic incentive structure in agriculture (*Kirsten, Van Zyl & Van Rooyen, 1994*). These two sets of policies are interlinked to some extent, and many researchers and analysts treat them as part of a wider set of apartheid policies.[2]

While these factors are interlinked and mutually reinforce each other, it is nevertheless important to consider each separately and to examine the impact of each on the environment. The reason for doing so is that, with the demise of apartheid and white domination in South Africa, there is a major change in the structure of incentives *vis-à-vis* large-scale farming, as well as in the property rights regime. Clearly, however, an immediate reduction of the impact of poverty among the rural black population on the environment will not occur. Consequently, while policy reforms aimed at improving allocative efficiency and equity and, in turn, environmental sustainability are necessary, they will not be able to address all of the environmental problems in rural areas.

The dominant form of agricultural production in South Africa is the white-owned, large-scale farm – accounting for more than 95 per cent of the value added and owning 86 per cent of the agricultural land. The current dominance of the large farm model of production (commercially-oriented and capital-intensive) is partly the result of more than a century of policy distortions, often racially based. At the heart of these distor-

2 These policies and their historical evolution are discussed in detail in Chapters 3 and 9. The main features are briefly summarized here again within the context of their influence on rural environmental issues.

tions were the discriminatory land acts, the first of which was passed in 1913. These acts drew a line between the areas of black and white land ownership by prohibiting the ownership of land by members of one racial group in areas reserved for the other group. By this and subsequent reinforcing mechanisms, blacks (approximately two-thirds of the population) were restricted to ownership of land in areas that comprised about 13 per cent of the country. Restrictions on land ownership were further reinforced by policies that severely restricted the provision of infrastructure and agricultural support services to those areas of the country where blacks were allowed to own land.

The culmination of this policy of geographical segregation of racial groups occurred in 1948 when the policy of apartheid was instituted which eventually created the self-governing homelands or bantustans for the African population. The bantustans (which were intended to function as labour reserves) were characterized by: (a) a dearth of income opportunities – on average about two-thirds of household income in the bantustans came from sources outside the bantustans, i.e. remittances and pensions; (b) a comparatively high population density – about 16, 2 hectares of agricultural land per rural resident in the white areas in contrast to 1,3 hectares per resident in the bantustans;[3] and, (c) a generally poor resource endowment. The poverty of the bantustans has resulted in considerable degradation of their natural resource endowment as households battle to make a living.

The complement to the apartheid policy of discrimination against ownership of land by blacks in most parts of South Africa was a set of policies that strongly favoured the white large-scale commercial farmers, e.g. subsidies on some input prices and credit, favourable tax policies, subsidized water for irrigation purposes and subsidized output marketing structures.[4] These policies have artificially supported the profitability of large-scale commercial farms and, apart from allocative and efficiency problems, have introduced distortions which have had profound environmental consequences. Examples of these consequences include the following:

3 In 1951 less than 5 million blacks lived in the homelands. This number peaked at 14 million in 1985 and declined to 13,1 million in 1988. The large majority of these people live in poverty (more than 70 per cent).

4 In addition to numerous single channel marketing schemes, the marketing system is also characterized by a domestic orientation in the case of foodcrops and livestock products where domestic producers have benefited from a wide variety of protectionist policies, mostly in the form of quantitative import controls and protective tariffs.

- Support of agricultural commodity prices, specifically for maize and wheat, at higher than market clearing levels, and the creation of a subsidized infrastructure, specifically storage facilities, contributed to an expansion of cultivation into marginal production areas;
- Subsidies for investments in irrigation facilities contributed to an expansion of irrigation into marginal lands and led to the accelerated depletion of groundwater aquifers;
- Subsidies for fertilizer and agricultural chemicals led to growth in fertilizer use of 7,7 per cent per annum between 1947 and 1979, while dips and sprays grew at 9,7 per cent per annum between 1947 and 1973 and at 20,6 per cent between 1973 and 1980;
- Subsidized water provision for irrigation purposes contributed to the treatment of water as an abundant resource being used in an inefficient manner;
- Favourable tax treatment of capital equipment combined with negative real interest rates has encouraged an over-mechanization of agricultural production that, in turn, has led to an extensification of production into marginal and often environmentally fragile land.[5]

Another feature of the South African policy environment that has impacted upon the degradation of the natural resource base is the framework of property rights.[6] An integral part of the concept of sustainable development is the need for property rights (and transaction rules) that create incentives for sustainable resource use.[7] A consequence of

5 For a detailed account of the incentives and the process that led to over-mechanization, see World Bank (*1994: 108–111*).

6 Property rights are social relations that define an individual (or a group) with respect to something of value (the benefit stream arising over time from an object) against all other individuals in the society. Property relations are triadic in that they involve: (a) benefit streams; (b) right holders; and, (c) duty bearers. A property right denotes a set of actions and behaviours that the owner may not be prevented from undertaking. A right – by definition – implies an obligation on the part of all others to respect certain actions or behaviours, and to refrain from preventing those actions or behaviours. Rightful actions are both permissible and inviolable, and therefore actions intended to interfere with rightful actions are not permissible.

7 The notion of sustainable development is concerned with ensuring natural resource use and management such that future residents will have access to resource services and income streams enjoyed by current residents. The central problem for any government is to create institutional arrangements that will ensure that natural resources are used in a manner that is sustainable. At the heart of these institutional arrangements are different property right regimes and transaction rules that govern the use of natural resources.

apartheid was to bestow private property rights (freehold) on those individuals deemed important to the success of the political and economic system (mines, factories, the energy sector, the commercial forestry sector, white commercial farmers), and then to retain everything else under state control. In forestry and agriculture, freehold tenure of land was granted to those sectors deemed essential to the economy. Tax laws, rules over the ownership of land, and rules over the partial alienation of the freehold estate, were largely motivated by the need to provide certain environmental resources (land, water, trees) to selected groups. The resulting property regimes, especially in the agricultural sector, are therefore often artifacts of apartheid.

As new environmental policy is formulated, it is essential that the existing property regimes in land be carefully assessed. The artificial economic incentives under apartheid and other policies encouraged agricultural enterprises to extend the freehold estate beyond the economically rational boundary (under most feasible economic policy regimes). While marginal lands were incorporated into the freehold estate under the past policy regime, new policies in South Africa will mean that many of these marginal areas will move back out of the freehold estate, driven by the institution of a more coherent taxing system on land and its economic value.

The result of these past policies on resource use has been contrasting: in the relatively small homeland areas, fairly scarce natural resources were used (abused) by millions of the poor to complement remittances and pensions from 'outside' as part of a survival strategy; and in the white commercial farming areas, relatively abundant natural resources were used (abused) by a small number of the 'privileged' to produce increasing surpluses. Further, the latter group was heavily supported by policies creating welfare transfers from taxpayers and consumers – particularly from the poor.

Total land resource depreciation and off-farm costs for various types of land degradation were calculated to be R1 898 million in 1992/93, some 14,6 per cent of agricultural GDP or 35 per cent of the total net national agricultural income. This amount is distributed as follows: annual depreciation of land – R1 041 million (arable land through water erosion – R240 million; wind erosion – R20 million; soil crusting – R80 million; soil compaction – R68,5 million; increased acidity – R228,5 million; salinization and water-logging – R85,7 million; and rangeland – R334,9 million); and off-farm costs of R857 million (sedimentation of dams – R285,7 million; increased costs of purification and maintenance

– R571,4 million). The total cost associated with water erosion – including on- and off-site costs – accounts for 55 per cent of total costs due to degradation. Of importance to policy formulation is that off-site costs of water erosion are far larger than on-site costs – some 71 per cent of total annual costs due to erosion are externalized. Of these off-site costs, roughly half can be ascribed to degraded rangelands (*McKenzie, 1994*).

Hence, the opinion that emerges is that many parts of South Africa's natural resources endowment are now significantly degraded, although measures of the magnitude of this degradation are imprecise. Two main causes of this degradation seem to be (a) the racially-based apartheid policies and, (b) policy distortions aimed at supporting large-scale white farming. These distortions have led to extreme poverty in the bantustans and the misuse and misallocation of resources in the large farm sector. It is hypothesized that both of these factors have played a critical role in the degradation of the natural resource base. Further, another consequence of the broader set of policy distortions (including apartheid) has been a set of property (and other natural resources, e.g. water) rights that have also contributed to this pattern of degradation.

Against this background, issues in natural resource management in South Africa are discussed. Most of these issues are interrelated and impact upon each other: land use – whether it is for commercial agriculture, subsistence, forestry or conservation; water rights; the legal environment; the institutional set-up and responsibility matrix. Nevertheless, the discussion below treats each of these issues separately. Their interrelatedness and joint nature, however, should always be taken into account.

Land resources

As indicated above, South Africa's current ownership and use of agricultural land is the inevitable outcome of decades of policies favouring white large-scale commercial farmers. Through the agricultural sector, the state has controlled the economic and political tenor of the rural areas. As pointed out earlier, the dominant form of agricultural production is the white-owned, large-scale farm which accounts for 90 per cent of the value added and owns 86 per cent of the agricultural land. At the core of racially-based policy distortion were the discriminatory land acts, the first of which was passed in 1913. It drew a line between the areas of black and white land ownership by prohibiting the ownership of land by members of one racial group in areas reserved for the other group.

At the same time that black agriculture was being suppressed, the large-scale white farming sector was being actively promoted through the use of a variety of mechanisms such as subsidies on capital equipment, subsidized credit for production and land purchases, pricing controls with prices above parity, and restrictions on where producers could sell. These measures seriously distorted the economic incentives in agriculture. Practices such as the extension of the freehold state beyond the extensive margin, a legal prohibition on the partition of agricultural holdings, expansion of the cultivated area into fragile lands, overuse of groundwater, distorted incentives on surface water use in agriculture, and the excessive use of agricultural chemicals, have not only been uneconomic, but also environmentally detrimental.

Arable land resources

The lack of complete and reliable data on land use makes it difficult to assess the availability of land in South Africa's white commercial sector. Nevertheless, the conventional wisdom is that arable land is generally fully utilized, albeit with varying degrees of efficiency. South Africa's total area is 122 million hectares. Estimates are that the land suitable for crop production (in the predominantly white areas) represents only 13,5 per cent of the surface area of the country (*Schoeman & Scotney, 1987*). While, theoretically, 14,3 million hectares of land are suitable for crop production, only 12,9 million were cultivated in 1986. Of the land used for farming, only 4 million hectares (or less than 3 per cent) are considered to be high-potential.

In particular, there appears to be room for expansion of the area cultivated in Mpumalanga (3,1 million hectares available versus 2,2 million cultivated). Such uncultivated arable land, as a general rule, is either under forest or pasture and expansion would entail its substitution with crop cultivation. However, in the North West Province some marginal land under maize monocropping should refer back to pastures.

The variability of South Africa's arable land is further reflected in the pressing need to reduce cultivation in other regions of the country, namely, the winter rainfall region (1,8 million hectares cultivated versus 1,5 hectares available) and the Free State (1,7 million hectares cultivated versus 1 million hectares available). The result of the past incentive structures has been the widespread use of marginal land for crop production. Examples are provided by the expansion of monocropped, mechanized maize and wheat cultivation in the North West Province and Free State (*World Bank, 1994*).

South Africa has relatively little irrigation potential. At present 1,2 million hectares are irrigated, while an additional 250 000 ha can be brought under irrigation. This brings the total irrigable potential to less than 1,5 million hectares, or less than 2 per cent of the total area farmed. Limited and unreliable water supply is a major constraint to irrigation – South Africa generally does not have large primary aquifers which can be exploited, and the majority of irrigation schemes are dependent on surface water supplies. Irrigation farmers are subsidized since irrigation water is generally priced at a fraction of its actual cost. Thus, irrigation water is frequently treated as an abundant rather than a scarce resource.

The extension of monocropping into marginal lands, aided by the artificial low price of fertilizers and irrigation, has brought thousands of hectares into the arable white-owned commercial estates. Serious soil erosion has been the result, generating both on-site costs such as reduced soil fertility that affects crop productivity and farm income, and off-site costs such as increased run-off siltation and water flow irregularities that affect water supplies, hydro-electric power facilities and fisheries. Annual soil losses in South Africa are about 300–400 million tons. The estimated loss of nutrients in South Africa is about 3 300 tons of nitrogen, 26 400 tons of potassium and 363 000 tons of phosphorus. If these nutrients were to be entirely replaced by commercial fertilizers, the cost has been estimated to exceed R1 000 million (*Huntley et al, 1989*). It is also noted that the net social losses from this run-off may be lower since at least part of the eroded soils (and nutrients) end up on the farms of neighbours and others.

Subsidies to agricultural chemicals have induced a movement away from labour use in agriculture. Moreover, tax laws and other incentives have encouraged large-scale mechanization of agriculture with the attendant soil compaction that implies. Estimates are that there are over 2 million ha in maize production areas where yields may be reduced by as much as 30 to 40 per cent on such affected soils (*South Africa UNCED Report*).

Most of these negative environmental effects, i.e. soil losses, are directly related to specific cultivation practices. Scotney and McPhee (*1990*) estimate that soil losses for selected tillage practices under average conditions in South Africa are as follows: maize, conventional tillage (20 tons/ha/annum); maize, mulch tilled (5 tons/ha/annum); bare soil (28,5 tons/ha/annum).

As mentioned earlier, total arable land resource depreciation and off-farm costs for various types of land degradation were calculated to be

R405,4 million in 1992/93: water erosion (R78,4 million); wind erosion (R7 million); soil crusting (R28 million); soil compaction (R24 million); increased acidity (R80 million); and salinization and water-logging (R30 million); with off-farm costs accounting for the balance (R150 million).

Non-arable and extensive land use resources

Non-arable land dominates most of the west and interior of the country where rainfall is below the 500 mm per annum needed for reliable cropping. Two broad modes of use can be identified: (a) about 69 million hectares are used for extensive commercial animal production; and, (b) about 11 million hectares are used for purposes of 'ecosystem services' such as water yield, biodiversity conservation, tourism and recreation. About half of this land is managed by the state, and the other half by private individuals and corporations.

Total non-arable land resource depreciation was calculated to be R117 million for on-farm costs and R150 million for off-farm costs, thus R267 million in total in 1992/93. These costs are mainly due to overgrazing and deterioration of rangeland through soil erosion.

Former homelands

Resource degradation is even more serious in the former homelands which are characterized by: (a) poor land; (b) politically enforced overpopulation; (c) labour shortages; and, (d) poverty:

(a) **Poor land.** From the beginning, blacks were generally allocated land with relative poor soil quality, scarce rainfall, and rocky and steep terrain. Borders were drawn and redrawn to exclude anything of value: industrial sites, transport lines, mineral resources and fertile land. In general, agricultural and other support services, as well as infrastructure, are also lacking. While there are a limited number of comprehensive surveys of land degradation in the former homelands, the Government's Ciskei Commission reported over a decade ago that 46 per cent of the land in the Ciskei reserves was already moderately or severely eroded, and 39 of its pastures were overgrazed. In addition to the poor land is the severity of the land shortages. Only in two former homelands did the ratio of arable land to rural population exceed 0,20 ha/person. These were in KaNgwane at 0,25 ha/person and Bophuthatswana at 0,27 ha/person. Five of the homelands registered 0,10 arable ha/person or less. This compares with a ratio in the white rural areas which ranges from 1,37 ha/person in Natal to 2,87 ha/person in the Cape.

(b) **Politically enforced overpopulation.** In 1951, less than 5 million people lived in the homelands. This number peaked at 14 million in 1985 and, in 1988, levelled off at 13,1 million. These changes resulted from the resettlement policies pursued by the government, particularly between 1960 and 1983, when millions of blacks were relocated to the homelands. Apartheid forced about two-thirds of South Africa's population to be confined to land in areas comprising about 13 per cent of the country. Apartheid also denied black people access to education, health care, family planning, and secure sources of livelihood, all of which are contributing factors to a birth rate which is higher in the homelands than anywhere else in South Africa.

(c) **Labour shortages.** With limited job opportunities in the former homelands, well over half of household income is derived from repatriated migrant earnings and pensions. This leaves the former homelands with few in their peak working years and the major share of the labour force made up of children, the sick, the elderly and women providing for their children and elders. These women, struggling to provide for so many, are too pressed to undertake land conservation measures.

(d) **Poverty.** In 1985, 8,4 million blacks in the homelands lived below the poverty line as defined by the Minimum Living Level of the Bureau of Market Research of the University of South Africa. Slightly higher incidences of black rural poverty – in the homelands, but also in the rural areas of the Republic of South Africa, are reported by the (unpublished) Family Surveys of the Central Statistical Service. An even higher proportion – 84 per cent – is estimated by other authors (*see,* for example, *Simkins, 1990; Bekker, Cross & Bromberger, 1992*). In contrast to the large-scale white farming sector, the farming sector in the former homelands is unable to meet the subsistence needs of its population with the result that the former homelands are net food importers. Income from agricultural production in these areas constitutes no more than 20 per cent of household income. Thus, the former homelands should be viewed as labour reserves for other sectors of the South African economy – not even as the subsistence sector of a highly dualistic agricultural system. Living day-to-day, homeland farmers lack the cash to make long-term investment in the management of arable and non-arable land. With an average disposable income of about US$150 per year, these farmers cannot afford to buy fencing supplies to control grazing, hire labourers to help terrace sloping fields, or invest in agroforestry to conserve soil (*Kirsten, 1994*).

Taken together, the elements of poor land, politically enforced overpopulation, labour shortages and poverty, form a vicious circle of economic impoverishment and environmental degradation. In comparisons between calculations of net present values of future income losses due to various factors, (if nothing is done to stabilize existing degradation), between farming areas in the former homelands and white commercial areas, it was concluded that, although degradation of natural resources is generally more severe in the homelands, the value of future losses is lower due to the lower income generation per hectare in these areas (*McKenzie, 1994*):

Table 10.1:

Net present value of future income losses (R/ha) due to:

	Former homelands	*Commercial*
1. Water erosion (including off site costs)		
severe	R1 142/ha	R1 362/ha
moderate	R391/ha	R560/ha
2. Soil compaction	R376/ha	R597/ha
3. Soil crusting	R425/ha	R756/ha

Forestry

South Africa's indigenous forest resources are very limited, comprising only about a quarter of a million hectares of land.[8] In contrast, about 1,2 million hectares are afforested under *eucalyptus* and *pinus*, and a further 30 000 hectares are being afforested annually to meet the demands of the pulp and paper, construction and mining industries. According to South Africa's UNCED Report, the country's forestry industry plans to

8 As comparisons, the total area under indigenous woodlands and plantations in Zimbabwe has estimates ranging from 11 million to nearly 17 million hectares. Malawi's forests cover 3,5 million ha, about 37 per cent of Malawi's total land area. Most of this forest land is indigenous woodlands.

double its current plantation area of 1 130 000 ha of commercial forest land over the next three decades. As the country develops its new forestry policy, it will require an assessment of the extent, ownership structure, and sustainability prospects of existing plantation forests. It will also be necessary to determine the extent, ownership structure, and sustainability prospects of existing commercial harvesting in indigenous forests.

The major limitation for forestry expansion is the scarcity of water; areas suitable for forestry should have a rainfall of at least 750 mm and a run-off of 100 mm per annum. Hydrological research has shown that afforestation can reduce run-off by 500 mm per annum, which severely impacts on downstream users. There is also evidence that eucalyptus utilize groundwater and also reduce run-off in the long term (*Gander & Christie, 1994*).

Commercial plantation forestry impacts further on the environment by enhancing erosion losses, soil compaction and acidification. Although using only 1 per cent of South Africa's total surface area, it often occurs in unique environments. Monocultural forestry practices lead to a reduction of biodiversity.

The present forestry structure favours the development of large plantations: 82 per cent of afforestation is conducted in plantations over 1 000 ha in size, and 98,5 per cent of plantations are over 100 ha in size. Little has been done to encourage commercial tree growing in mixed agriculture. The pricing structure further favours primary processing at the expense of timber production. The involvement of small-growers in the forestry sector will also enable disadvantaged sectors to benefit from the highly monopolized forestry industry.

Conservation areas

South Africa's national protection system includes 178 protected areas comprising 6 310 000 ha or 5,2 per cent of the land area. Moreover, there are a large number of privately-owned reserves which brings South Africa's total land area comprising national parks and nature reserves to 6 per cent. The relative size of the national protected land area is not uncommon when compared to other countries of the world. What is impressive is the absolute number of protected areas – South Africa's number is much higher than other middle income economies. Of all the low-income countries, only China at 289 and India at 359 – and with a considerably larger land mass – have a higher absolute number. In Africa, Kenya and Madagascar follow South Africa with the next highest number of protected areas at 36 (*World Resources Report, 1993, Table 20.1: 20*).

Internationally, South Africa is known for its high levels of biodiversity. The flora comprise some 20 300 species, about 8,4 per cent of the world total. The Cape Peninsula, at 47 000 hectares, has 2 256 known indigenous plant species. There are also 84 species of amphibians, 286 species of reptiles, 600 species of breeding birds, and 227 species of mammals (*South Africa's National Report to UNCED*). The country is also known for the high quality of management of the conservation areas that currently exist.

These natural resources, however, were managed for the benefit of the white minority, with local black communities largely excluded from either management or use. Of even greater concern is the emerging tendency to incorporate high-potential farm land into nature conservation areas (public and private), possibly to 'protect' them from any future land reform programmes.

A new environmental policy for South Africa should start with an understanding of the feasible policy direction in two aspects of natural resource conservation: (a) formal conservation areas now incorporated in national parks; and, (b) resource conservation mechanisms for all other lands.

In the formal conservation areas, several different models should be developed to allow Africans on the perimeter of existing nature reserves to become reincorporated into these areas or to benefit from the proceeds and activities. The much heralded Richtersveld National Park is an example. Its 30 year lease, agreed with the people in 1990, provides for 6 600 head of stock to graze inside the national park. The models to be developed would specify permitted land uses, the conditions and mechanisms under which those uses would be controlled, administrative and appeal procedures, and possible financial arrangements. The pricing of admission to existing nature reserves must also be explored, particularly with respect to several alternative pricing regimes for South Africans and foreign tourists.

Outside of the protected areas, natural resource conservation requires a clear understanding of the different policy instruments that will facilitate or induce improved resource conservation on freehold land. Exploring administrative alternatives for controlling land use will be needed, particularly in relation to alternative approaches which locate control at different levels (central, regional, local) within a national system.

Water resources

South Africa's water supply can be characterized by its scarcity compounded by past agricultural practices, its seasonal variability and poor distribution. The most arid 21 per cent of the country receives less than 200 mm of rainfall annually, with evaporation rates exceeding 1 800 mm. A further 65 per cent of the country receives less than 500 mm of rainfall annually (*South Africa's National Report to UNCED*).

Over the period 1970–1987, the annually renewable water resources amounted to 51 km^3, or 2 244 m^3 per capita per year. Surface run-off comprises 85 per cent of the total supply of water, and 15 per cent comes from underground supplies. Estimates are that about 60 per cent of available water can be exploited economically. However, actual withdrawal of internal, renewable water resources over the period 1970–1987 was 9,2 km^3 per year, about 18 per cent of the total of 51 km^3. Total per capita use was 404 m^3 of which 339 m^3 was for industrial and agricultural use, and 65 m^3 for domestic purposes. These figures refer to national averages and do not reflect the high variability in the regional distribution of water resources within the country. Natal, for example, occupies less than 7 per cent of the country's surface area but has 40 per cent of the available water (*World Bank, 1994*).

Estimates are that the exploitable surface, underground and return flow of water will be 38,666 million m^3 in the year 2010. Consumptive and non-consumptive use is expected to increase to 25 888 million m^3 in the year 2010, with the share of irrigation water being 11 885 million m^3 or 45,9 per cent of the total on average. Beyond the year 2020, the total utilization of water is expected to increase above the available supplies. Demand and supply of water is also not geographically balanced: 6 of the 24 primary water catchment areas of South Africa will be deficit areas in 2010.

The solutions proposed to the problem of the relative scarcity of water in South Africa are often structural, and require an increase in the supply of water through means such as dam construction and transfers between drainage regions. However, these solutions do not address whether the existing processes for the allocation and utilization of water among and by competitive uses and users in different regions and over time will solve the problem of water scarcity. For example, (a) the average irrigation water requirement for one hectare of land is enough to meet the domestic needs of almost 900 people for a year; and, (b) current water losses associated with the supply of irrigation water alone are around 30 per cent – a one per cent reduction (to 29 per cent) would be sufficient to meet the basic water needs of 9 million people.

According to South Africa's Water Act, there are only use rights and no property rights in water. Some of the main users of water are:

- **Human consumption.** South Africa's Standing Committee on Water Supply and Sanitation estimates that 12 million people lack an adequate water supply (8 million rural and 4 million urban), and 18 million lack adequate sanitation (11 million rural and 7 million urban). Just to service these gaps will have a significant impact on the current competition for water supplies. Significant demands will also occur with the further growth in the population. The role of groundwater to serve isolated rural communities is likely to become more important.
- **Agriculture.** It is the largest user of water, consuming as much as 50 per cent of the total demand. In the past, commercial farmers have benefited from substantial amounts of public assistance with respect to investments in water control. The low costs of irrigation water have encouraged both excessive use in agriculture (leading to increased loss through evaporation and leaching), and irrigation of low-value products such as grass meadows and forage crops. About 80 per cent of irrigable land is already irrigated. Many existing systems, however, are inefficient users of water, exacerbated by artificially low costs of water extraction and use. The development of new technologies and greater irrigation efficiency could generate savings in water to serve the needs of competing users.
- **Industry.** It is envisaged that industry will be a leading sector in South Africa's future economic growth, and the demands for water from this sector will increase proportionally.
- **Conservation and eco-tourism.** The Water Act still does not recognize any rights to water use in conservation and eco-tourism. The maintenance of biodiversity and the development of a potentially large eco-tourism industry will have large claims on water resources.
- **Waste quality.** This is also becoming an important issue and source of conflict among sectors. The external costs of pollutants created by urban, agricultural and industrial concerns must also be factored into decisions regarding water use efficiency.

A new environmental policy for South Africa must examine the misallocations of water in economic terms because these do not easily allow for the transfer of water rights to those sectors which derive high social and financial returns from water outside agriculture. This will involve an analysis of alternative water allocation mechanisms both within and among economic sectors. Water pricing mechanisms which reflect the real value of water could, at least potentially, play an important role since water for irrigation purposes is heavily subsidized at present. This has at least three effects: (a) it suppresses the price of agricultural products; (b) it increases farmers' profits; and, (c) it decreases the value of water, resulting in a smaller incentive to increase water use efficiency or conserve water. Effects of increased water prices on employment should also be considered.

Given the geographic disparities in water supplies and water demands, it will also be important to investigate the necessity (and feasibility) of future inter-catchment water transfers, including the international transfer of water among neighbouring countries such as the Lesotho Highlands Water Scheme and the Nkomasi Water Project, both projects partially funded by the World Bank.

Legal issues and the environment

South African environmental law does not stand on its own within the legal domain, but is rather an array of different laws ranging in several disciplines that seek to address the environment in a direct or indirect way. These laws seem to lack holistic rationale for their existence, and are ad hoc manifestations in response to problems that may have arisen in practice. The sources for environmental law, therefore, are diverse and are not embedded in a comprehensively codified statute; for example, one source is the common law, based on Roman-Dutch law. To define the exact parameters of environmental law is problematic; however, three broad areas are: (a) resource utilization and conservation; (b) waste management and pollution control; and, (c) planning and development.

South African law is embedded in private rights. While individuals can take neighbours to court for causing noise, air or water pollution, no such legal jurisdiction applies to the public interest. Since most environmental issues are public issues, the current legal framework does not afford public rights the same or similar privileges as private rights. Environmental interests, therefore, often do not enjoy adequate protection in law. Environmental law further lacks credibility because it is often seen to be the law for the privileged and has nothing to offer the poor, given its private law framework.

Water law and the allocation of water, in particular, is an important area that has been identified for further research because it also has important implications for rural restructuring in the wider sense, including land reform. The principles which underlie the water allocation mechanism of the 1956 Water Act are: (a) a distinction between public and private water; (b) a distinction between rights in respect of normal flow and surplus water; (c) 'riparian ownership'; and, (d) state control in the public interest.

A new water policy will thus have to address the present allocation of water resources. In this respect, the distinction between public and private water is impracticable for water allocation in a restructured rural system where the best possible distribution of all available water resources should be made. Further, the distinction between normal flow and surplus water ought not to be based on the capability of use with or without impoundment because, with modern technology, almost all water is impoundable and often best used after storage. Riparian ownership as fundamental to the allocation mechanism has also become obsolete, and consideration ought to be given to equal opportunities for the participation of all water user sectors within a basin in the use of the available water. Similarly, state control ought to happen as part of a system of basin representation. Groundwater is also part of an integrated water cycle, and thus ought not to have a separate legal status or set of allocation rules to surface water. Lastly, basin representation ought to underlie the water allocation mechanism from grassroots to government level: this will assist in decentralizing water matters without negating the necessity for state control.

Institutional responsibility

The present institutional framework and responsibility pertaining to natural resource use and policy, like the legal framework, is severely fragmented: the Department of Environmental Affairs is responsible for forestry and marine resources (forestry is in the process of being privatized); the Department of Water Affairs is responsible for water management; agricultural land and resource use is the responsibility of the Department of Agriculture; conservation either falls under the National Parks Board or the provincial administrations; tourism is handled by the Department of Tourism. This fragmentation has led to duplication and general inefficiency in handling environmental issues and the division of responsibilities.

The new interim constitution of South Africa allows for strong regional government, including natural resource issues. Thus, a new natural resource policy in South Africa should carefully consider institutional responsibility in executing this policy.

Specific issues to be addressed in a new environmental management policy

Rural land

A new environmental policy for South Africa pertaining to rural land must be concerned with understanding how the existing agricultural policies have led to unsustainable agricultural land use patterns. Land-use practices both in the commercial sector and the traditional agricultural areas of the former homelands (described below) need to be better understood. Assessing alternative agricultural policy reforms to change undesirable land use practices will be required. Moreover, special attention should be paid to analysing alternative institutional mechanisms for monitoring resource use in the future, particularly in the areas of soil erosion, water pollution, deforestation, and the management of arable and non-arable land.

Former homelands

Overgrazing and overstocking are problems that warrant particular attention in the former homelands. The added complexity of communally owned land, the role of cattle and other livestock in the agricultural economies of these areas, and other social and economic realities, should be specifically addressed by any future resource policy. These aspects should also be seen and analysed in the context of rural restructuring and the more extensive provision of support services.

Forestry

In commercial forestry, a new forestry policy must analyse and resolve problems related to the long-run sustainability of the sector, its environmental implications, the nature of contracts with landowners, the structural outlook for the industry under post-apartheid democratic reforms, and the general outlook over the next few decades in which rural land use and control will be in a state of transition. In social forestry, the issues will be the feasibility of extending the use of woody perennials into emerging smallholder agriculture, the role of agroforestry in emerging group farming schemes, extension programmes to provide information on forestry activities, and the use of trees to provide building materials and minor forest products to a newly enfranchised rural populace.

Economically, a new future forestry policy must understand the current pricing and subsidy regimes that drive commercial harvesting and afforestation practices both in the plantations and the indigenous forests. Work must also be undertaken on the economic status of South Africa's primary wood processing sector, its international competitiveness, and its long-run prospects in world markets. Gaining a better understanding of rural energy demand and supply, with special attention to the role of fuelwood in that system, will also be needed.

Of immediate concern is the serious depletion of fuelwood supplies in the rural areas. About two-thirds of South Africans, specifically the poor, use wood for fuel. Due to the extreme population pressure in the homelands, analysts at the Energy Research Institute of the University of Cape Town, by comparing the growth rates of woody biomass in the homeland ecosystems with fuel consumption rates, concluded that 4 out of the 10 homelands were already in fuelwood deficit in 1980 (*Durning, 1990*). A new environmental policy for South Africa should describe the prospects for enhancing the rural fuelwood situation, including the changes in legal and economic incentives to make fuelwood plantings feasible on freehold lands. It will also be important to understand the prospects for integrated forestry activities on both plantations and indigenous forests in which social forestry needs may be met with commercial wood needs.

Conservation areas

A new natural resource policy with respect to conservation areas should not see National Parks as the isolated islands of biodiversity they are at present, but rather as part of an overall land use policy. Unlike the existing policies, such a policy should consider large ecosystems and long-term trends, and various land use options should be developed through a holistic approach involving whole communities. Communities should gain direct economic and other benefits from wildlife and be empowered to take responsibility for the management of natural resources. The challenge of a new policy is to make the National Parks and reserves truly national assets which benefit all the people of South Africa.

Water

Specific issues which warrant further research and specific attention in a new water resource policy include issues centering around: (a) water law (support for rural restructuring; water rights; and constitutionally supported basic service rights); (b) water conflicts (water allocations; afforestation – thirsty green 'cancer' or economic saviour?; rural communities versus nature conservation; and falling ground water levels); (c) rural development (water shortages; efficiency of commercial irrigation; subsistence versus commercial agriculture; and vulnerability of rural communities to water shortages); (d) economics of rural water supply (water pricing – how much should water cost?; water subsidies; and availability of capital); (e) water quality (water pollution; non-point source pollution from agriculture; and sanitation); and, (f) international aspects (water demand of neighbouring countries; and international basin management).

Pricing of water resources needs particular attention. This will impact on most of the other decisions, i.e. allocation and utilization. Most of the existing controls and regulations on water and water rights effectively prevent the operation of a market in water rights. Even if deregulation takes place, trading of water rights and prices will not immediately reflect scarcity values since there are no property rights in water.

Legal issues

The aim of new legislation with respect to the environment should be the creation of the state's administrative capacity to manage and enforce the law, and to allow public participation in the law. Although the new constitution, to some extent, allows for this to take place, legal issues and how they can be addressed should form an important part of any environmental and resource use strategy.

Institutional responsibilities

As mentioned previously, the new interim constitution of South Africa allows for strong regional government, including natural resource issues. Thus, a new natural resource policy in South Africa should carefully consider institutional responsibility in executing this policy. The issues that specifically need to be addressed are: (a) the present fragmentation and general inefficiency of institutions involved in environmental management; and, (b) the appropriate level of decentralization of power and responsibilities to provincial and local government.

Conclusion

Unemployment and racially discriminatory legislation in South Africa closed off many possible avenues of making a decent livelihood for a large majority of rural inhabitants. As a result, these people lack sufficient incomes or access to credit to purchase appropriate tools, materials and technologies to practice environmentally sustainable agriculture and protect natural resources against degradation. The loss of entitlements could also be another cause for what Pinstrup-Andersen and Pandya-Lorch (*1994*) call the 'poverty-degradation relationship'. The forced removal from their land, as well as restricted access to land and natural resources, that was generally experienced by blacks in South Africa are prime examples of loss of entitlements. The condition of natural resources in the former homeland areas was largely caused by overpopulation, poverty and subsequent cultivation on marginal lands and the cutting down of vegetation for fuelwood.

Poverty poses the most serious environmental threat to the less developed areas of South Africa. The millions of people who live near the subsistence minimum will overexploit natural resources to survive. The eradication of poverty, perhaps through improved access to land, could be one of the important ways to prevent further environmental degradation. Once the issues of rights to natural resources have been adequately addressed, the other issues discussed above could be addressed in the process of formulating a coherent environmental policy.

References

Bekker, S., Cross, C. and Bromberger, N. 1992. Rural poverty in South Africa: a 1992 study using secondary sources. Durban: University of Natal.

Durning, A.B. 1990. Apartheid's environmental toll. *Worldwatch Paper 95*: May.

Gander, M. and Christie, S. 1994. Commercial and social forestry in South Africa. *Paper read at the LAPC Workshop on Environmental Issues*. Johannesburg: Mariston Hotel, 30–31 March.

Heath, J. and Binswanger, H.P. 1994. Making land and labour markets more effective. *Annex A to Colombia Agricultural Sector Review. World Bank unpublished document.* Washington, D.C.

Huntly, B., Siegfried, R. and Sunter, C.. 1989. *South African Environments into the 21st Century.* Cape Town: Human & Rousseau Tafelberg.

Kirsten, J.F. 1994. Agricultural support programmes in the developing areas of South Africa. *Unpublished PhD dissertation.* Pretoria: University of Pretoria.

Kirsten, J.F., Van Zyl, J. and Van Rooyen, J. 1994. South African agriculture during the 1980s. *South African Journal of Economic History*. Vol. 9 no. 2

September 1994.

McKenzie, C.G. 1994. Degradation of arable land resources: policy options and considerations within the context of rural restructuring in South Africa. *Paper presented at the LAPC Workshop on the Environment.* Johannesburg: Mariston Hotel, 30–31 March.

McKenzie, C.G., Van Zyl, J. and Kirsten, J.F. 1994. Agricultural issues relating to natural resources and the environment in South Africa. *Agrekon,* Vol. 33 (4): 222–224.

Pinstrup-Andersen, P. and Pandya-Lorch, R. 1994. Poverty, food security and the environment. *Unpublished IFPRI fact sheet prepared for the Union of Concerned Scientists/IFPRI Workshop.* Washington, D.C.: IFPRI, September.

Schoeman, J.L. and Scotney, D.M. 1987. Agricultural potential as determined by soil, terrain and climate. *South African Journal of Science,* Vol 86 : 395–402.

Scotney, D.M. and McPhee, P.J. 1990. The dilemma of our soil resources. Pretoria: Department of Agricultural Development.

Simkins, C.E.W. 1990. Black population, employment and incomes on farms outside the homelands revisited. *Paper presented at the IDASA Rural Land Workshop,* Houwhoek.

Wilson, F. 1991. A land out of balance. In Mamphela Ramphela and Chris McDowell (Eds), *Restoring the Land: Environment and Change in Post-Apartheid South Africa.* London: The Panos Institute.

World Bank. 1994. South African agriculture: structure, performance and options for the future. *Discussion Paper No. 6,* Southern African Department.

11

The farm size-efficiency relationship

Johan van Zyl

Introduction

South African agriculture has the appearance of being sophisticated and highly successful. A closer look at the present structure and performance of South Africa's agricultural sector, however, reveals that despite the appearance of efficiency, the sector has followed a pattern of growth that is far from normal.[1] Although agriculture is generally characterized by constant returns to scale and an inverse relation between farm size and productivity (*Binswanger, Deininger & Feder, 1993*), the sector is dominated by relatively large farms that are owned and operated by a comparatively small number of individuals. International evidence indicates that a large-scale mechanized farm sector is generally inefficient, especially when compared to small-scale family type farm models. Although there may exist very real economies of scale, they are mostly 'false' because they are usually the result of policies which favour larger farms over small farms.

At least two questions related to the productivity relations, both commercial and subsistence, in South African agriculture and the effects of size on these relations, have not been adequately addressed in South Africa before, and are important when considering land reform:

1 In its recent study, the World Bank (*1994*) , supported by several other studies (*see*, for example, *Van Zyl & Groenewald, 1988; Thirtle, et al, 1993; and Van Zyl, 1994*), concluded that agriculture in South Africa appears to be a highly sophisticated and successful sector, but this appearance hides severe distortions and inefficiencies. Evidence that is often cited in support of the former view is the fact that South Africa is self-sufficient with respect to most of its major agricultural commodity requirements. At the same time, the sector's relatively small and declining share of GDP is seen as indicating a pattern of secular decline of agricultural production that is consistent with a normal pattern of economic growth and development (*Van Zyl, Nel & Groenewald, 1988*). A closer look at the present structure and performance of South Africa's agricultural sector, however, reveals that despite the appearance of efficiency, the sector has followed a pattern of growth that is far from normal (*Van Zyl & Groenewald, 1988*).

- Are large mechanized farms and the present commercial white farms economically efficient relative to smaller holdings?
- What is the role of past policies in determining these observed productivity relations?

If larger farms are not efficient relative to smaller farms, then smaller farms and equalizing the ownership distribution would enhance both efficiency and equity; and if policy created artificial economies of scale, they should be adjusted.

This chapter aims to explore these issues by briefly reviewing the sources of economies of scale and international evidence on these issues, as well as analysing representative farm-level data in both the commercial and former homeland sectors. These analyses are conducted against the policy environment and changes therein, as well as against other factors which influence farm production, as discussed in Chapter 9.

There is a brief discussion of the sources of economies of scale with some international evidence. Then a picture of the structure of South African agriculture is provided, detailing distributions of farm sizes as well as some results of previous studies analysing farm size efficiencies. Thereafter, an analysis of the evidence on scale efficiency in the former homeland sector is provided, followed by an analysis of the farm size-efficiency relationship in commercial farming utilizing representative farm level survey data on the six major grain-producing areas and an irrigation area over the period 1975–1990. The role of policy in explaining these relationships is then discussed, and thereafter conclusions are drawn.

Values

'Efficiency for whom?' should be a central question in the determination of efficiency (*Schmid, 1994*). The issue can be conceptualized by the portrayal of two persons' indifference maps for two goods in a conventional Edgeworth box diagram. Any combination of goods held by the parties not on the contract curve is Pareto-inefficient, and any barrier to reaching the contract curve is inefficient. From any given starting place, the parties have a mutual interest in reaching the contract curve. But the portion of the contract curve that they can reach by mutually advantageous exchange is different for each starting place. Efficiency says nothing about the power question involved in the choice of starting place and the resulting equilibrium on the curve (*Schmid, 1987;*

Bromley, 1989). Furthermore, any voluntary agreement to trade does not clarify any agreement about the legitimacy of the starting place. If the original distribution of rights was illegitimate, any Pareto-improvement from it has no legitimacy either (*Calabresi, 1991*).

Since efficiency is always rooted in some distribution of rights, it can never be a basis for judging that distribution (*Schmid, 1992*). Rights are antecedent to efficiency calculations. In this context, it is neither useful nor meaningful to conceptualize policy issues as efficiency versus distribution. The issue is efficiency 1 versus efficiency 2, each with a different starting place that resolves the questions of power and rights.

Given the skewed land ownership in South Africa and the way in which these land rights were derived (see Chapter 3), the question 'Efficiency for whom?' is extremely relevant when comparing efficiency of different farm sizes and land distributions in South Africa. Because the validity of the efficiency argument depends on the legitimacy of the rights prior to the calculation, the efficiency calculations cannot be the only criteria for deciding on land reform when the very basis of these rights is in question. With this perspective, this chapter uses the existing distribution of rights to determine efficiency issues. It should be acknowledged from the outset, however, that this is not an attempt to justify the original distribution, but rather to show the efficiency impacts and issues given the existing distribution of power, wealth and rights.

International experience of economies of scale, farm size and productivity

In examining the relationship between farm size and productivity, it is necessary to look first at the sources of economies of scale which underpin the justification for the move towards large-scale production. In general these are: (a) lumpy inputs that cannot be used below a certain minimum level such as farm machinery and management skills; (b) advantages in the credit market and in risk diffusion arising from ownership of large holdings; and, (c) processing plants that transmit their economies of scale to farms, usually giving rise to wage plantations. A summary of the basic theoretical context is followed by a brief summary of these sources of economies of scale, and thereafter by a brief description of international empirical findings and related issues. For a more detailed discussion on these, see Binswanger, Deininger and Feder (*1993*) and Johnson and Ruttan (*1994*).[2]

2 These issues are also addressed in Chapter 4.

In theory, economies of scale are defined by a production function which exhibits a more than proportional increase in output for a given increase in magnitude of all inputs. In practice, the concept provides problems since there is rarely a situation when an increase in magnitude of some inputs does not imply a change in the factors of production (*Peterson & Kislev, 1991*). The general consensus of researchers on economies of scale in agriculture is that they do not exist, except under very special circumstances. Empirical studies typically find constant returns to scale (*see*, amongst others, *Johnson & Ruttan, 1994; Peterson & Kislev, 1991*), although lumpy inputs, credit and risk diffusion, and processing plants can be important sources of economies of scale.

Sources of economies of scale

Lumpy inputs

Farm machinery – threshers, tractors and combine harvesters – are lumpy inputs, and reach their lowest cost of operation per unit in relatively large areas. With the advent of agricultural mechanization, many people believed that the economies of scale associated with it are so large that it makes the family farm obsolete. Small owners would sell or lease their land to larger operators. However, it became quickly apparent that machine rental can permit small farmers to circumvent the economies of scale advantage associated with machines in all but the most time-bound of operations, such as ploughing and planting (seeding) in dry climates, or harvesting where climatic risks are high. In those situations, farmers compete for early service and therefore prefer to have their own machines. Thus, economies of scale associated with machines do increase the minimum efficient farm size, but by less than expected because of rental markets. The use of lumpy inputs leads to an initial segment of the production function that exhibits increasing returns with operational scale, but these technical economies vanish when farm size is increased beyond the optimal scale of lumpy inputs or when rental markets make the lumpiness of machines irrelevant.

Management skills, like machines, are an indivisible so, the better the manager, the larger the optimal farm size. Technical change strengthens this tendency. The use of fertilizers and pesticides, and arranging the finance to pay for them, require modern management skills. So does the marketing of high-quality produce. In an environment of rapid technical change, acquiring and processing information becomes more and more important, giving better managers a competitive edge in capturing the innovator's rents. Therefore, optimal farm sizes tend to increase with more rapid technical change. However, some management and technical skills, like machinery, can be contracted from specialized consultants and advisory services, or can be provided by publicly financed extension services. Contract farming for processing industries or bulk marketing companies often involves the provision of technical advice.

Access to credit and risk diffusion

Land, because of its immobility and robustness, has excellent potential as collateral, making access to credit easier for the owner of unencumbered land. On the other hand, rural credit markets are difficult to develop and sustain. Thus, there is often severe rationing of credit which can be partly relieved by the ability to provide land as collateral. The high transaction costs of providing formal credit in rural markets implies that the unit costs of borrowing decline with loan size. Many commercial banks do not lend to small farmers because they cannot make a profit. Raising interest rates on small loans does not overcome this problem since it eventually leads to adverse selection. For a given credit value, therefore, the cost of borrowing in the formal credit market is a declining function of the amount of owned land. Providing funds to overcome emergencies is a common function of informal rural credit markets. However, the amounts small farmers can borrow for consumption are usually tiny, and often only at high interest rates. Investigations into how farmers and workers cope with disaster show that credit finances only a small fraction of their consumption in disaster years. Access to formal commercial bank credit, therefore, gives large modern commercial farmers a considerable advantage in risk diffusion over small farmers without access. Establishment of a viable credit function for the family farm is a *conditio sine qua non* of modern commercial farming. Hence, emphasis on developing rural credit is needed, including co-operative banking and other savings-mobilization mechanisms.

Economies of scale in processing

Wage-based plantations continue to exist for typical plantation crops, for example, sugarcane, bananas and tea. This is not because of inherent economies of scale in producing these crops, but rather economies of scale which arise from the processing or marketing stage (and not the farming operation) and are transmitted to the farm. However, economies of scale in processing alone are not a sufficient condition for the explanation of the existence of plantations. The sensitivity of the timing between harvesting and processing is also crucial – sugarcane, tea or the fruits of the oil palm have to be processed within hours of harvesting. Plantation style production has never been established for easily stored products, such as wheat or rice, which can be bought at harvest time in the open market and stored for milling throughout the year. Even sugarcane can be contracted by millers with small farmers (e.g. in South Africa) as long as the logistics of harvesting and transportation can be solved. Thus, the superiority of the plantation depends on *a combination of economies of scale in processing with a co-ordination problem.* Plantations do not arise, or do not survive once labour coercion is abolished, unless both these conditions exist. In many cases, even where there is an even labour demand over the year, the plantation mode of production has declined sharply at the expense of smallholder production. This applies to commodities as diverse as sugarcane, tea, coffee, bananas, rubber and oil palm, as well as tobacco and cotton.

Wage plantations survive in areas where they were first established under conditions of low population density and with a large land grant. Where the same crops were introduced into existing smallholder systems, contract farming prevails. Processors seem not to have found it profitable to form plantations by buying out smallholders and offering them wage contracts. This suggests either that the co-ordination problem associated with plantation crops can be solved at a relatively low cost by contract farming, or that imperfections in the land sales markets are so severe that it is prohibitively expensive to create large ownership holdings by consolidating small farmers.

Evidence on the farm size–productivity relationship

Available literature clearly demonstrates that a systematic relationship between farm size and productivity is the result of market imperfections, and then only when more than a single market is imperfect. For example, if credit is rationed according to farm size, but all other markets are perfect, land and labour market transactions will produce a farm structure that equalizes yields across farms of different operational size. But, if there are imperfections in two markets – land rental and insurance, or credit and labour – a systematic relationship can arise between farm size and productivity.

In countries like South Africa, where markets facing small farmers for any combination of labour, land, credit, land rental, insurance, etc, are often imperfect or missing (at least for some farmers, in general small farmers), this may give rise to real economies of scale over the short-term. However, these economies of scale are 'false' in the sense that they are only temporary, and the result of deliberate elimination of, or restrictions on, these markets.[3] With development of these markets, economies of scale diminish and eventually disappear. Thus, the issue is not to pursue a farm structure that captures these benefits over the short-term but, over the long-term, gets a country locked into an inefficient and inequitable structure centering on large-scale mechanized farms.

Even without economies of scale, the question remains: Does size matter? Are larger farms more productive and/or profitable than smaller ones, even if an argument cannot be made for superior technical efficiency? The answer clearly is yes. Policies are rarely scale neutral, and external economies of scale are a reality. While these tend to favour larger farms, there are considerable transaction costs in the labour market, as well as supervision costs, which favour smaller farms. The issue is: What is the net effect of these factors?

Many studies on the farm size-productivity relationship reported on in the available literature suffer from severe shortcomings, for example, not accounting for differences in land quality or labour productivity; using physical yields; and not accounting for differences in operational holding size and ownership holding size. Proper measures of efficiency are the difference in total factor productivity between small and large farms, and the difference in profits, net of the cost of family labour, per unit of capital invested. Studies which apply these measures typically support the following generalizations (*Binswanger et al, 1993*):

- The productivity differential favouring small farms over large farms increases with the differences in size, implying that it is largest where inequalities in landholdings are the greatest, in the relatively land-abundant countries of Latin America and Africa, and smallest in land-scarce Asian countries where farm size distributions are less equal.

3 Under certain circumstances, such as those in South Africa, there are external economies of scale (*Johnson & Ruttan, 1994*). These occur when, as firms or farms increase in size, they experience advantages in terms of access to inputs, credit, services, storage facilities, or marketing and distribution opportunities relative to smaller farms. This gives large farms real advantages relative to small farms due to pecuniary economies or policy distortions rather than to greater efficiency. On the other hand, diseconomies of scale may also occur, for example when the labour market fails; or do not exist, when transaction costs in the labour market are high, or when the effort of hired labour is significantly affected by supervision (*De Janvry, 1987*).

- The highest output per unit area is often achieved not by the smallest farm size category, but by the second smallest farm size class, suggesting that the smallest farms may be the most severely credit constrained.

However, most of the empirical work on the farm size-productivity relationship has been flawed by methodological shortcomings, and has failed to deal adequately with the complexity of the issues involved. In general, studies which come to grips with some of the problems consistently show the superiority of smaller farms over large farms.

Numerous studies provide empirical evidence at the micro-level of the existence of an inverse relationship between farm size and the efficiency of resource use – as farm size increases, efficiency declines. This relationship is basically due to higher efficiency of family labour as compared to hired labour, in combination with commonly observed imperfections in credit and land rental markets (*Binswanger et al, 1993*). Berry and Cline (*1979*) found that the value added per unit of invested capital for the second smallest farm size group (10 to 50 ha) exceeded that of the largest farm size groups (200 to 500 ha) in a majority of zones that did not specialize in plantation groups.

A World Bank study (*World Bank, 1983*) on the higher efficiency of small versus large farms in Kenya, found that output per hectare was 19 times higher and employment per hectare was 30 times higher on holdings under 0,5 hectare than on holdings over 8 hectares. At the national level, this meant that a 10 per cent reduction in average farm size would increase output by 7 per cent and employment by over 8 per cent. Binswanger *et al* (*1993*) report similar results for many other countries. Chavas and Aliber (*1992*) found virtually no scale economies in dairy production in Wisconsin, and the very limited initial scale economies they observed were attributable to lumpiness of certain inputs.

Evidence is also available at the macro-level, but only in terms of physical yields – an imperfect indicator of efficiency. Prosterman and Riedinger (*1987*), using data from 117 countries, show that 11 of the top 14 countries in terms of grain yields per hectare are countries in which small-scale, family farming is the dominant mode of production.

However, theoretical models by Feder (*1985*) and Carter and Kalfayan (*1989*), demonstrate that the existence of market imperfections which tend to favour large farms (e.g. capital and insurance markets) may negate the inverse relationship between farm size and productivity. Carter (*1994*) finds that certain financial market disadvantages may render small farms non-competitive. Hence, whereas the small-scale farming strategy holds considerable promise from an efficiency perspective, this does not mean that its implementation is easy or can afford to ignore critical policy issues, such as resolving the usually constrained access of small farmers to credit markets.

Related issues: mechanization, labour organization and farm size

Also underlying the establishment and maintenance of large-scale farms is the misguided perception that there is a relationship between mechanization and large farms. This has been clarified in available literature (*see Johnson & Ruttan, 1994*). Capital intensity is explained by the substitution of capital for labour because of high wages. This substitution process, brought about by changes in relative factor prices (*Peterson & Kislev, 1991*), indirectly cause larger farms. Machinery allows farmers to work progressively larger units of land (*Hayami & Ruttan, 1985*).

In this respect, the work of Brewster (*1950*) on the influence of machinery on farm size is enlightening: mechanization in industry involves stationary machinery which implies that the number of workers can be increased substantially without increasing labour supervision costs. In agriculture, labour and machines are both mobile, making supervision expensive and increasing management costs. In addition, agricultural tasks are sequential in nature due to the annual cycle of production. This limits the opportunities for specialization and division of labour, which creates few advantages to expansion beyond the size of owner-operator.

Available literature clearly demonstrates (*see Berry & Cline, 1979; Binswanger & Rosenzweig, 1986; Binswanger & Kinsey, 1993; Binswanger & Elgin, 1992; Binswanger et al, 1993*) that family farms are generally more efficient and superior to other types of farming because of the way in which labour relations are organized. Family farms, by definition, are farms where the owner is the operator and where his/her family provides the large bulk of the regular labour requirements throughout the year. While the definition of family farms does not exclude the hiring of other people, especially in a part-time capacity when related to seasonal labour, it tends not to rely too much on such behaviour. In addition, in countries where capital is relatively scarce and expensive, the relationship between labour and capital should reflect this. Over-emphasizing modernization, restructuring, mechanization and other similar concepts, implying the use of more capital to labour than that dictated by economic realities, should be discouraged. This all implies farm sizes on the smaller side of the spectrum rather than larger sizes for family farms.

The structure of South African agriculture: issues related to size

Farm sizes in South Africa

Farm sizes in white South Africa began to increase in the 1950s and continued to increase until the 1980s. After steadily increasing until 1971, employment on these farms began to decline. Consequently, it can be argued that scale efficiencies appeared after 1950, and in particular after 1970, and were a main factor behind the steady decline of employment in agriculture *(Van Zyl et al, 1987*). Agriculture was the only major economic sector that experienced an absolute decline in employment between 1951 and 1985 – despite the fact that wages were rising at a slower rate in agriculture than in other sectors. This history suggests that, in South Africa, a number of interventions in the markets for land, labour and capital produced a structure of incentives which induced scale efficiencies, in particular since the 1970s.

From the beginning of the century until the 1950s, the number of farms and the total area cultivated increased, but the average farm size declined. After 1950 this trend was reversed; and farm size grew consistently, accelerating in the 1970s before levelling off in the late 1980s. Because the cultivated area remained the same, the number of farms declined – from 116 848 units in 1950 to 62 084 units in 1990 (*RSA, 1994*). The pattern seems to continue until the late 1980s, although there is some evidence of an increasing differentiation in farm sizes below the 100 hectare minimum which (in some areas) defines a farm in official statistics (*World Bank, 1994*).

Average farm size increased from 738 hectares per farm in 1953, to 867 hectares in 1960, to 988 hectares in 1971, and to 1 339 hectares in 1981, but declined to 1 280 hectares per farm in 1988. From 1955 to 1988, average farm size by province increased from 1 284 to 2 663 hectares per farm in the Cape Province; 471 to 998 hectares per farm in the Orange Free State; 403 to 629 hectares per farm in the Transvaal, and 390 to 609 hectares per farm in Natal. This data shows that the national average hides significant regional variations. In 1988, the median farm size was about 500 hectares, with farms in the high-potential areas significantly smaller. Such qualifications, however, should not distract from the fact that large-scale farms dominate South African agriculture, and that the average size of these farms is extraordinary by international standards.

Evidence of economies of scale

At present, there is mixed evidence for the existence of scale efficiencies in South Africa's commercial farm sector:

- The distribution of gross farm income in commercial agriculture is highly unequal: in 1988, 3 per cent of the farmers earned 41 per cent of the total gross farm income; 26 per cent earned 81 per cent, while the remaining 74 per cent of farmers earned a mere 19 per cent of total gross farm income (calculated from the 1988 agricultural census, CSS, 1993).
- Hattingh (*1986*) reports evidence of a direct relationship between farm size and efficiency in sheep farming in the Karoo and in cattle ranching in north-western Transvaal. Hattingh also reports that efficiency increased between small- and medium-sized irrigated farms at Vaalharts and dryland grain farms in the Orange Free State, before decreasing again on the larger farms (size ranges are not specific).
- Analysing the Department of Agriculture's Production Cost Surveys, Moll (*1988*) finds no significant economies of size[4] in maize-cattle regions (Western Transvaal, North-West Orange Free State and the Transvaal Highveld) or in wheat-sheep regions (Swartland). Using re-tabulated 1983 census data, however, Moll (*1988*) finds economies of size, but only in the maize areas in the 50–300 hectare range.

Conversely, there exists empirical evidence from South Africa to suggest an inverse relation between farm size and efficiency. Statistics from the 1988 Census of Agriculture (*CSS, 1993*) show that 50 per cent of farming units owning only 6 per cent of the farmland, with farm sizes of less than 500 hectares, were responsible for 30 per cent of gross farm income, 23 per cent of net farm income, 32 per cent of capital investment, and 29 per cent of farm debt. The larger farms (1000 ha +) comprising a third (33 per cent) of all farming units, collectively owed more than 50 per cent of the total farm debt. However, these farms were

4 Moll (*1988*) measures economies of size (all factors but operator labour changing) as opposed to economies of scale (all factors changing).

Table 11.1: Factor Intensities in Agriculture, 1988

Farming unit size groups (ha)	Number of farms	Gross margin (R/ha)	Employees (No./1 000ha)	Cash wage per worker (R/1 000ha)	Wages as % of gross (%)	Current expenditures (R/ha)
<2	142	5 096,77	2 779,6	10 534,31	16,4	28 210
2–4	1058	3 421,24	1 082,6	469,13	17,1	8 160
4–9	1525	2 517,84	673,5	84,56	15,3	3 133
9–19	1815	614,38	379,7	25,78	21,4	1 283
19–49	4837	986,41	335,3	6,49	14,0	1 942
49–99	4404	384,23	172,9	2,94	16,8	840
99–199	5690	344,41	133,6	1,35	17,3	680
199–299	4502	170,83	84,0	0,99	16,4	424
299–499	7044	91,45	45,8	0,35	12,5	277
499–999	10926	87,55	27,1	0,13	10,2	177
999–1 999	9230	74,55	17,7	0,08	10,1	117
1 999–4 999	7588	38,46	7,8	0,05	9,6	56
4 999–9 999	2573	18,06	3,0	0,07	8,6	25
10 000 +	1067	11,71	1,4	0,11	8,6	21
Total	62428	50,11	14,3	0,01	11,6	91

Source: 1988 Census of Agriculture data (CSS, 1993).

responsible for 53 per cent of total gross farm income. Table 11.1 provides evidence on this skew distribution.

Christodoulou and Vink (*1990*), based on data obtained from the Central Statistical Service which also covered the existing smallholdings in municipal areas, came to the following conclusions (Table 11.1):

- The gross margin per hectare was R1 514 for small farms (below 500 hectares), R87 for middle farms (500–1 000 hectares), and R36 for farms above 1 000 hectares. Moreover, small farms employed 632 workers per 1 000 hectares, compared to 27 and 29 workers per 1 000 hectares for middle and large farms respectively.
- The cash wage per 1 000 hectares paid by small farms was on average R1 189, compared to 13 cents and 7 cents paid by middle and large farms respectively.
- Farm workers earned 16 per cent of the gross income of small farms, but only 10 per cent and 9 per cent of the gross income of middle and larger farms.

- Smaller farms' total farm expenditures were nearly R5 000 per hectare, whereas middle and larger farms spent only R177 and R55 per hectare respectively.

The comparative efficiency of black, small-scale farming versus white, large-scale farming is very difficult to assess. More than a century of policy interventions has suppressed the profitability of black farming in order to protect white farmers from black competition and to assure the white farm sector of low-wage labour. The only areas where black farming was condoned were the former homelands. Given their location, lack of infrastructure and support services, generally poor soils, and extreme population pressure, it would be unfair to compare small-scale farming in the former homelands with farming in the white areas (see Chapter 3 for a more detailed discussion).

Nonetheless, a few cases exist in which small-scale farmers were given access to support structures roughly comparable to those of their white colleagues. The two case studies presented here in which small-scale, black farmers equalled or outperformed larger, white farms, come from tea and sugar farming.

In the tea industry, the case study illustrates that 'mini-farming' (where an individual leases a small area planted to tea from a tea estate and is remunerated according to the quantity of acceptable tea produced) shows an increase in yields, income and profitability for both the estate renting out the land and the mini-farmers. Compared to ordinary pluckers, mini-farmers obtained yields on their 0,5 hectare plots averaging 23 per cent more than that obtained by the large estate (*Van Zyl & Vink, 1992*).

The same applies to the sugarcane case study in the Eastern Transvaal, where black smallholders obtained 116,8 tons/hectare on their plots of 7,1 hectares (on average), while large-scale white farmers adjacent to these smallholders obtained 102,9 tons/hectare on 68,6 hectares (on average). Total costs amounted to R3 286/hectare for the smallholders and R3 448/hectare for the large-scale farmers (*unpublished data from representative samples gathered by the University of Pretoria*). Both these case studies confirm that, with the same support structures, small-scale farming is at least as efficient as large-scale farming in these specific areas and types of farming.

Evidence on causes of scale efficiency

The official definition of the viable farm in terms of size has had a profoundly negative effect on the relative profitability of farms smaller than the viable size. Given the high levels of official assistance and subsidies to farmers, the viability definition became almost a self-fulfilling prophecy because, under the Agricultural Credit Act, all farms below the viable size were excluded from assistance. Moreover, under the Subdivision of Agricultural Land Act of 1970, it is not possible to subdivide an existing title deed without ministerial approval. Permission is granted only with proof that a reasonable net farm income can be obtained with 'average' management. The subjectivity of this requirement, together with the lending criteria of the official funding agencies, precludes systematic empirical analysis of small farms in South Africa. Yet, it is interesting to observe that despite the lack of assistance for small farmers, official records of deed transfers show that the prices of small parcels of land increased more rapidly than the prices of large parcels since the 1960s.

Ironically, the benchmark for determining farm viability – farm size – has changed over time; during the 1960s and 1970s, expansion and mechanization were considered the solution to remain competitive with non-farm incomes. However, in the 1980s, the high debt loads from capital and land purchases reduced farm profitability and decreased returns to capital-intensive investment. Thus many farms once thought to be viable by the criteria set in the 1970s, were exposed as not viable in the financial crisis of the 1980s.

Farmers themselves seem to view consolidation of farms as a rational economic reaction capturing economies of scale. For instance, Moll (*1988*) reports that of 55 farmers surveyed in the Bredasdorp and Malmesbury regions who had bought land during the previous decade, 35 (or 64 per cent) indicated that they had done so partly to take advantage of size economies.

De Klerk (*1991*) attributes the process of farm consolidation to technical change, namely, mechanization. Generally, consolidation has caused a reduction in farm employment because the new mechanized farm did not need to employ workers from the more labour-intensive smaller farms that were acquired. While seasonal workers bore the brunt of mechanization, permanent workers were most directly affected by consolidation (*De Klerk, 1985*).

Sartorius von Bach, Van Zyl and Koch (*1992*) constructed an index of managerial ability based on indicators such as budgeting and the keeping of records, and found it to be highly correlated both with farm size and total farm income. By evaluating Cobb-Douglas production function coefficients with the managerial ability index included as an input, the authors found significantly increasing returns to scale among 34 farmers in the Vaalharts Irrigation Area. When managerial ability is excluded from the regression, however, results indicate constant returns to size. These results are confirmed by Van Schalkwyk, Van Zyl and Sartorius von Bach (*1993*) using non-parametric procedures to analyse the same sample, and adjusting land size for quality differences. The same patterns hold true for a sample of 100 farmers in the North-eastern Orange Free State.

Groenewald (*1991*) suggests that even beyond the indivisibility of capital and managerial inputs, economies of scale may result from scale efficiencies induced by the existing agricultural marketing system through volume discounts on the purchase of inputs and volume premiums on the sale of outputs. However, he ascribes most of the perceived economies of scale to management, with larger farms having better managers.

Roth *et al* (*1992*) econometrically tested a number of models explaining the reduction in the number of farms between 1972 and 1988. They found the number of farms to be positively correlated with the ratio of real machinery costs to real gross revenue, but negatively correlated with the ratio of farm requisites (mainly non-labour inputs) to output prices. This suggests that scale efficiencies in agriculture are strongly associated with a decline in machinery cost and an increase in the profitability of non-labour inputs. Both correlations suggest that the appearance of scale efficiencies in South African agriculture is rooted in the policy distortions that led to the reduction of the real cost of capital in the agricultural sector.

Chavas and Van Zyl (*1993*), using non-parametric analysis and accounting for quality differences in land, found a highly significant negative correlation between farm size efficiencies and debt burden, while size efficiency and managerial ability were positively correlated. The results show that the issue of scale efficiency is a complex one, and is influenced by a variety of factors of which managerial ability – the basic indivisible input in agriculture – seems to be dominant. A whole range of farm sizes, both in extensive and intensive commercial farming, was found to be scale efficient, depending on how farmers organize their specific variable and fixed input mix, as well as the combination of outputs they produce. Their results are consistent with the findings of Sartorius von Bach and Van Zyl (*1992*) who conclude that better managers have larger farms. It should be noted, however, that in general small farms will require less sophisticated management than large farms, which would explain why Chavas and van Zyl (*1993*) found efficient farms in all size categories. On the other hand, these results can be interpreted to mean that farm size is not really the central issue, but rather managerial ability.

Efficiency in the former homelands

The poor natural resource base and the continuous build up of demographic pressure since the beginning of this century through the racially segmented land, commodity, input and financial markets have made the former homelands 'functionally urban' or 'rural dormitories'. Given these conditions, future migrants from rural to urban areas are expected to come from the former homelands; but even the steady out-migration from these areas will not exceed population growth (*Urban Foundation, 1990*).

It is difficult to get a grip on efficiency issues in the former homelands; the existing data base is very weak and probably underestimates the importance of agriculture, even though it is extremely constrained by the overcrowded and poor resource base. For example, official estimates of homeland yields are consistently lower than case study data (for example, *Cairns, 1990*).

However, the available case study material does suggest that relatively few farmers are engaged in full-time agriculture in the former homelands. According to Nicholson and Bembridge (*1991*), the majority of households in South Africa's former homelands do not have enough land to provide for their subsistence needs. Thus most rural households engage in farming only part-time, and most of their output is kept for consumption at home.

Several studies have also explored the relationship between farm size and efficiency in the former homelands, where farms are in general much smaller than those in the commercial sector. Nieuwoudt (*1991*) notes that small farmers may use land much more intensively than do large farmers. Latt and Nieuwoudt (*1988*) used data on 140 households in the Umbumbulu district of KwaZulu to conduct a discriminant analysis of input use. They found that farms of less than 1 hectare applied inputs much more intensively than farms larger than 1 hectare; thus they suggested that smaller farms may maximize returns to land (their scarce resource), while larger farms maximize returns to labour or capital.

Moreover, even in comparison with the commercial sector, homeland agriculture is sometimes found to be more efficient. Case study material shows that dryland cotton smallholders in KwaZulu are more viable than large farmers (*Wheeler & Ortmann, 1990:251*). These trends are also confirmed by the two case studies mentioned earlier where small farmers, given the same support structures in tea and sugar farming, have done as well or outperformed larger farms (*Van Zyl & Vink, 1992*).

In a study of 60 farmers in Gazankulu, half of whom were identified by extension staff as commercially-oriented, Nicholson and Bembridge (*1991*) found that 'commercial farmers' cultivate more land than 'typical subsistence farmers' (12,2 hectares versus 2,3 hectares), own more cattle and equipment, are better educated, and more likely than their neighbours to keep records. Similar findings are reported for Transkei (*Bembridge, 1991a*), KwaZulu (*Bembridge, 1991b; Wheeler & Ortmann, 1990; Nieuwoudt & Vink, 1989*) and Lebowa (*Van Zyl & Coetzee, 1990*). The authors of these latter studies also note the importance of income from non-agricultural sources, such as wage employment, in providing working capital for the purchase of seeds, fertilizers and other production inputs.

Given the lack of rights to buy and sell land, much of the increase in operational holdings by successful farmers represents formal or informal acquisition of temporary use rights held by others. Such transfers of temporary use rights may be consistent with improvement in efficiency, because they combine idle land with surplus labour and other factors. Such transfers are limited by the extreme population pressure under which the tenure arrangements operate. Landowners fear they will lose long-term rights to their land if they permit others to use it.

Synopsis

The evidence on the farm size-efficiency relationship in South African agriculture is mixed. However, much of the evidence on scale efficiencies in South African agriculture cited above, with only a few exceptions, should be interpreted with extreme care because data was not standardized for differences in land quality or labour productivity, or for the particular commodity mix of farms and, in some cases, inappropriate analytical methods and measurement variables were used.

Evidence on scale efficiency in the homelands[5]

The neglect of agriculture in the former homelands extends to the availability of information. The data for commercial agriculture is on a par with that for developed countries, and includes detailed time series for inputs and outputs dating from the end of the Second World War. But,

5 This section builds on analyses conducted by Piesse, Sartorius von Bach, Thirtle and Van Zyl (forthcoming).

for the former homelands, there is a general paucity of data with only sample survey data being really reliable. Such data, collected for the Development Bank of Southern Africa, is analysed in this section to establish farm size-efficiency in these areas. This data is sufficient for estimation of efficiency of individual farms, which makes it possible to compare regions.

The methodology, including that of separating total from technical and scale efficiency, is discussed after the data and the last sub-section reports, and interprets the results, comparing efficiency levels for three of the former homelands.

The data

The data was collected from three of the former homelands – KaNgwane, Lebowa and Venda, all in the northern or eastern parts of Gauteng. Detail on the physical characteristics of the regions and the role of farmer support programmes (FSPs) can be found in Kirsten (*1994*).

A total of 23 FSPs were established by the Development Bank of Southern Africa (DBSA) in 1987 to provide participating farmers with services and training programmes, and to identify and address the major constraints to agricultural production (*Van Zyl, Fényes & Vink, 1992*). All of the farmers in the sample used in this analysis benefited from the local FSP, either as current members or as previous members, and have benefited from some level of education and training. Many aspects of the programmes are common to the three homelands, but one difference between the regions is the institutional arrangements for the provision of FSP services, since these are decided by the relevant homeland development corporation. However, the FSPs are similar to the extent that they all encourage the adoption of modern technology, including higher yielding varieties and chemical fertilizer (*Singini, Sartorius von Bach & Kirsten, 1992*). Extension and training relevant to new varieties and methods are also available.

The FSP data for KaNgwane, Lebowa and Venda covers the 1991 harvest which is regarded as a normal year with respect to rainfall and yields. The original sample groups are shown in Table 11.2, but attempts to include numerous minor crops that varied between regions were hampered by the lack of price data, making comparisons difficult.[6]

6 This data includes a range of socio-economic variables since the original objective was to monitor farmer attitudes to the FSP initiatives. Only agricultural production is inves-

Table 11.2: Population of Farms in Sample			
Area	Total number of farm households surveyed	Households producing only maize	Current members of FSP
Venda	60	30	27
Lebowa	84	64	56
KaNgwane	111	80	75
Total:	255	174	158

Thus, the only farms retained were those where maize production accounted for all the land utilization, and where inputs were maize-specific. This is considered to be acceptable since maize is by far the most important crop in the regions studied. Not surprisingly, labour was not reported as a crop-specific input. Therefore, a labour variable was constructed, using the total area cultivated and the total number of labour days spent on agricultural activities, with a distinction between family and hired labour. The land variable was hectares planted with maize, with some adjustment for land quality differences. These were small both within regions and between regions, which is the major reason for selecting these three surveys for the analysis. Seed, fertilizer and other variable inputs were reasonably well recorded. In cases where seed input was very low or zero, and still some output was reported, it was assumed that seed had been kept from previous years (subsequent inquiries indeed showed this to be the case). Most farms reported using hybrid seed, except for Lebowa where a majority of farms used traditional varieties. Output was maize production in metric tons.

Thus, the farmers in the sample are maize producers, some of whom follow the FSP recommendations and use fertilizer and hybrid seeds, while traditional farmers do not use modern inputs (although also belonging to the FSP). Both for outputs and some inputs – land and family labour – there is no reliable price information, and much of the maize is not sold, but consumed within the household.

Table 11.3 illustrates the differences in performance of the three regions in the sample. KaNgwane has the largest farms and the highest average output, while the farms in Venda produce far less output. This is

tigated in this paper which is not intended, in any way, to be an evaluation of the FSPs. Special care was taken not to repeat the mistakes that abound in the farm size-efficiency literature (*Binswanger et al, 1993*), also in South Africa.

Table 11.3: Mean Output and Selected Inputs in Maize Production, 1991

Region	Output (kg)	Land (ha)	Yield kg/ha	Seed/ha (kg)	Fert/ha (kg)
KaNgwane	4 317	3,93	986	18,91	161,1
Lebowa	2 360	2,13	1 399	9,12	143,6
Venda	320	1,15	273	11,08	112,3

partly because the farms in Venda are smaller, but also because yields are far lower. Lebowa has the highest average yields, followed by KaNgwane. An important point is that the production potential of these three areas – as defined by soil quality and rainfall in the 1990/91 production season – was essentially the same.

These basic statistics are useful, with yield being a partial productivity measure, when considering the relationship of output to the input of land alone. On the other hand, measures of total factor productivity relate output to the aggregate of all inputs, giving a better indication of the overall efficiency of the system, and allowing the more advanced producers to be compared with the traditional farmers who use little in the way of modern inputs. Total factor productivity can be measured using the programming methodology discussed in the next sub-section.

The measurement of productive efficiency

The method used to measure efficiency in the former homeland farms is Data Envelope Analysis (DEA).[7] This uses a linear programming procedure to minimize inputs per unit of output, to determine the frontier of best-practice farms, and then to determine the efficiency of all the production units relative to the frontier. This estimation approach is preferred to econometric modelling where the techniques impose a functional form, and where having a considerable number of zeros for some inputs can cause problems. The lack of price information limits econometric analysis to the estimation of production functions, precluding dual forms, such as the cost and profit function. Thus, the non-parametric efficiency measurement approach that was introduced by

7 See Chavas and Aliber (*1992*) and Piesse *et al,* (forthcoming), amongst others, for descriptions and applications of the DEA methodology in this context.

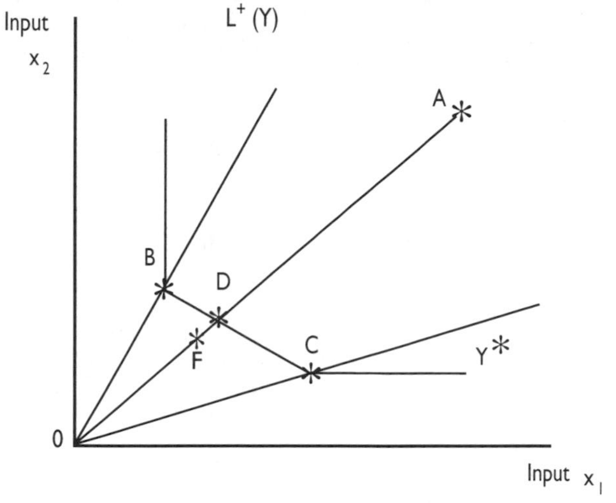

Figure 11.1: Farrel Efficiency Measurement

Farrell (*1957*), is used here largely because it does not require prices and leads naturally to simple efficiency comparisons. The efficiency frontier is expressed in terms of minimizing the input requirements per unit of output. The Farrell technical efficiency measure is defined so that the isoquant, which is the locus of the efficient points, uses the minimum required inputs to produce the unit level of output. The efficiency of the other farms is measured relative to this isoquant, as Figure 11.1 shows. In Figure 11.1, the efficiency frontier unit isoquant is determined by the linear combination of just two efficient farms, B and C, and is labelled Y*. The efficiency of a farm such as A, that is not on the frontier, is measured by the ratio OD/OA, since OD is the vector representing the lowest mix of inputs which farm A could use and still reach the isoquant, using its own factor combination.

The efficiency measures for 1991 which result from this analysis[8] are reported in Table 11.4 of the next section. This is an assessment of aggregate, or total efficiency, and includes both technical and scale effects, although these elements will be considered separately next. Since part of the current land reform policy in South Africa is based on

8 The equations for the programming problem are fully stated in Piesse *et al* (forthcoming).

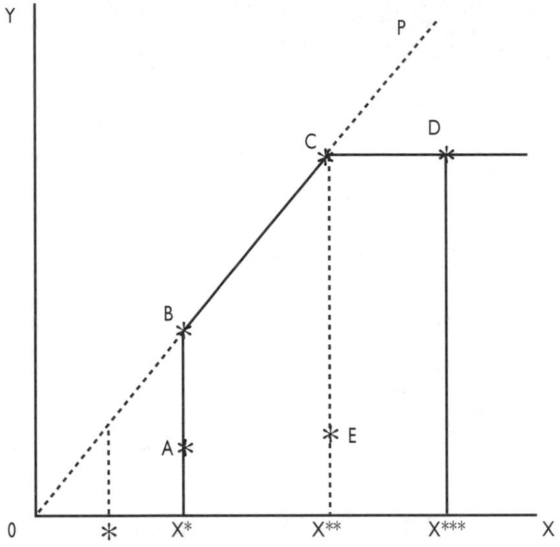

Figure 11.2: Decomposition of Technical and Scale Efficiency

the potential productive efficiency of small farms, it is important to measure the effects of farm size. Following Fare, Grosskopf and Lovell (*1985*), Figure 11.2 shows the effect of this decomposition.

In Figure 11.2, the constant returns to scale (CRTS) technology is denoted by the linear total product curve, OP, from the origin, through the efficient production units B and C. Units A and D, in this example, are inefficient as they are below the CRTS frontier. When non-constant returns to scale are allowed for, the frontier is concave, and the input-output combinations A, B, C and D are all technically efficient.

When technical efficiency is extracted from total efficiency, only the scale effect remains. Thus, farm A is scale inefficient by OX*/OX, due to being too small, but is technically efficient. Farm D is similarly technically efficient, but is too large and is scale inefficient by OX***/OX**. Finally, farm E is technically inefficient by OX**/OX* and scale inefficient by OX*/OX, giving a total level of inefficiency, relative to the CRTS frontier, of OX**/OX.

Results and interpretation

The objective of this section is to measure total efficiency, and to separate the scale effects resulting from farms being an inappropriate

Table 11.4: Total, Technical and Scale Efficiency Levels – 1991 (Frontier = 1,00)

	KaNgwane			Lebowa			Venda		
Item	Total	Tech	Scale	Total	Tech	Scale	Total	Tech	Scale
Mean eff. by region	0,358	0,703	0,487	0,427	0,794	0,547	0,476	0,671	0,698
# Efficient	6	35	6	4	31	5	6	12	7
% Efficient	7,52	43,8	7,52	6,3	48,4	7,8	20	40	23,3
Mean eff. pooled Sample	0,250	0,637	0,394	0,372	0,828	0,485	0,069	0,740	0,113
# Efficient	3	35	3	1	38	1	0	11	0
% Efficient	3,75	43,75	3,75	1,56	59,38	1,56	0	36,7	0

size from the technical efficiency of farmers. The top row of results in Table 11.4 shows the mean efficiency levels for the three homelands for the 1991 cropping year, decomposed into total, technical and scale efficiencies.

The results are in agreement with the previous analysis of the survey data. These efficiency levels of the farms are calculated with respect to the best-practice farms within each region, which means that inter-regional efficiencies are not compared. For KaNgwane, the mean level of total efficiency relative to the best practice farms is 35,8 per cent, which is lower than for Lebowa (42,7 per cent) or Venda (47,6 per cent). As the most advanced and commercialized region, KaNgwane has greater variation in input levels than the two other regions, as Table 11.3 suggested. This is particularly true of farm size; the small farms in KaNgwane are scale inefficient, relative to the larger units, giving the lowest average scale efficiency of 48,7 per cent. Conversely, Venda, which has the least variation in farm size, has a mean scale efficiency of 69,8 per cent, which more than compensates for the low level of technical efficiency (47,6 per cent) in the total efficiency calculations.

The next two rows show the number and percentage of farms that are on the efficiency frontier. For all three regions, at least 40 per cent of the farms are technically efficient, but only a little over 7 per cent are large enough to be scale efficient in KaNgwane and Lebowa, whereas 23,3 per cent are scale efficient in Venda. Farms with a wide range of characteristics determine the frontiers. For KaNgwane, there are six efficient farms which range in size from one to twenty hectares. Yields range from 0,19 to 4,6 tons per hectare, labour per hectare from 0,05 to 2 persons, seed from zero to 18 kgs per hectare and fertilizer from zero to 200 kgs per hectare. The other regions have almost as great a range of efficient farms.

Figure 11.1 shows why this occurs. The farms that define the isoquant in this figure are those that determine the extreme values of the factor ratios (OB and OC). Other farms will be included and, if there are enough efficient farms (such as F in the figure) with more moderate factor ratios, then the isoquant will begin to approximate the smooth neoclassical shape (the dotted line BFC). However, the extreme values will always play an important role in DEA, in contrast to linear regression analysis which tends to identify the characteristics of the average farm.

The isoquants for the three homelands have several dimensions, but the outstanding feature of the best practice isoquants can be described in the two dimensions of Figure 11.1. The two types of farm households that define the frontier may be called traditional and modern. If X1 is taken to represent the traditional inputs (land and labour) and X2 is modern inputs (hybrid seed and fertilizer), then the traditional farms can be represented by observations such as C in the figure, and modern farms that have followed the FSPs advice, by points like B, with a much higher ratio of modern to traditional inputs.[9] Thus, the DEA does not provide direct comparisons of the farms that have adopted the FSP practices and those that have not. Both are technically efficient in Farrell's terms, and further comparisons must be made on the basis of allocative efficiency which requires price information.[10] If an isocost constraint were to be added to Figure 11.1 (ignoring the hypothetical point F and the dotted frontier), C would be the allocatively efficient point[11] if the slope were less than that of the line BDC, meaning that modern inputs are expensive, relative to land and labour. If modern inputs were

9 In fact, the survey results are more extreme than this example; the most traditional farms use no improved seed and no fertilizer.

10 The difficulty is the lack of prices for family labour and for land. There are prices for seed and fertilizer.

11 The *economically* efficient point on the technically efficient isoquant.

cheaper, relative to traditional inputs, then the isocost line could be steeper than BDC, and B would be the economically efficient point. So, although technical efficiency can be analysed without price information, some crucial economic comparisons cannot be made without price data.

However, DEA does go further in accounting for the efficiency differences by separating the effects of farm size from pure technical efficiency. It seems that all three regions could considerably improve their efficiency levels if there were not constraints on farm size.

The bottom half of Table 11.4 compares between regions, by pooling the data and estimating inter-spacial efficiencies. In the fourth row, the efficiencies are calculated relative to a grand efficiency frontier, for the pooled sample of all three regions. There is no technical reason for not pooling the data for the three regions and measuring the efficiencies for the regions relative to a meta-efficiency frontier calculated from all the observations.[12] The results of this approach show that fewer farms lie on the combined frontier than on the separate regional frontiers. This is inevitable since, because the sample size is increased by pooling, a farm's efficiency can only decrease as its comparison set is augmented by new observations (*Nunamaker, 1985*).

The efficient farms are thus a subset of the regionally efficient units, with only three farms in KaNgwane and one in Lebowa remaining on the frontier. The three KaNgwane farms that were on the regional frontier and are not on the pooled frontier, were traditional producers, using few modern inputs. These are dominated by the one Lebowa farm on the pooled frontier which claims to use no improved seed and no chemical fertilizer.[13] None of the three efficient KaNgwane farms are current FSP members because they have evolved beyond the need for the FSP and are all large, commercial undertakings. However, they are following the FSP recommendations in that they reach the frontier by achieving high yields, using heavy applications of modern inputs.

12 Differences in land quality can be a problem when estimating meta-functions, and should be accounted for. In this case, the tree sample areas are relatively homogeneous both within and between regions. Furthermore, for the production year under consideration, rainfall and climatic conditions were, to a large extent, similar. This is the reason for analysing these three surveys.

13 The reader does not need to believe that the figures for this particular farm are strictly accurate. It may, or may not, be possible to get 3 200 kgs of maize per hectare without modern inputs in the most fertile parts of Lebowa; but if this farm were removed from the sample, there are several others very much like it, so the results are substantially unchanged.

The most striking feature of the pooled efficiency results is the scale efficiencies. Whereas Venda appeared to have the highest level of scale efficiency, this result depended on the fact that all the farms are small. Once the small farms in Venda are compared with the larger units in the other two regions, none are scale efficient and mean scale efficiency falls to 11,3 per cent, giving a mean total efficiency level of only 6,9 per cent. Thus, these results suggest that the small farms in Venda, which average only 1,15 ha, are too small to be viable.

The input-based DEA analysis measures efficiency relative to a best practice isoquant, as shown in Figure 11.1 but, as Ali and Seiford (*1993*) explain, this problem should be viewed as the first stage of a two stage model. Farms may be efficient in terms of being on the frontier, but one or more variables may be slack. If a variable is slack, it is not acting as a constraint on production in the programming problem so, to be fully efficient, a farm should have no slacks. Of the farms on the efficiency frontiers for the three regions, all but one are fully efficient in the sense of no variables being slack. For the farms that are not on the frontier, all have one or more slack variables, so the slacks provide an indication of the inputs that are in excess supply and those that are effectively constraining production. The results for Lebowa are perhaps the most enlightening; land is the main constraint, effectively limiting output for the vast majority of the farms while, at the other extreme, fertilizer in this sense was surplus to requirements for a large number of the farms. For Venda, seeds appear to be a binding constraint for the majority of farms, while a large number had an excess of fertilizer. Land appears to be slack for many farms, but this result should be disregarded, due to the problem of lack of variance of area planted in Venda. KaNgwane is far more mixed, with land being the most common constraint on production, and labour the least common.

Conclusions

This section uses recent farm-level data to study the regional differences in maize production efficiency in the former homelands. Non-parametric techniques allow estimation of total productive efficiency in the absence of prices, or when this data is unreliable. Regional differences do occur, and may be due to land quality variation, but the scale inefficiencies noted in Table 11.4 have implications for addressing the skewed distribution of land ownership in South Africa (*Van Zyl, Van Rooyen, Kirsten & Van Schalkwyk, 1994*). Relatively large efficiency gains can be achieved by redistributing other land to some farmers in order to increase farm size in the homelands.

Economies of scale in commercial agriculture

As noted already, the majority of the previous studies on the farm size-efficiency relationship in South Africa are flawed for a variety of reasons; therefore the results are not reliable. In particular, the studies generally suffer from the following shortcomings: only a minority of the studies adjust farm size for quality differences in land and other inputs; most of the studies use physical yields of specific crops or the value of agricultural output per unit of operated area, both imperfect (and at best only partial) measures of efficiency; differences in operational holding size and ownership holding size are sometimes not accounted for; and, managerial inputs from the farmer and his/her family and family labour, have not been included.[14]

This section makes use of three different methodologies to determine the farm size-efficiency relationship in commercial agriculture: first, total factor productivity differences between small and large farms are determined; second, non-parametric Data Envelope Analysis (DEA) is used to estimate scale efficiencies; and third, following the suggestions of Binswanger *et al* (*1993*), regression analysis is used to test the farm size-productivity relationship.

Data

The data used in these analyses comes from farm surveys conducted by the Department of Agriculture's Directorate of Agricultural Economics over the period 1974/75 to 1990/91. Farm surveys, covering a representative sample of between 65 and 85 individual farmers, were conducted in each of the six major grain production areas of South Africa. Two regions were surveyed per annum, implying that each region was surveyed every three years. These six areas involve rain-fed agriculture; subsequently, an irrigation area was also included in the analysis. The regions included in the analysis are representative of the relatively medium and high potential agricultural areas of South Africa, excluding perennial crops. More than 80 per cent of all maize, wheat and other grain are produced in these areas, while livestock (dairying, beef cattle and woollen sheep) is also important in most areas. Table 11.5 provides more information on the surveys included in the analyses. It was selected to represent all the regions in poor, normal and good rainfall years; thus, it selected years during the period 1974 to 1991.

14 See Binswanger *et al* (*1993*) for a discussion of these problems, as well as appropriate measures of farm size-efficiency.

The data from these surveys specifically allows for the elimination of the problems in previous studies. In particular, farm size is adjusted for differences in land quality within regions by using land value to normalise areas; differences in operational holding size and ownership holding size are incorporated into the analysis; and family labour is considered. Another important point is that, within a specific region, all farmers essentially face the same prices because they buy from the same input suppliers, and output markets for most commodities are controlled. This implies that monetary values of outputs and inputs (revenues and costs in the relevant categories) can be treated as quality adjusted quantities,[15] which greatly enhances the reliability of the analysis because it also normalizes input and output quantities by eliminating the effect of quality differences. The opportunity cost approach was used to derive the value of family labour.

Table 11.5: Surveys Included in the Farm Size-efficiency Analyses of Commercial Farming

Region	Type of farming (predominant)	Year covered by survey	Number of farmers surveyed
Eastern Free State	Summer rainfall (mixed): maize, wheat, cattle, sheep	1979/80 1982/83 1985/86 1988/89	92 83 76 72
Transvaal Highveld	Summer rainfall (mixed): maize, sorghum, cattle, sheep	1974/75 1983/84	71 77
Western Transvaal	Summer rainfall (grain): maize, sunflower, cattle	1981/82	78
North-western Free State	Summer rainfall (grain): maize, wheat, sorghum	1979/80	87
Ruens	Winter rainfall (mixed): Wheat, sheep, dairying	1978/79 1987/88	69 77
Swartland	Winter rainfall (mixed): Wheat, sheep, dairying, beef	1983/84	82
Vaalharts	Irrigation (annual crops): wheat, cotton	1990/91	34

All analyses were conducted separately for each region/survey. Because the analysis implicitly neglects possible production uncertainty (for example, due to weather effects), the underlying assumption is that all farmers within each survey face similar production uncertainty. This seems to be appropriate, given that the analysis is conducted for a given production year and one relatively homogeneous region at a time.

Table 11.6 provides a summary of the size characteristics of the farms in each of the surveys. From this information, it is clear that the surveys cover a relatively large range of farm sizes. While relatively small farms are also part of the data set, the average farm size indicates that the farms are in general large, specifically relative to world standards. The median farm size is smaller than the average in all the data sets, indicating a positively skewed size distribution.

The final data for each farm in the different samples involves inputs and outputs. These were aggregated to give two output series – crops and livestock, and seven input series – land, buildings, livestock and machinery represented the stock inputs, while labour, management (including family labour) and variable inputs[16] represented flows. All quantity measurements used in the analysis were annual flow variables. The stock variables were transformed into flow variables by calculating the equivalent annuities based on the relevant interest rate for that period and region, the average useful life of the particular assets, and the applicable tax rate.[17] Thus, the analysis presented below measures all inputs and outputs as annual flows expressed in monetary values.

15 This amounts to assuming that the corresponding implicit price indexes are unity. This approach has the advantage of being empirically tractable. Although it allows for price variation across years and areas, it has the disadvantage of neglecting price variations across farms within any particular survey. While the intuition is that these variations are small or even negligible, they cannot be ruled out. The 'rule of one price' (*Chavas & Aliber, 1992*), for example, does not take into account different transaction costs or market failures. However, the assumption that all farmers within a survey face the same prices seems too reasonable given the nature of the farm support system in these areas. An additional, but related point is that the 'rule of one price' implicitly accounts for commodities which are not of a homogeneous quality. Different farmers may face different prices because they purchase inputs or sell outputs of different quality. By using the monetary values of input and output as quantities, there is an adjustment for these quality differences, with an implicit assumption that the markets work fairly well.

16 Variable inputs represented all the other inputs, including seed, fertilizer, purchased animal feed, chemicals, etc.

Table 11.6: Summary of the Size Characteristics of Farms Analysed (adjusted ha)

Region	Year	Farm size characteristics (ha)–adjusted for quality differences			
		Average	Median	Maximum	Minimum
Eastern Free State	1988/89	993,2	763,0	3 418	32
	1985/86	1 375,4	943,7	9 221	108
	1982/83	1 154,8	885,5	4 287	41
	1979/80	1 019,5	860,4	2 504	162
Transvaal Highveld	1983/84	1 101,2	933,7	3 394	178
	1974/75	663,4	464,9	3 716	107
North-western Free State	1979/80	865,4	767,8	2 355	158
Western Transvaal	1981/82	474,6	361,1	1 461	118
Ruens	1987/88	1 501,0	1 167,3	6 638	230
	1978/79	1 435,3	1 054,2	4 706	187
Swartland	1983/84	793,4	704,5	2 675	259
Vaalharts	1990/91	50,4	50,4	97	17

Total factor productivity by farm size category

Total factor productivity (TFP) for different farm size categories is clearly a superior indicator of the farm size-efficiency relationship when compared to partial indicators, such as physical output or value of agricultural output per unit of operated area, since it fully accounts for differences in labour and input use. In this sub-section, TFP values for different farm size categories are compared for each of the surveys in Table 11.5. The Tornquist-Theil Index was used to calculate the comparative TFP index, while the farm with the highest TFP – the most efficient

17 To convert the stocks, namely land, buildings, livestock and machinery, into annual flows, the discount rates for these inputs based on the economic rate of depreciation (5 years for machinery and 20 years for buildings), the national price indices, the interest rate on the relevant annuities and the pertinent tax rate, were calculated and multiplied by the market value of each asset.

Table 11.7: Relative Total Factor Productivity and Labour/ Machinery Indices for Different Farm Size Categories*

Region	Year	Total factor productivity**			Labour/machinery ratio**		
		Small	Medium	Large	Small	Medium	Large
Eastern Free State	1988/89	129	126	100	192	113	100
	1985/86	115	107	100	136	116	100
	1982/83	104	101	100	115	106	100
	1979/80	102	99	100	129	98	100
Transvaal Highveld	1983/84	111	104	100	138	128	100
	1974/75	113	110	100	115	110	100
North-western Free State	1979/80	117	111	100	135	95	100
Western Transvaal	1981/82	103	91	100	122	91	100
Ruens	1987/88	128	110	100	125	97	100
	1978/79	112	110	100	132	92	100
Swartland	1983/84	106	102	100	118	104	100

Note: * Three farm size categories were defined for each data set: small represents the smallest third of the farms; medium represents the middle third of the farms; and large represents the largest third of the farms.

** TFP index and labour capital ratio of large farms are the norms (100) against which the other size categories were compared.

farm – was used as the reference point in these calculations.[18] The methodology in constructing the TFP index is described in detail in Thirtle *et al* (*1993*). Table 11.7 presents the results.

Both the results on total factor productivity and the labour/ machinery ratio per farm size category are enlightening when considering land reform (Table 11.7). Within the sample of relative large commercial farms, covering a range of farms sizes which all depend heavily on hired labour, the results are clear:

18 See Ball, Bureau and Butault (*1994*) for a review of the properties and recommendations on the selection of different index numbers based on the axiomatic and economic approaches. Following from this, the Tornquist-Theil methodology is appropriate for this analysis.

- It establishes that the negative relationship between farm size and efficiency also applies to South African commercial farming areas, in spite of a history of distortions and privileges to these farmers which particularly benefited the larger ones. Without exception, the relative TFP index of the smallest third of farms is higher than that of the largest third of farms. Efficiency differences are highest in the Eastern Free State for 1988/89 and 1985/86, where the small farms respectively performed 29 per cent and 19 per cent better than the large farms, and the Ruens for 1987/88, where small farms fared 28 per cent better. While, in most cases, these differences are not statistically significant at the 10 per cent level (with the exception of the three cases cited above) due to the wide variation of results between farms within a particular region, smaller farms are in general more efficient than larger farms.
- It seems that this negative relationship became more accentuated after 1985 when the movement towards the removal of distortions and the abolition of privileges to larger farms started taking effect. The three data sets covering the period after 1985 all yielded statistically significant differences in efficiency between small and large farms (at the 10 per cent level), while all the data sets covering farm operations before 1985 yielded statistically insignificant differences (at the 10 per cent level). This aspect needs further investigation to fully confirm these observations. However, the result is fully compatible with prior expectations.
- Smaller farms consistently have a higher labour/machinery ratio than larger farms in all the areas and for all the periods covered, indicating that they are relatively more labour intensive. Differences between these ratios are statistically significant between small and large farms for most of the areas at the 10 per cent level of significance (with Eastern Free State in 1982/83, Transvaal Highveld in 1974/75 and Swartland in 1983/84 being the exceptions).

Thus, the conclusion is that, in general, smaller farms are not only more efficient than their larger counterparts, but are also relatively more labour intensive in their mode of production. However, these general results derived from averages within groups mask the wide variability between specific farms. Figures 11.3 and 11.4 provide an indication of this variability, respectively in the Ruens (1978/79) and the North-western Free State (1979/80). From these figures, it is obvious why efficiency differences between small and large farms were not statistically significant for these two areas.

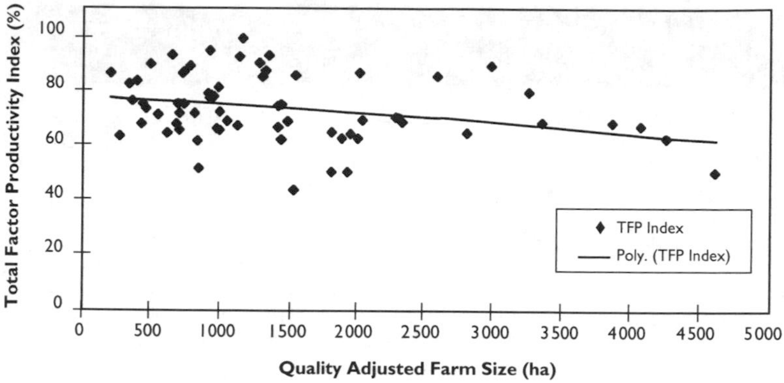

Figure 11.3: Farm Size-Efficiency in the Ruens, 1978/79

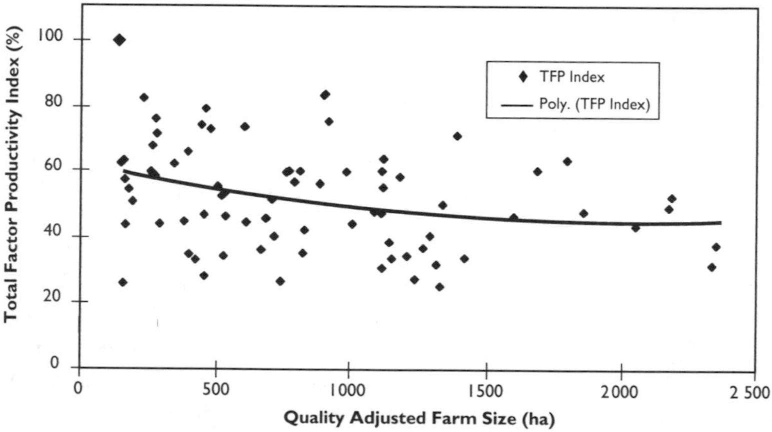

Figure 11.4: Farm Size-Efficiency, North-western Free State, 1979/80

Using market prices to measure productivity assesses differences in private efficiency, while the use of social opportunity costs as a measure eliminates the impact of distortion and measures differences in social efficiency. Few studies, none of them in South Africa, have made this distinction in the analysis of the farm size-efficiency relationship. During the period under consideration, the price of capital was distorted by several factors, including tax benefits and interest rate subsidies. This contributed, amongst other things, to over-capitalization of specifically larger farms (see Table 11.7). On the other hand, output prices were also

Table 11.8: Social Relative Total Factor Productivity for Different Farm Size Categories*

Region	Year	Average social/private TFP ratio	Social total factor productivity**		
			Small	Medium	Large
Eastern Free State	1988/89	0,86	138	133	100
Ruens	1987/88	0,91	135	135	100
Western Transvaal	1981/82	0,78	118	118	100
Eastern Free State	1979/80	0,75	121	97	100

Note: *Three farm size categories were defined for each data set: small represents the smallest third of the farms; medium represents the middle third of the farms; and large represents the largest third of the farms.

**TFP index and labour capital ratio of large farms are the norms (100) against which the other size categories were compared.

distorted due to protection and market price support.[19] Most of these privileges went to relatively large farms. Thus, accounting for these distortions is important when looking at farm structure and production relations from a social point of view.

Social efficiency estimates were calculated for four of the data sets analysed above, namely the Eastern Free State (1988/89) and Ruens (1987/88), as well as the Western Transvaal (1981/82) and Eastern Free State (1979/80). These four surveys respectively represent those with the two largest differences and the two smallest differences in the average TFP between small and large farms in Table 11.7. Alternatively, they can also be regarded as representative of the beginning and the end of the decade of the 1980s – thus pre-reform and just after the first reforms started taking effect. They are also representative of all the areas in the analysis. Social opportunity costs for capital, labour, variable farm inputs and farm outputs (both crops and livestock) were obtained from previous studies and were incorporated into the analysis.[20] The TFP analysis of each farm was repeated using these social opportunity costs rather than the actual private costs. The results of these social efficiency estimates are summarized in Table 11.8.

19 See Chapter 6 for a synopsis of these policies and their effects.

20 *See*, for example, Helm and Van Zyl (*1994*), Van Heerden and Van Zyl (*1992*) and Meyer and Van Zyl (*1993*).

The results from the social TFP analysis should be interpreted with care. Farmers react to the incentive structure facing them and, if capital is relatively cheaper, they should use more of it and vice versa. For this reason, the social TFP calculations are more indicative of the distortions than the actual social costs or efficiency losses. Strictly, changing the values from private to social prices does nothing to the physical input and output ratio, and the TFP stays essentially the same, although the weighting of the inputs and outputs changes. However, the point here is to determine to what extent farm size influences the farmer's ability to capture benefits and use the structure of incentives. The results obtained from the social TFP analyses, which are summarized in Table 11.8, indicate that:

- Average social TFP is lower than average private TFP in all the regions. The difference is much more accentuated at the beginning of the 1980s than later in the decade when some of the privileges were already removed. The reason for this is that, because all farmers face the same prices, the value of outputs and inputs can be treated as quality adjusted quantities. While these differences are meaningless in terms of efficiency, they indicate to what extent policies have been distorted.
- Larger farms are less efficient relative to smaller farms when social opportunity costs are used to determine the value of input and output instead of actual market prices. The reason for this stems mainly from the differences in the relative importance of labour and capital in the input mix of large and small farms (see Table 11.7). The value of output of small and large farms is generally affected in a similar manner because the ratio of livestock to crops does not differ significantly between these groups, but the input mix varies considerably, with large farms being relatively more capital intensive and small farms being relatively more labour intensive. Because the social opportunity cost of labour is lower than the actual wage rate (due to massive rural unemployment), and the social opportunity cos for capital is higher than the subsidized prices farmers face, the total value of inputs increases more for large farms than for small farms.
- The positive effects of removal of distortions on small farms (or negative effects on large farms) are relatively greater where the distortions have been large. For example, the analysis shows that small farms gain more in relative efficiency (compared to the private analysis in Table 11.7) under such situations.

Non-parametric efficiency estimation using Data Envelope Analysis (DEA)[21]

The analysis of efficiency has fallen into two broad categories: parametric and non-parametric. The parametric approach relies on a parametric specification of the production function, cost function or profit function (*see*, for example, *Forsund et al, 1980; Bauer, 1990*). Alternatively, production efficiency analysis can rely on non-parametric methods (*see*, for example, *Seiford & Thrall, 1990*). Building on the work of Farrell (*1957*) and Afriat (*1972*), the non-parametric approach has the advantage of imposing no *a priori* parametric restrictions on the underlying technology (*see*, for example, *Färe et al, 1985*). Also, it can easily handle disaggregated inputs and multiple output technologies. As the non-parametric approach develops, its applications to production analysis have become more refined (*Chavas & Aliber, 1993*) which provides some new opportunities for empirical analysis of economic efficiency. This sub-section uses this non-parametric or DEA approach to estimate the farm size-efficiency relationship in the Eastern Orange Free State (1979/80, 1982/83, 1985/86 and 1988/89) and the Vaalharts irrigation area (1990/91). In particular, the scale efficiency of each of the farms is determined relative to that for the whole data set.

Non-parametric scale efficiency (SE) measures were developed in response to earlier work on technical efficiency (TE) and allocative efficiency (AE) (*see Baumol et al, 1982*). While TE and AE take the output level as given, SE is concerned with choosing the output level itself. The key question becomes whether firms are operating under decreasing, increasing or constant returns to scale. The SE takes on values between 0 and 1, where SE=1 identifies scale efficiency under (local) CRTS. Finding SE<1 means that the firm is not scale efficient, i.e. does not produce at a scale exhibiting local CRTS. In this context, (1–SE) can be interpreted as the relative decrease in average cost obtainable from rescaling outputs to the point of (locally) constant returns to scale. Tables 11.9 and 11.10 provide the results for the *Eastern Orange Free State* (all four surveys).

To a large extent, the results are similar to those obtained with the TFP analyses. In addition, the methodology used here isolates scale

21 The non-parametric analysis of efficiency benefited from discussions with, and suggestions by, Jean-Paul Chavas from the University of Wisconsin-Madison. The GAMS code for the initial analysis was provided by him, while Paula Despins did some of the initial calculations.

Table 11.9: Summary of Efficiency Results, Eastern Free State (1.00 = efficient)

Efficiency	Year	Average	St Dev	Median	Mode	Min	Max
Technical (TE)	1979/80	0,85	0,23	0,91	1,00	0,38	1,00
	1982/83	0,85	0,18	0,91	1,00	0,38	1,00
	1985/86	0,90	0,15	1,00	1,00	0,39	1,00
	1988/89	0,89	0,16	1,00	1,00	0,48	1,00
Scal (SE)	1979/80	0,87	0,23	0,93	0,98	0,24	1,00
	1982/83	0,84	0,14	0,89	1,00	0,51	1,00
	1985/86	0,79	0,21	0,86	1,00	0,08	1,00
	1988/89	0,83	0,16	0,88	1,00	0,33	1,00
Allocative (AE)	1979/80	0,73	0,23	0,72	1,00	0,21	1,00
	1982/83	0,74	0,17	0,76	1,00	0,25	1,00
	1985/86	0,78	0,15	0,78	1,00	0,39	1,00
	1988/89	0,66	0,20	0,66	1,00	0,30	1,00

Table 11.10: Average Farm Size of Efficient versus Inefficient Farms (scale, technical and allocative efficiency), Eastern Free State (ha)

Item*	1979/80	1982/83	1985/86	1987/88
SE-eff	1 208,3	766,4	697,3	475,3
SE-ineff	1 369,1	349,1	1 624,6**	1 221,8
TE-eff	1 265,2	1 205,3	1 312,1	1 226,5
TE-ineff	1 283,9	1 293,9	1 329,7	1 215,4
AE-eff	1 258,2	1 246,3	1 285,1	1 387,3
AE-ineff	1 267,5	1 253,2	1 309,4	1 188,8

Notes: *SE = scale efficiency; TE = technical efficiency; and AE = allocative efficiency
**The average is relatively high due to two very large farms

efficiency from technical and allocative efficiency, while TFP measurements do not differentiate between them. The results yielded statistically significant differences between average farm sizes of scale efficient (SE-eff) and inefficient (SE-ineff) farms ($p<0,10$) for 1982/83, 1985/86 and 1988/89. No similar trend was encountered for technical efficiency

(TE) and allocative efficiency (AE), implying that there is no meaningful relationship between TE or AE and farm size. This implies that the farm size-efficiency relationship has its origin in scale efficiency and not technical or allocative efficiency. While technical and allocative efficiency do not differ significantly across farms sizes, scale efficiency does differ significantly. The conclusions are as follows:

- Differences in scale efficiency are relatively more important in explaining efficiency differences between small and large farms than differences in technical or allocative efficiency.
- These results even more clearly establish an inverse relationship between farm size and efficiency.
- Another trend which emerges from these results, which also confirms the previous observations, is that the average farm size of efficient farms declined over time. This correlates with the abolition of tax and credit policies favouring relatively large farmers more than small farmers.

In order to identify where and why the diseconomies of scale set in, the inverse of the SE measure was investigated. Following Chavas and Aliber (*1993*), this inverse can be interpreted as something akin to an average cost function, i.e. it is a declining function of outputs under increasing returns to scale (IRTS), and an increasing function of outputs under decreasing returns to scale (DRTS). In all years, there is very little difference between the average scale inverses for the largest 25 per cent of the farms. Between the largest 10 per cent and the rest, however, there are diseconomies of scale emerging in most years. These appear to be driven largely by diseconomies of scale in crop production with very little noticeable difference between farm sizes and the inverse for livestock production, which is mostly extensive ranching. Since most farms derive a very large percentage of their gross farm income from crop production, this would appear to explain the low number of perfectly efficient farms. Given that only 10 per cent of the farms are exhibiting strong diseconomies of scale, the relatively reasonable performance of individual farms is also not surprising.

The analysis of scale efficiency in the *Vaalharts irrigation area* differs in two complementary ways from the above analyses:

- Farm sizes are much smaller in the Vaalharts irrigation area than in the grain producing regions (Table 11.6).

- Scale efficiency is also related to the managerial ability of farmers (apart from quality adjusted farm size) in the Vaalharts area.

Managerial skill in the Vaalharts area was measured explicitly according to the method proposed by Burger (*1971*) who developed and validated a scale of 'managerial aptitude' of farmers. This scale is based on six different factors: vision, planning, record keeping, labour management, budgeting and maintenance tasks. This scale was found to be positively associated with size related variables, including farm size and return to assets (*Groenewald, 1991; Van Schalkwyk, et al, 1993; Sartorius von Bach & van Zyl, 1992*).

Scale efficiency of the individual farms in the Vaalharts area was determined using the methodology described on pages 294–296.[22] Figures 11.5 and 11.6 present the results of the analysis by plotting the inverse of the scale efficiency index (1/SE) against quality adjusted farm size and managerial ability. This inverse (1/SE) can be interpreted in a similar way to an average cost function: (1/SE) is a decreasing function of outputs under increasing returns to scale, and an increasing function under decreasing returns to scale.

Figure 11.5 shows that a whole range of farm sizes is efficient, from the smallest to the largest. However, there is evidence of economies of scale for very small farms, with the majority of inefficient farms being on the small side. This result differs from the previous findings where no economies of scale were found. However, farm sizes in Vaalharts are much smaller, making supervision of labour easier and less expensive. On the other hand, machinery is the slack variable for many of the small farms that are scale inefficient. Indivisibility seems to play some role here, particularly with respect to tractors. To perform the necessary tasks on even the smallest farms in the sample, at least two tractors are required because some of the tasks have to be performed simultaneously. While combine harvesters and other large machinery can be contracted or hired, this is not possible for tractors within the present farm support structure. Two small tractors can work areas of up to 50–60 ha easily if managed properly; therefore, the lumpy input argument seems to apply for small farm sizes. But, this argument applies only to some farms: roughly one third to one half of the small farmers are scale efficient. Upon further investigation it seems that these farm managers found a

22 The procedure is similar to that used by Chavas and Aliber (*1993*) who analysed efficiency of grain and dairy farmers in Wisconsin.

way around the lumpy machinery input problem by entering into sharing arrangements and/or diversification of their operation to minimize the need for simultaneous use of machinery. Management seems to be important in this respect.

Figure 11.6 establishes a strong relationship between scale and managerial ability. Managers with better managerial ability are generally

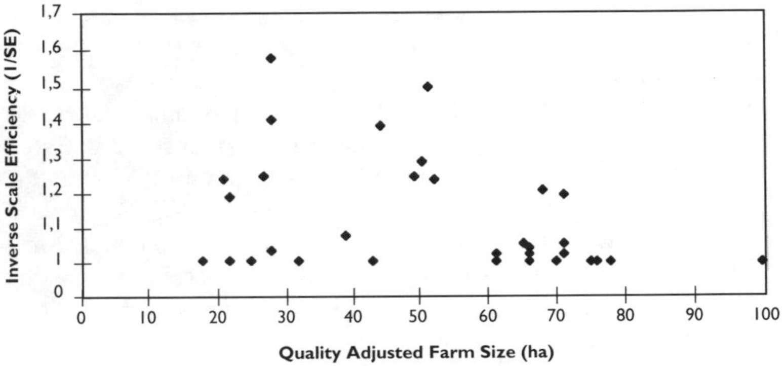

Figure 11.5: Farm Size-Efficiency in the Vaalharts Irrigation Area, 1990/91

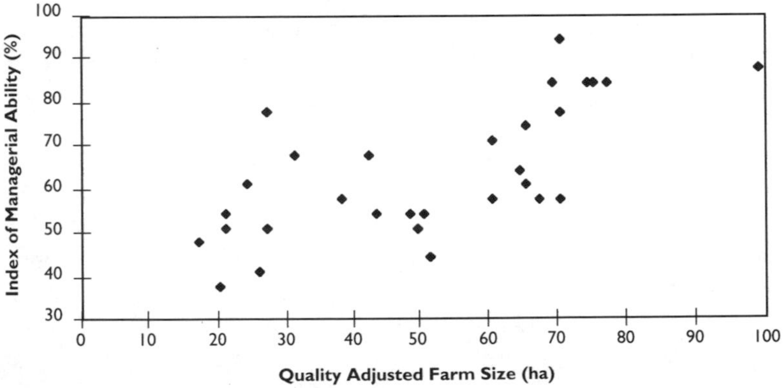

Figure 11.6: Managerial Ability and Farm Size, Vaalharts (1990/91)

more scale efficient. This relationship is also statistically highly significant: the correlation coefficient between farm size and inverse scale efficiency is –0,334 (p=0,062); between managerial ability and inverse scale efficiency is –0,589 (p<0,0001); and between farm size and managerial ability is 0,686 (p<0,0001).

Results obtained from the analysis are mixed, and clearly demonstrate the complexity of the issues involved. A whole range of farm sizes seem to be scale efficient (Figure 11.5), depending on how farmers organize their specific variable and fixed input mix, as well as the combination of outputs they produce. In this respect, a number of relatively small farms are scale efficient, although there seems to be a bias towards a larger number of relatively small farms being scale inefficient. On the other hand, the relationship between scale efficiency and managerial ability seems to be much stronger. This emphasizes the importance of management: results support the notion that better managers operate on larger farms than less skilled managers.[23] In this respect, Groenewald (*1991*) is of the opinion that 'returns to management' is a more appropriate concept than returns to scale.

Econometric estimation of the farm size-efficiency relationship

Another appropriate way in which to measure relative efficiency of small and large farms is to investigate the difference in profits, net of the cost of family labour, per unit of capital invested (*Binswanger et al, 1993*). The following test of the farm size-productivity relationship, proposed (and discussed) by Binswanger *et al* (*1993*), was used to take most of the methodological considerations into account:

$$\mathbf{P/K = g\ (OP, OW, H, Z)\ \text{with expected signs}\ g_1 < 0;\ g_2 > 0;\ \text{and}\ g_3 > 0;} \qquad \mathbf{(1)}$$

where K is assets; L is labour; P is private or social profits net of private or social cost of family labour; OP is operated area or value of operated land; OW is owned area or value of owned land; H is the number of household workers; and Z is a vector of exogenous land quality, distance from infrastructure, and land improvement variables. Also, g1 should be negative because of rising supervision costs; g_2 should be positive

23 This result is consistent with earlier findings reported by Van Schalkwyk *et al* (*1993*) for Vaalharts, and Sartorius von Bach and van Zyl (*1992*) for the Aberfeldy district in the Eastern Free State.

because ownership provides better access to credit; and g_3 should be positive because family members have incentive to work and can supervise. Equation 1 does not describe a casual relationship, but a multiple correlation.

In addition to this specification, managerial ability of the farmer (M) is an important explanatory variable which needs to be considered. This clearly also impacts on efficiency (see results of DEA analysis) and, because it varies for different farms, should also be included as part of the multiple correlation depicted by equation 1.

The equation was estimated for only Ruens in 1987/88 and the Vaalharts area in 1990/91. These surveys included information which allows for specification of the variables in equation 1, with some adjustments to capture the specific set of circumstances particular to the local situation. The specification of P/K and OP in equation 1 is straightforward from the available data. Most of the land is owned by the operator, with the results that OW is correlated with OP. To avoid estimation problems, OW was specified as the percentage of land owned by the operator. Due to the situation in South African commercial agriculture, H was not specified as the number of household workers, but as remuneration of management (including the owner and household members, and people other than those employees of the owner who benefit from revenue sharing arrangements)[24] expressed as a percentage of other (hired) labour costs. Lastly, Z was specified as the fixed land improvements per unit area, because most of the other land quality characteristics were already considered when adjusting farm size for land quality differences.

In addition, the estimation for Vaalharts was done both with and without inclusion of managerial ability (M) as the independent variable in equation 1. The argument for inclusion is similar to that for incorporating differences in land quality. The estimation of equation 1 assumes that all factors not specified are fixed and similar for all farms. Factors which differ between farms, and which may have an influence on the results (such as managerial ability), should therefore be accounted for explicitly. In this respect the intuition is that the coefficient of M should be positive, since better managers should get a higher return on their investment.

24 This approach, in essence, broadens the concept of family members engaging in and benefiting from supervision tasks. Only a relatively small number of farms (16 per cent) had a value other than zero for H.

Table 11.11: Results of the Regressions on Farm Size-Efficiency in Vaalharts, 1990/91

Regression/ Region	Variable					Equation	
	OOP	OW	H	Z	M	F Value	Adj R^2
Ruens	–0,0067 (0,093)	0,0233 (0,114)	0,0142 (0,223)	0,0025 (0,002)	—	45,28	0,62
Vaalharts 1	–0,0428 (0,259)	0,0574 (0,008)	0,0811 (0,042)	0,0002 (0,2875)	—	37,53	0,55
Vaalharts 2	–0,0735 (0,182)	0,0239 (0,094)	0,0756 (0,66)	0,0001 (0,365)	0,0533 (0,050)	74,25 (0,000)	0,69

Notes: 1. Value in brackets indicates the significance of the coefficient or value.
2. R^2 of a regression model without an intercept, measures the proportion of variability in the dependent variable about the origin explained by the regression, which cannot be compared to the R-square for models which include the intercept.

Table 11.11 shows the results of the regression analyses which were used to determine the relationship between profit (net of family labour) per unit of investment, and the other variables.

Table 11.11 shows that, within a framework that accounts for many of the shortcomings of previous analyses on the farm size-efficiency relationship, farm profit (net of family labour) per unit of capital invested is negatively related to operated farm size (OP), (although it is not statistically significant in Vaalharts), but is positively related to the percentage of the operated area owned (OW). This further complements the findings obtained from the TFP and non-parametric analyses: farm size tends to be inversely related to efficiency, however, larger owned areas are positively related to higher farm profits per unit of capital. In addition, the supervision component (H) is also positively related to farm profitability (although it is not statistically significant for the Ruens).

Inclusion of the managerial ability variable (M) yielded interesting results. As a whole, a better fit was obtained. Managerial ability is significantly positively related to efficiency. On the other hand, the Z variable yielded mixed results. As expected, it has a highly significant positive relationship with efficiency for the Ruens, but is insignificant for Vaalharts.

In general, the results presented in Table 11.11 are strictly according to prior expectations regarding the signs of the different coefficients. Some specific coefficients, in particular operated area (OP), however, are not significant.

Explaining the results: policy, technology and management

The different analyses of the farm size-efficiency relationship in the grain-producing areas (which represent approximately 60 per cent of all cultivated areas) and the irrigated areas (for crops only) in South Africa yield consistent and complementary results from which it can be concluded that:[25]

- Farms in the former homelands seem to be scale inefficient. This is not surprising, given the history of lack of access to support services and infrastructure, of policies discriminating against them, and the extremely fragmented and small land use rights of farmers. In addition, credit, information, insurance and labour markets are missing or imperfect.
- There is an inverse relationship between farm size and efficiency in the commercial farming areas for the range of farms analysed, regardless of the methodology used.
- This inverse relationship in commercial farming seems to become stronger and more accentuated as policy distortions, which largely favour large farms relative to smaller farms, are removed.
- Large farms use relatively more capital intensive methods of production, while smaller farms are more labour intensive.
- Managerial ability seems to be closely related to farm size, with better managers having larger farms.

From these results, it is clear that the policy framework is crucial since it has an important impact on the farm size-efficiency relationship. However, even in South Africa where a small group of large commercial farmers have captured most of the benefits from the extremely distorted policy regime which heavily supported them, this was not enough to offset the disadvantages brought about by higher labour supervision costs and transaction costs associated with labour, and imperfect labour

25 High value crops, such as export fruit, were not included in the analyses and, as a result, all of the conclusions would not necessarily also apply to these crops.

markets. In addition, for the range of commercial farms analysed, advantages large farms have in access to inputs, credit, services, marketing and distribution opportunities were also negated. The conclusion is that even a policy environment favouring large farms over small ones, resulting in huge social opportunity costs, was not enough to make large farms more efficient than relatively smaller farms.

In addition, it seems that larger farms have better managers than smaller farms. However, even this 'advantage' was offset by the supervision and transaction costs associated with hiring a large number of labourers in the more extensive and lower potential rain-fed areas, although it seems to have a bigger impact in the higher potential irrigation areas where some economies of scale were observed in small farm sizes.

These results apply to the existing technologies used on South African farms. These technologies essentially originated in the United States, where labour is relatively expensive and capital abundant, and were adapted to the local situation (*Van Zyl & Groenewald, 1987; Van Zyl et al, 1987*). In addition, research and extension concentrated on encouraging the adoption of such technologies, many of which are inappropriate given South Africa's factor endowment. The argument is that small farms, even smaller than the range of farm sizes evaluated in the analyses presented in this chapter, will be even more efficient than larger farms if there were more appropriate technologies available; if these technologies were properly supported by research and extension; and if the policy environment in general was more friendly towards small farmers.

Finally, the results provide some insights on how to think about the farm size-efficiency relationship in general. They support the idea that economies of scale arise because of missing or imperfect markets, or distortions and pecuniary economies favouring large farms over small farms. They show, however, that the costs associated with labour supervision, as well as other labour-related transaction costs are huge, and outweigh many of the advantages of being large. Even in the South African commercial farm sector, where relatively larger farms have benefited substantially more from a comprehensive range of policies and privileges, this was not enough to compensate for these costs, and an inverse farm size-efficiency relationship is observed. However, markets do exist in these areas and they function fairly well for even the smaller commercial farmers. On the other hand, where they are missing or imperfect, for example, in the former homelands where the situation is further compounded by a lack of support systems and infrastructure, small farms are less efficient than the larger farms (although all farms are relatively small due to over-population and often extreme fragmentation of use-rights).

The farm size-efficiency relationship thus seems to be determined by the relative importance of the factors benefiting smaller farms and those benefiting larger farms. On balance, how these factors impact on the relationship, and the net outcome of their effects, are influenced by several factors, both individually and jointly. These include the production relations and technology utilized on the farms, the relative factor endowment facing the broader society, and the managerial ability of the farm manager. For example, managerial ability seems to have a smaller impact where there are other factors which are more restrictive, or where there are no alternative technologies available. This is the case in the dryland areas as opposed to where irrigation is available. In the latter situation, the upper efficiency boundary of the individual's farm is more reliant on managerial ability than on some exogenous factor such as rainfall. In addition, in an economy where the factor endowment (and relative prices) favour the use of labour, farm size should be smaller because the disadvantages of using labour kick in at smaller farm sizes. Thus, production relations and factor endowment (which includes management) together determine the impact of pecuniary economies and distortions on farm size-efficiency on the one hand, and supervision and transaction costs associated with labour on the other.

Implications for land reform

The inverse farm size-efficiency relationship, which is present in South African agriculture despite a history of policies favouring relatively large mechanized farms, implies that significant efficiency gains can be made if farm sizes in the commercial sector become smaller. An important element in such a process would be the removal of all policies and distortions favouring larger farms relative to smaller farms. The basic principle should be to make markets work by removing distortions and privileges favouring large farmers, and by creating markets to service small farmers in areas where they are missing, without entrenching new privileges. Imperfect markets should be made to work better.

Although the efficiency argument cannot be a judge of the present distribution of land rights given the history of how these rights were acquired, it does provide a strong argument for land reform in light of the inverse farm size-efficiency relationship observed in South African commercial agriculture. However, a precondition is the removal of all privileges to the farm sector because they tend to favour large farms over smaller ones, as well as the addressing of missing and imperfect markets for small farmers. Thus, the playing field should be levelled.

The results on management and farm size also have important implications for land reform. They further support the call for flexibility in policies regarding farm size and structure of agriculture, while also showing the value of proper training and extension aimed at increasing the individual farmer's managerial ability. The results clearly support the notion of a farm structure with smaller farms opening the way for land reform.

The results obtained particularly support the abolition of the Act on the Subdivision of Agricultural Land (Act 70 of 1970), and especially the way in which it is applied. Apart from prohibiting the creation of more efficient small farmers in South Africa's commercial areas, applications for the subdivision of agricultural land are based on the notion of 'average management' which is clearly inappropriate. The point is to make training and extension an integral part of the land reform programme, and not legislation controlling minimum farm sizes.

References

Afriat, S.N. 1972. Efficiency estimation of production functions. *International Economic Review,* 13 : 568–598.

Ali, A. and Seiford, L. 1993. The Mathematical Approach to Efficiency Analysis. In Fried, H., Lovell, K. and Schmidt, S. (Eds), *The Measurement of Productive Efficiency, Techniques and Applications.* Oxford University Press.

Ball, V.E., Bureau, J.C. and Butault, J-P. 1994. Intercountry comparisons of prices and real values. *Unpublished Paper.* Washington, D.C.: Economic Research Service, USDA.

Bauer, P.W. 1990. Recent developments in the econometric estimation of frontiers. *Journal of Econometrics,* 46 : 39–56.

Baumol, W.J., Panzar, J.C. and Willig, R.D. 1982. *Contestable Markets and the Theory of Industry Structure.* New York: Harcourt Brace Jovanovich, Inc.

Bembridge, T. 1991a. *The Practice of Agricultural Extension: A Training Manual.* Halfway House: Development Bank of Southern Africa.

Bembridge, T. 1991b. Crop farming system constraints in Transkei: implications for research and extension. *Development Southern Africa,* 4(1) : 67–152.

Berry, R.A. and Cline, W.R. 1979. *Agrarian Structure and Productivity in Developing Countries.* Baltimore: Johns Hopkins University Press.

Binswanger, H.P., Deininger, K. and Feder, G. 1993. Power, distortions, revolt and reform in agricultural land relations. *Discussion Paper.* Washington, D.C:

World Bank. (Forthcoming in Srinivasan, T.N. and Behrman J. (Eds), *Handbook of Development Economics,* Vol III.)

Binswanger, H.P. and Elgin, M. 1992. What are the Prospects for Land Reform? In Maunder, A. and Valdez, A. (Eds), *Agriculture and Governments in an Interdependent World.* Buenos Aires: International Association of Agricultural Economists.

Binswanger, H.P. and Kinsey, B. 1993. Characteristics and Performance of Resettlement Programs: A Review. *World Development,* 21(9) : 1477–1494.

Binswanger, H.P. and Rosenzweig, M.R. 1986. Behavioural and Material Determinants of Production Relations in Agriculture. *Journal of Development Studies,* 22(3) : 503–539.

Brewster, J.M. 1950. The machine process in agriculture and machinery. *Journal of Farm Economics,* 32(1) : 69–81.

Bromley, D. 1989. *Economic Interests and Institutions.* New York: Basil Blackwell.

Burger, P.J. 1971. The measurement of managerial inputs in agriculture III: The construction and evaluation of a sale. *Agrekon,* 10(4) : 5–11.

Cairns, R.I. 1990. An agricultural survey of subsistence farmers in the Nkandla district of KwaZulu. *Development Southern Africa,* 7(1) : 77–104.

Calabresi, G. 1991. The pointlessness of Pareto: carrying Coase further. *Yale Law Journal,* 100 : 1211–1237.

Carter, M.R. 1994. Sequencing capital and land market reforms for broadly based growth. *Paper No. 379.* Madison: Department of Agricultural Economics, University of Wisconsin.

Carter, M.R. and Kalfayan, J. 1989. A General Equilibrium Exploration of the Agrarian Question. *Unpublished Research Paper.* Madison: Department of Agricultural Economics, University of Wisconsin.

Chavas, J-P. and Aliber, M. 1993. An analysis of economic efficiency in agriculture: A nonparametric approach. *Journal of Agricultural Resource Economics,* 18 (1) : 1–16.

Chavas, J-P. and Van Zyl, J. 1993. Scale-efficiency in South African grain production: a non-parametric analysis. *Unpublished Research Paper.* Washington, D.C: World Bank.

Cristodoulou, N. and Vink, N. 1990. The potential for black smallholder farmers' participation in the South African agricultural economy. *Paper presented at the Conference on Land Reform and Agricultural Development in South Africa.* United Kingdom: Newick Park Initiative.

CSS. 1993. *Census of Agriculture, 1988.* Pretoria: Central Statistical Services, Government Printer.

De Janvry, A. 1987. Farm Structure, Productivity and Poverty. *Working Paper No. 432.* Berkeley: Department of Agricultural and Resource Economics, University of California.

De Klerk, M. 1985. Technological change and employment in South African agriculture: the case of maize harvesting in the Western Transvaal, 1968–1981.

Unpublished MA Economics dissertation. Cape Town: University of Cape Town.

De Klerk, M. 1991. The accumulation crisis in agriculture. In Gelb, S. (Ed), *South Africa's Economic Crisis.* Cape Town: David Philip.

Fare, R., Grosskopf, S. and Lovell, C.A.K. 1985. *The Measurement of Efficiency of Production.* Boston: Kluwer-Nijhoff.

Farrell, M.J. 1957. The measurement of productive efficiency. *Journal of the Royal Statistical Society,* A 120, Part 3 : 253–81.

Feder, G. 1985. The relation between farm size and productivity: the role of family labour, supervision and credit constraints. *Journal of Development Economics,* 18 : 297–313.

Fényes, T., Van Zyl, J. and Vink, N. 1988. Structural imbalances in South African agriculture. *South African Journal of Economics,* 54(2&3) : 181–194.

Forsund, F.R., Lovell, C.A.K. and Schmidt, P. 1980. A survey of frontier production functions and their relationship to efficiency measurement. *Journal of Econometrics,* 13 : 5–25.

Groenewald, J.A. 1991. Returns to size and structure of agriculture: a suggested interpretation. *Development Southern Africa,* 8(3) : 329–342.

Hattingh, H.S. 1986. The skew distribution of income in agriculture. *Paper presented at the Agricultural Outlook Conference.* Pretoria.

Hayami & Ruttan 1985. *Agricultural Development: An International Perspective.* Baltimore: Johns Hopkins University Press.

Helm, W.E. and Van Zyl, J. 1994. Quantifying agricultural support in South Africa. *Agrekon,* 33(4).

Johnson, N.L. and Ruttan, V.W. 1994. Why are farms so small? *World Development,* 22(5) : 691–706.

Kirsten, J.F. 1994. Agricultural Support Programmes in the developing areas of South Africa. *Unpublished PhD thesis*. Pretoria: University of Pretoria.

Latt, E.A. and Nieuwoudt, W.L. 1988. Identification of plot size effects on commercialization of small-scale agriculture in KwaZulu. *Development Southern Africa,* 5(3) : 371–382.

Meyer, N.G. and Van Zyl, J. 1993. Comparative advantages in Region G: an application of a sectoral linear programming model. *Agrekon,* 31(4) : 307–312.

Moll, P.G. 1988. Economies of size in 'white' South Africa: implications for land reform. *Unpublished Research Paper.* Cape Town: University of Cape Town.

Nicholson, C. and Bembridge, T. 1991. Characteristics of black commercial farmers in Gazankulu. *South African Journal of Agricultural Extension,* 20 : 7–17

Nieuwoudt, W.L. 1991. Efficiency of land use. *Agrekon,* 29(4) : 210–215.

Nieuwoudt, W.L. and Vink, N. 1989. The effects of increased earnings from traditional agriculture in Southern Africa. *South African Journal of Economics,* 57(3) : 257–269.

Nunamaker, T.R. 1985. Using Data Envelopment Analysis to measure the efficiency of non-profit organizations: a critical evaluation. *Managerial and Business Economics,* 6(1) : 50–58.

Peterson, W. and Kislev, Y. 1991. Economies of scale in agriculture: a re-examination of the evidence. *Staff Paper P91-43.* St. Paul: Department of Agricultural and Applied Economics, University of Minnesota.

Piesse, J., Sartorius von Bach, H., Thirtle, C. and Van Zyl, J. Agricultural efficiency in the Northern Transvaal homelands. *Journal of International Development*. (Forthcoming).

Prosterman, R.L. and Riedinger, J.M. 1987. *Land Reform and Democratic Development.* Baltimore and London: The Johns Hopkins University Press.

Roth, M., Dolny, H. and Wiebe, K. 1992. Employment, efficiency and land markets in South Africa's agricultural sector: opportunities for land policy reform. *Unpublished Review paper.* USA: Land Tenure Centre, University of Wisconsin-Madison.

RSA. 1994. *Abstract of Agricultural Statistics, 1994.* Pretoria: Department of Agriculture, Republic of South Africa.

Sartorius von Bach, H.J., Koch, B.H. and Van Zyl, J. 1992. Relating perceptions and associated economic criteria to economic survival in commercial dryland farming in South Africa. *Agrekon,* 31(4) : 210–215.

Sartorius von Bach, H.J. and Van Zyl, J. 1992. Comment: Returns to size and structure of agriculture – a suggested interpretation. *Development Southern Africa,* 9(1) : 75–79.

Schmid, A.A. 1987. *Property, Power and Public Choice.* New York: Praeger.

Schmid, A.A. 1992. Legal foundations of the market: implications for formerly socialist countries of Eastern Europe and Africa. *Journal of Economic Issues,* 26 : 707–732.

Schmid, A.A. 1994. Institutional Law and Economics. E*uropean Journal of Law and Economics,* 1 : 33–51.

Seiford, L.M. and R.M. Thrall. 1990. Recent developments in DEA: The mathematical programming approach to frontier analysis. *Journal of Econometrics,* Vol 46 : 7–38.

Singini, R.E., Sartorius von Bach, H.J. and Kirsten, J.F. 1992. One to four hectares: what effects can the farmer support programme strategy have on such households? *Agrekon,* 31(4) : 228–234.

Thirtle, C., Sartorius von Bach, H.J. and Van Zyl, J. 1993. Total factor productivity growth in South African agriculture, 1947–1991. *Development Southern Africa,* 10(3) : 301-318.

Urban Foundation. 1990. Report on Agricultural Development. Johannesburg.

Van Heerden, W.R. and Van Zyl, J. 1992. The aggregate measure of support for maize in South Africa. *Unpublished Research Paper*. Pretoria: University of Pretoria.

Van Schalkwyk, H.D., Van Zyl, J. and Sartorius von Bach, H.J. 1993. Management and returns to farm size: results of a case study using parametric and non-parametric methodology to measure scale efficiencies. *Agrekon,* 32(4): 252–256.

Van Zyl, J. 1994. Market-assisted rural land reform in South Africa: efficiency, food security and land markets. *Agrekon,* 33(4) : 105–116.

Van Zyl, J. and Coetzee, G.K. 1990. Food security and structural adjustment: empirical evidence on the food price dilemma in South Africa. *Development Southern Africa,* 7(1).

Van Zyl, J., Fényes, T.I. and Vink, N. 1987. Labour-related structural trends in South African maize production. *Agricultural Economics,* 1(3) : 241–258.

Van Zyl, J., Fényes, T.I. and Vink, N. 1992. Effects of a Farmer Support Programme and changes in marketing policies on maize production in South Africa. *Journal of Agricultural Economics,* 43(3) : 466–76.

Van Zyl, J. and Groenewald, J.A. 1987. Economical farm technology for different levels of managerial ability: the case of maize cultivar selection. *Quarterly Journal of International Agriculture,* 26(1) : 46–57.

Van Zyl, J. and Groenewald, J.A. 1988. Effects of protection on South African commercial agriculture. *Journal of Agricultural Economics,* 39(3) : 387–401.

Van Zyl, J., Nel, H.G. and Groenewald, J.A. 1988. Agriculture's contribution to the South African economy. *Agrekon,* 27(2) : 1–9.

Van Zyl, J., Van Rooyen, C.J., Kirsten, J.F. and Van Schalkwyk, H. 1994. Land reform in South Africa: options to consider for the future. *Journal of International Development,* 6(2) : 219–239.

Van Zyl, J. and Vink N. 1992. The mini farmer approach: a case study of the South African tea industry. *Development Southern Africa,* 9(4) : 493–500.

Wheeler, M.W. and Ortmann, G.F. 1990. Socio-economic factors determining the success achieved among cotton-adopting households in two magisterial districts of KwaZulu. *Development Southern Africa,* 7(3) : 323–333.

World Bank. 1983. *Kenya: Growth and Structural Change.* Washington, D.C: Basic Economic Report, Africa Region.

World Bank. 1994. South African agriculture: structure, performance and options for the future. *Discussion Paper 6, Informal Discussion Papers on Aspects of The Economy of South Africa.* Washington, D.C.: Southern African Department, The World Bank.

12

The land market

Herman van Schalkwyk and Johan van Zyl

Introduction

International experience with land reform and rural restructuring, as well as that of South Africa, suggests a land reform programme design that relies as much as possible on the existing land market. This stems from the observed weaknesses of non-market oriented programmes that typically vest too much control in public sector bureaucracies which develop their own set of interests that are in conflict with the rapid redistribution of land (*Binswanger & Deininger, 1993*). Nonetheless, non-market interventions may be necessary to ensure successful implementation of the programme (*World Bank, 1994*). In this context, it is important to analyse the South African land market so that forces driving the market and their effects are identified. Understanding these issues is necessary before embarking on any market-assisted land reform process.

Against this background, this chapter addresses two questions:

1. If land reform is a key to a more optimal distribution of operational holdings, to what extent can the land market and price mechanism be used to effect the desired changes?
2. If large ownership holdings are inefficient, why do their owners not split them up and rent or sell them to family farmers, or what prevents the land market from bringing ownership holdings in line with the optimal distribution of operational holdings?

The first question is addressed by analysing the land market through the use of a structural land price model developed for this purpose; the second question is addressed by analysing the gap between the market and productive value of land, i.e. to what extent has the land market been inflated by the capitalization of privileges to farmers and other distortions? In this context, all the information, tables and analyses in this chapter refer to farmland owned by whites outside of the former homeland areas.

This chapter is structured as follows: the next section provides an overview of rural land transactions and land transfers in South Africa, as well as a description of land price movements in relation to key economic indicators. Then a model for, and results of, simulating land price changes is described. These results are subsequently used to analyse the gap between market values and productive values of farmland. Thereafter some conclusions are provided.

The South African land market: historical background

Land transactions constitute an important element of the land market. Since 1964, between 7 561 and 14 889 deeds, and between 3,1 million to 5,5 million hectares of rural immovable property, have been transferred annually (Table 12.1). The total area of transfers has remained remarkably constant at around 4 per cent of the total surface area in the commercial sector.

The average size of land transfers has risen over time. The number of transfers dropped in the 1980s, both nationally and for most size categories, but there does not appear to be a corresponding drop in the total area transferred. Particularly in the upper size ranges of the market for rural land, transfers have remained relatively constant in number, while areas transferred have increased. Transactions involving smaller parcels of land dominated: of the 8 852 parcels transferred in 1990/91, 26,8 per

Table 12.1: Average Annual Rural Land Market Transactions in South Africa, 1964–1991

Region	Transfers: total number	Area transferred (ha)	% of land transferred (%)	Average area transferred (ha)
Cape Province	2 942	1 944 641	4,02	661,2
Natal	1 183	254 545	4,28	215,2
Transvaal	5 438	1 112 089	4,19	204,5
Orange Free State	1 358	402 457	4,09	296,4
Total: South Africa	10 921	3 713 732	4,15	328,7

Source: Registrar of Deeds (1992).

cent were less than 19 hectares in size, 26,7 per cent between 10–99 hectares, 17,9 per cent between 100–299 hectares, 10,4 per cent between 300–499 hectares, 9,4 per cent between 500–999 hectares, 5,1 per cent between 1000–1999 hectares, and 3,6 per cent above 2000 hectares. However, the frequency of recorded deed transfers of the smallest parcels is declining in relative terms, which may suggest that progressively more transfers are taking place off the record. Transactions involving larger parcels, on the other hand, dominated the total area transferred. Of the 3,2 million hectares of land transferred in 1990/91, 0,6 per cent were parcels less than 19 hectares in size, 3,2 per cent between 20–99 hectares, 9,2 per cent between 100–299 hectares, 11,2 per cent between 300–499 hectares, 18,1 per cent between 500–999 hectares, 19,5 per cent between 1000–1999 hectares, and 38,3 per cent above 2000 hectares.

Figure 12.1 shows the relationship between the real land prices and the percentage land transfers in South Africa. It is evident from this figure that real land price is not the only factor influencing land transfers. For instance, a low percentage of land transfers is associated with

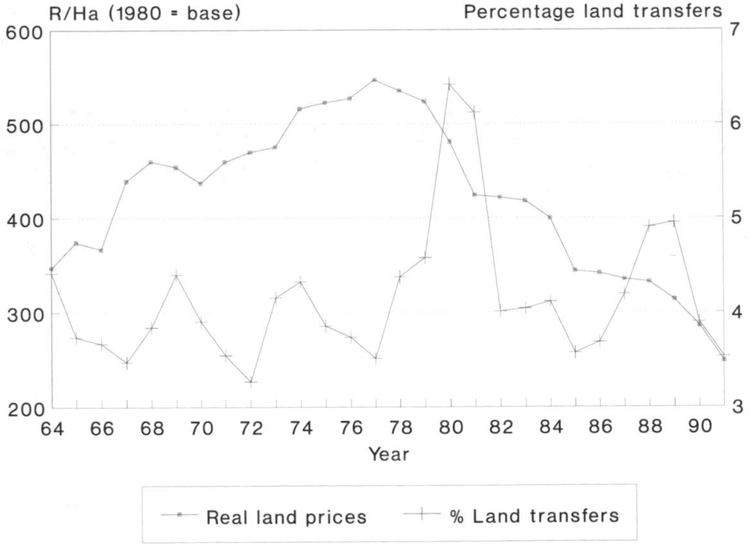

Figure 12.1: Percentage Land Transfers and Real Land Prices (1964–1991)

high land prices in 1977, while a low percentage of land transfers is associated with lower land prices in 1990.

In 1963, total leased land represented only 13,1 per cent of total land area; but in 1988 rented, leased and sharecropped land represented 19,5 per cent of the total surface area, with considerable regional variation: 26,9 per cent in the Orange Free State, 22,9 per cent in the Transvaal, 17,3 per cent in the Cape, and 15,7 per cent in Natal. Hattingh and Herzberg (*1980*) found that those who lease land are mainly farmers who already own land. Moreover, although the official statistics point to a relatively high rental rate of nearly 20 per cent of total area, in fact most rentals are between the older and younger generations of the same white family. Such rental arrangements are *de facto* pension schemes, and the proportions of genuine rentals can be as low as 5 per cent. It has been suggested that the low rate of genuine rentals at least partly reflects owners' fear that renters will 'mine' and destroy the fragile land (*Van Zyl et al, 1994*).

Historic movements of average South African farmland prices since 1955 are subsequently compared to several important variables. First, it is important to see how price movements differed between regions. Figure 12.2 shows that, except for the winter rainfall region, price movements over the last decade were fairly similar for all the regions. Nevertheless, important variations still exist.

Underlying the research on farmland values in South Africa are some interesting historical patterns of land prices, returns, rents, interest rates, financing and inflation. These patterns are presented in Figures 12.3 to 12.8 to facilitate evaluation of competing hypotheses suggested by previous research. These patterns provide insight pertinent to model discrimination and are discussed here to provide background for later analysis. The most widely accepted explanation of farmland prices is based on expected returns or rents. Because expected returns are unobservable, an intuitive comparison of real land values with current and lagged real returns is informative.

Recent studies that find returns to be the major explanation of land prices, explain land prices by complicated distributed lags on returns (*Alston, 1986; Burt, 1986*). In contrast with the USA, as illustrated by Just and Miranowski (*1993*), Figure 12.3 reveals that real land values in South Africa follow an almost parallel pattern to current real returns and that land values appear positively related to recent changes in returns as plausible expectations schemes would require. The major trends in returns and real land prices, however, have been in opposite directions during most of the period since 1983.

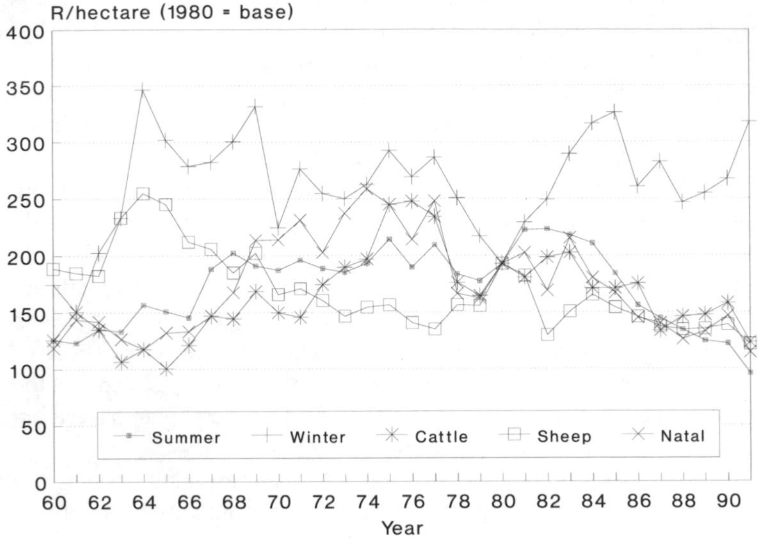

Figure 12.2: Real Index of Land Prices in Different Agro-economic Regions (1960–1991)

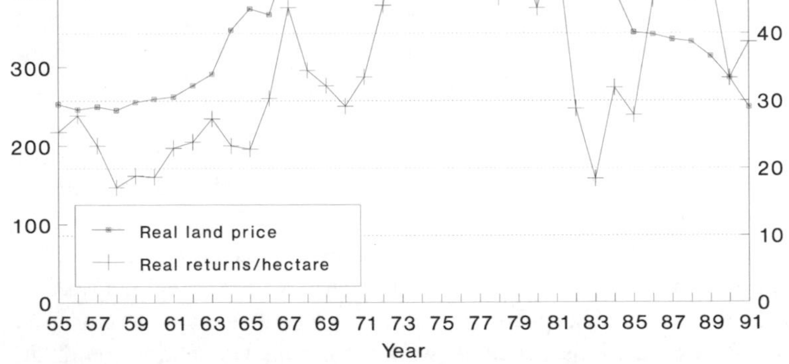

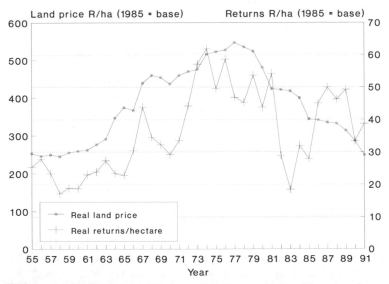

Figure 12.3: Real Land Prices and Returns per Hectare in South Africa (1955–1991)

Rapid rates of inflation are also used to partially explain land price increases of the late 1960s and early 1970s. Inflation not only reduces the rate of capitalization of future returns, but land serves as a hedge against inflation. Figure 12.4 relates real land prices to the inflation rate. Although less volatile than the rate of inflation, land prices follow a similar pattern with a short lag. Thus, the inflation explanation is appealing, even though the mechanism by which inflation affects land values is far from clear.

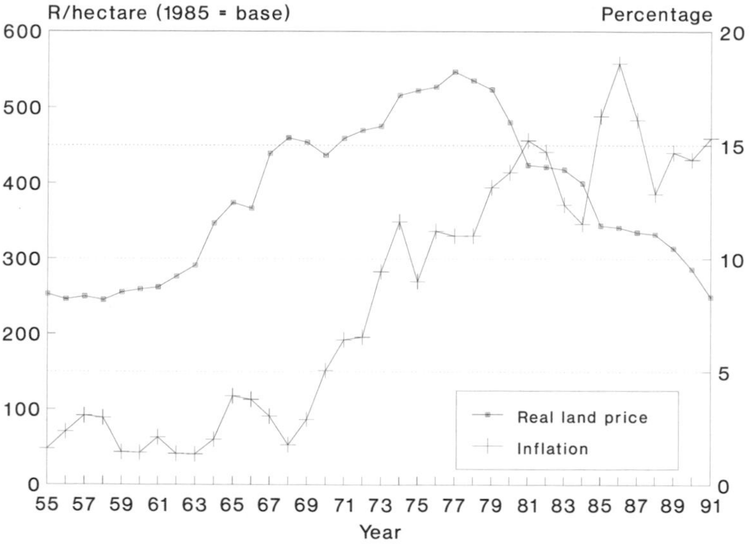

Figure 12.4: Real Land Prices versus Inflation (1955–1991)

Explaining land values by access to credit and credit control is supported by the similar pattern which per hectare farm real estate debt and land prices follow (Figure 12.5). The lag between the 1976 real land price peak and the 1985 debt peak is a direct result of the expectations of agricultural financiers that real farmland prices would increase. The problem, however, as shown in Figure 12.6, is that farm real estate debt as a percentage of land value remained stable while land prices increased (1955–76), and then increased rapidly when land prices headed into a decline (1979–85). These observations suggest that the farm debt bubble may have occurred more as a consequence of high land values rather than as a causal factor.

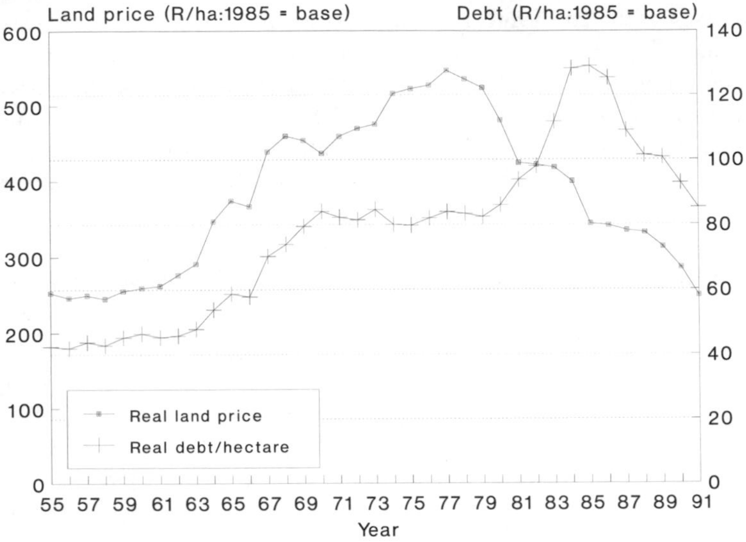

Figure 12.5: Real Estate Debt and Land Prices in South Africa (1955–1991)

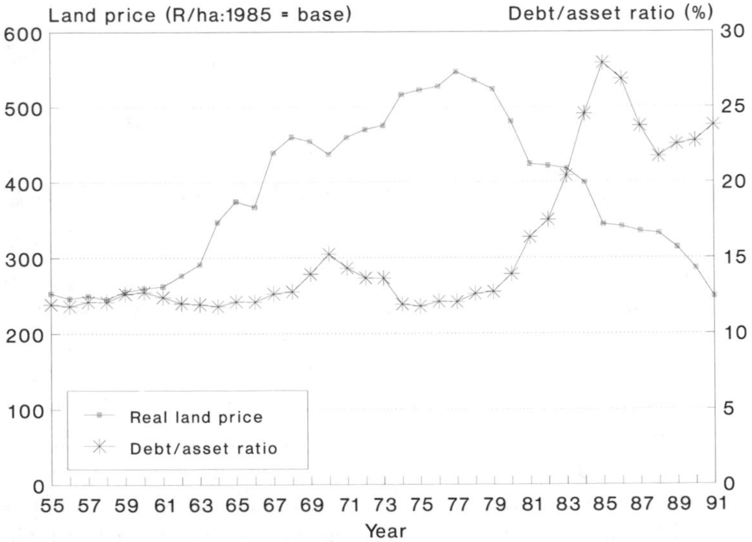

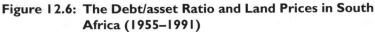

Figure 12.6: The Debt/asset Ratio and Land Prices in South Africa (1955–1991)

Alternatively, the real interest rate on farm real estate debt can be used as an indicator of debt constraints. During the 1970s, low and even negative real interest rates displayed an inverse relationship with land values while high real interest rates in the 1980s were associated with declining land values. Also, the sources of credit changed significantly during the 1970s and 1980s possibly reflecting easier credit. However, traditional sources of credit may have tightened in the 1980s as debt-asset ratios declined which, in turn, motivated a shift to the Land Bank and agricultural co-operatives for financing. Financing by the Land Bank and agricultural co-operatives went from 29 per cent of total financing in 1970 to 47 per cent in 1991. Van Schalkwyk and Groenewald (*1993*) established that there is a negative relationship between the amount of debt financed by agricultural co-operatives and solvency ratios. The increased financing of the Land Bank and agricultural co-operatives therefore leads one to accept that solvency ratios and land prices should have declined. This is illustrated in Figure 12.7.

Treating the real interest rate as the opportunity cost of capital, rather than as a measure of credit tightness, the opportunity cost of capital

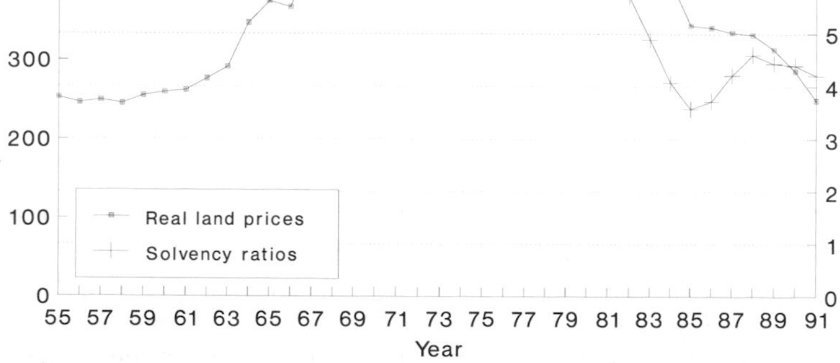

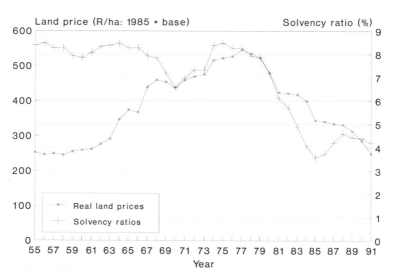

Figure 12.7: Real Land Prices and Solvency in South Africa (1955–1991)

appears to be a more important explanation of land values than credit availability (Figure 12.8). The savings interest rate closely parallels the debt interest rate, so empirical distinction is difficult. Because farm real estate debt does not vary sharply in response to changes in the real interest rate on debt, the opportunity cost explanation appears more plausible. The tight credit explanation thus applies to a small component of the land market while the opportunity cost explanation applies to the whole market.

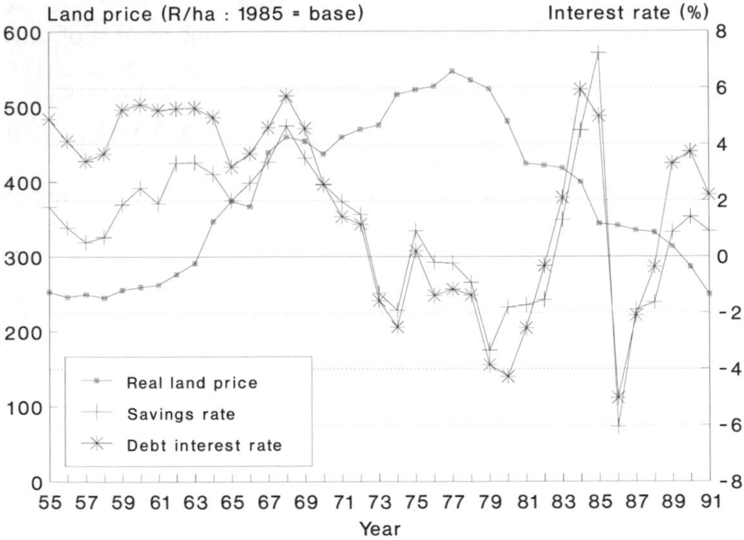

Figure 12.8: Saving and Debt Interest Rates versus Land Prices (1955–1991)

After reviewing the historical data, many of the factors hypothesized to affect farmland values appear to have correlations (Table 12.2) that suggest validity and reflect the results obtained by studies examining each individually. These relationships explain why empirical results based on ad hoc and partial analyses are conflicting, and imply that a comprehensive and theoretically defensible framework is needed to identify the relative importance of each.

Table 12.2: Correlation of different variables with the average South African land price

	Inflation rate	Net farm income	Debt load	Interest rate on savings	Interest rate on debt	Debt asset ratio
r	0,4397	0,6209	–0,3188	–0,6548	–0,2433	–0,4770
p-value	0,0678	0,0060	0,1972	0,0032	0,3306	0,0453

Modelling land price changes

Modelling the land market

The traditional ad hoc econometric approach to empirical analysis has the advantage of tailoring results closely to observed data, but is vulnerable to misleading results due to spurious correlations and an inability to identify proper functional forms. Typically, it can 'identify' only a few factors, so information on interaction with other variables (possibly subject to large changes outside the sample) is not obtained. Alternatively, theoretical analysis has the advantage of maintaining plausible relationships among variables, but suffers from the need for stringent assumptions to obtain unambiguous results (*Just & Miranowski, 1993*).

This analysis draws on the advantages of both approaches. Economic theory is used to impose plausible relationships among variables so that econometric identification is possible with more variables. While some restrictive assumptions are required for tractability, the assumptions are arguably as general as unknown implicit restrictions imposed by arbitrary choices of functional forms for ad hoc investigations. Additionally, the resulting model contains several unknown parameters for which good extraneous information exists – parameters that can be identified more accurately from alternative information than econometric estimation. After imposing these coefficients, the remaining parameters are estimated conventionally.

The structural model of land prices used for this analysis includes the multi-dimensional effects of inflation on capital-erosion, savings-return erosion, and real debt reduction; it also develops the effect of changes in the opportunity cost of capital. The method of approximation and procedure is largely based on that followed by Just and Miranowski (*1993*) in their computation of farmland price changes in the USA, which was specially adapted by Just (*1993*) for the South African land market. This model is shown below. It provides a comprehensive framework for analysing the relative importance of factors determining farmland prices over the past two decades. Free-form econometric investigations cannot estimate coefficients on all variables with sufficient precision to resolve the important issues. The model was estimated for different agro-economic regions and for South Africa as a whole:

$$\bar{p}_t = f_t \frac{\rho(1-\tau_t \nu_t \psi_g)\bar{P}^* + (1-\tau_t)\bar{R}^*_t - \beta\phi^2 \bar{A}\Sigma_t}{1-\tau_t \nu_t \psi_g + \chi_t (1-\tau_t) + \psi_s Z_t + \psi_t - \psi_d f_t (1-\Delta) Z_t} \qquad (1)$$

where

$Z_t \;=\; -(1-\tau_t ([\chi_t - r_t - (1+\chi_t)\Delta]/1-\Delta)$

$S_t \;=\; (1-\tau_t \nu_t \psi_g)^2 \rho^2 \omega_t + (1-\tau_t)^2 \sigma_t + 2(1-\tau_t \nu_t \psi_g)(1-\tau_t)\xi_t,$

the variables are:

$\bar{p}_t$ = average land price resulting from transactions at the beginning of period t

f_t = 1 plus the current rate of inflation at time t

τ_t = the average tax rate on current income

v_t = the proportion of capital gains taxed in period t

$\bar{P}^*_t$ = average land price expectation for the end of period t held at the beginning of period t

$\bar{R}^*_t$ = average expected net returns to farming per hectare (including subsidies) for period t

$\bar{A}_t$ = average farm size in period t

Σ_t = perceived variance of end-of-year wealth per hectare against beginning-of-year expectations

χ_t = rate of interest earned on savings in period t

r_t = rate of interest paid on debt in period t

Z_t = effective cost of debt

ψ_t = property tax per hectare on real estate in period t

ω_t = perceived variance of end-of-year land price

σ_t = perceived variance of net returns from farming per hectare (including subsidies)

ξ_t = perceived covariance of land price and net returns per hectare

the unknown parameters are:

β = coefficient of absolute risk aversion on profit
ϕ = $\beta^*/(\beta^* + \beta)$ where β^* is the absolute risk aversion coefficient on short-run variations in wealth
ρ = 1 minus the rate of sales commissions on land transactions
Δ = rate of finance charges and other transactions costs on new debt

and the indicators of strength of various regimes and phenomena are:

ψ_g = proportion of current land value attributable to capital gain
ψ_s = proportion of farmland in farms with a binding minimal savings constraint
ψ_d = proportion of farmland value financed by debt

While the model appears rather complicated, the intuition is straightforward (*Just & Miranowski, 1993*). First, if all the complications of inflation ($f_t = 1$), taxes ($\tau_t = 0$, $\psi_t = 0$), credit market imperfections ($\chi_t = r_t$), transactions costs ($\Delta = 0$, $\rho = 1$) and risk aversion ($\beta = 0$) are eliminated from the model, then this equation reduces to the standard discounting equation:

$$\bar{p}_t = \frac{\bar{P}^*_t + \bar{R}^*_t}{1+\chi_t} \qquad (2)$$

which in equilibrium ($\bar{p} = \bar{P}^*_t$) yields $\bar{p}_t = \bar{R}^*_t/\chi_t$. Adding simple inflation considerations multiplies the right-hand side of the discounting equation by f_t obtaining $\bar{p}_t = f_t\,(\bar{P}_t{}^* + \bar{R}_t{}^*)/(1+\chi_t)$ which, in long-run equilibrium, reduces to the same basic equation as does the model developed by Feldstein (*1980*). All the additional effects in the model are justified as a modification of this equation. To see this, note that the numerator represents the value of holding a hectare of land while the denominator represents the opportunity cost of channelling a Rand's worth of wealth into land. In this context, the terms in the model can be examined and interpreted one by one (*see Just & Miranowski, 1993*).

Estimating the model

In this section, the farmland model is estimated for different agro-economic regions and for the country as a whole. The results are used to decompose farmland price changes, beginning with the boom of the 1960s. The results show that inflation and changes in the real returns on capital are major explanatory factors in farmland price swings, in addition to returns to farming. Additionally, the effects of credit market constraints and expectation schemes are considered explicitly in the analytical model. Data for the period 1955 to 1991 were used for estimation. The results are reported for the summer grain region and South Africa as a whole. The model was estimated by the non-linear, seemingly unrelated, regression (SUR) method to take advantage of the high correlation of disturbances that exists among regions. Predictions fit the 1955 to 1990 data very closely (see Figures 12.9 and 12.10 for South Africa and the summer grain region respectively).

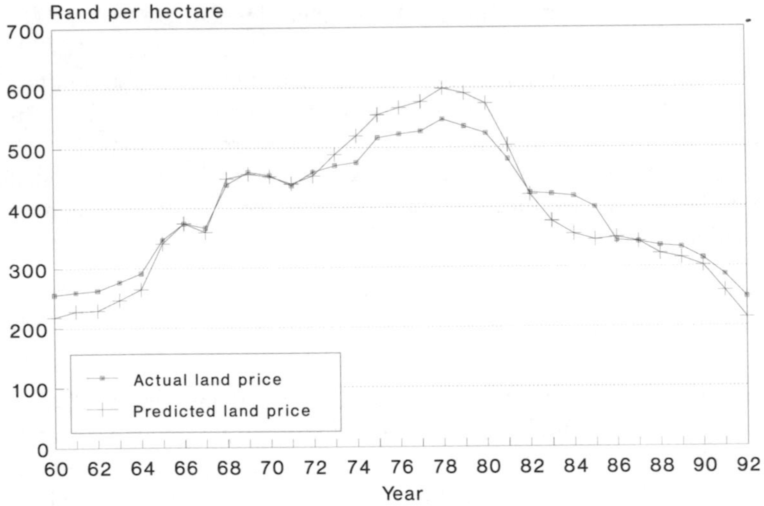

Figure 12.9: Actual and Predicted Land Prices in South Africa (1955–1991)

Decomposition of price movements

To understand the source of land movements, this section decomposed predicted annual land price changes among all the effects represented in the model. That is, the price changes are decomposed according to the effects represented by the various terms of the numerator and denominator. The decomposition of predicted price changes is reported in Table 12.3 for South Africa by effect for the years of land price vitality, 1970–90. Note that the inflation effect is on real prices rather than nominal prices (the inflation effect on the numeraire is removed). The

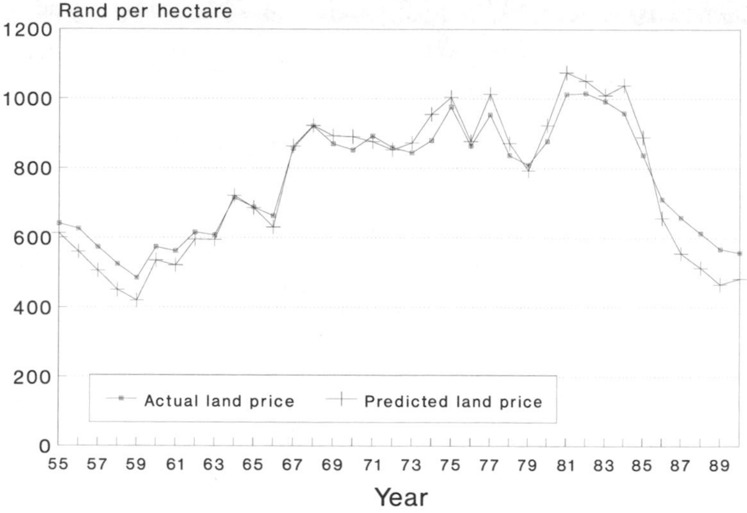

Figure 12.10: Actual and Predicted Prices in the Summer Grain Region (1955–1991)

predicted price change and its components are reported in real 1985 Rands.

Land price expectations are the most important explanatory force in every agro-economic region. However, the change in land price expectations is explained by changes in previous prices and, thus, indirectly by previous changes in other variables. With extrapolative expectations, the change in price expectations for period t is explained by the change in price expectations and all other variables in period t-1, the change in price expectations in t-1 by price expectations and all other variables in period t-2, etc. Thus, the relative role of variables other than price expectations is crucial in understanding the wide swings in the South African land prices. The contribution of price expectations in each year is primarily important in understanding the dynamic effects of the other variables.

For the remaining variables, the most striking effect is the dynamic role of inflation and the opportunity cost of capital. These two effects are each roughly as important as increased returns to farming. This is well illustrated in Table 12.3 for the South African 1971 land price take-off period and the 1975 surge. From 1971 to 1973, the inflation rate increased from 6,4 per cent to 9,4 per cent (as measured by the consumer

Table 12.3: Decomposition of Predicted Real Land Price Changes by Effect for South Africa, 1970–1990

Year	Total	Expectations			Tax rate paid	Opportunity		Inflation
		Price	Returns	Risk		Saving	Debt	
1970	–2,661	–4,425	–0,068	–0,079	–0,034	–0,445	0,207	1,292
1971	2,925	1,573	1,284	–0,385	–0,040	0,338	0,048	0,108
1972	7,650	1,719	1,232	–0,630	0,002	2,100	0,513	2,715
1973	6,114	1,810	1,650	–0,081	0,096	0,471	0,158	2,009
1974	7,248	12,204	0,252	–0,233	–0,024	–2,162	–0,430	–2,358
1975	2,159	0,259	–1,023	–0,355	0,103	0,867	0,270	2,038
1976	1,719	0,560	1,769	–0,291	–0,113	0,024	–0,038	–0,193
1977	3,931	4,531	–0,967	–0,294	0,085	0,532	0,035	0,009
1978	–1,139	–4,796	–0,072	–0,431	–0,089	1,896	0,439	1,916
1979	–2,831	–3,081	0,705	–0,056	0,085	–1,156	0,085	0,588
1980	–12,513	–10,877	–1,204	–1,211	0,038	–0,055	–0,435	1,232
1981	–16,016	–16,024	1,430	–0,743	0,490	–0,135	–0,599	–0,435
1982	–10,443	–0,185	–4,759	–0,620	–0,400	–1,787	–0,682	–2,011
1983	–5,737	–0,604	–2,422	1,063	0,012	–1,788	–1,253	–0,745
1984	–2,335	–3,693	3,746	0,942	–0,018	–1,378	0,347	–2,281
1985	3,933	–16,058	–0,856	–0,445	0,617	7,445	4,535	8,696
1986	–3,581	–0,820	2,721	0,138	0,096	–2,506	–1,120	–2,090
1987	–2,991	–2,343	–1,150	0,761	–0,256	–0,156	–0,609	0,763
1988	–3,813	0,112	0,957	–0,262	0,110	–1,462	–1,309	–1,959
1989	–4,101	–3,876	0,089	0,306	0,108	–0,316	–0,149	–0,263
1990	–12,733	–11,896	–2,727	0,212	–0,083	0,442	0,505	0,813

price index). This increase in the rate of inflation explains 35 per cent of the predicted land price increase in 1972 in South Africa. This effect is the direct result of capital erosion, i.e. the opportunity cost of a Rand invested in any activity declined because it would be worth 9,4 per cent (rather than 6,4 per cent) less in real terms after one year of use (apart from the rate of return it earns).

Another major force in the 1971 take-off period is the opportunity rate of returns on capital. From 1968 to 1974, the real rate of return on savings dropped from 4,6 per cent with 6,5 per cent. This caused investment in land to become more attractive by comparison. This effect explains 27 per cent of the predicted land price increase in 1972 for South Africa as a whole. Note that the effect of the rate of interest on debt has a minor effect.

By comparison, the increase in returns to farming explains 16 per cent of the predicted change in South African land prices in 1972. Over the five-year period from 1971 to 1975, the rate of inflation and the real rate of return on capital had similar effects to those of farming returns. Following the 1971 take-off period, much of the ensuing land price appreciation was due to the 1968–1974 effects working through the system and culminating in price expectation effects. To understand this explanation, note that an initial price increase due to inflation or opportunity cost had a positive effect the following year on price expectations; these higher price expectations, in turn, caused a higher price the following year, which then caused higher price expectations to be transmitted to a third year, and so on. While, on the surface, this explanation may suggest that land price changes are being explained tautologically with land price changes, the adjustment process actually works much like a Nerlovian model. Each external shock has a declining distribution of effects over time, reflected through the land price expectation which is the lagged land price. Apart from higher expected returns to farming, inflation and opportunity costs are the only major explanatory forces behind the increased price expectations of 1971–1977. By 1979, inflation and opportunity cost had returned to pre-1968 extremes. Land prices started to drop in 1977 – a direct effect of high inflation. Furthermore, the land price volatility in the 1980s led to large increases in perceived risk, tending to decrease prices further.

The model predicts the price turn around in 1977 very well. The 1982 shock is primarily due to perceived risk, opportunity-cost and farming returns. From 1973 to 1983, farming returns decreased while the rate of inflation increased. The associated opportunity-cost effect explains about 40 per cent of the predicted decline in land prices for South Africa as a whole.

Conclusion

The structural model of land prices includes the multi-dimensional effect of inflation associated with capital erosion, savings-return erosion and real debt reduction, as well as the effect of changes in the opportunity cost of capital. In spite of the imposition of substantial *a priori* theoretical structure and extraneous information, the model fits the data well, compared to ad hoc econometric models. The results show that the large price swings are mainly explained by inflation rates and changes in real returns on alternative uses of capital. These effects caused substantial appreciation in 1971 and substantial depreciation in 1978. The large shock of 1971 tended to continue as indirect effects worked their way through land price expectations. The lagged effects of later changes were moderated or offset by changes in other causal variables.

The results described above have important implications for a market-based rural land reform in South Africa. They clearly demonstrate that: (a) the rural land market is active enough, and also (b) stable enough to be used as a transfer mechanism for substantial amounts of rural land to the people disadvantaged and excluded by the apartheid policies of the past. In addition, real land prices have declined considerably because some of the distortive agricultural and other policies favouring farming in general, and specifically large-scale mechanized farming have been abolished. This issue is subsequently addressed in more detail.

To further illustrate these conclusions, the land-prices model developed on pages 319–321 is used to illustrate the effect of new liquidity in the land market. The effect is measured by imposing four land redistribution objectives (1 per cent, 2 per cent, 3 per cent and 4 per cent of total land per annum, in addition to normal land market transactions which

Table 12.4: The Effect of New Liquidity on Real Land Prices in South Africa (1980–1991)

Year	Predicted land prices (R/ha)	Increase in land price (%) Annual redistribution objective			
		1%	2%	3%	4%
1980	479,36	1,80	3,35	4,66	5,70
1981	576,17	2,34	4,47	6,37	8,03
1982	460,55	4,98	9,68	14,11	18,23
1983	315,03	10,14	19,94	29,39	38,48
1984	289,74	10,51	20,81	30,88	40,71
1985	206,58	13,74	27,51	41,28	55,04
1986	377,21	7,37	14,84	22,39	30,02
1987	217,70	10,97	21,99	33,04	44,14
1988	341,24	6,18	12,35	18,52	24,71
1989	370,51	5,85	11,65	17,44	23,23
1990	267,20	8,08	16,04	23,91	31,72
1991	297,75	8,04	15,97	23,80	31,56
Average		7,50	14,88	22,15	29,30

occurred), using 1980 as the starting point for the analysis to illustrate the different paths land prices would have followed. The results are presented in Table12. 4.

Table 12.4 reveals that real RSA land prices will rise on average by 7,5 per cent, 14,88 per cent, 22,15 per cent and 29,3 per cent if 1 per cent, 2 per cent, 3 per cent and 4 per cent of the land was bought for redistribution purposes respectively. However, this is the worst case scenario because two important assumptions were made to be able to make the calculations: (a) additional liquidity leaves the agricultural sector in the same proportion as observed during the period 1980–91 and, (b) these rates of transfer represent additional transfers over and above the actual transfers during the period 1980–91. In practice, the land price increase will thus be considerably lower.

It is clear from the above that a market-assisted approach to land distribution will have different effects on land prices according to different redistribution objectives. The magnitude of these effects is important for future policies regarding land redistribution.

Is South African agricultural land overvalued?

In most countries, the major advantage of ownership of land has been the price appreciation of land over time. Unlike most resources used in farming, land does not depreciate or deteriorate if managed properly. Although the farmer has not received the financial benefits of price appreciation in a cash form that is available for direct consumption, appreciation has increased net worth. This increased net worth can be used as a financial base for borrowing funds to expand the farm operation, as well as a cushion or reserve against short-term financial losses that may require refinancing. Thus, land ownership has important income, capital appreciation and risk-reduction dimensions for the farm operator, as well as the social and family dimensions of a permanent home and residence for the farm family.

The price that must be paid for these attributes of ownership is the substantial capital outlay needed to purchase land. Most farmers, and particularly emerging farmers, do not have sufficient capital for the down payment required for land acquisition as well as enough funds left for machinery, equipment purchases and working capital. The financial requirements of purchasing land can drain valuable funds away from other investment alternatives. The basic question, therefore, becomes one of which method of land acquisition has the highest financial payoff compared to alternative uses of the farmer's funds, and which alternative is 'financially feasible' or within the financial capability of the farm operator (*Boehlje & Eidman, 1984*).

The difference or gap between the market and agricultural value of land does not contribute to the farmer's ability to repay a loan made to acquire land. Often, however, this contributes to the ability of the farmer to obtain credit (*Binswanger & Deininger, 1993*). Van Schalkwyk and Groenewald (*1993*) found that non-farm factors like policy distortions, policy and institutional expectations get capitalized into market values, hence the difference between the market and agricultural value of land. The non-farm factors, for example, also represent expectations of present landowners that their land can be sold for non-farm purposes. Land in the vicinity of cities is usually more expensive than similar land further afield, not only because of the mentioned expectations but also because of cost savings on transport. They also found that high gross revenues – partially a result of price supports – become capitalized in land values. This tends to lend some support to arguments by Paarlberg (*1962*) and Groenewald (*1978*), that the profitability gains the present farming generation receives because of price supports become a cost of doing business for the next generation.

The size of the gap between the agricultural and market value of land is of major importance for land reform purposes, especially if the affordability of a basically market-oriented land reform is taken into account. It is therefore important to understand the forces underlining the difference between the market and land use value of agricultural land in South Africa. This section aims to identify these forces and to quantify the gap between the market and agricultural value of land.

Alternative agricultural value estimates

According to Boehlje and Eidman (*1984*), there are generally three methods by which land can be appraised, namely, the market, cost and income approach. The market approach to valuing real estate essentially attempts to determine what the property would bring if sold. The basic philosophy of the cost approach is to inventory the various resources of the farm, estimate their cost, and then sum these costs to obtain a total value. Because of the extremely difficult task of associating a cost with land, this approach is quite difficult to use for unimproved land. In essence, the income approach to valuation determines the long-run profitability of a land investment (*Boehlje & Eidman, 1984*). The income approach to land valuation was subsequently used because of its consistency with the net present value method of evaluating investments.

The income-capitalization approach is based on the logic that the market value of a piece of land should equal the present value of the stream of all future incomes. In its most simple form (where income is assumed to accrue in perpetuity), earnings value $V = I/r$, where I is the average yearly return to land and r is the discount or capitalization rate (*Locken et al, 1978*). This simple formula does not consider income taxes; both the income stream and the capitalization rate are calculated on a before-tax basis. If taxes are included as a cash expense, then the capitalization rate must also be reduced to an after-tax rate. A number of refinements can be made to this approach to account for changes in the income stream or discount rate, taxes or any other changes that may affect the income generated from a parcel of land over time (*Locken, 1976*). While these refinements are not difficult to deal with conceptually, empirical implementation requires knowledge of the future income streams and other changes affecting agricultural value. Failure to incorporate these changes by capitalizing current rather than future income streams certainly has an impact on estimates of agricultural values. However, one can argue that agricultural values based on recent performance may be the only acceptable alternative for empirical estimation of the earning value of land. By comparing these agricultural values with those developed through a market approach, one can argue that market participants setting land market values have just as much difficulty in perceiving the future as any researcher. They, too, may only have crude estimates of the future income potential of land, and they may rely most heavily on the recent performance of land as their basis for appraising its future productivity (*Locken et al, 1978*).

One of the most difficult decisions required in using the income approach to valuation is choosing the appropriate capitalization rate. From a conceptual viewpoint, the capitalization rate should reflect the cost of capital or the cost of funds committed to the purchase of land. However, adjustments are necessary to reflect differences in the risk associated with land compared to alternative investments.

Reynolds and Timmons (*1969*) have suggested that the capitalization rate should reflect the rate of return on other farm inputs, thus representing the opportunity cost of investing in farmland. Scofield (*1964*) argues that one should employ rates of interest or rates of return on non-farm investments which represent the opportunity cost of investing in any farm inputs. He argues that non-farm income producing real estate (such as apartment buildings and office complexes) or common stock has similar liquidity and risk characteristics, and is analogous to farmland in an investment sense. He objects, however, to the use of interest rates on real estate mortgages as a capitalization rate because they are a fixed monetary (Rand) investment.

Although Scofield (*1964*) argues that fixed monetary investments have a lower risk than farmland, rates of return on alternative investments may still be useful. It has been suggested that farmers as a group may use a lower capitalization rate because of a propensity for farming and a preference to live in a certain area (*Reynolds & Timmons, 1969*). On the basis of these arguments, the annual return on government bonds was selected as the capitalization rate. This is in accordance with the suggestions of Locken *et al*, (*1978*).

Empirical results

Refinements were made to the numerator of the mentioned income-capitalization formula in order to measure other important factors which also influence the agricultural value of land. The refined formula involves $V = (I^* + S - E - L - i)/r$, where I^* = total expected cash farm receipts, S = services received by holding land, E = total cash farm expenses, L = the value of the operator's remuneration and unpaid family labour, i = interest on capital, and r = the capitalization rate.

Data on average agricultural income streams, total cash farm expenses and interest on capital ranging from 1970 to 1992, was obtained from the Directorate of Agricultural Economic Trends (*1994*), while the interest rate on government bonds was obtained from the Central Statistical Service (*1994*). Using this data, alternative regimes for expectations on returns per hectare were used to postulate future income streams. Extrapolative expectations on net returns per hectare were specified by extending a four-year trend. This approach gave the best results in previous research on land markets in South Africa (*Van Schalkwyk, 1995*), and are therefore also used here. Only the results of the extrapolative expectations are shown since they provided the best results. Average salaries for all employees as reported by the Central Statistical Service were used to measure the value of operator's and unpaid family labour because actual figures were not available. Land provides its owner with free housing and water, cheaper food, etc (*Binswanger & Deininger, 1993*). The mentioned services, received by owning land, were measured by calculating the actual cost of these services if the operator had to pay for them.

Figure 12.11 compares these calculated agricultural values of land with the market value of agricultural land.

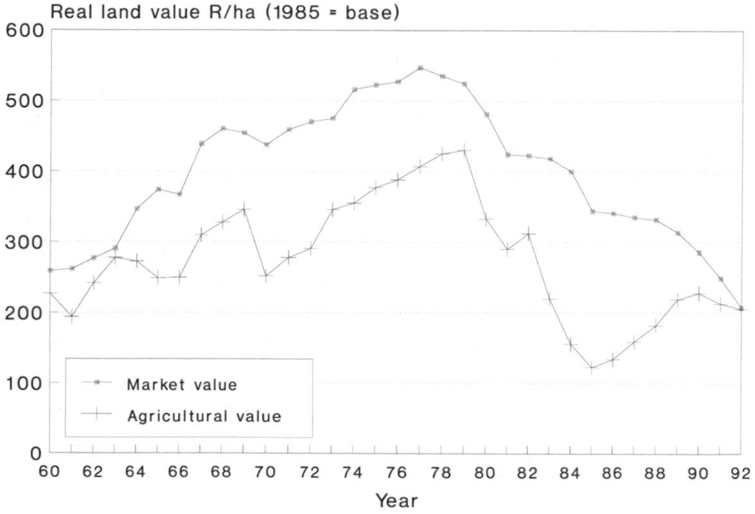

Figure 12.11: Real Market Value and Expected Agricultural Value of South African Land (1960–1992)

Figure 12.11 shows that the market and agricultural value of land followed almost the same trend since the 1960s. Agricultural values rose up to the mid seventies and then gradually declined. Figure 12.11 emphasizes Van Wyk's (*1976*) finding that the difference between the market price and agricultural value in general increased during the period 1960–1969. However, it does also reveal that the difference between the market and agricultural value of land reached its maximum in 1984 after which it plummeted and reached a minimum in 1992, where the difference was insignificant. The agricultural value of land declined over the long-term. The market value of land, however, declined at a much faster pace which caused the gap between the agricultural value and the market value of land to decrease.

Discussion

Inflation has become a major consideration in any investment or disinvestment decision. If buyers expect land to appreciate at a rate similar to the rate of inflation, they can expect to pay more for the same land at some future date. Consequently, if they have adequate financing and want to expand their land base, it may be desirable to make the land purchase now rather than to wait. For the seller, inflation is also an important consideration. Sellers must be careful not to lock themselves into fixed or constant income investments where the income stream and the investment principle do not adjust with inflation or increase with the general price level.

According to the analysis on pages 323–325, the major force in the 1971 take-off period in land market prices of land was the opportunity rate of returns on capital. From 1968 to 1974, the real rate of return on savings dropped from 4,6 per cent with 6,5 per cent to –1,9 per cent. This caused investment in land to become more attractive by comparison. Following the 1971 take-off period, much of the ensuing land market price appreciation was due to the 1968–1974 effects working through the system and culminating in price expectations effects. Again, while on the surface, this explanation may suggest that land market price changes are being explained tautologically, the adjustment process actually works much like a Nerlovian model. Each external shock has a declining distribution of effects over time reflected through land market price expectations (which is a lagged form of market land price). As mentioned previously, by 1979, opportunity cost had returned to pre-1968 extremes, but this time coupled with an added high inflation rate. Land market prices started to drop in 1977 – a direct effect of the high inflation rate. Furthermore, the land market price volatility in the 1980s led to large increases in perceived risk tending to decrease market prices further.

The agricultural value of land on the other hand is affected by the ability of land to generate profits. Van Zyl *et al* (*1993*) showed that profits are mainly affected by changes in productivity and price recovery: from 1947 to 1991 total factor productivity increased rather slowly at 1,3 per cent per annum; there was no growth until 1965; then 2,15 per cent until 1981 and fairly rapid growth of 2,88 per cent per annum since 1981. They also showed that land productivity increased at 3,13 per cent per annum since 1947. The increasing rate of growth over the period is in accordance with Van Zyl and Groenewald's (*1988*) perception that farmer's profits came under increasing pressure as inflation gathered pace.

Since 1974, highly inflationary conditions prevailed. Input prices have risen faster than product prices and a cost-price squeeze has been experienced. This cost-price squeeze obviously exerts considerable pressure on the income and therefore also on the agricultural value of land. Real net farm income has increased by nearly 181 per cent since 1947. This has been ascribed by Van Zyl *et al* (*1993*) to the growth in total factor productivity of nearly 161 per cent which countered the decline of 27 per cent in terms of trade. However, real net farm income declined by 1,06 per cent per annum from 1973 until 1991, and by 8,14 per cent from 1973 to 1983. This decline is a direct result of the unfavourable growth rate in terms of trade.

It is evident from the above that inflation had a negative effect on both the market value and the agricultural value of land. This, coupled with the withdrawal of some of the major support services and policy distortions from the state to the farming community, led to the general misconception that the difference between the market and agricultural value of land did not decline, but that at best it stayed the same. However, the effect of the fairly rapid growth in productivity which countered the negative effect of the terms of trade on profits and hence on agricultural values, were never taken into account. The growth in productivity did in fact push up net farm incomes, and hence also agricultural values, which resulted in a declining market/agricultural land value gap.

Conclusion

This chapter analysed agricultural land prices in South Africa over time, including the sources of change and the difference between the market and agricultural value of land in South Africa. From the analysis, it is clear that the gap between the average market and agricultural value of South African land showed a general decline since 1984. The decline is attributable to the withdrawal of some of the major privileges benefiting the commercial farming community, and inflationary conditions which had a negative influence both on sellers and buyers, as well as an annual growth in productivity of 4,63 per cent since 1983. This had a positive effect on agricultural land values, thus closing the gap between the market and agricultural value of land.

The decreasing gap between agricultural land values and market values of land makes land more affordable and enhances repayment ability because buyers of land will now find it easier to repay a loan from the productive capacity of the land itself. This scenario also affects land reform in the sense that emerging farmers will not have to pay much above the agricultural or productive value of land in a market related land reform. This will enable them to have a better repayment ability or, alternatively, reduce the size of grants required to make new farms financially viable to beneficiaries of a market-assisted land reform programme.

In summary, this chapter illustrates that:

- the land market has very real possibilities as a major mechanism for redistributive land reform;
- present land price levels are, on average, relatively near to the capitalized value of farm profits;
- the land market is active and stable enough to handle relatively large quantities of new liquidity and additional land transfers;
- non-market interventions may be necessary, specifically in some regions, to ensure a well-functioning market.

References

Alston, J.M. 1968. An analysis of growth of US farmland prices, 1963–82. *American Journal of Agricultural Economics,* Vol. 68 : 1–9.

Binswanger, H.P. and Deininger, K. 1993. South African land policy. The legacy of history and current options. *World Development,* Vol. 21 (9) : 1451–1476.

Binswanger, H.P., Deininger, K. and Feder, G. 1993. Power, distortions, revolt and reform in agricultural land relations. *World Bank Discussion Paper.* Washington, D.C.: World Bank.

Boehlje, M.D. and Eidman, V.R. 1984. *Farm Management.* New York: John Wiley and Sons.

Burt, O.R. 1986. Econometric modelling of the capitalization formula for farmland prices. *American Journal of Agricultural Economics,* Vol. 68 : 10–26.

Central Statistical Service. 1994. *South African Statistics.* Pretoria: Department of Statistics.

Feldstein, M. Inflation, portfolio choice, and the prices of land and corporate stock. *American Journal of Agricultural Economics,* Vol. 62 : 910–16.

Groenewald, J.A. 1978. Production quotas in agriculture: Comment. *South African Journal of Economics,* Vol. 46: 62–65.

Hattingh, H.S. and Herzberg, A. 1980. Ownership or leasing of agricultural land: production economic aspects. *Agrekon,* Vol. 19 (2): 1–7.

Just, R.E. 1993. A model of farm investment and land prices. *Unpublished Working Paper.* University of Maryland.

Just, R.E. and Miranowski, J.A. 1993. Understanding farmland price changes. *American Journal of Agricultural Economics,* Vol. 75 (1): 156–168.

Locken, G.S. 1976. Alternative methods of estimating the use-value of farmland in New York. *MS thesis.* New York: Cornell University, Ithaca.

Locken, G.S., Nelson, L.B. and Richard, N.B. 1978. Estimating agricultural use values in New York State. *Land Economics,* Vol. 54: 50–63.

Paarlberg, D. 1962. Discussion: contributions of the new frontier to agricultural reform in the United States. *Journal of Farm Economics,* Vol. 44: 1179–1183.

Republic of South Africa. 1994. *Abstract of Agricultural Statistics.* Pretoria: Department of Agriculture.

Reynolds, J.E. and Timmons, J.F. 1969. Factors affecting farmland values in the United States. *Agriculture and Home Economics Experiment Station Research Bulletin 566.* Ames: Iowa State University, February.

Scofield, W.H. 1964. *Land Prices and Farm Earnings.* Farm Real Estate Market Developments. USDA-ERS, CD 66 (Oct): 39–47.

Van Schalkwyk, H.D. 1995. Modelling South African agricultural land prices. *Unpublished PhD thesis.* Pretoria: University of Pretoria.

Van Schalkwyk, H.D. and Groenewald, J.A. 1993a. Solvency, entrepreneurial action and the economic environment: lessons from the recent past. *Agrekon,* Vol. 32 (4): 270–275.

Van Schalkwyk, H.D. and Groenewald, J.A. 1993b. Agricultural land price and quality. *Development Southern Africa,* Vol. 10 (3) : 401–410.

Van Schalkwyk, H.D. and Van Zyl, J. 1993. The South African land market: an analysis of land prices. *Unpublished Research Report.* Halfway House: DBSA.

Van Schalkwyk, H.D., Van Zyl, J., Van Rooyen, C.J. and Kirsten, J.F. 1994. Market based rural land reform in Southern Africa – real possibility or pipe dream? The case of South Africa's commercial agricultural sector. *Paper delivered at the 22nd IAAE conference.* Harare, 23–29 August.

Van Wyk, S.P. 1967. Trends in land values in South Africa. *Agrekon,* Vol. 6 (1) : 23–30.

Van Zyl, J. and Groenewald, J.A. 1988. Effects of protection on South African commercial agriculture. *Journal of Agricultural Economics,* Vol. 39 (3): 387–401.

Van Zyl, J., Van Schalkwyk, H.D. and Thirtle, C. 1993. Entrepreneurship and the bottom line: how much of agriculture's profits is due to changes in price, how much to productivity? *Agrekon,* Vol. 32 (4) : 223–229.

Van Zyl, J., Van Rooyen, C.J., Kirsten, J.F. and Van Schalkwyk, H.D. 1994. Land reform in South Africa: options to consider for the future. *Journal of International Development,* Vol. 6 (2) : 219-239.

World Bank. 1994. South African agriculture: structure, performance and options for the future. *Discussion Paper 6.* Washington, D.C.: Southern Africa Department, World Bank.

13

Agricultural growth linkages: international experience and some South African estimates

Johann Kirsten and Johan van Zyl

Introduction

This chapter pursues a further motivation for the introduction of a land reform programme: the existence of rural growth or farm–non-farm linkages and the belief that smaller farms have stronger linkages, mainly due to production being more labour intensive, and a more egalitarian distribution of benefits resulting in smaller leakages. It stresses the argument that a land reform programme could lead to increased farm income and, through the existence of these linkages, contribute to growth in the rural economy and, in particular, the rural non-farm economy. This positive indirect contribution associated with a land reform programme could be one of the major benefits of, and thus motivation for, land reform[1].

The link between agriculture and the rural non-farm economy is a crucial aspect in this respect. Farming is central to the rural economy and, as such, differences and changes in agriculture will explain much of the variation in rural non-farm activity. On the demand side of the rural non-farm economy, agriculture exerts a large influence, since non-farm enterprises depend primarily on the farm input and consumption demand of agricultural households. Driven largely by agricultural earnings, rural income levels determine the extent of consumer diversification into non-foods. Moreover, land distribution affects income distribution and hence the share of incremental expenditure allocated to rurally supplied, as opposed to imported, non-foods (*Haggblade & Hazell, 1989*). Agriculture also influences the supply side of the rural non-farm economy, primarily through the labour market.

1 Some of the authors in this volume (*see Aliber* – Chapter 23, and *Van den Brink, De Klerk & Binswanger* – Chapter 17) make use of the existence of these linkage or multiplier effects, to quantify the benefits of land reform (Chapter 23) and to estimate the additional livelihoods (Chapter 17) created by such a programme.

The international literature on agricultural growth linkages gives considerable evidence on the linkages generated by technological change in agriculture and investment projects in agriculture – the next section provides a summary of this evidence. However, as Lipton (*1993*) rightly points out, very little evidence on the linkage effects of land reform programmes on farm income and output, on secondary growth and poverty reduction emerged from this debate. On the other hand, it could be argued that a land reform programme dramatically changes the production relationships in much of the farm economy, for example a change to more labour intensive agriculture, thus implying some form of technological change which suggests that much of the discussion in the first part of the chapter could be relevant in the context of a land reform programme. In this sense, this chapter is exploratory, to a large extent, but nevertheless it makes a strong case in support of the existence of such linkages from a land reform programme. It stresses the importance of research programmes to quantify the extent of the linkages.

After initially reviewing the theory and evidence of agricultural growth linkages in the next section, the following section reviews South African studies and evidence to provide a sense of the extent of possible growth linkages.

Theory and international evidence on agricultural growth linkages

Many of the earlier studies on agricultural linkages to economic growth focused largely on production linkages arising from the establishment of new production activity. These linkages were classified into backward and forward linkages. Backward linkages consist mainly of derived demand for inputs from the new activity. Forward linkages consist of the induced creation of new productive activities from having a new intermediate product on the market. Earlier writers in development theory (*Prebisch, 1959; Hirschman, 1958*) argued that agriculture has weak linkages and that it should not be a big priority in fostering growth in developing countries. This generated the 'agricultural pessimism in development theory' during much of the 1950s and 1960s (*see Eicher & Staatz, 1992*).

The writings of Johnston and Mellor (*1961*), Nicholls (*1964*), and Mellor (*1966*) again encouraged economists to view agriculture as a positive force in economic growth and development. These authors also stimulated debate on the interdependence of agricultural and industrial growth which, in turn, led to a growing interest in the measurement of intersectoral resource transfers.

The case for agriculture as an engine for economic growth was enhanced through increased focus on the impact that growth in the agriculture sector has on incomes, and hence, rural demand for consumer goods and services from the non-farm economy (*Delgado & Alfano, 1994*). Against the background of India's experience with the green revolution, Mellor (*1966, 1976*) points out that although production linkages from the agricultural sector may in fact be weak – especially in the case of subsistence agriculture – consumption linkages from the agricultural sector do have major indirect effects on the rest of the economy. According to Mellor's view, the impact of agricultural growth on other sectors in the economy depends on how much of the extra wage income is spent on non-agricultural goods and services, leading to a new round of spending, and how much is spent on the increased food production itself, or leaks out into savings.

The 'linkages paradigm' (after *Delgado & Alfano, 1994*) was formulated in the 1960s and 1970s in response to the writings referred to earlier, but also in response to the concern over rapid urbanization and the growth in urban unemployed as a result of the inability of the more capital intensive urban economy to provide enough employment opportunities for the growing population. The perceived need was to find a way to create jobs outside agriculture and outside cities. This, according to Delgado and Alfano (*1994*), led to a focus on growth processes boosting demand for rural non-farm activities. Much of the literature during the 1970s and early 1980s, therefore, stresses the advantages of creating demand in rural areas for locally-produced non-food goods and services.

The theory of agricultural growth linkages (or the 'linkages paradigm') rests on the premise that agricultural growth can lead to substantial indirect growth in non-farm incomes and employment. These effects arise partly as a result of increases in the use of farm inputs, and in processing, marketing and transport services to handle the larger output. The indirect effects, however, also arise from increases in household expenditures on consumer goods and services as a result of increased farm incomes (*Hazell & Roëll, 1983*). Earlier studies by authors such as Johnston and Kilby (*1975*), highlighted the importance of production linkages or agriculture's demand links to the non-farm economy. Johnston and Kilby, together with authors such as Mellor and Johnston (*1984*), also argue the case for a small farmer development strategy on the basis that it will generate rapid and equitable growth because of more labour intensive linkages with the rural economy. Johnston and Kilby (*1975*) in particular, point to small farmer demand for fertilizer, construction inputs, equipment and repair services typically provided by small labour-intensive enterprises. Consequently, the indirect effects of agricultural growth have the potential to help alleviate rural underemployment and to contribute to the reduction of rural poverty and malnutrition (*Hazell & Roëll, 1983*).

Apart from the existence of production linkages, Hazell and Roëll (*1983*) and Mellor (*1976*) show the importance of consumption linkages in stimulating second round growth effects. Strong consumer expenditure linkages between agricultural households and the non-farm economy are important, firstly, because the income and employment generated by these linkages is predominantly concentrated in rural areas; and secondly, because the kinds of goods and services demanded are typically produced by small, labour intensive enterprises (transport, distributive trades, health, and housing and residential construction). Hazell and Roëll (*1983*) argue that households which spend the largest share of incremental income on goods and services not traded inside the regional economy (non-tradables), would contribute the most to the promotion of labour-intensive growth. These non-tradables typically include locally produced goods and services – often non-foods – that are consumed entirely within the region.

The categorization of commodities in the linkage literature into tradables and non-tradables is largely attributed to Siamwalla (*1982*), who has shown that leakages from net additional demand occur as a result not only of expenditures on imported goods, but also on exportables. The consideration here is that only new expenditure on non-tradables has the potential to create additional local income since they are the only goods that are demand-constrained.

Given the discussion above, a number of factors are listed that will determine the size of the indirect benefits induced by agricultural growth (*see also Hazell, 1990; Delgado & Alfano, 1994*):

- Agriculture should account for a large share of aggregate employment.
- The amount of extra income generated by farmers as a result of increased agricultural output. The greater the increase in farm incomes, the greater the incremental expenditure by farm households on consumer goods.
- The nature of agricultural growth must be such that the benefits of growth are widely spread, allowing the effective demand for goods and services of a broad base of rural people to increase.

- An important factor is the structure of rural household expenditure patterns. Expenditures of the direct beneficiaries of agricultural growth must be such that large shares of increments to income are spent on labour-intensive local non-agricultural goods and services, typically non-tradables, thereby stimulating demand for sectors that employ large numbers of rural people outside the tradable agricultural sector. This implies that indirect (or consumption) growth linkages are more likely to be of major importance where a major share of local economies is non-traded. The key determinant here is the propensity of rural households to consume non-tradable goods and services out of additional income.
- It is also important that there should be a supply of under-used local resources. If supply of local production factors and non-tradables are inelastic, then increased household demand for these goods and services will simply increase prices rather than employment and real incomes of labour.

These arguments can be summarized by arguing that growth that only benefits a small number of very large farmers, or a relatively small agricultural sector, would not have large rural consumption linkages for locally produced goods and services, the production of which would provide a great deal of local employment. On the other hand, growth that stimulates the incomes of large numbers of small farmers is likely to lead to widespread increase in demand for local consumer goods and services. The more these goods are demand-constrained, the greater the growth impact (*Delgado & Alfano, 1994*).

Given the fact that South African agriculture is characterized by an unequal distribution of land, assets and income,[2] a few large farmers will capture all the benefits of any growth in agricultural income. Thus there is the notion that any growth in agricultural income could not generate large farm – non-farm linkages resulting in very little growth in the rural economy. This argument is pursued in more detail towards the end of the chapter.

In conclusion, it should be stressed that agricultural growth linkages cannot be the basis for a development strategy themselves, since they have no sustained engine. They arise only in response to initial income flows from sales to outsiders (*Delgado & Hopkins, 1994*) or from

2 In Chapter 11 it was indicated that 16 100 farmers, or roughly 26 per cent of the total number of farmers, earn 81 per cent of the total gross farm income.

technological and structural change. Growth linkages occur because underemployed resources, such as labour or vacant land, are drawn into production by new local demand. This can only occur if there are underemployed resources, which is often the case in countries in Africa. Resources are assumed to be underemployed due to insufficient local demand to purchase the end products; non-local demand is not a factor, due to high transport costs. This situation typically arises because of remoteness and poverty, and may be associated with visible or hidden underemployment of labour or land.

Estimates of agricultural growth linkages

The magnitude of agricultural growth linkages is usually estimated as regional growth multipliers which measure the amount of extra income generated in a region from stimulating new production of goods and services through the increased consumer and intermediate spending that arises from an initial boost in household income (*Delgado & Hopkins, 1994 ; Delgado et al, 1994b ; Haggblade et al, 1989*). A growth multiplier of 1,80, for example, would imply that one unit of household income from higher yields or new cash cropping would eventually add 0,8 of a unit of income from new production of non-tradables, induced by household spending of the initial unit – thus implying a total increase in income of 1,8 units (*Delgado & Hopkins, 1994*).

Many of the studies on agricultural growth linkages over the last two decades focus largely on the linkages or multiplier effects of technological change, in particular the green revolution and agricultural investment projects.[3]

Most of the studies that estimated agricultural growth multipliers used fixed-price models including economic base models, input-output models, and semi-input-output tables (or social accounting matrices (SAMs)). To compute agricultural growth multipliers, the fixed-price models classify economic activities as tradable or non-tradable. Furthermore, the fixed price models share three common assumptions (*Haggblade, Hammer & Hazell, 1991*):

- Increased production of tradable goods is viewed as the driving force in regional economic growth.

3 *See* for example: King and Byerlee, 1978; Bell and Hazell, 1980; Bell, Hazell and Slade, 1982; Haggblade, Hazell and Brown, 1987; Haggblade and Hazell, 1989; Hazell, Ramasamy and Rajagopalan, 1991.

- Production is modelled as fixed-coefficient Leontief technology.
- Constant prices both for tradable and non-tradable output are assumed.

The fixed price models are attractive to practitioners because they are relatively easy to estimate and solve. But Haggblade, Hammer, and Hazell (*1991*) argue that the underlying assumption of a perfectly elastic supply of non-tradables is troublesome, which means that fixed-price models could overestimate the amount of indirect income generated within the regional economy. To correct this bias, Haggblade, Hammer, and Hazell (*1991*) developed a price-endogenous model which allows for increasing cost of non-tradable supply and less extreme assumptions about input substitution. Using data sets from previous fixed price model studies, the estimates of agricultural growth multipliers by the Haggblade, Hammer, Slade-model indicate that, as a rough rule of thumb, the Asian multipliers lie in the range of 90 per cent of the levels projected by the semi-input-output models (or 45 to 65 per cent as esti-

Table 13.1: A Summary of Estimates of Agricultural Growth Multipliers

Author(s)	*Locality*	*Methodology*	*Multipliers*
Rangarajan (*1982*)	India	Macro-economic model	1,70
Bell, Hazell and Slade (*1982*)	Muda, Malaysia	Semi-input-output model	1,80
Hazell (*1984*)	Muda, Malaysia	Semi-input-output model	1,82
Hazell, Ramasamy and Rajagopalan (*1991*)	North Arcot, India	Semi-input-output model	1,87
Hazell and Haggblade (*1991*)	India	Econometric models Semi-input-output model	1,46 – 1,93 2,37 – 2,54
Rogers (*1986*)	Mauritania	Semi-input-output model	1,27
Haggblade, Hazell and Brown (*1987*)	Sierra Leone and Nigeria	Semi-input-output model	1,50
Delgado, Hopkins and Kelly (*1994a*)	Burkina Faso Zambia Senegal Niger	Semi-input-output model	2,88 2,48 1,97 1,96

mated by Hazell & Haggblade, 1991), while the African multipliers stand at closer to 75 per cent of the fixed-price estimates.

A brief summary of the multiplier estimates of selected studies is provided in Table 13.1. However, the results should be interpreted in the light of the findings of Haggblade, Hammer and Hazell (*1991*), as discussed above.[4]

The extent and nature of farm–non-farm linkages in sub-Saharan Africa

In order to understand the nature of the farm–non-farm linkages in Africa, it is necessary to briefly discuss the study by Haggblade, Hazell and Brown (*1989*) in which they show the strength of intersectoral linkages in sub-Saharan Africa. In the context of land reform in South Africa, the appropriateness of a more detailed discussion of the linkage effects in sub-Saharan Africa is obvious. The important linkages are factor market linkages, capital and labour, and product market linkages through backward and forward production linkages, as well as consumption demand linkages.

Factor market linkages

It is generally believed that the capital flow from agriculture is substantial and often more than the reverse flow from non-farm activity to agriculture. The evidence from studies cited by Haggblade *et al* (*1989*) suggests that surpluses have consistently been transferred out of agriculture through fiscal, crop pricing and trade policies. But there is also evidence that surpluses generated from non-farm activities are used to acquire productive agricultural assets, especially land. The labour market in sub-Saharan Africa is characterized by substantial seasonal labour flows between the rural farm and non-farm sectors. It is estimated that 20 to 40 per cent of the labour force works in both farm and non-farm activities, thus resulting in considerable flows of labour moving back and forth between the rural farm and non-farm sectors.

Forward and backward production linkages

Agriculture's demand for production inputs is the reason for the existence of backward linkages between agriculture and the manufacturing and

4 For more detailed reviews of these studies, readers are referred to Haggblade, Hazell and Brown (*1989*), Haggblade, Hammer and Hazell (*1991*), and Delgado and Alfano (*1994*).

trade sectors. Often these inputs are supplied by rural enterprises that are considerably more labour intensive than their urban counterparts. The type and magnitude of backward linkages vary, depending on agricultural technology, size of holding, type of crop and whether production is irrigated or rain-fed. To some extent, they also depend on infrastructural and locational aspects within the rural economy. The lack of good roads and other infrastructure could influence the location of rural enterprises; it could happen that many of the backward linkages do not accrue to the region itself, but to the nearest metropolis or regional centre.

Haggblade *et al* (*1989*) argue that, in general, the backward linkages for agriculture in sub-Saharan Africa appear to be weaker than those measured in Asia. In particular, it is Africa's topography and hydrology that severely limit irrigation potential, and thereby reduce the demand for inputs, pumps and other irrigation equipment. The current farm technology in Africa is also responsible for the fact that backward linkages appear far smaller than the forward processing linkages from agriculture.

Food processing is by far the most prominent of the forward linkages, and virtually all processing activities involve transformation of local agricultural production. After food processing, distribution of agricultural products generates the second largest of the forward linkages from agriculture.

Consumption linkages

An increase in per capita farm incomes typically leads to increased demand for local services, housing, durables, livestock and horticultural products. Evidence from Asia suggests that the production of these commodities and services is labour intensive, which could thus lead to increased rural employment. Haggblade *et al* (*1989*) present data that shows that African spending patterns support far less rural non-farm activity than do those in Asia. Marginal budget shares are important predictors of consumption linkages to be anticipated from growing incomes. In African countries, such as Nigeria and Sierra Leone, it has been found that consumers spend only between 11 and 18 per cent of incremental income on rural produced non-foods, while the comparable Asian figures vary between 26 and 31 per cent. According to Haggblade *et al* (*1989*), this difference arises because African consumers spend far more of their average and marginal income on rural produced foods. It appears that transportation networks and proximity to rural towns contribute to the much higher Asian incremental consumption on rural produced goods and services.

The evidence cited by Haggblade *et al* (*1989*) suggests that lower farm–non-farm linkages in Africa are largely due to the high share of non-marketed goods and services in total consumption. Because they are not marketed, many rural African goods and services are not measured and, if values for consumption of home produced goods are not imputed, it is highly likely that farm–non-farm linkages can be underestimated. A study by Delgado *et al* (*1994a*) confirms that these consumption linkages cannot be ignored and that they are often underestimated and considerably stronger than suggested by the earlier studies. They found that the share of growth linkages attributable to consumption alone vary between 42 per cent in Senegal, 93 per cent in Burkina Faso and 98 per cent in Zambia.

A study by King and Byerlee (*1978*) in Sierra Leone found enough evidence to support the hypothesis that low income households consume goods and services which require less capital and foreign exchange and more labour than do the goods embodied in the consumption patterns of higher income households.

The impact of agricultural technology on the size of agricultural growth linkages

To conclude this section it is necessary, given the context of land reform, to briefly discuss the effects of the choice of technology and farm type on the size of the agricultural growth multipliers. Evidence from the literature suggests that different agricultural technologies generate different patterns of non-farm linkages to the extent that the input demand and the consumption patterns they generate differ. The input intensity, consumption profiles of targeted farms and processing characteristics of the farm output all affect the size and composition of non-farm spin-offs. The multiplier effects also vary across countries as a result of differing institutions, population densities, spatial settlement patterns and policy environments.

Haggblade and Hazell (*1989*) estimate that amongst the high-yielding-variety rice farms in Asia, the multiplier is the greatest when technological change is focused on medium- rather than small-sized farms. Given an assumption of identical savings rates, the multiplier is 1,74 for medium-sized farms and 1,55 for small-sized farms.[5] It should be

5 It should be kept in mind that the farms referred to here are in general extremely small, with the largest farms not covering more than 10–15 hectares. The arguments made here by Haggblade and Hazell should thus be seen against this background, also taking account of the fact that most of the studies analysed by Haggblade and Hazell mainly included rice farms in several countries in Asia.

remembered that these results also assumed that small- and medium-sized farms used the same technology. The higher multiplier for medium-sized farms is the result of stronger household consumption linkages due to the higher income levels of these households, as found by Hazell and Roëll (*1983*).

In their analysis of one region each in Malaysia and Nigeria, Hazell and Roëll (*1983*) came to the conclusion that households on the larger farms in both regions have the most desired expenditure patterns for stimulating secondary rounds of growth in the local economy. It should be qualified, however, that the farms analysed in the two case studies are not particularly large by most standards, and also excluded wealthy estate owners. In the Muda region of Malaysia, the largest farms are 28 hectares on average while, in the Gusau region of Nigeria, the largest farms are 9 hectares on average, which is not particularly large in terms of South African commercial agriculture, but considerably larger than the allotted pieces of land cultivated by black farmers in the former homelands.

Haggblade and Hazell (*1989*) also argue that large agricultural estates in Africa and Asia – similar to the commercial farms in South Africa – generate smaller multipliers due to their more urbanized household expenditure patterns with low marginal budget shares for rural non-tradables. Only if their expenditure patterns approach those of smaller-sized farms do estates generate more favourable multipliers.

It has been found that farms in Asia growing high-yielding varieties of rice show a larger multiplier than those farms growing traditional varieties (1,56 versus 1,38). Farms growing traditional varieties, on the other hand, require fewer tradable inputs, and their values to gross output are higher. But this positive contribution toward the multiplier is more than offset by lower household demand linkages as a result of lower incomes and smaller marginal budget shares for non-tradables (*Haggblade & Hazell, 1989*).

Finally, Haggblade and Hazell (*1989*) found that the choice of mechanization package does not affect the multiplier on Asian rice farms, although oxen cultivation does produce higher indirect income increments than hoe cultivation amongst Africa's smallholders. It is argued that greater mechanization increases the demand for non-tradable intermediates in production, which adds to the size of the multiplier. However, at the same time, the size in the multiplier is reduced due to a decline in the value added generated per unit of output. The size of the multiplier will therefore be determined largely by consumption linkages. If income is increased as a result of mechanization, and households have a high propensity to consume non-tradables, the multiplier will be larger.

Linkages and multipliers in South African agriculture

In this section, we review studies that estimated the intersectoral linkages between South African agriculture and the rest of the economy on a national, regional and household level. In the latter case it is particularly the expenditure patterns of rural households that are reported. These are valuable in estimating the possible impact on the rural non-farm economy through agricultural growth linkages (consumption linkages) in the aftermath of a land reform programme. However, this section will not attempt to estimate the growth linkages.

Intersectoral linkages

Agriculture's current and future importance to the South African economy lies, to a large extent in the sectoral linkages that exist, rather than in the direct contribution of agricultural production to GDP. In terms of forward linkages, agriculture has supplied raw materials for an expanding number of agricultural processing industries that create output and jobs (381 000 in 1976, according to *Van Zyl et al, 1988*).

Agriculture also creates a derived demand for goods and services through backward linkages. In 1992/93 farmers spent R503 million on packaging materials, R1 419 million on fuel, R1 069 million on fertilizers, and R882 million on dips and sprays. They further invested R931 million in tractors, machinery, and implements, and another R835 million in fixed improvements (fencing, buildings) (*RSA, 1994*).

In order to be able to provide an estimate of the forward and backward linkages of agriculture, Van Zyl *et al* (*1988*) calculated sectoral multipliers using the input-output methodology. Table 13.2 indicates selected sectoral multipliers calculated according to the 1978, 1981 and 1985 production structure of the South African economy. These multipliers reflect the overall effects, direct and indirect, produced by forward and backward linkages in the economy. According to the 1985 production structure, agriculture has the highest employment/production multiplier of the selected sectors, while the income/production multiplier is the second lowest. This suggest that when there is a general increase in production, the associated increase in agricultural production will create the greatest number of job opportunities throughout the economy, but that the accompanying rise in income will be comparatively low.

Table 13.2: Selected Sectoral Multipliers (direct and indirect effect) according to the Production Structures for the years 1978, 1981 and 1985

Year	Sector	Capital/ production multiplier (per unit)	Employment/ production multiplier (labourers per R mil)	Income/ production multiplier (per unit) per R mil)	Income/ capital multiplier (per unit)	Employment/ capital multiplier (labourers)
1978	Agriculture	2,4	244,1	0,90	0,38	101,7
	Gold mining	1,7	189,6	0,94	0,55	111,5
	Fertilizer and pesticide industry	2,1	133,1	0,70	0,33	63,4
	Agricultural machinery	1,3	127,9	0,86	0,66	98,5
	Electricity and gas	5,1	94,2	0,94	0,18	18,5
	Construction	1,7	263,1	0,87	0,51	154,8
	Trade	1,7	140,1	0,93	0,55	82,4
	Transport	4,0	122,7	0,89	0,22	30,7
	Services	2,3	27,7	0,97	0,42	12,0
1981	Agriculture	2,1	136,8	0,87	0,41	65,1
	Gold mining	1,5	88,3	0,94	0,63	58,9
	Fertilizer and pesticide industry	1,8	64,6	0,58	0,32	35,9
	Agricultural machinery	1,0	58,8	0,74	0,74	58,8
	Electricity and gas	5,6	56,1	0,93	0,17	10,0
	Construction	1,5	140,6	0,83	0,55	93,7
	Trade	1,6	78,2	0,93	0,58	48,9
	Transport	3,8	73,7	0,88	0,23	19,4
	Services	2,3	16,3	0,96	0,42	7,1
1985	Agriculture	2,8	100,9	0,88	0,31	36,0
	Gold mining	1,8	51,1	0,95	0,53	28,4
	Fertilizer and pesticide industry	2,2	43,2	0,76	0,35	19,6
	Agricultural machinery	1,3	43,9	0,86	0,66	33,8
	Electricity and gas	6,5	29,2	0,94	0,14	4,5
	Construction	1,6	82,8	0,88	0,55	51,8
	Trade	1,7	46,5	0,94	0,55	27,4
	Transport	4,4	45,2	0,90	0,20	10,3
	Services	2,2	6,4	0,98	0,45	2,9

Source: Van Zyl *et al* (1988).

From the multipliers provided in Table 12.2 it can be deduced that, with the exception of building construction, agriculture created the greatest number of job opportunities in 1981 and 1985 per unit increase in capital.

However, it should be noted that treating agriculture as a single sector in the national input-output tables, may partly obscure the relative importance of agriculture on a regional level, and the fact that a large number of other production activities, especially manufacturing sub-activities, are depending on agriculture as the destiny of their outputs and as a source of their inputs. It is possible in this regard to think of fertilizer and agricultural equipment as sub-activities of manufacturing, as well as the various food processing activities such as grain milling, meat processing, and canning of fruits and vegetables. It is against this background that Van Seventer *et al* (*1992*) undertook an analysis of 'agribusiness' in an attempt to decompose the backward and forward linkages of agriculture in a more detailed manner.

The results of this study show that agribusiness processing activities have higher backward linkages than the agricultural sub-activities. The opposite seems to apply with regard to the forward linkages, which show higher values for the agricultural sub-activities compared to the processing activities. A comparison of agribusiness and non-agricultural activities shows that the former have relatively larger backward linkages in terms of income and employment. Non-agricultural activities show relatively higher forward income and employment effects. Mixed results are obtained in respect of the combined linkages. Agribusiness scores higher with regard to employment, and non-agricultural activities with regard to income.

Linkages between commercial irrigation farming and the rural non-farm economy

In this sub-section, we report on an earlier study by Kirsten (*1989*) which estimates the farm–non-farm linkages between the irrigation areas and the regional economy of the south-western Orange Free State. The results from this study (reported in *Kirsten & Van Zyl, 1990*) are discussed to give some indication of the farm–non-farm linkages and multipliers of smaller commercial farms on irrigated land. Most of the farms included in the study consisted of an irrigation plot that varied in size from 11 to 120 hectares plus, on average, an additional 200 to 1 000 hectares or more in natural grazing land. The irrigation technology also varied between farmers with the smallest farms using flood or sprinkler irrigation (both of which are more labour intensive), and the larger plots using centre pivots.

The profile of the non-farm economy of the south-western Free State provides an indication of the interaction between the irrigation farmers in the region and the non-farm enterprises. This is summarized in Table 13.3.

The bulk of economic activity in the region is aimed at servicing the primary sectors. It can be seen from Table 13.3 that little or no inputs for agriculture are manufactured within the region. The bulk of the products produced by irrigation farmers are also processed and consumed elsewhere. However, agriculture is still an important sector in the regional economy, contributing 25 per cent to the Gross Geographical Product. Irrigation farming in the region, comprising only 1 per cent of the total surface area, provides 40 per cent of the total output of agriculture. In estimating the linkages of irrigated agriculture with the regional non-farm economy, use was made of the input-output methodology. Output, income and employment multipliers were calculated, using a regional input-output model.

The output multiplier for irrigation agriculture is 1,7012 which means that, for every R1 of output in irrigation agriculture, an additional output of 70,12 cents is generated in the intermediate sectors and households within the region. It is important to note that the output multiplier for irrigation agriculture is among the highest of all the sectors in the region. The output multiplier is comprised of the following:

Initial impact	:	1,0000
First round effects	:	0,4647
Secondary effects	:	0,1818
Consumption-induced effects	:	0,0487
Total output multiplier		1,7012

It is worth noting that the consumption-induced effects are considerably lower than estimated in other studies reported earlier, contributing only 7 per cent to the indirect linkages. This could be attributed to the relatively high income levels of the irrigation farmers and the skew distribution of income amongst the minority white farmers. These households do have consumption patterns which are characterized by high marginal propensities to consume for imported goods. These farmers are able to

Table 13.3: Profile of the Farm and Non-farm Economy in the South-western Orange Free State, 1985

	Magisterial District				
Item	Fauresmith	Jagersfontein	Philippolis	Jacobsdal	Koffiefontein
Number of farmers	205	40	107	189	31
Land utilization:					
Irrigation	5 588 ha	365 ha	1 125 ha	7 749 ha	2 218 ha
Dryland cultivation	115 111 ha	0	0	10 000 ha	12 000 ha
Natural grazing	463 163 ha	127 840 ha	341 532 ha	242 546 ha	176 496 ha
Type and number of businesses in region:					
Co-operatives	3	1	1	2	1
General dealer	7	5	4	11	5
Café	3	5	3	3	–
Dairy	–	1	–	–	1
Chemist	–	2	–	2	1
Retail: Clothes	1	2	1	–	1
Retail: Furniture	–	–	1	–	1
Filling stations/ garage	3	1	3	4	3
Tractor dealer	–	1	–	1	2
Hotel	2	–	1	2	1
Repairs and services	1	2	1	–	1
Petroleum product/ dealer	1	–	–	–	3

Marketing of products by irrigation farmers within the region (%)

	Fauresmith	Jagersfontein	Philippolis	Jacobsdal	Koffie-fontein	Outside region
Wheat	1,5	0,0	0,0	66,7	6,1	25,7
Maize	0,0	0,0	0,0	27,3	0,0	72,7
Cotton	0,0	0,0	0,0	80,0	0,0	19,2
Potatoes	23,1	0,0	0,0	0,0	0,0	76,9
Vegetables	0,0	0,0	0,0	1,0	2,0	97,0
Grapes (wine)	0,0	0,0	0,0	100,0	0,0	0,0
Lucerne	11,0	0,0	0,0	33,0	0,0	55,6

Purchasing of inputs and equipment by irrigation farmers (%)

	Fauresmith	Jagersfontein	Philippolis	Jacobsdal	Koffie-fontein	Outside region
Seed	13,5	0,0	0,0	62,2	8,1	16,2
Seed potatoes	23,5	0,0	0,0	23,5	5,8	47,1
Fertilizer	11,4	0,0	0,0	64,5	8,9	15,2
Packaging	22,6	0,0	0,0	41,9	19,4	16,1
Fuel	5,1	0,0	0,0	50,0	10,2	34,6
Chemicals	6,7	0,0	0,0	62,2	8,1	23,0
Tractors	3,7	0,0	0,0	33,9	9,4	52,8
Implements	10,0	0,0	0,0	46,9	10,9	31,3
Irrigation equipment	18,5	0,0	0,0	38,5	3,1	40,0
Parts	10,4	0,0	0,0	55,8	3,9	29,9

Source: Kirsten and Van Zyl, 1990.

drive to the nearest city (Bloemfontein or Kimberley) to purchase consumables, durables and often some of their inputs. Thus, very little benefit will flow to the regional economy if these farmers experience an increase in income. These leakages through expenditure on regional imports are responsible for the low consumption linkages. This result confirms the findings and observations of Hazell and Roëll (*1983*) and Haggblade and Hazell (*1989*) discussed earlier. The farm–non-farm linkages are further highlighted by the sectoral distribution of the indirect multiplier effects as shown in Table 13.4. It is important to note that the strongest linkages are with the trading sector which captures around 42 per cent of all the indirect effects.

The income multiplier for irrigation agriculture which measures the change in income resulting from the initial output, was estimated at 15,3 cents. On the other hand, the employment multiplier for irrigation agriculture was estimated as 89,64. This means that for every R1 million output produced by the irrigation areas in the south-western Free State, 89,64 persons are employed, of which 67,8 persons are employed by the irrigation farmers. The rest are employed in other sectors of the regional economy as a direct result of agricultural production under irrigation.

Similar results on the linkages and multipliers of irrigation agriculture were found in later studies using the input-output methodology (*Botha, 1991 and Du Plessis, 1994*).

Table 13.4: Sectoral Distribution of Indirect Effects

Sector	Indirect effects (cents)	Percentage of total (%)
Rest of agriculture	4,04	5,76
Irrigation agriculture	5,34	7,62
Mining	0,00	0,00
Manufacturing: Food	0,77	1,09
Manufacturing: Liquor	0,07	0,09
Manufacturing: Textiles	0,01	0,01
Manufacturing: Non-metal minerals	0,57	0,81
Electricity	9,88	14,23
Construction	0,25	0,35
Trade	29,66	42,29
Transport and communication	9,00	12,84
Finance	4,72	6,73
Services	5,71	8,14
Total:	70,12	100,00

Source: Kirsten and Van Zyl, (1990).

Agricultural growth linkages: expenditure patterns of rural households in the former homelands

In the spirit of King and Byerlee (*1978*) and Hazell and Roëll (*1983*), this sub-section endeavours to describe rural households' expenditure patterns, average budget shares and expenditure (income) elasticities to get some feel for the extent and nature of consumption linkages in rural South Africa. Given the assumption that land reform could lead to increased household incomes, it is important to know how rural households would spend their increased income. Knowledge of expenditure patterns of rural households, therefore, is important in order to be able to estimate the impact of a land reform programme on the rural non-farm economy and the rural economy as a whole.

This section provides some evidence on the expenditure patterns of rural households based on case studies in two of the former homelands – Lebowa and Venda, reported in Dankwa *et al* (*1992*), and Van Zyl and Vink (*1992*), as well as KwaZulu, as reported by Nieuwoudt and Vink (*1989*).

Haggblade, Hammer and Hazell (*1991*) argue that marginal budget shares for non-tradable goods drive the estimates of agricultural growth multipliers. The studies reviewed here have only calculated average budget shares and expenditure elasticities for selected commodities. However, these estimates would provide a good indication of consumption demand linkages given the relationship between expenditure elasticity (ξ), marginal budget share (MBS) and average budget share (ABS), as identified by Hazell and Roëll (*1983*):

$$\xi = \text{MBS} / \text{ABS}$$

Table 13.5 presents the expenditure patterns (average budget shares) among emerging and subsistence farmer households in the two localities.[6]

Table 13.5: Average Budget Shares for Farmer Households in former Lebowa and Venda, 1991

Item	Lebowa		Venda	
	Emerging farmers	Subsistence farmers	Emerging farmers	Subsistence farmers
Food/groceries:				
Maize meal	2,95%	8,62%	–	–
Other food	15,26%	9,93%	–	–
Total food	–	–	17,26%	19,60%
Household expenses	6,33%	2,49%	–	–
Personal (cosmetics, soap, etc)	–	–	7,98%	4,22%
Transport	4,39%	2,39%	4,88%	5,43%
Clothing	12,19%	9,35%	17,23%	15,61%
Savings	21,85%	–	16,81%	15,88%
Durable household expenditures	18,27%	11,19%	4,24%	8,89%
Farm expenses	12,19%	14,82%	–	–
Instalments	–	–	9,42%	13,53%
Education	6,58%	41,21%	15,31%	9,45%
Medical	–	–	6,87%	7,39%

6 Emerging farmers are farmers who make use of support services to increase agricultural production. In many cases, these farmers rely to a larger extent on income from farming, while the subsistence farmers produce less and rely more on remittances and pensions. For a more detailed discussion on the performance of farmers under this programme, *see* Kirsten (*1994*).

Lebowa households in the group of emerging farmers spend more on other food (15,26 per cent) but very little on maize meal (2,99 per cent). Expenditures on clothing, savings, farm expenses and household durables are relatively high. However, expenditures on transport (4,39 per cent) and education (6,58 per cent) are fairly low. The relatively low expenditure on maize meal supports the perception that these households produce comparatively more maize on their farms as compared to the subsistence group.

In Venda it was found that the group of subsistence households spend relatively more on food/groceries (19,60 per cent) than the other group (17,26 per cent). The emerging farmers spend relatively more on education and personal items (cosmetics, soap, etc), and vice versa for all other expenditure items.

The study by Dankwa et al (*1992*) also provides some results on expenditure elasticities for selected commodities. Table 13.6 summarizes the elasticities for both emerging and subsistence farmer households in Lebowa and Venda based on a log-linear model. Each equation was estimated using Ordinary Least Squares (OLS) procedures. The table also provides estimated elasticities of the effect of household size on the expenditure of selected items.

Both the individual group and pooled results confirm Engel's Law. The estimated food expenditure elasticities are less than one. The results show that the emerging farmers are able to spend more on other food as compared to the subsistence households. However, the estimated expenditure elasticities for maize of the subsistence households (–0,72) were higher than the emerging farmers (–0,08), which indicates that the households in the latter group spend less on maize meal than the former. The expected positive relationship between household expenditures, farm expenses and education is confirmed by the estimated elasticities of 0,75 and 0,89 respectively. The elasticity estimate on the effect of family size on other food was negative and highly significant. The income effect therefore dominates, meaning that an increase in family size of the subsistence households makes the group relatively poorer to the extent that other food becomes luxurious items to buy.

Table 13.6: Estimates of Elasticities for Selected Expenditure Items for Rural Households in Lebowa and Venda, 1991

Item	Lebowa			Venda		
	Emerging farmers	Subsistence farmers	Pooled	Emerging farmers	Subsistence farmers	Pooled
Maize meal						
β_1	–0,08	–0,72	–0,24	0,99*	0,22	0,91**
	–(0,24)	–(0,63)	–(0,71)	(5,07)	(0,42)	(5,49)
β_2	0,65	0,15	0,72	1,22**	1,59	1,29**
	(0,78)	(0,03)	(0,87)	(3,48)	1,57	(4,09)
R^2	0,10	0,15	0,11	0,45	0,20	0,39
Other food						
β_1	0,94**	0,72	0,88**	0,78**	0,80**	0,80**
	(40,92)	(1,64)	(4,88)	(8,99)	(7,09)	(12,06)
β_2	–0,41	–5,83**	–0,72	0,17	0,22	0,06
	–(0,91)	–(3,47)	–(1,61)	(1,14)	(0,97)	(0,44)
R^2	0,24	0,62	0,36	0,60	0,84	0,64
Household expenses						
β_1	1,29**	0,17	1,14**	1,10**	1,37**	1,14**
	(4,79)	(0,29)	(4,62)	(7,53	(4,00)	(9,11)
β_2	0,17	0,28	0,29	–0,69	–0,14	–0,53*
	(0,26)	(0,13)	(0,28)	–(2,64)	–(0,20)	–(2,25)
R^2	0,33	0,02	0,29	0,64	0,62	0,49
Transport						
β_1	0,94**	–0,18	0,79**	1,58**	1,81**	1,66*
	(4,72)	–(0,66)	(4,38)	(6,19)	(3,41)	(8,08)
β_2	–0,72	0,12	–0,68	0,45	–1,26	0,09
	–(0,53)	(0,11)	–(1,52)	(0,99)	–(1,21)	(0,25)
R^2	0,43	0,05	0,29	0,42	0,56	0,44
Clothing						
β_1	0,80**	0,92	0,84**	1,10**	1,66**	1,15**
	(3,49)	(1,66)	(3,99)	(4,73)	(4,43)	(6,03)
β_2	0,02	1,93	0,11	–1,16**	1,05	–0,75*
	(00,04)	(0,90)	(0,22)	(–2,78)	(1,42)	–(2,06)
R^2	0,24	0,36	0,26	0,27	0,69	0,30

Item	Lebowa			Venda		
	Emerging farmers	Subsistence farmers	Pooled	Emerging farmers	Subsistence farmers	Pooled
Clothing						
β_1	2,39**		2,14**	0,77**	0,98*	0,92**
	(60,99)		(6,69)	(3,62)	(2,43)	(5,26)
β_2	0,22		0,05	–0,54	1,27	–0,34
	(0,28)		(0,07)	–(1,42)	(1,60)	–(1,02)
R^2	0,39		0,34	0,17	0,47	0,24
Durable household expenditure						
β_1	2,33**	2,17*	2,32**	0,79**	0,25	0,64**
	(6,29)	(2,47)	(6,85)	(6,64)	(1,65)	(6,27)
β_2	–1,11	3,19	–0,84	0,18	0,05	0,18
	(–1,27)	(0,95)	–(1,01)	(0,85)	(0,16)	(0,93)
R^2	0,34	0,52	0,35	0,44	0,22	0,33
Farm expenses						
β_1	0,75**	0,58	0,75**	0,94	1,11*	0,94**
	(4,79)	(0,84)	(4,79)	(4,65)	(3,92)	(5,92)
β_2	–0,03	1,75	0,04	–0,89	–0,25	–0,81**
	–(0,08)	(0,66)	(0,10)	(2,46)	–(0,45)	–(2,69)
R^2	0,23	0,15	0,21	0,26	0,61	0,30
Education						
β_1	0,81	1,28	0,89**	1,27**	1,30*	1,34**
	(4,69)	(1,43)	(4,91)	(4,54)	(2,60)	(5,81)
β_2	1,79**	0,60	1,69**	0,53	1,89	0,61
	(4,41)	(0,18)	(3,79)	(1,06)	1,92	(1,39)
R^2	0,37	0,22	0,32	0,29	0,52	0,32

* = coefficient is significant at 5 per cent level

** = coefficient is significant at 1 per cent level

β_1 = Expenditure elasticity

β_2 = Household size elasticity

Numbers in parentheses are t–statistics

Estimated elasticities of the effect of family size on education were significant for the pooled results (1,69) and the emerging group (1,79). The positive sign indicates that the 'specific effect' dominates, which means there is an increase in the need for education expenditures as family size increases. Since the estimated elasticities are greater than one, education is also considered a luxury good. None of the estimates of the effect of family size on the other expenditures are significant.

In Venda, Engel's law is again confirmed by the individual group and pooled results. The estimated elasticities for all food products are less than one and significant. The pooled expenditure elasticity was 0,80, also highly significant (t = 5,49) at both levels. The pooled elasticity estimates for clothing, savings and transportation for Venda households were all significant. These estimates were: clothing 1,14 (t = 9,11); savings 1,66 (t = 8,09) and transportation 1,15 (t = 6,03). The fact that the elasticity estimates for clothing, savings and transportation exceed one indicates that these are luxury items.

The elasticity estimates of a number of studies in South Africa's rural areas are compared in Table 13.7 to provide an overview of the nature of expenditure patterns in rural areas. The results from the Hazell and Roëll (*1983*) study in Gusau, Nigeria were also included to provide further comparison. A striking aspect of the information included in Table 13.7 is the similarity in the size of the elasticity estimates across regions and countries.

The results from the discussion above provide a good starting point for the estimation of consumption linkages. The next step would be the classification of consumption items into tradables and non-tradables. None of the studies discussed above have done this classification, and without further knowledge about the locality aspects of consumption, no classification can be done. Given the nature and extent of the former homeland economies, it can be expected that many of the items listed in the tables would be tradables and would normally be imported from urban areas or from neighbouring, previously 'white' controlled areas. But, with the removal of apartheid and the formation of new regions and provinces, one could view this from a different perspective and consider imports from neighbouring formerly 'white' rural towns as non-tradables. This, however, still needs some debate. Non-tradables would usually include staple food (home-produced and exchanged with neighbours), and education. Consumption of most of the other items represents linkages to other regions. To enable proper results and estimates of consumption linkages, it would be necessary to embark on further research of rural expenditure patterns. The studies listed above, however, are a good start to enable researchers to estimate the agricultural growth linkages to be expected from increased household income resulting from a land reform and rural restructuring process.

Agricultural growth linkages and land reform: a synthesis

The lessons from the review of the literature on agriculture's linkage effects, within the context of land reform, can be summarized as follows: agriculture is the key sector in the rural economy, and usually has strong backward (production) and forward (production and consumption) linkages with the non-farm sectors. Two factors are crucial for a viable and vibrant rural non-farm economy:

- a healthy agricultural sector;
- strong linkages between agriculture and the non-farm economy.

Backward linkages are generally the least important of the different linkage effects. Higher input use in more intensive agriculture associated with modern biological technology produces stronger linkages than traditional less intensive practices. The effect of mechanization on these linkages is inconclusive, and can strengthen or weaken them. The type and magnitude of backward linkages vary depending on agricultural technology, farm size, crop and agricultural potential (land quality, climate and water availability).

In large-scale agriculture, food processing is the most prominent of the forward linkages, while consumption linkages are small. On the other hand, in small-scale agriculture, consumption linkages are far more important than the production linkages due to expenditure patterns of smallholders who generally support the local economy. Total forward linkages (production and consumption) are stronger in small-scale agriculture.

Growth that only benefits a small number of very large farmers does not have large rural consumption linkages for locally produced goods and services, the production of which would provide a great deal of local employment. On the other hand, growth that stimulates the incomes of large numbers of small farmers is likely to lead to widespread increase in demand for local consumer goods and services.

Large-scale farms generate smaller multipliers due to their more urbanized household expenditure patterns with low marginal budget shares for rural non-tradables. Only if their expenditure patterns

Table 13.7: A Comparison of Elasticity Estimates for Rural Households in South Africa

Expenditure Item	Venda (*Dankwa et al*, *1992*)		Lebowa (*Dankwa et al*, *1992*)		Venda (*Van Zyl & Vink*, *1992*)		KwaZulu (*Nieuwoudt & Vink*, *1989*)		Gusau, Nigeria (*Hazell & Roell*, *1983*)	
	ABS@	Elasticity**	ABS	Elasticity**	ABS	Elasticity*	ABS	Elasticity*	ABS	Elasticity**
Staple food	–	0,91	2,9%	0,24	4,5%	0,414	N/A	0,40	–	–
Other food	17,3%	0,80	15,3%	0,88	19,6%	0,956		0,868	0,6%	0,94
Household expenditure	–	1,14	6,3%	1,14	7,1%	0,967		0,93	4,3%	1,02
Transport	4,9%	1,66	4,4%	0,79	3,6%	1,039		0,86	1,9%	1,41
Clothing	17,2%	1,15	12,2%	0,84	15,9%	1,266		1,27	7,2%	1,24
Savings	16,8%	0,92	21,8%	2,14	11,5%	0,894		2,61–		
Durables	4,2%	0,64	18,3%	2,32	10,9%	1,078		1,24	1,1%	1,25
Farm expenses	–	0,94	12,2%	0,75	14,3%	0,935		0,87#	–	–
Education	15,3%	1,34	6,6%	0,89	12,5%	1,061		0,93	1,1%	1,42

Notes: @ Average Budget Share
* Income elasticity
** Expenditure elasticity
\# Not an aggregate estimate – only for the Mbongalwane region of KwaZulu

approach those of smaller-sized farms do they generate larger multipliers.

These findings provide a strong argument for a land reform programme in South Africa. The creation of smaller farms with more people getting access to agricultural income will strengthen consumption linkages which will be beneficial for rural economic growth. In addition, the choice of technology on these smaller farms, given the crushing rural poverty and unemployment, should favour labour-use. Policy should also explicitly consider strengthening the farm–non-farm linkages, since this will provide a strong impetus for rural growth, employment creation and poverty alleviation. In this respect, it is important that a land reform programme span wider than the redistribution of agricultural land. Not everyone will benefit directly from land redistribution. Since the landless and the near-landless households depend on non-farm earnings, support of rural non-farm enterprises should therefore also feature high on the agenda to ensure additional rural livelihoods.

References

Bell, C.L.G. and Hazell, P.B.R. 1980. Measuring the indirect effects of an agricultural investment project on its surrounding region. *American Journal of Agricultural Economics,* 62 (1) : 75–86.

Bell, C., Hazell, P. and Slade, R. 1982. Project evaluation in regional perspective. Baltimore: Johns Hopkins University Press.

Botha, S.J. 1991. Die direkte en indirekte ekonomiese gevolge van waterbeperkings vir gebruikers van Vaalrivierwater oor die tydperk 1983 tot 1987. (*The direct and indirect economic impact of water restrictions on water users in the Vaal River region during 1983 to 1987.*) *Unpublished MCom dissertation.* Bloemfontein: Department of Agricultural Economics, University of the Orange Free State.

Dankwa, K.B., Sartorius von Bach, H.J., Van Zyl, J. and Kirsten, J.F. 1992. Expenditure patterns of agricultural households in Lebowa and Venda: effects of the Farmer Support Programme on food security. *Agrekon,* 31(4) : 225–230.

Delgado, C.L. and Alfano, A. 1994. Agricultural growth linkages: relevance and research issues for Africa. In Delgado, C.L., Hopkins, J.C. and Kelly, V.A. 1994. Agricultural growth linkages in sub-Saharan Africa. *Unpublished IFPRI report.* Washington, D.C.

Delgado, C.L. and Hopkins, J. 1994. The impact of changing export sector incomes on local rural economies in sub-Saharan Africa. *Paper prepared for the Workshop on 'The Role of Agriculture in promoting Sustainable Economic Development in Africa*. Switzerland: Hotel Chaumont, Neufchatel, 30 October – 3 November.

Delgado, C.L., Hopkins, J.C. and Kelly, V.A. 1994a. Agricultural growth linkages in sub-Saharan Africa. *Unpublished IFPRI report.* Washington, D.C.

Delgado, C.L., Hazell, P., Hopkins J.C. and Kelly, V.A. 1994b. Promoting intersectoral growth linkages in rural Africa through agricultural technology and policy reform. P*rincipal paper prepared for the Annual Meetings of The American Agricultural Economics Association.* California: San Diego, August 7–10.

Du Plessis, L.A. 1994. Die ontwikkeling van verliesfunksies en 'n rekenaarmodel vir die bepaling van vloedskade en vloedbeheerbeplanning in die benede Oranjeriviergebied. (*The development of loss functions and a computer model for the determination of flood damage and flood control in the lower Orange River.*) *Unpublished MSc Agric dissertation.* Bloemfontein: Department of Agricultural Economics, University of the Orange Free State.

Eicher, C.K. and Staatz, J.M. (Eds), 1992. *Agricultural Development in the Third World.* (2nd edition). Baltimore: Johns Hopkins University Press.

Haggblade, S., Hammer, J. and Hazell, P. 1991. Modelling agricultural growth multipliers. *American Journal of Agricultural Economics,* 73 : 361–374, May.

Haggblade, S., Hazell, P. and Brown, J. 1987. Farm–non-farm linkages in rural sub-Saharan Africa: empirical evidence and policy implications. *World Bank Discussion Paper, No. 67.* Washington, D.C: Agriculture and Rural Development Department, World Bank.

Haggblade, S., and Hazell, P. 1989. Agricultural technology and farm–non-farm growth linkages. *Agricultural Economics,* 3 : 345–364.

Haggblade, S., Hazell, P. and Brown, J. 1989. Farm–non-farm linkages in rural sub-Saharan Africa. *World Development,* 17(8) : 1173–1201.

Hazell, P.B.R. 1984. Rural growth linkages and rural development strategy. *Paper prepared for the fourth European Congress of Agricultural Economics.* Kiel: 3–7 September.

Hazell, P.B.R. 1990. Agricultural growth linkages and the alleviation of rural poverty: importance and implications for agricultural and macro models. In Pinstrup-Andersen, P. (Ed), Macro-economic policy reforms, poverty and nutrition: analytical methodologies. *Cornell Food and Nutrition Policy Program. Monograph No. 3.* February.

Hazell, P.B.R. and Ramasamy, C. 1991. *The Green Revolution Reconsidered: The Impact of High-yielding Rice Varieties in South India.* Delhi: Oxford University Press.

Hazell, P.B.R., Ramasamy, C. and Rajagopalan, V. 1991. An analysis of the indirect effects of agricultural growth on the regional economy. In Hazell, P.B.R. and Ramasamy, C. 1991. *The Green Revolution Reconsidered: The Impact of High-yielding Rice Varieties in South India.* Delhi:Oxford University Press.

Hazell, P.B.R and Roëll, A. 1983. Rural growth linkages: household expenditure patterns in Malaysia and Nigeria. *Research Report No. 41.* Washington, D.C.: IFPRI.

Hirschman, A.O. 1958. *The Strategy of Economic Development.* New Haven: Yale University Press.

Johnston, B.F. and Kilby, P. 1975. *Agriculture and Structural Transformation: Economic Strategies in Late-developing Countries.* New York: Oxford University Press.

Johnston, B.F. and Mellor, J.W. 1961. The role of agriculture in economic development. *American Economic Review,* 51 (4) : 566–593.

King, R. and Byerlee, D. 1978. Factor intensities and locational linkages of rural consumption patterns in Sierra Leone. *American Journal of Agricultural Economics,* 60 (2) : 197–206.

Kirsten, J.F 1989. Die Ekonomiese Impak van Besproeiingslandbou in die Suidwes Vrystaat. (*The Economic Impact of Irrigation Agriculture in the Southwestern Free State*). *Unpublished MSc Agric dissertation.* Pretoria: Department of Agricultural Economics, University of Pretoria.

Kirsten, J.F. 1994. Agricultural Support Programmes in the developing areas of South Africa. *Unpublished PhD dissertation.* Pretoria: University of Pretoria.

Kirsten, J.F. and Van Zyl, J. 1990. The economic impact of irrigation agriculture: methodological aspects and an empirical application. *Development Southern Africa,* 7 (2) : 209–222.

Lipton, M. 1993. Land reform as commenced business: the evidence against stopping. *World Development,* 21(4) : 641–657.

Mellor, J.W. 1966. *The Economics of Agricultural Development.* New York: Cornell University Press, Ithaca.

Mellor, J.W. 1976. *The New Economics of Growth: A Strategy for India and the Developing World.* New York: Cornell University Press, Ithaca.

Mellor, J.W. and Johnston, B.F. 1984. The World Food Equation: interrelationships among development, employment and food consumption. *Journal of Economic Literature,* XXII : 531–574.

Nicholls, W.H. 1964. The place of agriculture in economic development. In Eicher, C.K. and Witt, L.W. (Eds), *Agriculture in Economic Development.* New York: McGraw-Hill.

Nieuwoudt, W.L. and Vink, N. 1989. The effects of increased earnings from traditional agriculture in South Africa. *South African Journal of Economics,* 57(3) : 257–269.

Prebisch, R. 1959. Commercial policy in the underdeveloped countries. *American Economic Review,* 49 : 251–273.

Rangarajan, C. 1982. Agricultural growth and industrial performance in India. *Research Report No. 33.* Washington, D.C.: IFPRI.

Rogers, G.R. 1986. The theory of output-income multipliers with consumption linkages: an application to Mauritania. *PhD dissertation.* USA: University of Wisconsin-Madison.

RSA. 1994. *Abstract of Agricultural Statistics.* Pretoria: Department of Agriculture.

Siamwalla, A. 1982. Growth linkages: A trade-theoretic approach. *Mimeo.* Washington, D.C.: IFPRI.

Van Seventer, D.E.N., Faux, C.S. and Van Zyl, J. 1992. An input-output analysis of agribusiness in South Africa. *Agrekon,* 31(1) : 12–18.

Van Zyl, J., Nel, H.J.G. and Groenewald, J.A. 1988. Agriculture's contribution to the South African Economy. *Agrekon,* 27(1) : 1–9.

Van Zyl, J. and Vink, N. 1992. Effects of Farmer Support Programmes on consumption and investment. *Development Southern Africa,* 9(4) : 493–499.

Options for Land Reform

PART IV

14

Overview of land reform issues

Robert Christiansen

Introduction

This chapter is concerned with the broad issues of rural restructuring and development, but with a special emphasis on the potential of smallholder agriculture and the changes necessary to achieve that potential. The thesis that underpins this chapter is that greater reliance on smallholder production offers South Africa the opportunity of achieving significant gains in terms of social justice, employment and agricultural potential. The central, but not exclusive, ingredient in this strategy is a land reform process. The balance of this chapter is concerned with examining options for how that process might work in South Africa. As a starting point, the chapter briefly reviews lessons from international and South African experience.

Land reform: lessons from experience

Historically, land reform programmes have been associated with changes in political regimes. The most thorough and rapid programmes have followed revolutions (e.g. China, Eastern Europe and Cuba) or occupation after military defeat (e.g. Japan, Taiwan and Korea). More benign land reform programmes and ones with less fundamental changes have followed the end of colonial rule, the shift of power from an immigrant settler community or relatively peaceful ideological shifts (e.g. Algeria, Kenya, Zimbabwe, Bolivia and Mexico).

The five salient lessons to emerge from international experiences in land reform are listed below:

1. **The speed of implementation** of the programme. One characteristic of a successful programme is rapid implementation. In the absence of fast-paced programmes, a combination of bureaucratic inertia, legal challenges, and the power of present is likely to render the programme ineffective.
2. **Economic viability of the farm models**. Before a reform programme

is implemented, there must be a careful assessment of the models or livelihood options available to rural households. That is, the models should indicate whether the persons resettled on the land have sufficient land size and quality to provide at least the target income. Further, in computing the costs and benefits, other assistance and infrastructure necessary to generate the income should be planned.

3. **Political acceptability and legitimacy** of the programme. There must be a consensus across the spectrum of political opinion that the programme is both necessary and the most acceptable way of achieving the stated goals. Land reform programmes are not irreversible, particularly where this consensus has not been achieved. Thus, for instance, in Chile, the overthrow of the Allende regime resulted in the reintroduction of the former skewed landownership patterns during the Pinochet regime.
4. **Clear definition of the role that the public sector can and will play**. The proposed programme must be evaluated in the light of an understanding and acceptance of the roles that the public sector can and must play, and what should be best left to the non-governmental sector. Programmes that have relied entirely on the public sector in the belief that it is the only sector capable of maintaining integrity, delivering services, determining needs, and managing the process, have been failures.
5. Land reform is only one part of **a comprehensive programme of economic reconstruction**. The redistribution of land is necessary, but not sufficient to guarantee the success of a development programme. There is the need for additional services – infrastructure, markets, incentives, health – to be considered and access provided. These considerations are necessary both to sustain higher productivity consequent on reform and to include others who may not benefit from the direct provision of land.

Land reform: invasion, restoration and redistribution

In the context of a land reform or restitution programme, there are three basic means for landless people to gain access to land in any country: (a) land invasion; (b) restoration; and, (c) redistribution. Of these three broad mechanisms, this chapter advocates a combination of restoration and redistribution; however, it is recognized that where there is uncertainty about future access to land, invasion is often seen as the only option available to the landless population. This section examines each of these mechanisms and develops options that may be able to provide guidance to a South African government faced with the necessity of designing and implementing a land reform programme.

Invasion

When people are not given access to land either through restoration or redistribution options, one alternative is land invasion. The process of land invasion would present a new South African government with an exceedingly difficult problem. The credibility of existing legal institutions will have been undermined because of their inability to implement the land restitution process effectively. Consequently, people who have invaded land are unlikely to respond to court orders to vacate, and forced evictions may become necessary.

Given the likely inability of the government to undo land invasion, it is essential that an expeditious, transparent, and thorough land restoration and redistribution process be implemented as a means of discouraging the landless from resorting to land invasions. Although squatting and trespass are likely to be decriminalized, landowners will continue to have remedies to protect their property rights. In the event of invasion, the landowner will retain rights to seek a court order requiring the squatters to vacate the property. Under these circumstances, landowners would have the right to demand that a court order be implemented, failing which they may claim that the state had effectively confiscated their property and must pay compensation in accordance with the constitutional protection against expropriation without compensation. Invasion, therefore, will not usually result in lower fiscal cost of land acquisition than expropriation in public interest or purchase of land. However, decriminalization will make it necessary to develop additional disincentives to invasions that could include the denial of grants under the redistributive process to groups participating in invasions, once the restoration and redistribution programmes are functioning.

An outstanding issue concerns the status of land invasions that occurred prior to the implementation of new programmes. One option is to subject these invasions to judicial review. If the invaded land is state land, the problems may be resolved through the fast-track procedure of administrative land allocation. However, if the invaded land is private property, the status may be resolved either through adjudication or, alternatively, the state may expropriate and offer compensation. However, the communities who have settled on land prior to the launching of the programme will not be subject to the penalties envisaged against those who invade once the programme is underway.

Restoration

The role of a land claims process in restructuring existing patterns of land occupation and use in South Africa will paradoxically be both limited and all important. It will interact closely with the broader process of rural restructuring and the land redistribution process. If the redistributive process is inadequate or too slow, people are likely to frame their needs in terms of claims, which could get bogged down in legal processes.

The land restoration (or land claims) process involves the return – by means of an administrative or adjudicative process – of specific parcels of land to individuals or communities who were unjustly removed in pursuance of racially-based land legislation or policies.[1] This process has the benefit of expediting the restoration of specific land; however, it raises important gender and class issues which will need to be addressed. Restoring the *status quo ante* could reimpose the original patterns of land holdings, in some cases, including a partriarchical ownership structure to the exclusion of women and in some cases tenants.

Eligible claimants

In order to facilitate the process of recognizing and identifying claimants, it is possible to define categories of removals which sought to further a particular apartheid ideology. These qualifying acts may be identified through the differentiation of four types of legal rules which characterized racial land law in the period since 1913: (a) rules dividing the land surface into race zones; (b) removal provisions implementing resettlement in terms of an apartheid legislative map, e.g., s 5(1)(b) of the Native Administration Act 38 of 1927 and s 46(2) of the Group Areas Act 36 of 1966; (c) other removal provisions employed to achieve apartheid objectives, e.g. Prevention of Illegal Squatting Act, trespass, expropriation, slums acts, etc; and, (d) provisions which had the effect of denying persons of a particular race group access to land in certain areas by criminalizing their 'unlawful' presence, e.g. pass laws.

Of these four types, only the specific removal provisions implementing resettlement are employed to achieve apartheid objectives; categories two and three would be deemed to be the basis of claims for lost land.

Categorizing the rural claims that are likely to arise from these two sets of rules, it is possible to identify four broad categories of rural forced removals to which restoration will apply: (a) 'black spot' removals; (b)

1 For a full discussion of the land claims process, readers are referred to Chapter 15. Only a number of important issues regarding the restitution process are discussed in this section.

'homeland' consolidation removals; (c) labour tenants and squatters; and, (d) betterment schemes.

(a) **'Black spot' removals.** These were the removals of blacks from pockets of black-owned land (or land to which blacks at least had legally recognized rights of some kind) which remained in the 'non-black' rural race zones after the coming into operation of the Natives Land Act of 1913 and the Development Trust and Land Act of 1936. Since 1913, and particularly from the mid-1950s onwards, it was the policy of the South African government to force people living in these areas to move to one or more of the self-governing areas or newly 'independent' states. Given the clear statutory evidence available and the more recent nature of these claims, they provide the clearest cases for unrestricted restoration.

(b) **Homeland consolidation.** Using the National States Constitution Act of 1971 and the Borders of Particular States Extension Act of 1980, the South African government, in co-operation with the national state or self-governing territory concerned, amended the borders of the various homelands so as to include areas previously excluded from the system. It thereafter became possible for the homeland governments concerned both to ignore any registered rights that the affected communities might have had in respect of the land in question, and to remove such people under the guise of privatization.

(c) **Labour tenants and squatters.** This class of persons poses difficult questions as to the appropriate form of redress which should be given. Most of the evictions of labour tenants took place under the provisions of Chapter IV of the Development Trust and Land Act, until the abolition of labour tenancy in 1979. After the repeal of Chapter IV in terms of the Abolition of Influx Control Act, No. 68 of 1986, a new section (Section 6F) was inserted into the Prevention of Illegal Squatting Act, providing for the ejection of persons occupying land outside the jurisdiction of a local authority who were not employed by the owner or legal occupier of such land. There are several questions raised as to the nature of the property rights to be given to such labour tenants and what compensation if any should be paid to the existing owner for the loss of part or all of his/her land.

Despite the questions, it is essential to recognize the birthright of long-standing land occupiers in the restoration process. Labour tenants and squatters who have suffered eviction will have a claim to restoration. However, given the difficulties of determining the extent of such claims, and the relationship between tenancy under Roman-Dutch law and birthright under indigenous conceptions of the land ethic, a number of remedies must be available to this group. First, if claimants wish to continue farming and are still resident or recently evicted, a portion of the farm may be awarded to them collectively. Second, if there is concern that restoration will lead to unresolvable conflict with the owners, claimants may obtain alternative land or monetary compensation. Finally, in cases where the evidence is inadequate to demonstrate a birthright to a particular portion of an existing farm, but the claimants want access to land, such claimants should be directed to the redistributive process where they should be given priority.

(d) **Betterment planning**. The policy of betterment planning, whereby many black rural areas were completely restructured in a misguided attempt to promote agricultural production, was first introduced in 1939. In order to give effect to the policy, hundreds of thousands of people in the homelands were relocated from their dwellings to closer settlements, while arable land and commonage were arranged into larger, contiguous units. Very little compensation was paid. However, due to the extent of the claims and the potential for intra-community conflict, it would be advisable to direct claimants away from the restoration process to the redistribution process.

From the choice of particular apartheid rules and the intention to provide restitution for past discrimination, a valid limitation emerges. This is the exclusion of claims from landowners who may have been expropriated by this policy, but who are not members of communities who suffered legal discrimination under apartheid. Justification for this approach is based on the fact that white property owners were adequately compensated and, furthermore, had access to the political process denied to the victims of apartheid.

Land claims forums and process

Because the majority of the claims are likely to be by groups or communities, it is necessary that a suitable form of class action be recognized. The existence of multiple claimants requires the establishment of clear notice requirements. Further, given the nature of some of the claims, there will be a need to relax certain formal evidentiary rules, to enable the land claims court to obtain the information necessary to make its decisions. This may require some form of sliding scale of evidentiary burden, i.e. the older the case the greater the burden, so as to balance the uncertainty and lack of living witnesses while, for more recent cases, there would be a lesser evidentiary burden. Closely related to this is the cut-off date of 1913. However, to avoid discrediting the process, it is important for the land claims court to have the power to relax this barrier, as a matter of equity, in cases of clear demonstrated injustice.

Redistribution

It is important that the reader understand what this section is intended to accomplish. The options and ideas put forward here should not be seen as prescriptions or a blueprint. Instead, they should be regarded as a starting point and a set of tools that can be useful in designing a programme for redistributing land in South Africa. Although these options cover many aspects of land redistribution, they do not address all the questions that will need to be answered. For example, while it is recognized that in an environment of rising land prices speculators may try to enter the programme, no solutions are offered to this problem.

The salient points that emerge from the options put forward in this section are that:

- market-assisted land redistribution programmes tend to perform better than those administered and operated by the public sector;
- the role of the public sector in a land redistribution programme centres on ensuring adequate supplies of land in the market and monitoring the overall operation of the programme;
- criteria for participation are necessary and must be discussed and agreed upon in advance;
- welfare objectives can be met by including a grant component in the programme;
- a matching grant scheme that forces participants to use some of their own resources in order to gain access to land will help to assist in self-selection of participants and encourage the productive use of land;
- the grant elements of the programme are essential in order to accomplish a redistribution of assets and to ensure that beneficiaries emerge from the programme with a net increase in their asset position and low debt/asset ratio as a means of ensuring viability and sustainability of their enterprises;
- in addition to addressing the fundamental issue of social justice, these options are likely to increase net rural employment significantly and ensure that the cost of the programme is very reasonable;

- a redistribution programme will not be able to provide land for everyone, and the programme will need to be complemented by a rural safety net and by programmes for urban groups;
- although the discussion concentrates on agriculture and small farmers, these are not the limits of the programme. It is envisioned that the programme will be a vehicle for supporting a wide range of land use activities, including trading activities and small-scale enterprises.

It is anticipated that in South Africa the majority of land reform activities will occur under the heading of land redistribution rather than restoration. Models of land redistribution can be thought of as consisting of two stages – acquisition and distribution of land. The central issue in the way these two stages are implemented is the role of the public sector. Although there are numerous variations in the role that the public sector plays in these stages, it is possible to think of each stage as being either market-assisted or publicly administered. Figure 14.1 depicts the four general models of redistribution that emerge from this classification, and some examples.

	Market-assisted acquisition	Administrative acquisition
Market-assisted distribution	Both stages handled by market mechanisms, such as a willing buyer–willing seller model. The role of the state is limited to monitoring and facilitating the process.	In this case, the state acquires the land (through expropriation or direct purchase) and relies on a market device for distribution, e.g. a bidding process.
Administrative distribution	In this case, the state acquires land in the market and then administers the settlement programme, e.g. homeland consolidation and betterment schemes.	An example of an administratively handled land redistribution is the process of directed resettlement on state-owned or expropriated land.

Figure 14.1: Models of Land Redistribution

International and South African experience clearly indicates that the performance (as measured by pace and extent of land reform as well as the performance of settlers) of models that are dominated by the public sector – through various administrative devices such as centralized decision-making – is typically unsatisfactory. In contrast, reliance on market forces improves the performance of the model, but requires government intervention or guidance to ensure that certain social objectives are achieved. The redistribution options described in this chapter (for both the acquisition and distribution stages) rely on market mechanisms, combined with a recognition of the need for government to ensure that the broader social objectives of the land redistribution programme are achieved.

Options for redistribution

Apart from the role of the public sector in land reform, one of the central tensions in designing a land redistribution model is the tension between the desire to address welfare and asset transfer objectives through the redistribution of land (essentially transferring an asset), and the desire to promote the productive use of land. Some individuals who qualify for land or assistance under welfare objectives of a programme, may have little experience in agriculture. In contrast, the most experienced and well-qualified farmers are unlikely to qualify to receive land under welfare objectives. Some redistribution models have taken the view that those individuals qualifying under welfare guidelines will acquire the necessary farming skills, but because of the time required for such skill acquisition, this approach is more costly. However, this difficulty should not be cause for sacrificing welfare and asset redistribution objectives in a land redistribution programme.

To the extent that a South African land redistribution programme wishes to address welfare objectives, one option is for the land redistribution programme to include a basic grant. The basic grant would be available to all individuals who meet the requirements for participation in the redistribution programme. The size of the individual grants would be determined largely by budgetary considerations and the needs of the participants, but would likely be sufficient to pay for a rural housing site. In the context of rural restructuring, a basic grant scheme would provide the South African government with a very flexible tool. If the government opts for a large welfare (grant) component to the land redistribution programme, the size of the basic grant could be increased – subject to budget constraints – thereby allowing grant recipients access to more land. For such a basic grant scheme to operate effectively, the government will need to ensure that the land market provides opportunities for beneficiaries to gain access to land. Options for the government to fulfill this responsibility are described below.

In order to support increased access to land by individuals or groups who will use land for production, a matching grant option can be added to the programme. Under this option the individual or group who wishes to gain access to land (through purchase or lease arrangements) would provide a portion of the purchase price as would the redistribution programme. Therefore, the resources for the purchase (or lease) price of the land would consist of a matching grant from the government – which can be separated or combined with the pure grant component discussed above – and the beneficiaries' co-payment. The size of the matching grant could be designed in several ways; however, it is assumed that the higher the value of land the smaller the matching grant share. (In terms of Figure 14.1, this option corresponds to relying on market-assisted devices for both acquisition and distribution.)

An example of how the matching grant option would work is provided in Figure 14.2. Any purchase of land up to the value of point A would be entirely paid for by a grant (i.e. 100 per cent grant financing). In designing the programme, the value of point A would be determined largely by budgetary considerations. Referring to line I – which provides a 50 per cent subsidy above the basic grant – in Figure 14.2, any purchase of land between points A and B would receive the basic grant (point A) plus 50 per cent of the additional land value. For example, if A = R5 000 and B = R20 000, and an individual or group wanted to purchase a piece of land valued at R17 000, the total grant available to the individual would be R11 000 (R5 000 + (.5)R12 000). An alternative formulation of the grant function (line II) provides a decreasing subsidy as the value of land to be purchased increases. For example, referring to line II, if an individual wishes to purchase a piece of land valued at R16 000 (point C), the total grant available to the individual would be R9 400 (R5 000 + (.4)R11 000).

Apart from budgetary considerations, there are other advantages and disadvantages associated with a matching grant or co-payment requirement. Among the advantages are the following: (a) co-payment acts as a self-selection or rationing device and reduces frivolous requests for land from people who are not really interested in living in rural areas or operating a farm; (b) co-payment increases the incentives for using land in a productive and sustainable manner; (c) co-payment may encourage prudent thrift and capital accumulation among the beneficiaries, and familiarity with financial instruments; and, (d) co-payment makes it less attractive for the wealthy to claim land, especially if the co-payment requirements increase progressively with the amount of land claimed. Therefore the programme will have much fewer built in incentives for arbitrary allocations of the land or corruption.

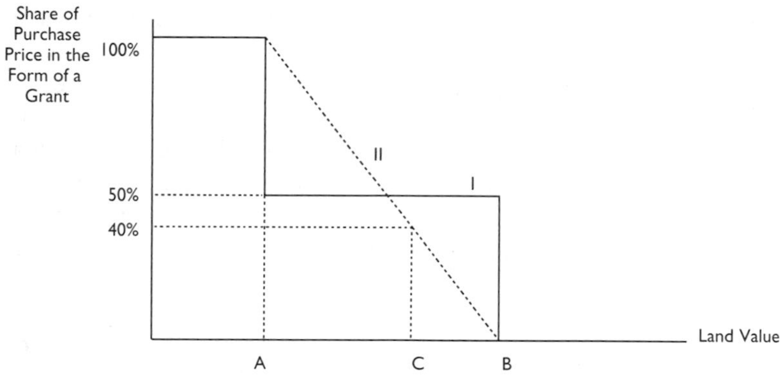

Figure 14.2: A Grant Function Based on Land Value

At the same time, however, a matching grant scheme has disadvantages. First, such a requirement may be inconsistent with the prevailing ideology that the right to land for purposes of shelter and food production should be free, especially in view of past abuses in South Africa. Second, if a co-payment is required, low income individuals and households are likely to be excluded from the programme, thereby subjecting the programme to charges of elitism. Clearly these disadvantages centre on the tension between the welfare objectives and the desire to have land used productively. It is for this reason that the options presented in this chapter are based on two components – basic grants and co-payments. This approach allows some land to be available (through the grant component) for housing (or other uses) even if the beneficiary is unable to participate in the co-payment component of the programme. Any additional access to land through the redistribution programme would require a co-payment. Within this context, government would have the option to make the programme more equitable, and to reduce the incentives for corruption, by placing a ceiling on the size of the government's co-payment. In order to further reduce the burden of co-payment requirements of the programme, poor beneficiaries might also be allowed to pay for their co-payment portion in kind, i.e. in the form of labour or materials for the construction of the basic infrastructure which would be part of the programme.

Another question that needs to be considered is who will receive the matching grants and how much land (or grant value) will they receive? International experience clearly indicates that the characteristics of those who participate in a land redistribution programme are an important factor in determining the success or failure of the programme. In the context of a land redistribution programme, there are two broad methods for determining who will participate in the programme: (a) self-selection; and, (b) criteria based on certain characteristics of the beneficiaries. These two methods can operate individually or be combined in a variety of ways. The basic self-selection model is one in which any individual or group could participate in the programme. (Even in this option some simple criteria are likely to be applied, e.g. participants need to be historically disadvantaged and South African citizens.) In terms of the matching grant model discussed above, an individual or group would have their contribution to the purchase price of the land automatically matched in accordance with the particular co-payment or grant function selected for the programme. For example, an individual (or group) wishing to purchase a piece of land would compute the value of the matching grant they were eligible for (typically based on the value of land) and their required payment. Once their required payment was deposited in an escrow account and their application to the programme was validated, the matching component would be deposited in the same escrow account. The account could then only be used for a land purchase.

The application of criteria to the programme simply means that individuals or groups need to be certified as having the necessary qualifying characteristics before being allowed to participate in the programme. For example, points could be assigned for various characteristics that are judged to be desirable, e.g. relevant experience, social circumstances, and/or gender. On the basis of total points, an individual (or group) would qualify for a particular level of grant financing. In Figure 14.3, an individual or group receiving more points (reflecting the characteristics valued by the programme) would be eligible for a higher share of grant up to some limit, say 50 per cent (or 100 per cent if a full grant component was judged necessary for the programme). The example shown in Figure 14.3 would have the characteristics that individuals with points less than level D would receive the land on a grant basis, while individuals with more points would obtain a grant, but with a co-payment requirement. No additional grant would be provided to individuals with a very large number of points above point E.

A variation of the points system could involve qualifying individuals with certain characteristics, e.g. tenants, landless people, or victims of betterment planning, for immediate participation in the programme. Alternatively, criteria could be based in large measure on means testing. In such an option, it needs to be stressed that decisions regarding who participates in a land redistribution programme must be accepted by society in order for the programme as a whole to have credibility.

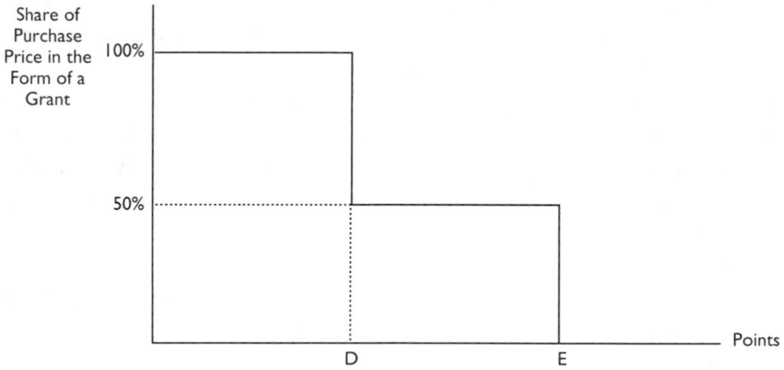

Figure 14.3: A Grant Function Based on Points Criteria

Ensuring the availability of land

As stated earlier, a central theme of the land redistribution options described in this chapter is a reliance on market forces for facilitating both the acquisition and redistribution of land (Figure 14.1). At the same time, it is recognized that the government must accept ultimate responsibility for ensuring the success of the redistribution programme. A critical requirement for a market-assisted redistribution programme is the assurance that sufficient land will come onto the market, and that land prices will not be driven up beyond levels that correspond to the productivity and profit potential of the land. Among the options for achieving these objectives are the following:

- Elimination of policies and programmes that favour large-scale agriculture will reduce the land price (which is simply a capitalization of these privileges into the land price). Much of this has already happened over the last few years and has contributed to the decline in real land prices.

- State land can either be sold or directly provided to particularly needy beneficiary groups such as former labour tenants or victims of dislocation associated with betterment schemes. Sale of state land will increase the supply of land, while direct distribution will reduce the demand for land by beneficiaries. Since low quality land is unsuitable for direct settlement, only the unoccupied, high quality, state-owned land should be used for direct distribution. The remaining land can be sold, with the profit going to purchase more appropriate land. Of the total amount of state-owned land, it is estimated that only about 600 000 hectares are suitable for crop production, of which only about 320 000 hectares are unoccupied.
- If land supply is still a problem, another option would be to encourage landowners, through a variety of means, to sell their land. For example, what amounts to a pension scheme (based on the net worth of the farm) could be designed to provide current owners with a secure income stream based on the net worth of their farms. If the current owners perceive the pensions as providing a secure income stream, rather than risky farm profits, they are likely to be willing to sell the land for a present value of pension benefits which is lower than the market price of the land. The pension scheme has the additional advantage that it allows for the deferral of the fiscal cost of the land purchase over time. To reduce the perceived risk of the pensions, a portion could be guaranteed by the South African state, while another portion of the pensions could be guaranteed by an international donor consortium against political risk. A time limit on the pension offers would encourage additional land supply for the critical first few years in which much land needs to change hands to satisfy the land hunger.
- Another option for increasing the supply of land would be to encourage the Agricultural Credit Board and other financial intermediaries to accelerate the foreclosure on farms with non-performing loans. For a certain number of years the banks would be recapitalized for a certain percentage of their loan losses, say 30 per cent, on farms which change hands from current owners to any of the beneficiary groups, either in a direct sale, or as a consequence of foreclosure. Finally, the state could pay brokers a fee for bringing additional land onto the market.
- On the demand side, the credit system is likely to place limits on the mortgage loans that beneficiaries could apply to their land purchase.

Even today, for example, the Land Bank's mortgage finance is capped with an upper limit defined by the productive value of the land. Thus, by limiting the price which beneficiaries are able to bid, such a lending policy will help to avoid price wars between competing beneficiary groups.

- If all the above measures fail to stimulate an adequate supply of land in a given district, an institution such as a district land committee could be empowered to recommend expropriation of specific farms to a national land committee under the constitutional provision for expropriation in the public interest. In order for the benefits of a market-assisted system to be realized, this option needs to be used only rarely, if at all. Such action, therefore, should be triggered by a clearly specified set of indicators, such as an excessive rise in the real land price or an excessively low level of transactions relative to beneficiary demand. A recommendation to activate this option would have to be reviewed by the national or regional land committee to ensure that common policies with respect to the trigger indicators are satisfied before the district land committee could expropriate with compensation. If an expropriation order is issued, the land committee would then be able to distribute the land through the normal market-assisted mechanisms to programme participants.

Summary

The operation of the land reform (or land restitution) programme is summarized by the flow diagram in Figure 14.4. Beginning at the top of the diagram, individuals (or groups) that wish to gain access to land have three choices: (a) seek land through the restoration process; (b) acquire land through the redistribution channel; or, (c) purchase land without assistance from the programme. The criteria for participation in the restoration process will be well-known, and any individuals (or groups) that qualify would submit an application to the appropriate institution (e.g. a land claims court, a land commission or a land committee). Having entered the claims process, there are four possible outcomes: (a) a land award; (b) a cash award; (c) the denial of the claim; or, (d) conversion of the claim to the redistribution process. The latter outcome would be an option for a claimant who felt that the claims process would be too time-consuming and who was willing to accept compensation through automatic participation in the redistribution process. In designing the programme, it may be decided that special consideration should be given to such individuals (or groups) in the redistribution process.

Hence, claimants who relinquish their claim to restoration could be put in the front of the queue or awarded additional points if that is the relevant model.

The individuals or groups wishing to participate in the redistribution process (or channel) would apply to the relevant institution. If the applicants meet the criteria (which would need to be set at the national level after consultation with regions), they would be responsible for identifying and assessing land. The beneficiaries would be responsible for identifying a piece of land, but would be eligible for assistance in this activity. This assistance could be used to find and assess land, as well as provide advice on various land use options. Once a suitable piece of land is identified, financing could be arranged from three possible sources: (a) the grant component available from the programme; (b) the beneficiaries' own resources; and, (c) a bank loan. Once the sale (or lease) is completed, the beneficiaries would then be eligible for support services.

Land reform: outstanding issues

The various options described above leave unanswered several questions about redistribution and how it will operate. The present section attempts to address two of these questions: (a) is it appropriate to compensate current landowners for assets acquired under the apartheid model; and, (b) recognizing that land redistribution will not provide agricultural land for the entire rural population, what safeguards and/or alternative programmes are necessary?

Should current landowners be compensated?

Whether or not current landowners will be compensated is of concern because these landowners have acquired land with the assistance of racially motivated apartheid policies. If they are compensated at the prevailing market price, some would argue that they have benefited from apartheid. Ultimately, the government will take the decision on this issue, but there are several points that might inform the decision.

First, the current owner is likely to have acquired the land in a bona fide transaction, typically as a purchase in the market. In such cases, the previous owners were the primary beneficiaries of apartheid, since many of the privileges associated with apartheid policies would have been capitalized into the prevailing land prices.

Second, in addition to the former landowners, apartheid policies benefited much of the white population who had no direct interest in agricultural land. This includes not only the providers of agricultural input, marketing and financial services, many of whom captured rents from anti-competitive restrictions at the expense of farmers and consumers, but also the mining, manufacturing and service sectors, which benefited from the cheap labour and restrictions on entry of black entrepreneurs and competitors. Therefore, failure to compensate current owners of farmland for the price of land would impose the cost of a land redistribution programme on only one group of those who have benefited from apartheid, namely the current owners.

Third, many of the current landowners are heavily indebted, with the most severely indebted ones being most likely to leave the farm sector.

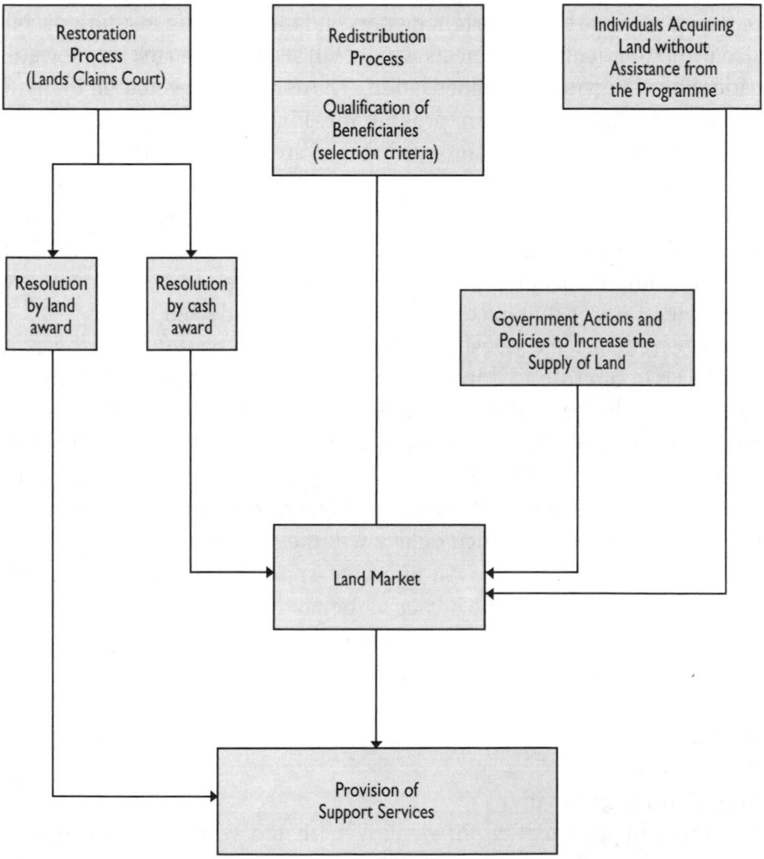

Figure 14.4: Land Reform or Restitution Process

Confiscation would mean that the capital loss would be shared between the net worth of the current landowner and the debt claim of the concerned lender. Consequently, a large-scale programme of land redistribution without compensation would not only eliminate the net worth of the current owners, but could also bankrupt many of the agricultural lending institutions. If, as is likely, these institutions are needed to finance the new owners in the future, the new government will find it necessary to refinance the losses of the lending institutions associated with the confiscation of the land. Thus, even if compensation is not paid, the state will probably need to pay, at least, for the unrecoverable farm debts of the farms it takes over.

Fourth, failure to pay compensation will be viewed as politically unacceptable, not only by the current owners and their lending institutions, but also by non-agricultural interests which will see this as a first step towards nationalization without compensation. These interests would be likely to oppose the entire land reform programme. Finally, external investors and South Africans with capital invested abroad are likely to react by reducing their investment in or repatriation of capital to South Africa.

Even though there are many reasons to regard the past acquisition of farmland as being of doubtful legitimacy, it is the inequities and the lack of practicality associated with any confiscation or compensation at less than market prices that makes such an action ill-advised.

There remains the question whether the compensation to be paid should be at or close to a market price. While there are legitimate reasons to tax the current landowners for their share in the gains from apartheid, taxation in the form of a reduction in the compensation paid to them may not be advisable. Such efforts will be viewed as expropriation without full compensation and will be resisted and contested, possibly with lengthy delays. Such delays will inevitably be at the expense of the potential beneficiaries. An alternate approach would be to tax the capital gains of the current owner associated with the compensation at market prices, or to include the payment of the current owners in a broader restitution levy on all wealth.

Rural poverty

Scope of rural poverty

The most important ways of dealing with the widespread poverty in South Africa's rural areas will be increasing the incomes of poorer people through distribution of productive assets and accelerated, labour-intensive growth of the economy. The first of these includes the land redistribution and supplementary rural sector initiatives, described earlier in this chapter. The welfare of the poor will also be enhanced if agricultural policies help to lower food prices in real terms. While these various initiatives will help to address rural poverty, there will continue to be people who need additional assistance. This section examines options for addressing the needs of these people.

An Urban Foundation study measured the proportion of households whose incomes fell below a minimum living level, and showed that about 42 per cent of households in the country were in poverty in 1990, the majority in rural areas. The study estimated that, in rural areas, 5 per cent of white households, 8 per cent of Asian households, 40 per cent of 'coloured' households and 68 per cent of black households had living levels below the poverty line. In rural areas of the former bantustans, as many as 83 per cent of black households fell below the poverty line. Studies carried out between 1990 and 1992 in KwaZulu, Transkei and Bophuthatswana, showed that the poorest groups lacked access to wages, remittances or social transfers. They accounted for about 4 per cent of the population, and had monthly per capita incomes of R32 (male-headed households) and R23 (female-headed households). For this poorest group, more than half of their income came from agriculture; in no other group did more than 10 per cent of their income arise from agriculture.

The poorest groups described above lack the wherewithal to benefit from greater access to land. The proportion of female-headed households is higher for these groups than for others. It is likely that better access to employment, to activities of the informal sector, and to welfare payments, will be more important for raising the incomes of these groups than access to land alone.

Welfare

The poorer groups receive substantially less of welfare expenditure per capita than others. The majority of the budgets are allocated to old age pensions (68 per cent for whites, 40 per cent for 'coloureds', 39 per cent for Indians, and 77 per cent for blacks). The next largest category is disability payments, followed by family and child care, and finally very small amounts spent on poor relief. The family and child care category may become very significant for blacks in the future, since such provisions (especially for children's health) are increasingly regarded by UNICEF as vital for the welfare of children and, on the demand side, the large number of single parent households among the black population, to whom such welfare would apply most particularly, are only beginning to become aware of the service and apply for it.

The old age pension is an especially important source of income for blacks in rural areas. It provides regular payments, probably going disproportionately to women (who most need it), and the administrative structures for delivery are already in place. Serious efforts have been made by NGOs to increase the proportion of eligible persons who are actually enrolled. In 1991, it is estimated that around 71 per cent of eligible rural black families were receiving pensions. There remain 29 per cent who are not, none of whom receive pensions from private funds. However, at the same time, as more full coverage is being sought, there may be a need to decrease the level of the old age pension in order to provide funds for other programmes which target poverty more directly in families who do not have family members old enough to be eligible for the pension.

National Nutrition and Social Development Programme

The NNSDP, which began in 1990, is funded by government, administered by the Department of National Health and Population Development, and implemented by NGOs. The programme had difficulty developing appropriate working relationships between government and the NGOs, and the arrangements for this vary greatly among the regions. The programme relied initially on food parcels whose contents were inappropriately formulated to address under-nutrition. There have been concerns about corruption in the commercial arrangements for assembling the parcels, as well as with the competition the parcel system provides for trading stores, especially in remote areas.

In the absence of an effective nutrition surveillance system, allocation of food parcels has not been effective, and the programme has developed a marked urban bias. There have also been problems of coordination between the NNSDP, run by the Department's Welfare Division with a poverty focus, and the Protein Energy Malnutrition scheme, run by the Department's Nutrition Division with a focus on malnourishment. Ideally, there should be close collaboration between such schemes so that, when a child receives treatment for malnourishment, the family can also receive assistance on grounds of poverty. While a start was made during 1993 on addressing the various problems of these food programmes, much more work needs to be done to increase their effectiveness. The establishment of a nutrition surveillance system, in particular, has yet to be started.

The country's welfare system will need to be widened over the next few years to increase access by households without employable adults, where adults cannot work away from the home without reducing necessary care to others, and for households in temporary distress. This may need some reduction in levels of old age benefits as their reach is extended towards universal coverage. No reduction in levels should be countenanced, however, until other programmes are in place. Such other programmes would include more direct efforts to provide employment, for example, through public works programmes. When the latter have been established, it may be possible to phase out the NNSDP. There is a need, however, for regional experimentation over thc next few years, including food vouchers and other feeding schemes, to develop a mix of initiatives to address rural poverty and food security problems more effectively.

Employment programmes

The main way of dealing with the country's massive unemployment will be economic growth which is employment-intensive (in contrast to that of the past decade which, in the formal sector, has generated extra jobs for as few as 6 per cent of additions to the economically active population). In support of such an employment-intensive growth path, maintaining low food prices in real terms will be crucial. The key factors for that are continuing improvements in agricultural technology, and substantially increasing the efficiency of food processing and marketing. But food security in rural areas will also need to be assisted by income generating programmes outside agriculture, and special employment creation schemes will be an important part of this. The enormous disparities in regional economic activity, welfare and infrastructure, provide the setting for such special employment schemes to have a valuable impact on economic growth while creating jobs.

A rural Public Works Programme (PWP) is proposed, which would create productive economic infrastructure and build human capacity, thus simultaneously addressing two central needs of the rural areas. The proposed PWP would be developed around the following principles:

- The programme would be national in scope, and would require a consensus among government, trade unions and local communities about employment conditions, political and financial commitment.
- While careful engineering design would be needed, local communities would be closely involved in project identification and planning

in order to ensure that they would 'own', manage and maintain the assets constructed under the programme; local management and maintenance would be critical to success.

- Pilot projects would help accelerate decisions about the nature and design of infrastructure.
- Among others, projects undertaken would include roads and tracks, water supply, rural electrification, sanitation, schools, clinics, training facilities, housing, irrigation and tree planting.
- Labour-based construction methods would be employed which have been found to be compatible with acceptable engineering quality of assets in most cases.
- Specific training to increase the skills of participants would be undertaken.
- A national fund is proposed, underpinned by a long term (at least ten years) commitment by government, with clear, efficient guidelines for access, and mechanisms to ensure that benefits would not be captured only by the better organized and informed.
- The proposed national fund would be formed by consolidating in one place many of the budget allocations for 'employment creation', 'emergency relief', 'training', and so on, which are scattered throughout the government; the total of such programmes in the 1992/93 budget is estimated to be around R6,2 billion, or 5 per cent of the total budget; such a consolidated fund, it is argued, would be likely to have more impact than the many individual fragmented programmes.
- Financing would be from the national budget initially, with the consolidation proposed above, but a variety of other funding mechanisms would be explored, including from the private sector, international aid, and contributions from the beneficiary communities, possibly in kind in some instances (although the emphasis would be on remunerative employment for poorer persons).
- Some models for joint private and public sector initiatives are already available in the National Housing Forum and the Independent Development Trust – (the latter's Relief Development Programme in 1992/93 expended about R49 million to generate 4,7 million days of work, a cost per equivalent full-time job of about R2 600), to construct roads, dams, water supply systems, clinics, classrooms and crèches).

Summary

In the context of a land reform or restitution programme, there are three basic means for landless people to gain access to land in any country:

(a) land invasion; (b) restoration; and, (c) redistribution. Of these three broad mechanisms, this chapter advocated a combination of restoration and redistribution and examined options for how such a process of land reform might work in South Africa. In the light of this favoured approach to land reform in South Africa, individuals (or groups) who wish to gain access to land have three choices: (a) seek land through the restoration process as outlined in the chapter to follow (Chapter 15); (b) acquire land through the redistribution channel; or, (c) purchase land without assistance from the programme. Individuals (or groups) who qualify would submit an application to the appropriate institution (i.e. a land claims court, a land commission or a land committee).

Due to its superior performance, a market-assisted redistribution process is favoured. In this process, the beneficiaries would be responsible for identifying a piece of land, but would be eligible for assistance in this activity. Once a suitable piece of land is identified, financing could be arranged from three possible sources: (a) the grant component available from the programme; (b) the beneficiaries' own resources; and, (c) a bank loan. The grant elements of the programme are essential in order to accomplish a redistribution of assets and to ensure that beneficiaries emerge from the programme with a net increase in their asset position. A matching grant scheme that forces participants to use some of their own resources in order to gain access to land will help to assist in self-selection of beneficiaries.

It is accepted that a redistribution programme will not be able to provide land for everyone, and that the programme will need to be complemented by a rural safety net and by programmes for urban groups as well as a rural public works programme.

15

Historical claims and the right to restitution

Heinz Klug

Introduction

The historic relationship in South Africa between legal process and dispossession makes it imperative for any new legal dispensation both to acknowledge its past failures and to address the consequences of the past. While the 1993 Constitution and the adoption by Parliament of the Restitution of Land Rights Act provide the basis for achieving this reconciliation, it is important that an analysis of these developments remains rooted in the historical context. In discussing these developments, this chapter makes use of comparative experiences to highlight different aspects of this process.

The historical role of legal processes in land holding and dispossession

Social relations in the rural areas of South Africa may be represented by a series of formal legal relationships describing people's relationship to the land or to each other. Despite a formal legal approach which presumes that these relationships are politically neutral and unchanging over time, it can be clearly demonstrated that in South Africa they are the product of a history of dispossession in which the law and legal practitioners have played a significant role.

The 1913 Natives Land Act is often described as having 'turned into law the process of dispossessing Africans of their land that had been going on for the past 200 years'.[1] However, a closer examination shows that law was applied in the process of dispossession long before 1913 and continued to be used to justify and implement the apartheid regime's policy of forced removals.

Although it is true that military dominance was central to the process of dispossession, it is important not to ignore evidence of a history of

1 L. Platzky and C. Walker, 1985. *The Surplus People: Forced Removals in South Africa.*

coexistence and betrayal in which legal mechanisms were applied to secure expropriation of the land. Analysts who focus on the primacy of military conquest see legal mechanisms as only the *final* stage in a lengthy process. However, the statement that 'law, not war, was the final means of conquest'[2] is misleading. The idea that law in the form of the Land Acts of 1913 and 1936 gave a legal basis to the expropriation of African land rights involves an antiquated colonial idea that military conquest gives rise to legal rights. Seeing the operation of legal mechanisms as only the *final* stage, also ignores the significance of different conceptions of land ownership and use in the law and custom of African and settler communities. Permission to use and settle on land granted by African communities to white settlers was later held to be a grant of ownership, upheld by the colonial legal system as the basis for denying land rights to these same African communities. Colonial grants of rights to settlers and the simultaneous denial of African communities' existing rights to the land were and remain central to understanding the process of dispossession.[3]

The 1913 Land Act, moreover, not only consolidated colonial occupation – preventing the African people even from buying back their own land – but further reduced African access to the land. 'Extensive areas of African freehold land and of unsurveyed state land which had long been regarded as African areas were left out of the schedule. In 1916 it was estimated that over one and a half million hectares of African land was excluded in this way, not counting the far larger area of rented white land that was also lost.'[4] Thus by excluding areas of land and prohibiting Africans from renting land outside the boundaries of the reserves in the future, the Land Acts actively dispossessed millions of South Africans of their rights to the land.

In contrast to the law's role in the direct expropriation of land through the 'legalization' of military conquest and colonial settlement, the legal process also functioned indirectly to further the process of dispossession. Increasing evidence indicates a process of gradual rural impoverishment leading to landlessness. This process was strongly supplemented and encouraged both by political forces demanding an increased

2 *Id.*

3 *See* H. Klug, Defining the property rights of others: political power, indigenous tenure and the construction of customary land law. (Draft, January 1995.)

4 L. Platzky and C. Walker, 1985. *Surplus People: Forced Removals in South Africa*, pp.84–85.

labour supply and a pattern of commercial and legal practices amounting to corruption and fraud, if not a conscious process of expropriation.

Three examples will serve to illustrate the use of the law, particularly the law of mortgage, to this end. First, in 1884 the Boers annexed Barolong lands in the Thaba'Nchu area of the Orange Free State; after taking the bulk of the territory for themselves, they 'recognized two tiny African locations and 95 farms belonging to individual Barolong landowners'.[5] By 1900 only 54 of the original 95 farms remained intact, the other Barolong landowners having 'lost their land to the whites by taking out mortgages which they could not pay back'.[6]

Second, after East Griqualand was annexed in 1874, whites were recognized as owning 63 of approximately 505 farms granted by the Griqua polity. Five years after restrictions on the sale of Griqua titles were lifted, however, approximately half of the Griqua titles had changed hands.[7] 'The titles passed very largely into the hands of local merchants and speculators to whom many Griqua, in trying to establish themselves on the land, had become indebted. Some Griqua landowners undervalued their only substantial asset and used it directly to pay debts. But the white traders also enforced payment in land rather than offering long term mortgages.'[8] In a short time the Griqua, a 'landed community, allied to the colony, became … a collection of landless people with a deep-seated sense of grievance.'[9]

In a final example, the lawyer and later parliamentarian V.G. Fenner-Solomon used 'legal' sleight of hand to hasten the dispossession of the Kat River people. Following the passage of the Boedel Erven Act in 1905 which defined who had legal rights to land among the Kat River people, 'Solomon extended credit to them on a lavish scale … Or so he claimed.'[10] He did not ask for receipts and he kept no records. When transfer deeds arrived from Cape Town, Fenner-Solomon informed his clients that they owed him amounts, including government charges, legal

5 *Id,* p. 74.

6 *Id.*

7 Beinart, W. 1986. Settler accumulation in East Griqualand from the demise of the Griqua to the Natives Land Act. In Beinart, W., Delius, P. and Trapido, S. (Eds), *Putting a Plough to the Ground: Accumulation and Dispossession in Rural South Africa 1850–1930,* pp. 264–265.

8 *Id,* p. 265.

9 *Id,* p. 260.

10 Peires, (1987). The Legend of Fenner-Solomon. In Bozzoli (Ed), *Class, Community and Conflict: South African Perspectives,* p. 73.

fees and personal loans, many times what they were expecting. Although he 'offered them time to pay, he demanded that they immediately sign power of attorney for him to pass bonds payable to himself ... Co-heirs were made jointly indebted and severally liable for the payment of the bond, so that the whole property could be expropriated for the default of a single heir.'[11]

It is possible to argue that, when he foreclosed on the mortgages, 'Fenner-Solomon was only expediting processes which capitalist agriculture, working through the laws of contract and land tenure, would ultimately have executed anyway.'[12] However, this fails to recognize the particular colonial form that this process took in South Africa. The process of dispossession, in this context, represents both a consolidation of colonial domination and the emergence of capitalist agriculture.

It was the weight of this history and the potential consequences for the legitimacy of property rights in general that kept the issue of restitution alive despite a concerted effort to argue that the new South Africa should look towards the creation of a just future and not dwell on the conflicts of the past. While many argued that the distribution of economic benefits and the allocation of property rights should ideally be left to the democratic process, the insistence that property rights be included as a fundamental right forced negotiators to address the legacy of dispossession.

The 1993 Constitution's recognition of a right to restitution

The 1993 Constitution provides a clear framework for the recognition and adjudication of land claims. First, a right to restitution is recognized in section 8(3)(b) as part of the constitution's commitment to addressing apartheid's legacy. Second, sections 121, 122 and 123 of the constitution provide a framework for the establishment of a land claims process.

That property relations have been historically shaped by the very processes of 'untold suffering and injustice' which the 'postamble' to the 1993 Constitution suggests the interim constitution is designed to remedy, is recognized in these sections which provide a constitutional basis for the restitution of land rights. Although criticized as 'a cheap

11 *Id*, pp. 73–74.

12 *Id*, p. 88.

13 *See* Marais, Snatching defeat from the jaws of victory. *Work in Progress 95*, p. 26 (February/March 1994).

crook'[13] in that they make the remedy of restitution dependant upon state certification that 'the restoration of the right in question is feasible,'[14] these clauses do constitutionalize a process of land claims which the state is compelled to address.

A right to restitution

Section 8(3)(b) creates a right to the restitution of land dispossessed under any law which would have violated the 1993 Constitution's prohibition on discrimination. This creation of a constitutional right to restitution is significantly incorporated within the equality clause of the bill of rights and is directly linked, with the notion of affirmative action, to the constitutional mandate to address apartheid's legacy.

Recognized in both civil and common law, restitution is the act of restoring anything to its rightful owner, of making good or giving equivalent for any loss; it requires a person who has been unjustly enriched at the expense of another to make restitution to the other.[15] In Roman law, restitutio was based on injury to the applicant's proprietary interests resulting from a transaction or event, and could be asserted on the basis of a good cause, significantly including a person's inability to make a claim for a particular period of time.[16] By analogy, it may be argued that during South Africa's period of colonial domination the colonized were unable to make a claim for the restitution of their dispossessed land rights but will, with the emergence of a democratic South Africa, be in a position to assert their claims.

Although agrarian reform programmes are described, at times, as being comparable to affirmative action programmes in industrial countries, and the property redistributed during land reform is considered to be redress, 'just as job preference in affirmative action is explicitly given to minorities and women as recompense for years of maltreatment', only rarely have agrarian reform programmes been explicitly based on principles of restitution.[17] One notable exception, however, was the Mexican Revolution which, by the time the Morelos revolutionaries won official recognition in 1920, had 'forced on its regime a policy of concern for the

14 South African Constitution 1993. Section 123(1).

15 H. C. Black. 1979. *Black's Law Dictionary* 1180. (5th Ed.)

16 J. A. C. Thomas. 1976. *Textbook of Roman Law,* p. 113.

17 Thiesenhusen, W. 1989. Introduction: searching for agrarian reform in Latin America. In W. C. Thiesenhusen (Ed), *Searching for Agrarian Reform in Latin America,* pp. 1 and 6–7.

nation's rural poor'.[18] Proclaimed in November 1911, the Plan de Ayala, the Zapatistas' revolutionary manifesto, gave notice that regarding 'the fields, timber, and water which the landlords, *cientificos,* or bosses have usurped, the *pueblos* or citizens who have the titles corresponding to those properties will immediately enter into possession of that real estate of which they have been despoiled by the bad faith of our oppressors, maintaining at any cost with arms in hand the mentioned possession; and the usurpers who consider themselves with a right to them [those properties] will deduce it before the special tribunals which will be established on the triumph of the revolution'.[19]

The restitution of dispossessed lands was institutionalized in the Zapatistas' Basic Law of Agrarian Reform which in Article 1 decreed: 'to communities and to individuals the fields, timber, and water of which they were despoiled are [hereby] restored, it being sufficient that they possess legal titles dated before the year 1856, in order that they enter immediately into possession of their properties.' The agrarian law acknowledged, however, that restitution alone would not adequately address the problem of landlessness and, in Article 4, recognized 'the unquestionable right which belongs to every Mexican of possessing and cultivating an extension of land, the products of which permit him to cover his needs and those of his family; consequently and in order to create small property, there will be expropriated, by reason of public utility and by means of the corresponding indemnization, all the lands of the country, with the sole exception of the fields belonging to pueblos, rancherias, and communities, and those farms which, because they do not exceed the maximum which this law fixes, must remain in the power of their present proprietors'.[20]

The legal structure created by the Mexican agrarian law thus incorporated both the notions of restitution and affirmative action in order to address both issues of dispossession and landlessness. The effect of a similar incorporation in the South African context will be to provide a basis not only for the redistribution of resources to ensure equal participation in the context of the Freedom Charter's ideal that the land should be shared among those who work it, but also a basis upon which to

18 *See* J. Womack. 1968. *Zapata and the Mexican Revolution*, p. x.

19 The Plan de Ayala, para. 6, reprinted in J. Womack. 1968. *Zapata and the Mexican Revolution,* p. 402.

20 Arts. 1 & 4 of the Agrarian Law. Reprinted in J. Womak. 1968. *Zapata and the Mexican Revolution,* pp. 406–407.

resolve existing claims and to address issues of compensation in the context of land reform.

Creating a framework for land claims

The restitution of land rights clauses form a sub-chapter of the 1993 Constitution which lays out a framework for the implementation of the right to restitution granted by section 8(3)(b). These three clauses broadly define the nature of the claims that are to be considered for restitution; the administrative process for identifying and investigating claims; and the process through which a court exercising jurisdiction over land claims must reach its decision.

While they fail to grant a right to the restoration of lost lands by dispossessed communities, sections 121–123 do require Parliament to pass a law establishing a land claims process and empowering the courts to grant a range of remedies to claimants.[21] Two significant limitations on restitution are included in these clauses, with the intention of facilitating those land claims which the negotiating parties agreed were central to the process of addressing apartheid's immediate legacy. The first deals with the question of a cut-off date, and establishes 19 June 1913 as the earliest date that the legislature may set for the recognition of claims. Based on political expediency in the negotiations, and the rationale that it would be very difficult to obtain the necessary evidence to ascertain the circumstances and identities of beneficiaries for claims of dispossessions that dated back more than two generations, the choice of 19 June 1913 was justified in terms of the symbolic importance of the passage of the 1913 Land Act. While the traditional leaders have already expressed their dissatisfaction with this cut-off of historical land claims, it is not clear that if even a portion of the approximately 3,5 million persons who were forcibly removed from their land under discriminatory laws since 1913 were to lodge their claims, that the institutions and processes established would have the resources to ensure that these claims are adequately addressed.

The second limitation imposed by the restitution clauses excludes all claims to any rights in land that was expropriated under the Expropriation Act where just and equitable compensation was paid at the time. Although this exclusion would seem unjust in circumstances where the expropriation was directed at furthering the implementation of a racial policy, it is distinguishable in cases where those whose land

21 South African Constitution 1993, section 123.

was expropriated had political rights under the apartheid system to challenge these policies. While this effectively excludes white landowners whose land was expropriated for the purpose of consolidating the bantustans, it also serves the purpose of excluding those with the legal resources to dominate the process. This exclusion may be further justified by situating this process of restitution within the framework of the equality clause of the bill of rights as one of the mechanisms to address past discrimination.

The constitution provides a clear set of remedies in order to address claims of restitution. However, restoration of the land, which is the essence of the notion of restitution, is made subject to the state certifying that restoration of the right in question is feasible. Even if it is held to be feasible by the state, the Court is still required, in the case of land in the possession of a private owner, to find that it is just and equitable to restore the right, taking into consideration all relevant factors including: 'the history of the dispossession, the hardship caused, the use to which the property is being put, the history of its acquisition by the owner, the interests of the owners and others affected by any expropriation, and the interests of the dispossessed'.[22] Even if the court finds on this basis that restoration of the privately held land is just and equitable, the expropriation of the land for the purpose of restoration shall be subject to the payment of compensation in terms of the property clause of the constitution.

Failure to grant a right to restoration, and particularly the power given to the state to certify whether restoration is suitable in particular cases, pose the greatest challenges to the legitimacy of this framework. If in implementing this framework and addressing land claims, the state repeatedly fails to allow the courts to grant restoration, it is likely to be confronted with an increasing pattern of land invasions as communities resort to alternative strategies to return to their land. Although no idle threat, given the growing effective organization of rural communities and their demands for land reform,[23] the prospect of land invasions forcing confrontations between communities, the government and the courts, is likely to ensure that the state is careful not to discredit, by constant denial of court endorsed claims to restoration, the orderly claims process established in the constitution. It is only this restitutionary process, coupled with a programme of land reform designed as a

22 S. Afr. Const. 1993, section 123 (2).
23 *See*, *Work in Progress* No. 95, 1994, p. 3.

corrective measure in terms of s 8(3)(a) of the constitution to give access to those who have been historically denied the right to enter the land market, which will avoid increasing contestation over property rights.

The Restitution of Land Rights Act

The adoption by Parliament of the Restitution of Land Rights Act in November 1994 provided the detailed legal framework and institutional structure for the implementation of the constitutional mandates of land restitution. The Act deals with three main elements of the restitutionary process guaranteed in the constitution. First, the Act clarifies who may lodge a valid claim to restitution by defining a number of key terms and specifying the conditions under which a valid claim may be brought. Second, the Act implements the constitution's mandate to establish institutions for the processing of claims to land rights. The Act provides for a Commission on Restitution of Land Rights to administer and facilitate claims, while it establishes a Land Claims Court to exercise specific jurisdiction over the restitutionary process. Third, the Act provides clear guidelines to both the Commission and the Court for the processing of claims from their initial lodgement through their negotiation and final adjudication.

Defining eligible claimants

In debating the issue of land restoration prior to the elections of April 1994, there was a great deal of anguish and concern over the question of reach. How far back should we go in time in recognizing claims to the restoration of land? How do we define eligible claimants or types of dispossession so as to ensure that all those with morally valid claims are able to obtain justice? How do we prevent any claims process from being dominated by better resourced urban claimants to the detriment of those impoverished rural communities in the greatest need?

To meet these challenges, participants in the debate produced an elaborate categorization of forced removals which sought to further a particular part of apartheid ideology, and attempted to give different weightings to these situations so as to facilitate the process of recognizing and identifying claimants. These 'qualifying' acts were identified through the differentiation of four types of legal rules which characterized racial land law in the period since 1913: (a) rules dividing the land surface into race zones; (b) removal provisions implementing resettlement in terms of an apartheid legislative map, e.g. s 5(1)(b) of the Native Administration Act 38 of 1927 and s 46(2) of the Group Areas Act 36 of 1966;

(c) other removal provisions employed to achieve apartheid objectives, e.g. the Prevention of Illegal Squatting Act, trespass, expropriation, slums acts, etc; and, (d) provisions which had the effect of denying persons of a particular race group access to land in certain areas by criminalizing their 'unlawful' presence, e.g. pass laws.

Of these four types, only the specific removal provisions implementing resettlement or employed to achieve apartheid objectives, categories two and three were, considered to provide a basis of a claim which would be able to identify a specific act of removal, or threat of removal in terms of which the claimants lost their land. (See also Chapter 14.) Categorizing the claims that are likely to arise from these two sets of rules, it was thought possible to identify four broad categories of rural forced removals and to indicate what type of weighting a restitutionary process may give claims falling into the different categories: (a) rural 'black spot' removals, where the availability of clear statutory evidence and the more recent nature of these claims provide the clearest cases of unrestricted restitution requiring the full restoration of the land; (b) homeland consolidation, providing another example where removals would have been carried out according to statutory scheme and therefore ensuring that it should not be difficult to restore the land; (c) labour tenants and squatters, whose clearly established rights will pose difficult questions both with regard to evidentiary issues and to the appropriate form of redress which should be given. In the case of labour tenants, unless claimants assert historic title to the land, their award would remain subject to the rules of tenancy and, if these were interpreted to be transferable, then the landowner may very well argue that the court has effectively violated his/her property right by imposing a form of co-ownership where none had previously been recognized; and, (d) betterment planning, a situation where, due to the extent of claims and potential for intracommunity conflict, it would be advisable to recognize claims but to direct claimants away from the restoration process to the redistributive process.

Fears were expressed, however, that a process of categorization may become a means of disempowerment through which government officials could engage in the prior identification of valid claimants; thus it was suggested that this categorization should serve merely to validate or exclude claims, subject to appeal, in the initial phases of the process. However, this typology did provide a valid limitation following from the choice of particular apartheid rules and the intention to provide restitution for past discrimination, in excluding claims from landowners who may have been expropriated by the government in order to further apartheid zoning or homeland consolidation, but who were not members of communities who suffered legal discrimination under apartheid. Although this 'affirmative action' type approach would not function in an inclusive race-neutral way, it would serve the function of preventing those with legal and financial resources from clogging the restitution process with claims that they were not adequately compensated or had land expropriated in the furtherance of apartheid policy although they received adequate compensation. Justification for this approach was based on the fact that white property holders had access to the political process denied the victims of apartheid.

Instead of relying on a categorization of apartheid laws and actions, the Restitution of Land Rights Act is based on the 1993 Constitutional framework which provides a broad claim to restitution of land rights under two conditions. First, if the dispossession took place after 19 June 1913 and, second, if the dispossession 'was effected under or for the purpose of furthering the object of a law which would have been inconsistent with the prohibition of racial discrimination' contained in the equality clause of the constitution's chapter on fundamental rights.[24] This inclusive definition of eligible claimants is enhanced by the broad definitions given to the terms 'community' and 'rights in land' in section 1 of the Act. Community is defined for the purposes of the Act as 'any group of persons whose rights in land are derived from shared rules determining access to land held in common', while rights in land are defined as 'any rights in land whether registered or unregistered'. The definition of a land right also explicitly includes interests not formally recognized or substantially discounted by South African common law, including: 'the interest of a labour tenant and sharecropper, a customary law interest, the interest of a beneficiary under a trust arrangement and beneficial occupation for a continuous period of not less than 10 years prior to the dispossession in question.'[25]

While the only formal exception to this broad definition of claimants flows from the constitution – claimants whose rights were expropriated under the Expropriation Act and who received just and equitable compensation[26] – the requirement that the act of dispossession was in furtherance of a discriminatory law may have dire consequences for some

24 S. Afr. Const. 1993, sections 121 (2)(a) & (b).
25 Sections 1(iv) & (xi), Restitution of Land Rights Act 1994.
26 S. Afr. Const. 1993, section 121 (4).

claimants. Despite a generally inclusive approach, a question hangs over the claims of labour tenants and farm workers who were paid for their labour in part by being allowed access to some land on the farm for residential and domestic production, and who were subsequently forced off 'white' farms by the 'private' action of the farmer. In such cases, the communities or individuals will be faced with the difficult task of demonstrating that their dispossession was under or in furtherance of the object of a discriminatory law. However, the white farmer need merely assert that his action was based solely on economic considerations or changing use patterns on his property. While communities may argue that their removal as labour tenants was in fact in furtherance of the 1936 Land Act which moved to prohibit African labour tenancy, it will be difficult to distinguish economic from political or legal reasoning in the actions of private land holders.

Further constraints on claimants stem from the provisions determining the conditions under which claims may be enforced. While parliament adopted 19 June 1913 as the earliest date set in the constitution for the recognition of claims, all claims must be lodged, according to section 2(b) of the Act, within three years of a date to be fixed by the Minister of Land Affairs, but not prior to the commencement of the Act which was 25 November 1994. This fairly short timeframe before the imposition of a statute of limitations barring all further claims should not present a major obstacle to claimants, given the degree of political mobilization around land claims. Such a limited window of opportunity for individuals and communities to bring their land claims may also be justified by the need to finalize the status of land in order to avoid a clash between a future redistributive land reform and continuing land claims. However, the legitimacy of this limitation will rest on the amount of effort which will have to go into ensuring that the vast bulk of obviously colourable claims are registered. The exclusion of even a small minority of clearly justified claims by the imposition of the statute of limitations will severely undermine the credibility of the process and would only serve to rekindle and repoliticize wider claims of restitution in the name of historic claims and communal dispossession.

The structure and powers of land claims

Early discussions over the institutional structure for a land claims process debated the merits of an administrative versus a judicial process, and expressed concern over possible violations of strict notions of the separation of powers if an administrative body were to exercise a quasi-judicial function. The Act resolves these debates by creating an investigative commission, to make an initial determination, with a final disposition being endorsed by a special land claims court.

The Commission on Restitution of Land Rights is given extensive powers to:

- receive all claims for the restitution of land rights;
- assist claimants in the preparation and submission of claims;
- inform the public of the land claims process;
- encourage mediation of conflictual claims and to report the terms of any negotiated or mediated settlement to the court;
- define disputed issues between claimants and other interested parties with a view to expediting the court process.

In addition to these wide-ranging functions aimed at establishing and settling claims, the Commission is empowered to:

- monitor and make recommendations concerning the implementation of Court orders providing a wide range of remedies to recognized claimants;
- advise the Minister of Land Affairs on the most appropriate forms of relief, if any, for those who do not qualify for restitution;
- refer questions of law and interpretation to the court;
- administer the process so as to ensure that priority is given to claims impacting upon substantial numbers of people, or cases of pressing need, or where claimants have suffered substantial loss.

Once claims are received by the Commission, it is granted specific powers to investigate the claim and to ensure that the property in dispute – including all improvements, use and the residence of people on the land – remains undisturbed until the claim is settled. While the investigation of a registered claim is facilitated by the requirement that the commission has access to all information in the hands of government departments and the power to demand the production of documents from any person, in cases where a claimant is unable to provide the initial information to register a claim, the Chief Land Commissioner is required to direct officers working for the Commission to make available the information necessary for a claimant to formulate a claim. In order to conserve resources and facilitate the resolution of claims, the Commission is also empowered in the course of an investigation to give the public notice that all claims in respect of a particular property must be brought within a specified period so that the claims process with respect to that particular parcel of land may be finalized. The Commission is allowed, however, on a showing of good cause, to accept further claims against a particular property after the expiry of such a period, but no later than the three years set as a statute of limitations for all land claims.

The powers and functions of the Commission are reinforced by a limited criminal sanction which can be imposed on anyone refusing to provide the Commission with relevant information and documents or attempting to frustrate the workings of the Commission. While this is an attempt to prevent resistance by property holders to the processing of claims, the maximum sentence of three months imprisonment will not deter determined objectors. Accountability for the Commission's progress lies with the legislature which, according to the Act, will appropriate monies for the Commission's functions and receive an annual report on the Commission's activities.

The creation of a special Land Claims Court with a status equivalent to a provincial division of the Supreme Court and exercising national jurisdiction over land claims, is both in accordance with the constitutional mandate and characteristic of the innovative style of the Restitution of Land Rights Act. The President of the Court and any additional judges are to be appointed by the President on the advice of the Judicial Service Commission and may either be seconded from the existing Supreme Court judiciary or appointed for a fixed term. The Court, in order to be accessible to claimants, may conduct hearings at any place in the Republic, and shall be presided over by a judge and at least one accessor who shall have an equal vote to the judge, except in relation to questions of law. Significantly, the Act specifically holds that the application of the factors which have to be taken into account by the court under section 33 of the Act, and the decision over whether the restoration of land rights may be excluded as a remedy before the final determination of any claim, are not to be deemed questions of law.

The Act provides for extremely broad and flexible standing and evidentiary rules designed to meet the special circumstances of the land claims process. While the Act specifically suggests that hearsay evidence and expert anthropological and historical evidence shall be admissible, the weight given to such evidence shall be as the court deems appropriate. Although locus standi is granted to all parties listed by the Commission or their legal representatives, and the state is granted the right to be heard in all cases, any other person or community has the right to apply to the court for permission to appear. In addition, the Act provides that the Chief Land Claims Commissioner may arrange legal representation for a party which cannot pay for legal representation itself.

Another innovative element in the Act is the introduction of a pre-trial conference procedure with the purpose of clarifying the issues in dispute and identifying issues on which evidence will be necessary. This opportunity for the Court to become engaged in pre-trial negotiations will provide a powerful tool for informal settlements, saving court time and bolstering the Commission's role in mediating conflicting claims. The Court's oversight role is further strengthened by being given exclusive jurisdiction to exercise the Supreme Court's powers of review over the activities of the Commission and its functionaries. The only avenue of appeal against the decisions of the Court, on both procedural and substantive grounds, will be to the Constitutional Court or to the Appellate Division of the Supreme Court.

Processing a claim to land rights

The Restitution of Land Rights Act establishes a clear procedure for the submission and processing of a claim to the restoration of a land right. The first stage in the process is the lodgement of a claim which may be made by any person or member of a community or their representative. In lodging a claim, the claimant must describe the land and the nature of the rights being claimed on forms to be supplied by the Land Claims Commission. On receiving a claim, the Regional Land Claims Commissioner (RLCC) must decide whether the claim has been lodged in the prescribed manner and that it is not precluded by section 2(1) which defines which claimants are eligible and the statutory period in which claims may be brought. If the RLCC finds that the Land Claims Court has already issued an order with respect to the particular parcel of land being claimed, then the claimant may make an application to the Court for permission to lodge a claim.[27]

Once the claim is successfully lodged, the Commission must place notice of the claim in the government gazette and publicize the claim in the district where the land is located.[28] If there is reason to believe that any damage may be caused to the land, the RLCC may enter the property

27 Restitution of Land Claims Act 1994, section 11 (5).
28 Section 11 (1).

for the purpose of drawing up an inventory of property and persons resident on the land. The Commission must also conduct an investigation into the merits of the claim. In terms of section 13 of the Act, the Chief Land Claims Commissioner (CLCC) may direct the parties to engage in mediation over the claim in circumstances where there are two or more competing claims, or competing groups within the claimant community, or if the holder or owners of the land are opposed to the claim. If, as a result of mediation, the parties reach agreement and the RLCC is satisfied that the agreement is appropriate, then the agreement may be submitted to the Court for its endorsement. The Act requires that the Court be given the result of the Commission's investigation into the claim, a copy of the relevant deed of settlement and a signed request from the parties endorsed by the CLCC, asking the Court to make the agreement an order of the court.

If the Chief Land Claims Commissioner is not satisfied with the agreement,[29] or if the Court refuses to endorse it, the claim would now be referred to the Court for a hearing. If there was no settlement between the parties, then the parties are required to agree in writing that it is not possible to settle the claim by mediation or negotiation. The RLCC must also certify that it is not feasible to resolve the claim by mediation and negotiation, and that the claim is ready for court. When referring the claim to the Court, the Commission must provide the Court with documentation setting out the results of the Commission's investigation into the merits of the claim, as well as the Commission's recommendation as to the most appropriate resolution of the claim. The Commission must also supply the Court with a list of parties who ought to have the right to make representations to the Court, as well as a report on the failure of any party to accept mediation in the earlier attempts to settle the claim.

Simultaneous with its submission of the claim to the Court, the Commission must request the Minister of Land Affairs to certify whether: (a) in the case of state land, the restoration of the right in the land is feasible; or, (b) in the case of private land, it is feasible for the state to acquire the right; or, (c) it is feasible to designate alternative state-owned land. Within 30 days, the Minister's decision must be made taking into consideration whether: (a) the zoning of the land has been altered since the dispossession; or, (b) any relevant urban development plan has been made with respect to the land; or, (c) if there is any physical or inherent defect in the land which would cause it to be hazardous

29 Section 14 (4).

for human habitation. If the Minister considers the feasibility of alternative state land, the Minister must take into account what land is owned by the state, particularly in the area where the dispossession occurred and the suitability of such land in meeting the needs of the claimant. While all the parties have a right to make submissions to the Minister on the issue of feasibility,[30] the Minister must give reasons for the decision which are subject to review by the Land Claims Court, and may be done simultaneously with the Court's consideration of the merits of the claim.

In deciding on the merits, the Land Claims Court must take into account the following factors:

- the desirability of providing for restitution of rights in land or compensation;
- the desirability of remedying past violations of human rights;
- requirements of equity and justice;
- the desirability of avoiding major social disruption;
- any provision which exists for dealing with the land in question in a manner designed to achieve the constitutional goals of equality through affirmative action;
- any other factor relevant and consistent with the spirit and objects of the constitution and in particular the equality clause of the constitution.

The Land Claims Court is also empowered to make a decision excluding the restoration of a particular parcel of land before any determination is made on claims for restitution in regards to that land. This process may be initiated by any national, provincial or local government body, but the Court may only issue such an order precluding restoration if it is satisfied that it is in the public interest not to restore the rights or that the public or a substantial proportion of the public will suffer substantial prejudice if restoration is not precluded. While this process is designed to enable government to obtain certainty over the future allocation and use of particular parcels of land for development or other public purposes, its use as a foreclosure on the possibility of restoration could undermine the legitimacy of the claims process if communities feel that clear cases of restoration are being precluded by government initiative.

If a claimant is denied restoration but is offered alternative relief –

30 Section 15 (5).

state land or compensation – the claimant may instead apply to the Minister to be registered as a preferential claimant to benefit from any state support programme for housing or rural land allocation and development.[31] If a claimant is to be paid compensation, the amount has to be determined in terms of section 123(4) of the constitution which requires that it be just and equitable, taking into account the circumstances at the time of the dispossession, including any compensation that was paid at that time. If the Court orders the restoration of the right to the land in terms of sections 123(1)(a) or (b) of the constitution, the Court may determine conditions which must be fulfilled before the right is in fact restored to the claimant, including: (a) the amount, manner and time of any payment required by the claimant before the right is restored; (b) if the claimant is a community, then the manner in which the rights are to be held or the compensation is to be paid or held; and, (c) the Court's recommendation to the Minister that the claimant be given priority access to state resources. In the case of a community, the Act states that these conditions should ensure that all dispossessed members of the community concerned have fair and non-discriminatory access to the Court's award, particularly in the case of women and tenants. Provisions should also be made according to the Act for the accountability of persons who hold the land or compensation on behalf of the community.

Although the Land Claims Court is empowered to set conditions on otherwise successful claims, this power must be used very restrictively. Significantly, neither the fact that the land has been transformed by development, (for example, urban development on formerly rural land), nor the question of future use, affects the claimants' entitlement to restoration. Only the nature of the Court's award may be effected, either through the granting of alternate land or compensation. However, it is important to recognize that, with a restitutionary approach, claimants who have a recognized right will be placed in an extremely strong bargaining position with respect to the existing land holders.

Historical claims and land restoration in comparative perspective

After 65 years of legal conflict over Indian land claims in United States courts, the United States Congress established a 'special commission to handle exclusively Indian cases under a broad new jurisdiction and with

31 Section 35 (1).

the firmly expressed goal of finality'.[32] The United States Indian Claims Commission (ICC) was created on 13 August 1946, and operated from July 1947 until 30 September 1978 when the remaining matters before it were transferred to the United States Court of Claims.

The Commission was granted very broad jurisdiction in an attempt to ensure that all possible claims could be resolved through this process. Section 2 of the Indian Claims Commission Act of 13 August 1946 states that 'The Commission shall hear and determine the following claims against the United States on behalf of any Indian tribe, band, or other identifiable group of American Indians residing within the territorial limits of the United States'. It then goes on to list five categories of claims: (a) claims in law or equity arising under the constitution, laws, treaties of the United States, and executive orders of the President; (b) all other claims in law or equity, including those sounding in tort, with respect to which the claimant would have been entitled to sue in a court of the United States if the United States was subject to suit; (c) claims which would result if the treaties, contracts, and agreements between the claimant and the United States were revised on the grounds of fraud, duress, unconscionable consideration, mutual or unilateral mistake, whether of law or fact, or any other ground cognizable by a court of equity; (d) claims arising from the taking by the United States, whether as the result of treaty or cession or otherwise, of lands owned or occupied by the claimant without the payment for such lands of compensation agreed to by the claimant; and, (e) claims based upon fair and honourable dealings that are not recognized by any existing rule of law or equity.

Another aspect of the ICC experience of particular interest was the imposition of a five year filing period. All claims had to be filed within a 'period of five years after the date of approval of this Act and no claim existing before such date but not presented within such period may thereafter be submitted to any court or administrative agency for consideration, nor will such claim thereafter be entertained by the Congress'.[33] At the beginning of 1951, the last year of filing, only 200 claims had been filed; 25 cases had been decided, of which only two had won awards totalling US$3,5 million. However, as a result of a number of

32 United States Indian Claims Commission: 13 August 1946–30 September 1978, Final Report. U.S. Government Printing Office, 1979, p. 3.

33 Section 12, Indian Claims Commission Act, Pub.L. 726 (60 Stat. 1049), 13 August 1946.

factors, (including the awards granted, the difficulty some tribes experienced in getting legal representation and the task of preparing a case), the last six months of the filing period saw the number of petitions double, resulting in a total of 600 claims.[34]

The ICC Act also provided for very open rules of standing. Section 10 stated that any claim within the provisions of the Act could be brought by 'any member of an Indian tribe, band, or other identifiable group of Indians as the representative of all its members; but wherever any tribal organization exists, recognized by the Secretary of the Interior as having authority to represent such tribe, band, or group, such organization shall be accorded the exclusive privilege of representing such Indians, unless fraud, collusion or laches on the part of such organization be shown to the satisfaction of the Commission.'

Another aspect of interest is the criteria used by the ICC to establish valid Indian title. The three factual elements which appear most commonly in formulations of the requirements for proving title are: (a) 'the extent of the use and occupancy, (b) the exclusiveness of the use and occupancy, and (c) whether the use continued "for a long time" '.[35] The extent of use and occupancy was determined in *United States vs Seminole Indians*[36] as that 'which usually coincides with what is shown to be the area used for subsistence by all the members of the tribe. Not only agricultural use, but religious use, hunting and gathering use, even sporadic irregular hunting use, may, if not in conflict with another tribe's use, bring an area within the compass of a tribe's Indian title lands.'[37] Exclusive use and occupancy did not require a tribe to 'show formal political hegemony over an area, or assertion and exercise of power to exclude all members of other tribes. Usually a more intensive and persistent use of an area as compared to a casual and incidental use by another tribe will be sufficient for Indian title.'[38] In *Creek Nation vs United States*,[39] the Commission held that a tribe may permit guests to use its lands without defeating its title. Finally, although there was an unwillingness to find ownership of a specific area of land in a nomadic tribe, 20–50 years of occupation was found to be judicially acceptable as meeting the requirement of 'a long time'.

34 United States Indian Claims Commission, Final Report, p. 5.
35 Final Report, Indian Claims Commission 1979, p. 128.
36 180 Ct. Cl. 375 (1968), aff'g 13 Ind. Cl. Comm. 326, 1964.
37 *Id.*
38 *Id,* p. 129.
39 23 Ind. Cl. Comm. 1, 1970.

Claimants were permitted to introduce evidence from a wide range of sources to establish the three factual criteria required to show Indian title. Not only did the 1946 ICC Act require that all government departments make all documentary evidence in their possession available to the Commission as evidence in the claims process, but the claimants also made extensive use of enthographic evidence, both documentary and as provided by expert witnesses. The most extensive documentation of forms of evidence obtained for this chapter, was used in the claim brought by the California Indians for the whole of California.[40] The evidence was brought to support the argument that California Indians owned and defended their tribal territories and were permanently attached to their lands. The evidence was divided into four main sections: (a) testimony regarding the land-owning unit; (b) documents showing various kinds of native ownership of lands and resources; (c) evidence of ownership, among which was reaction to trespass of boundaries, the existence of boundary or ownership markers, seasonal use of tribal areas and instances of 'sale or rent'; and, (d) arguments for the permanence and continuity of occupation, supported by showing that the traditional territories of some tribal groups were occupied by them at the time when they first had contact with European settlers. Additional arguments for permanent occupation were supported by: (a) early maps which showed the location of tribes; (b) archaeological sites known to have been occupied in late prehistoric times by the ancestors of groups occupying the area in the historic period; (c) the practice of bringing home ashes of persons who died at a distance so that their remains could be buried in their birthplace; (d) the inheritance of resources (tracts of land, fishing spots, seed areas, etc); (e) conscious conservation or use in rites of food or other natural resources existing within tribal territories; (f) mentions by ethnographers of native persons living before the American period who had participated in or witnessed some known historical event – the argument here being that some members of identifiable tribes were living at the time in their traditional areas; (g) the Indian's feeling of attachment to the land of their ancestors; (h) traditional use of certain spots or traditional ceremonies held at particular places; (i) intimate knowledge of the features of one's tribal territory; and, (j) the stay-at-home attitude and practice of most California Indians – meaning their stable pattern of settlement.

40 *The California Indians vs The United States of America,* 8 Ind. Cl. Comm. 1, 1959, aff'd 167 Ct. Cl. 886, 1964.

Significantly, this process of Indian Claims was founded on the notion – created within the common law tradition – of aboriginal title. As stated in the Final Report of the Indian Claims Commission: '[h]owever often ignored in practice, the legal doctrine of every European colonizer of the Americas acknowledged that the Indians had certain rights to the peaceful possession of their land. England acknowledged such Indian rights, and in general her American colonies expanded through purchase or lands from the tribe claiming them'. In terms of Roman-Dutch common law at the time, it should be noted that 'the colonies established by the Dutch ... were all founded on lands which were purchased from the Indians. Dutch policy is indicated in the 1629 "New Project of Freedoms and Exemptions" [in which] the patroons are required by article 27 to purchase lands from the Indians: "The Patroons of New Netherland, shall be bound to purchase from the Lord's Sachems (i.e., Indians) in New Netherlands the soil where they propose to plant their colonies and shall [have] such rights thereunto as they will agree for with the said Sachems".'[41]

A severe limitation of the ICC's jurisdiction, and the reason why its decisions are still being challenged before the US courts and why it was unable to achieve its aim of finality, was its inability to grant the restitution of land. The only remedy open to the commission was a financial reward representing payment for the land, to be determined by the value of the land at the time of the taking. However, at the time of its dissolution on 30 September 1978, the ICC had made a total of 342 awards totalling US$818 172 606,64, reflecting a judicially established recognition that over half the land mass of the continental United States had been Indian land.[42]

Conclusion: land restitution and the new constitution

The role of a land claims process in changing existing patterns of land occupation and use in South Africa's rural areas will paradoxically be both limited and all-important. A land claims process will interact closely with the broader process of land reform. As a matter of empowerment, it will provide an alternate route for those previously dispossessed who wish to regain access to land.

If the redistributive process is inadequate, people are likely to frame their needs in terms of claims which, if great enough in volume, will

41 P.A. Cumming and N.H. Mickenberg. 1972. *Native Rights in Canada,* p. 16.
42 See map in pocket, Final Report, Indian Claims Commission, 1979.

threaten to swamp the claims process. Alternatively, if the claims process is bogged down in legal processes, appeals and claims for review, etc, the redistributive process may provide an escape valve in that claimants with an initial determination of right from the Commission on the Restitution of Land Rights, may trade that claim for preferential access to the redistributive process.

Although they will be closely interrelated in practice, there is a fundamental distinction between land restoration and land redistribution. The restoration of land rights, by means of administrative and adjudicatory processes involving specific parcels of land, directly empowers the successful claimants and guarantees their access to some resource. However, unless these processes of land restitution and redistribution are perceived as delivering access to land – for land claimants, land hungry rural dwellers and urban housing needs – the question of property rights may once again haunt the constitution-making process. Although the constitutional principles contained in Schedule 4 ensure the inclusion of 'universally accepted fundamental rights, freedoms and civil liberties', this does not guarantee the inclusion of property rights in the final constitution. While property rights are increasingly recognized, they were not included in either of the United Nations Covenants on human rights of 1966 or even, initially, in the European Convention on Human Rights and Fundamental Freedoms of 1950. Even the requirement that due consideration be given to 'the fundamental rights contained in Chapter 3 of the interim constitution,'[43] does not guarantee the inclusion of property rights in the final constitution. The legitimacy of property rights will rise or fall in large measure with the fortunes and failures of the process of land restitution guaranteed by the constitution and implemented through the Restitution of Land Rights Act.

43 South African Constitution 1993, Schedule 4 (II).

16

Market-assisted rural land reform: how will it work?

Johan van Zyl and Hans Binswanger

Background: agriculture in crisis

South African agriculture in general is going through a period of crisis associated with change. Many privileges to large-scale agriculture, such as taxation benefits, subsidized credit and price supports have been withdrawn (*Meyer & van Zyl, 1992; Van Zyl, 1994a; 1994b*).[1&2] Combined with the effects of increasing international agricultural competition under the GATT agreement, it will and has already placed agricultural profits and land prices under severe pressure, especially in the grain and livestock sub-sectors. For example, full market and trade liberalization will cause the producer surplus in the Western Cape to decrease by 21 per cent from 1988 levels (*Van Zyl, 1994b*).

In the present situation, a market driven (or market-assisted) land reform programme provides a unique opportunity to alleviate some of the problems caused by market liberalization and removal of privileges, especially in grain and livestock production. The details of such a proposal are discussed below.

The elimination of privileges in marketing and tax policies, and the high real interest rates associated with macroeconomic stabilization and the demands of the Reconstruction and Development Programme (RDP), will result in:[3]

- an investment pause in agriculture;
- numerous farm bankruptcies;
- very low demand for land;

1 See Chapter 9 for a detailed discussion.

2 See Helm and Van Zyl (*1994*) on the quantification and composition of agricultural support over the period 1988 to 1993.

3 An additional (well-documented) example of the process described below is the adjustment in the US farm sector in the early 1980s following high real interest rates – more than 30 per cent of all US farmers left the farm sector, despite a well-developed financial support system (*Tweeten, 1986*).

- the collapse of farmland prices;
- demand for blanket debt relief from both the farm and bank sector;
- aggravation of already high urban and rural unemployment;
- worsening of rural poverty and the nutrition situation.[4]

The social impact will be most severe in the rural areas where poverty is already crushing – 75 per cent of the total number of poor are from the rural areas (*Kirsten et al, 1994*). Market-assisted land reform can help solve some of these difficult problems:

- It can assist in resolving the financial crisis of the commercial farm sector by creating a market for land.
- It can assist in solving the employment problem by generating self-employment on the new farms and the associated non-farm economy at a low cost per job.

Design criteria

It is clear from the literature cited in earlier chapters[5] that a land reform programme should avoid the following:

- maintenance of privileges to the farm sector, most of which are captured by the large-scale mechanized farms, for example, tax shelters and credit subsidies;
- general debt bail-outs and blanket debt relief to farmers;
- confiscation, expropriation or any other administrative/ legal 'cheap option' to get land;
- land acquisition by state or local land reform or administration bureaucracies;
- administrative beneficiary selection;
- settling land reform beneficiaries on low quality land.

Desirable design features include:

- making the law consistent with the objectives and process of the reform;

4 This has already happened and is still the case in many areas in South Africa where large-scale bankruptcy has been prevented or prolonged in general through blanket debt relief and credit subsidies (see Chapter 9).

5 *See*, among others, Binswanger and Deininger (Chapter 4), and Kinsey and Binswanger (Chapter 5).

- using a market-assisted approach involving willing buyers and willing sellers, with targeted financial assistance to poor buyers;
- using self-selection by communities and individuals through the rules and incentive structures of the programme;
- using grants to target the poor and to provide appropriate self-selection incentives;
- limiting restrictions on the beneficiaries, for instance on renting or selling farms, or choosing alternative enterprise and community models;
- involving beneficiaries in the planning of infrastructure and services;
- making use of a decentralized institutional structure;
- decentralizing supervision;
- emphasizing ex-post supervision rather than ex-ante approval at each step, combined with suitable penalties.

How will land reform contribute to resolving the debt crisis?

The workings of the proposed market-assisted land reform process in an impending debt crisis situation are described in Figure 16.1. The processes described above exert serious downward pressure on farm profitability, specifically in the large-scale commercialized and mechanized grain sectors, causing farm debts to increase, farm bankruptcies to increase, and farmland prices to fall dramatically. Although all farms are affected, they can be classified into three categories, namely:

- farms which are not viable under the new policy regime and are likely to go bankrupt;
- farms which are viable, provided they shed assets and restructure their debt;
- good farms which can survive the crisis without major change or assistance.

Since all farms are affected by the crisis in one way or another, they will join together to request blanket debt relief, regardless of their category. The banking sector is not keen to declare farms bankrupt since it knows that the farms will be nearly impossible to sell. The banks, therefore, will also join the political alliance for blanket debt relief (Process 1).[6]

6 As shown in Chapter 9, this has already happened several times over the last decade, with high costs to taxpayers.

This approach is expensive and will keep inefficient farmers alive. A better approach is based on differential assistance of the specific categories of farms (Processes 2 and 3).

Farms in financial trouble, but which can make it with some restructuring of assets and liabilities, could receive selective financial assistance conditional on their restructuring, including the shedding of

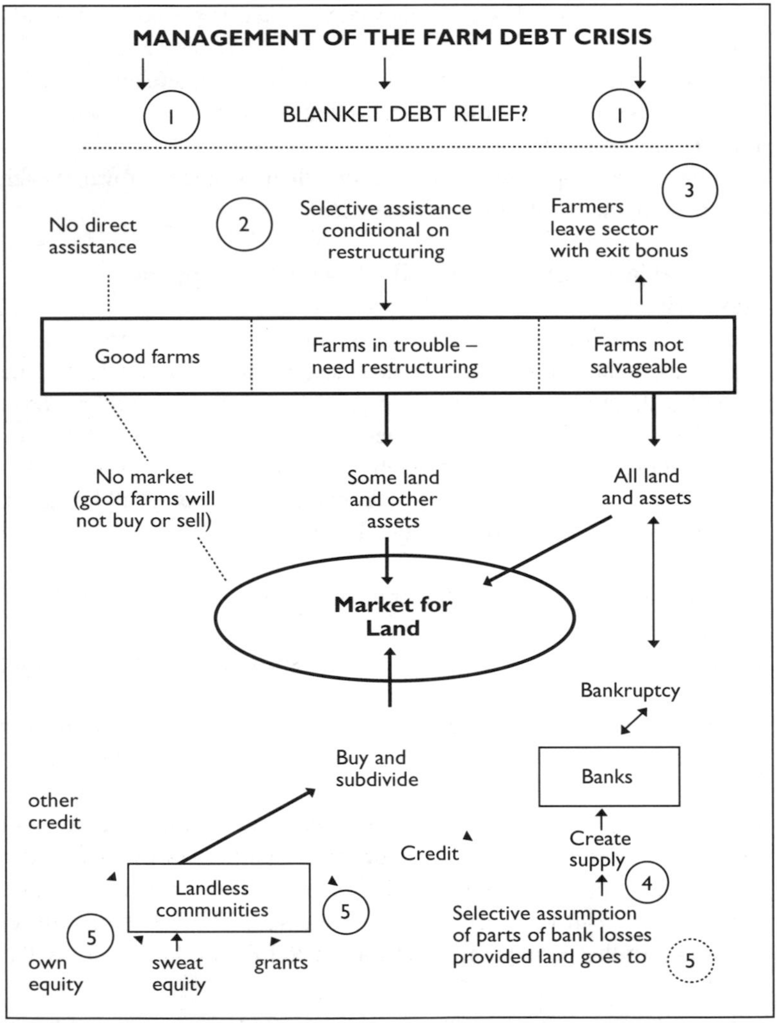

Figure 16.1: The Process of Market-assisted Land Reform

some land and other assets (Process 2). The decision about which farms should be restructured, and how, is based on their future viability and is made by the banking sector.

Farmers whose businesses are not viable could receive an exit bonus or other package in order to increase the supply of land and other assets to the market (Process 3). These could include relocation allowances, or assistance with retraining, or setting up a new business. Good farms need no financial assistance, but will also benefit since the process of market-assisted land reform will stabilize their land prices and the value of their assets. The number of farms in each of the categories is determined by the severity of the debt crisis, and the criteria and amount of selective assistance to farms. Land and other assets will become available on the market from both the non-viable and the restructured farms. However, financially sound farms are unlikely to buy the land and assets because they are already large enough, and because of high real interest rates and decreasing farm profitability.

The above process of increasing the supply of land and farm assets can further be aided by incorporating the bank sector as an active partner and beneficiary (Process 4). Banks can be encouraged to participate by the selective assumption of part of their losses when they foreclose on non-viable farms and assist in the restructuring of others, provided that the land and other farm assets go to the intended beneficiaries of land reform.[7]

On the demand side of the land market, there are the intended beneficiaries of the land reform process, namely the present large number of landless communities and individuals, often farm workers, in the farm and rural sector in general (most of them 'coloured' or black). A land market is created by providing eligible beneficiaries with a grant and access to credit. They can combine this with their own (limited) equity in the form of assets and financial contributions, and in the form of labour contributions to the farm and infrastructure development (Process 5). Credit sources to the beneficiaries include the former owners of the land, the former lenders (banks) and contractors (e.g. factories). Credit could also be linked to the eligibility of part of a bank's loss being taken over (Process 4) if it forecloses on former farmer clients. Communities who buy land can subdivide it according to their own negotiated agreements. They, or individuals, should be eligible for partial grants to create the necessary infrastructure or for other approved purposes, on condition that they have good quality land.

7 The bank sector could partly finance these beneficiaries if they meet normal bank standards (at market rates and conditions).

Policy environment and other issues

A prerequisite for any market-assisted land reform programme is the removal of policy distortions favouring relatively large farms over smaller farms (levelling the playing field), and driving up land prices to levels well above the capitalized value of future farm profits. This is important for several reasons:

- High land market prices relative to the capitalized value of future farm profits will increase the cost of a reform programme, will serve as an incentive for the selling of farms by beneficiaries to get windfall profits,[8] will limit the supply of land to beneficiaries, and will encourage land purchases by groups other than beneficiaries, for example, good farmers and business people looking for tax shelters.
- Measures favouring relatively large farms over smaller farms lower the reservation prices of small farmers who are often poor, thereby encouraging the selling of land by beneficiaries and the purchase of land by groups other than beneficiaries.

Coalitions

A key to market-assisted land reform is to break the coalition between farmers (good and bad) and the banks, in favour of blanket debt relief. Instead, a coalition between good farmers, banks and beneficiaries is necessary to make the process work. Limited exit bonuses to non-viable farmers and benefits to participating banks are important in this respect.

Exit bonuses

Exit packages for non-viable farmers who are likely to go bankrupt have as their objective the increase of the supply of land to the market. These packages can take various forms, including exit bonuses, training for other careers, alternative employment, increased foreign currency allotments and pension schemes.

Cost of the programme

Initial calculations indicate that market-assisted land reform will cost between 55 and 78 per cent of a blanket debt relief approach, depending on the assumptions of the analysis (*Van Zyl, 1994c*). This represents considerable savings to taxpayers.

8 See Carter and Mesbah (*1993*) on the reservation prices of large and small farmers.

Grants and grant sizes

Small farmers cannot purchase land and other inputs and pay for them from the proceeds of their farming operation. Even when the difference between the market price of land and the capitalized value of future farm profits is small, farmers may find it difficult to borrow money to purchase land and to pay back the loan from the proceeds of farming. In this respect, experience has shown that, depending on the type of farming, farmers with a debt/asset ratio of more than 0,3 are unlikely to be able to shoulder the high risks of farming and be financially viable and creditworthy in the long run.

Market-assisted land reform therefore requires some financial assistance to the beneficiaries of such a programme so that the desirable debt/asset ratio can be reached and maintained, and debts can be repaid. In this respect, grants are superior to subsidies because they are immediate, transparent, can be targeted and their distortive effects are small.

Grants can be justified on other grounds as well. Poverty reduction objectives can justify targeted interventions to promote access to rural finance for particular groups of the population. Financial interventions must be shown to be cost-effective means to achieve poverty reduction compared to other targeted interventions. Externalities, such as possible risk diversification and nutrition benefits associated with improved access to land, are an additional justification. Covariance, seasonality and risk averseness may require additional equity to serve as a safety-net against disaster. Infant industry considerations justify limited initial grants to allow inexperienced farmers to face the special inherent difficulties in farming.

Two aspects are important when considering grants, namely: (a) a comprehensive benefit-cost analysis should demonstrate the private, economic and social efficiency of the proposed grants; and (b) the results from such a benefit-cost analysis should be used to determine the size of grants which are required (see Chapter 23). Two calculations are recommended:

- *Total grant* needed to achieve target ratios with respect to liabilities and wealth of farmer beneficiaries. A target ratio of 0,3 (liabilities/wealth) is proposed.
- *Base grant* necessary to provide at least a 10 per cent social rate of return (including nutrition, risk diffusion benefits and other social benefits).

The size of grants required to achieve the 30 per cent debt/asset ratio beyond which farms become financially risky and not viable, depends on several factors which can be categorized into those independent of the reform programme, and those related to the programme specifications.

In the last category are farm policy and macroeconomic variables, on-farm investment requirements (land, machinery and equipment, livestock and working capital) and profitability, as well as interest rates. These factors are exogenously given in the construction of farm models. A major issue is the difference between the market prices of land and other assets on the one hand, and the capitalized value of farm profits on the other.

Factors impacting on grant sizes which are dependent on programme specifications are:

- the larger the farms of the beneficiaries, the larger the grant required per beneficiary to ensure that the farm is financially viable;
- equity that the beneficiaries of the programme have to contribute from their own resources.

The target farm size and own equity requirements are related to the target income, to give the farm household an income comparable to that which can be earned outside of agriculture (with the necessary adaptations for risk, use of farm products for own consumption, on-farm housing, etc). In addition, the selection of beneficiaries is significant. The really poor cannot contribute substantial amounts from their own resources beyond their own labour. So it matters whether targeting of the poorest of the poor is important.

Redirecting agricultural credit: starting the process

As shown in Chapter 9, farmers continued to increase their indebtedness in the 1980s to all government supported credit sources, including the Land and Agricultural Bank, Agricultural co-operatives (obtaining credit from mainly the Land Bank), and the Department of Agriculture (specifically the Agricultural Credit Board).

At present, the government and specifically the Department of Agriculture (through the Agricultural Credit Board), directly controls an 'agricultural credit book' of well over R1 billion. The Agricultural Credit Board (ACB) serves more than 20 000 of the approximately 60 000 commercial white farmers in South Africa. The large majority of these farmers are unbankable in the private commercial banking sector.

The credit provided by the ACB distorts agricultural credit markets because it is given at less than half the market rate. There is little incentive to repay ACB credit – roughly 60 per cent of all creditors are either defaulting on their loan or breaking some of the repayment conditions. There is little risk involved in such behaviour – it is not the ACB's policy to call up bad loans. The general practice is that other creditors should sue for payment of their debts first, before a farmer is foreclosed.

The credit provided by the ACB costs South African taxpayers in excess of R200 million per annum. It crowds out private sector involvement. It breeds inefficiency by using state subsidies to keep non-viable farmers on the land. It inhibits the working of the land market and sustains land prices at relatively high levels. This makes it difficult for new entrants to start farming through open market purchases of land. Credit provided by the Land Bank is subject to the same criticism, but to a significantly lesser degree.

Agricultural credit provided or guaranteed by the government should be used as an instrument to correct some of the imbalances caused by previous policies, rather than perpetuating their consequences. It could be a powerful tool in achieving the objectives of the Reconstruction and Development Programme (RDP). Existing credit lines can be retargeted and restructured to benefit new small-scale family-type black farmers. Calling-up bad and overdue debt could be used to start the land reform process.

Other support services are equally important to ensure viability and sustainability of a land reform programme over the longer term. It is clear that the redirection of credit and other support services is politically sensitive, and therefore requires a strong political will.

Conclusion

Management of the debt crisis following market liberalization and the withdrawal of other privileges can be done by facilitating market-assisted land reform rather than through a much more costly, inefficient and inequitable blanket debt relief programme. This will also have the advantages of increasing employment and adding to the rural safety-net. In addition, the process outlined above provides a mechanism for land reform without the problems and excessive procedures associated with the usual state or parastatal approach.

References

Binswanger, H.P. and Deininger, K. 1993. South African land policy: the legacy of history and current options. *World Development,* 21(9): 1451–1477.

Carter, M.R. and Mesbah, D. 1993. Can land market reform mitigate the exclusionary effects of rapid agro-export growth? *World Development,* 21 (7) : 1058–1100.

Helm, W. and Van Zyl, J. 1994. Domestic agricultural support in South Africa from 1988/89 to 1993/94: a calculation. *Paper read at the 32nd Annual Conference of AEASA.* Pretoria: 19–20 September.

Kinsey, B.H. and Binswanger, H.P. 1993. Characteristics and performance of resettlement programmes: A review. *World Development,* 21(9) : 1477–1494.

Kirsten, J.F., Van Zyl, J. and Hobson, S.D. 1994. Food security in South Africa with particular emphasis on rural households. *Paper presented at a Workshop on Employment and Aids in Southern Africa.* Zimbabwe: 22nd Conference of the IAAE, Harare.

Meyer, N.G. and Van Zyl, J. 1992. Comparative advantages in Development Region G: an application of a sectoral linear programming model. *Agrekon,* 31(4): 307–312.

Tweeten, L. 1986. Macroeconomic policy as a source of change in agriculture. *Agrekon,* 25(3) : 5–11

Van Zyl, J. 1994a. Modelling the South African animal feed sector: the example of yellow maize. *Unpublished Research Report.* Pretoria: Protein Research Advisory Committee, Agricultural Research Council.

Van Zyl, J. 1994b. Regional implications of the restructuring of agriculture in the Western Cape: effects on the farm sector. *Unpublished Research Report.* Halfway House: Development Bank of Southern Africa.

Van Zyl, J. 1994c. Productivity, efficiency and land markets in South African Agriculture. *Research Report.* Halfway House: Development Bank of Southern Africa.

17

Rural livelihoods, fiscal costs and financing options: a first attempt at quantifying the implications of redistributive land reform

Rogier van den Brink, Mike de Klerk and Hans Binswanger

Introduction

If redistributive land reform is implemented in South Africa and beneficiaries are given access to land, what are the types of property relations and land use activities that they are likely to develop? What would a typical farm look like in terms of area cultivated, area used for pastoral activities, herd size and machinery employed by the household? Moreover, since farming will typically only be a part-time activity for the household, how much of the household's income should be expected to be derived from farming? How would the household spend its income in the rural economy, and what would that imply for the creation of non-farm jobs in the local economy? What would be the net gain or loss of jobs, if one also takes account of the jobs lost on the large, commercial farms if some of them are replaced by smallholder farms? What would be the likely fiscal cost of a market-assisted programme which redistributed a significant amount of land at market prices and provided basic infrastructure and support services to the beneficiary households? How would such a programme be financed?

These questions can only be answered in an indicative way. However, the policy debate on land reform and rural restructuring cannot avoid addressing these questions upfront since the feasibility of the entire exercise hinges on the answers. In 1993, through a collaborative effort between the Land and Agricultural Policy Centre and the World Bank, a team of South African and international researchers was formed to provide a first set of preliminary answers to the above questions.[1]

1 The team responsible for the farm models case studies consisted, inter alia, of Johann Kirsten, Julian May, Agnes Nyamande, David Tapson, Rogier van den Brink and

The first set of issues to be addressed concerned the modelling of a number of typical post-reform households. Four regional case studies were undertaken, yielding a range of typical household models, but only four models were ultimately selected for quantitative analysis. The four models are derived from available empirical data, sometimes limited, on land use, net farm incomes, household size and agricultural income shares. The models refer to the four agro-ecological zones for which case studies were undertaken. The case studies cover a roughly representative cross-section of farming areas, and the aggregate totals derived from the four models reflect approximately average national agricultural potential and cost of land and farm capital acquisition per hectare.

The second set of issues to be addressed concerned the estimation of the fiscal costs of a rural restructuring programme which would attempt to induce the type of rural economy indicated by the above household models. The various cost components were estimated, options to reduce the fiscal burden outlined, and an indication of the likely pace of a programme which relied on restitution and market-assisted redistribution was given.

The main hypothesis emerging from the analysis in this chapter is that large numbers of poor, but often potentially commercial, farm households can obtain reasonably attractive household and employment levels through fiscally affordable land redistribution. This process can be stable, sustainable, consistent with urban, rural and household food security, and help to restrain the speed of migration from rural to urban areas.

Further research is needed to test this hypothesis. The paucity of comprehensive empirical data on small-scale agriculture in South Africa and the limited number of agro-ecological zones studied, pose obvious limitations to the generalization of the results which were generated. However, the results presented in this chapter tentatively predict that a redistribution of land (say 30 per cent in five years) from the large-scale, commercial sector to small-scale, part-time farmers would create a substantial number of rural livelihoods at acceptable income levels. When land moves from large to small holdings, there is a substantial increase

Wilfred Wentzel, who were, in turn, assisted by a number of others. The team coordinator was Michael de Klerk and the team's advisers were Michael Lipton and Daniel Bromley who both contributed to this chapter. Hans Binswanger provided the theoretical framework for the exercise. The team responsible for the programme cost estimates included Nick Vink and Johan van Rooyen.

in the number of rural livelihoods. First, this is because farming becomes more labour-intensive. Second, it is because a new type of rural economy is created in which there is stronger linkage to labour-intensive rural or small-town production of non-farm goods and services. These two effects greatly outweigh the loss of jobs on the larger farms as they are restructured. Moreover, the cost per livelihood in terms of land, livestock and machinery – priced at current market values[2] – as well as institutional and infrastructural investments needed, is relatively small when compared to the likely costs of job creation in other major sectors of the economy (e.g. industry and mining).

The organization of this chapter is as follows. First, a qualitative description is given of the range of potential beneficiaries, property relations and farming options that could be contemplated under a rural restructuring programme. Second, a quantitative analysis of four farm models – one for each major agro-climatic zone – is outlined. Third, an estimate of the potential for rural livelihood creation is provided. Fourth, the fiscal costs and financing options of a rural restructuring programme are estimated. Finally, research recommendations are given.

Property relations and beneficiaries

Under the political and economic policies of apartheid, white settlers in rural areas were to be provided secure economic opportunities. Economic policy in general – and property tax policy in particular – created an incentive for white farmers to hold freehold title in their entire agricultural estate. Given that the fixed cost to the farmer of owning extensive freehold land approached zero, there was a strong incentive to expropriate large tracts of land (that in other similar regions of the world carry the name 'wasteland') into the private domain. When this agrarian structure based on large land holdings was exposed to a policy framework which heavily subsidized capital costs, labour intensity dropped below its economic optimum and effectively depopulated the commercial farm areas. Redistributive land reform in South Africa means the creation of new property relation – a new agrarian structure – which will repopulate the commercial farm areas with a rural population living in sustainable social, economic and environmental conditions.

In the higher potential areas of Natal, Gauteng and the Free State, for example, most farms would likely consist of private property. This private estate could be held by an individual farm family, or by several farm

2 Unless otherwise indicated, 1993 Rands are used throughout this chapter.

families who hold a group title and farm the land under several alternative enterprise forms, such as individually, cooperatively or as a corporation. In the more arid regions, such as the Northern Cape and the Karoo, the superior arable lands would surely be held individually, again by individual families or under group title by several families. However, the extensive rangelands that are necessarily associated with these private holdings could be held under several managerial forms of common property. In the medium potential areas, a wide range of different combinations of private and common property would probably emerge.

Hence, a combination of ecological reality and the preferences of beneficiaries will probably lead to different property rights and relations. The reform process should respect this fact. It may also reduce the cost of resettlement: households could decide to rent grazing rights instead of purchasing the full set of land use rights.

The following broad categories of potential beneficiaries emerged from the various case studies – (see also Chapter 18 for further discussion on potential beneficiaries): (a) communities that have traditional tribal origins, or are based on voluntary associations of people such as in the church, labour unions, neighbourhoods; (b) farm workers currently or recently employed on commercial farms; and, (c) individuals. For each of the above categories of beneficiaries, the following range of typical property relations and farming options was identified. The list is not exhaustive and models will adapt to local circumstances.

Communities

- **Common property model.** Under this model (which should be distinguished from production co-operatives or collectives discussed below), one can envisage a range of arable/pastoral mixes as determined by community preferences and the natural resource base. The common area can be held as freehold or leasehold with the option to buy under group title. Secure usufruct or freehold rights for individual households can be established for residential and arable land within the area, and common access to pastoral land can be regulated by community rules. Farm sizes will vary considerably and a shift towards more labour-intensive technology will be likely to occur. Appropriate levels and forms of support services are essential to accompany this shift. Especially in arid/semi-arid areas, communities may develop more flexible grazing arrangements with other communities. This will extend the effective pastoral range over larger areas than covered by existing individual commercial farms.

The scope for increasing the flexibility in stocking rates under well-managed common property regimes could substantially improve the efficiency and environmental sustainability of livestock production in South Africa. For instance, though ecologically distinct from South Africa, the summer pastures of Switzerland are held in common property and used under a variety of managerial regimes. In South Africa, these property regimes would be likely to bear a strong resemblance to customary tenures known in southern Africa since time immemorial.

There can be little doubt about the breadth of support for this type of property relations model. 'Many of the communities who are re-establishing themselves (in the Transvaal) after the return of their land, e.g. Mogopa, Goedgevonden, Monnakgotlas and Doornkop, have made communal control over land one of the mainstays of the new systems ... The apartheid policy of land dispossession has created the need in African people for security of tenure but not necessarily in the form of individual ownership of a registered piece of land. Generally people want legally respected property rights in the form of title deeds to land (held by the community as a whole), but they still wish to retain community involvement in internal land allocation.' (*Small & Winkler, 1992:33*).

- **Irrigated garden plot model.** This model applies to irrigated perimeters that can be adapted to small-scale use. The irrigated perimeter can be held under group lease or freehold by the community. Individual households obtain secure usufruct, freehold or sub-lease rights to their garden plots, with sizes ranging from 0,05 to 0,25 hectares or smaller. Experience suggests a maximum community size of about 100 households. Initially, farm size and activities would primarily emerge from the food security needs of households. In the longer run, more commercialized activities may evolve. An important pre-condition for the model would be to establish the community's property rights to water and ensure sustainable water use.

  Anecdotal evidence suggests that this model – which is a departure from existing irrigated farm systems – accords well with community preferences. In one locality in the Eastern Cape where a community was settled on irrigated land, elaborate plans for cash crop production were disregarded and the land was used along the above lines instead. The model would have most appeal to dense settlement, small town and mine village communities.

Farmworker groups

- **Improved conventional employment model.** For many farmworkers, rural restructuring does not necessarily mean land redistribution, other than in a limited sense. Rather, it means improvement in their social conditions and their relations with their employers. This model assumes continued wage employment of the farmworkers on an existing commercial farm. However, farmworkers are assumed to hold secure rights to residence, arable or garden plots, and the pastoral land of the farm. Increased labour participation in management decisions and improved access to social services seem to be prerequisites for the social sustainability of this model. The model is also known as the 'normalized' industrial relations model. Whereas the state could probably enforce a minimum wage package without substantial loss of employment, a more significant improvement of working conditions of existing farmworkers crucially depends on the opportunity costs of labour. Substantial land redistribution and the resulting emergence of a more employment-intensive rural economy would raise the opportunity costs of labour, thereby increasing the bargaining power of farmworkers *vis-à-vis* their employers. Hence, the success of the 'improved conventional employment model' would still hinge on the success of rural restructuring.
- **Common property model.** Under this model, farmworkers establish common property relations as indicated in the common property model for traditional communities outlined above. For instance, employees on an existing commercial farm would buy out the owner and operate the farm under a common property model.
- **Production co-operative model.** This model is based on joint management of the commercial farm enterprise by workers, and joint ownership or lease of land and equipment. Freehold land is held by the members under a group title, or by a corporation in which the members are shareholders. Provision of support services focusing on management skills and the promotion of a shift towards more labour-intensive production is essential. An extensive theoretical and empirical literature on production co-operatives, collectives and labour-managed farms suggests that these farms often suffer from complex incentive problems and are frequently converted to individual or common property models after a few years.

    However, production co-operatives will have an appeal in a society emerging from a history of inequality as draconian as South Africa's. Some groups of farmworkers or others may have the motivation, resources, social cement and good fortune to make them successful. But even in traditional communities, co-operation in production is generally confined to short reciprocal exchanges of labour during peak periods. Moreover, abysmal experiences in Africa with production co-operatives suggest that this will not be an easy option.

- **Equity-sharing option.** Under this option, continued wage employment of farmworkers and continued participation of the former sole owner in management and equity is assumed. Participation by farmworkers in management decisions is a prerequisite for profit and equity sharing with respect to land and/or operating company. The model is described in detail by McKenzie (*1993*), who advocates its potential importance in, for instance, horticulture.

Whereas the broadening of the ownership base in agriculture is an important goal in itself, the equity sharing model is unlikely to generate a significant number of livelihoods – and could even do the opposite if 'labour aristocracy' tendencies emerge. The case for the model becomes stronger when economies of scale exist. However, even in horticulture, international experience does not indicate that small-scale farmers would be less efficient than large-scale farmers.

Moreover, the bargaining process implied by this model again depends on the opportunity costs of labour. For farmworkers, the existence of a feasible alternative – independent farming – will greatly increase their bargaining power in this process.

Individuals

- **Irrigated market garden model.** This model is likely to emerge in peri-urban areas, in dense rural settlement or on existing irrigation schemes. It assumes freehold or lease by individuals or groups of market garden plots, and a wide range of plot sizes, production technologies, income levels and degrees of commercial orientation. Little reliable information is to be found on small-scale urban and peri-urban farming in South Africa (no national estimate of employment in this sector exists), and the potential is barely appreciated. However, May (*1993*) empirically investigates a wide range of detailed farming options and stresses the great potential of this model. An important condition for the success of this model would be the access to property rights in water, and the establishment of incentive-compatible rules for sustainable water use.
- **Outgrower model.** This model applies to subdivided or expanded large-scale horticultural or sugar farms. Outgrowers gain access to

land under individual or group freehold or lease, while contracting with the 'core' farm for the provision of services (e.g. input supply, output marketing, financial management). Where households have accumulated little capital and few management skills, the model has considerable advantages. Once households accumulate capital and management skills, however, a more independent type of operation will probably evolve.

Anecdotal evidence suggests there is substantial demand for this model, while some large-scale farmers (who would be the 'core' of the model) have expressed support. Glover (*1993*) notes the success of outgrower models in east and southern Africa, while the expansion of the outgrower model in sugarcane farming in Natal provides a South African success story of this model. Cousins (*1993*) describes a promising experiment in the Western Cape.

- **Commercial family farm model.** This model is based on conventional private freehold or lease. A wide range of plot sizes, production technologies, income levels and degrees of commercial orientation is envisaged. Support service packages should be developed to assist the farmer in efficient and sustainable farming.

The success of all the above farming options in terms of social welfare and employment creation crucially hinges on the adoption of the appropriate agricultural policy framework. In the most general of terms, it is suggested that a wide range of policy distortions, which currently favour large-scale farming and discriminate against small-scale producers, will need to be eliminated in order to achieve a socially and economically optimal mix of farm sizes. For instance, an obvious policy distortion which needs to be removed is the Subdivision of Agricultural Land Act, which impedes the subdivision of land and the establishment of private property rights for small-scale farms.

Moreover, in order to increase net income per hectare and restrain debt burden, post-reform farmers should be helped to adopt processes that raise output by applying increasing levels of labour with acquired skills, rather than selecting products or methods that displace labour by purchased inputs (e.g. herbicides, hired combine harvesters). For the sector as a whole, this involves removing existing incentives to increase capital intensity and radical changes in support institutions.

Stability of rural income will be essential if post-reform communities are to be food secure and creditworthy. Hence restructuring should encourage multiple income sources; rotations, crop and/or livestock mixing, and other techniques to reduce farm risk; employment in slack seasons/years in public works to increase rural infrastructure; and appropriate water use (including farmer- or community-controlled irrigation).

Quantifying four farm models

Some of the richness in terms of the range of the property relations and farm options outlined above is lost in the effort of quantifying, undertaken in the following section. However, in order to obtain a first estimate of the potential number of jobs, and the costs of creating these jobs, it is necessary to simplify the above variety into a restricted number of relatively simple and robust models. Four models were selected: the summer grains area of the southern Transvaal and the northern Free State; the livestock-dominated areas of the Eastern Cape; the fruit/vegetable/wine areas of the Western Cape; and the peri-urban vegetable areas around Durban, Johannesburg/Pretoria and Cape Town. The case studies cover 28 per cent of agricultural land and 34 per cent of existing farm jobs (excluding the peri-urban areas).

Several important agro-ecological zones and crops were excluded. Some, like the Karoo, are areas of low agricultural potential. Others, e.g. the subtropical areas of Natal and Mpumalanga (fruit, sugar and timber), and the irrigated perimeters, have high agricultural potential. The case studies therefore cover a roughly representative cross-section of farming areas. Some of the excluded crops, notably sugar and some sub-tropical fruits, are proven by South African and international experience to be especially appropriate for efficient small-scale farming. Overall, the aggregate totals derived from the four models reflect approximately average national agricultural potential and cost of land and farm capital acquisition per hectare.

The two main principles on which the models are based are the following: beneficiary households are (a) neither expected to generate unduly high agricultural incomes, (b) nor required to rely wholly on farming. For instance, today in South Africa, the typical income of a full-time farm labourer appears to be approximately R3 000 per year. Hence, in the modelling exercise here it is assumed that, depending on local conditions, an indicative annual income, including the value of own consumption, would lie between R5 000 and R12 000[3] for a typical

3 The relatively wide range of incomes assumed reflects current local conditions and relatively imperfect labour mobility between zones. It may be argued, however, that post-reform equity objectives should attempt to equalize incomes across regions. More uniform indicative incomes would then be appropriate.

range of post-reform households of six persons. More importantly perhaps, while only between 25 per cent and 75 per cent of this income is assumed to be derived from the household's agricultural resources, this is a higher proportion than is now typical of small-scale farmers in South Africa. If higher income levels are modelled, or higher shares of agricultural incomes, then the number of beneficiaries will be smaller. Of course, some households will seek much higher incomes (or proportions coming from agriculture) at once, and incomes should grow. No explicit risk premium is assumed, but both yield estimates and debt-asset ratios take frequent drought conditions into account.

The indicative income levels and shares of agricultural incomes for which models have been constructed are the following:

- In the Eastern Cape, the income level assumed is R5 000 per household per year, half of which is derived from agriculture. The other half comes from equal shares of non-farm and other sources. Current farmworkers' annual wages, as mentioned above, appear to be in the order of R3 000.
- In the Western Cape, the income level assumed is R12 000 per household per year, three quarters of which is derived from agriculture. A quarter of income comes from non-farm activities. No other source of income is assumed to exist. Current semi-skilled farmworkers' annual wages appear to be about R6 000.
- In the summer grain area, the income level assumed is R6 400 per household per year, roughly the current rural household subsistence income level, with half derived from agriculture. The other half comes from equal shares of non-farm and other sources.
- In peri-urban areas, the income level assumed is R9 000 per household per year – reflecting greater income-earning opportunities in peri-urban areas, and a quarter of income is derived from agriculture. An additional quarter of income comes from non-farm sources, while half of income derives from other sources.

Empirically estimated values for basic parameters were used to calculate net farm income for each of the four models.[4] Net farm income was then

4 The definition of net farm income used is the following. It consists of gross revenues less purchased inputs, purchased labour, depreciation and capital service charge on machinery, and the rental rate on the portion of capital that is borrowed. Net farm income can thus be interpreted as the economic return to the household's owned

compared to indicative net farm incomes. Household models were then rescaled – up or down – to conform to indicative net farm incomes. This is particularly important in the case of labour requirements. If the rescaled model would yield labour requirements that vastly exceed the household's own labour capacity, the household would need to hire additional labourers. Only in the case of the peri-urban areas did such a situation occur, and it was necessary to iterate back to the empirical case and add costs for hired labour. In the other regions, rescaled values seemed well within the household's labour capacity, taking into account some underemployment.

The value of a number of basic model parameters was obtained as follows. Average household size is six persons, comprising two adult equivalents. The rate of return on livestock assets, taking into account all cash and non-cash forms of income, is assumed to equal 34 per cent per year (*Tapson, 1990:147–167*). Stocking rates used vary from 5 ha/LSU in the summer grain area to 6 ha/LSU in the Eastern Cape. Average physical yields of unirrigated maize range between 0,48 tons/ha in the Eastern Cape and 1,36 tons/ha in the summer grain area. These conservative estimates take into account the frequent occurrence of drought.

Farm households' sustainable debt-burden on assets (land, livestock and machinery) is assumed not to exceed a 30 per cent debt-asset ratio, indicative of a relatively risky agricultural environment. This is consistent with international and local norms (*SAAU, 1984:56*). The debt-asset ratio is applied to total asset value of land, capital and machinery. Given a 30 per cent debt-asset ratio, the remaining 70 per cent of asset value represents the capital cost of the models. This is assumed to come from a combination of own equity or grants – no assumption on the actual source of capital cost is made in the modelling exercise.

The imputed cost of servicing the 30 per cent of assets that are encumbered by debt are calculated using current market prices. For land and livestock, which account for the bulk of farm capital, empirically observed rental rates from South Africa are used (*Kirsten, 1993; & Nieuwoudt, 1993*). They vary between 5 and 7 per cent of the price of land, depending on the region. In the case of machinery, currently prevailing commercial rates (15 per cent) are applied. Land prices used are those currently prevailing (see note to Table 17.1).

resources (labour and the portion of capital assets unencumbered by debt). This figure is different from farm profits, which would be arrived at by subtracting the opportunity costs of owned resources from net farm income.

Table 17.1: Main Characteristics of the Four Indicative Farm Models

Variable	E. Cape	Summer grain	Peri-urban	W. Cape
NFI ®	2 500	3 200	2 250	9 000
NFI/ha	56	180	12 341	6 178
Farm size (ha)	45,0	17,8	0,2	1,5
Cultivated (ha)	5,5	3,0	0	0
Irrigated (ha)	0	0	0,2	1,5
Pastoral (ha)	39,5	14,8	0	0
Herd size (LSU)	6,6	3,0	0	0
Machinery (R)	0	0	456	8 740
Main activities	Cattle Maize	Maize Cattle	Vegetables Maize	Winegrapes[5] Vegetables

Note: Land prices (1993 Rands per ha) used in the models are the following: (a) Eastern Cape cultivated land R700, and pastoral land R500; (b) Summer grain cultivated land R1 000, and pastoral land R500; (c) Peri-urban irrigated land R14 000; and (d) Western Cape irrigated land R18 000.

Capital cost does not include provision for the cost of housing and rural infrastructure. Households are assumed to build their own houses or to occupy existing houses. The costs of provision for public rural infrastructure (e.g. water supply and roads) are estimated below.

The basic characteristics of the rescaled models are presented in Table 17.1. The models capture a wide variety of farm types. For instance, net farm income per ha varies dramatically from R56 per ha in the mixed farm model of the Eastern Cape to R12 341 per ha in the peri-urban vegetables/maize model. Farm sizes necessary to achieve the indicative incomes range from 0,2 to 45 ha.

Estimating rural livelihoods gained

Assumptions

- **Rural livelihoods.** As mentioned above, the models do not assume that 100 per cent of a rural household's income is required to come from

5 Winegrapes were selected given data limitations. However, deciduous fruit holds considerable promise for small-scale farming, and appears to be more profitable than winegrapes.

agriculture. The rural restructuring envisaged should neither attempt to create small replicas of current commercial farmers, nor should its benefits be measured only by the number of such replicas created.[6] Hence, the modelling attempts to go a step further than merely replacing commercial farms with small farms, and calculating the net farm job gains due to the increase in labour intensity. The modelling roughly estimates the full economic benefits of a new, more densely populated and dynamic rural and peri-urban economy in which farm incomes, supported by incomes from other sources, create non-farm jobs through forward and backward linkages. It is assumed that income from other sources (business income, wages, remittances and pensions) diminishes in relative importance as farm income grows.

To capture job creation in this new setting, the concept of rural livelihood is used. A rural livelihood is defined as a full-time equivalent worker living in a rural household who derives an income from farming activities, from non-farm activities directly related to farming (e.g. farm labour, input supply, output processing and marketing, farm capital production and maintenance), and from other sources (e.g. business activities not related to farming, wages, remittances, pensions).

Two full-time equivalent workers are assumed per household of six persons. For instance, such a household could consist of two adults working full time, or of a single head of household working full time, an older relative working half time, and the balance of labour requirements provided by older children during weeks and days that school is not in session.

6 Hertz (*1994*) points out that a large portion of the employment effects are due to the assumed levels of remittance (including pension) incomes. His position is that only farm income, and not farm income plus remittances, should be used to estimate the non-farm linkages, because he doubts that these remittances will emerge spontaneously in the resettled communities. We believe that such remittances will indeed emerge, but will become less important than is currently the case in the rural homelands where they account for about 35 per cent of household income. Since both farm income and remittances are likely to be spent in the restructured rural economy, we applied the multiplier to the sum of both incomes. Hertz also asserts that the emergence of remittances in the resettled areas should not be attributed to the change in agrarian structure. It is true that a cost-benefit analysis would evaluate the opportunity costs of these remittances, and correct for the jobs lost by transferring these remittances from, say, the current rural homelands or urban squatter camps to the newly resettled areas. We doubt, however, that there would be a one-to-one trade-off. Hence, Hertz overstates his case. The current exercise, moreover, is not a cost-benefit analysis, but an estimate of the employment potential and fiscal costs of a restucturing of the commercial farm areas. Only jobs – and their linkages – lost on commercial farms are counted.

- **Multiplier.** The multiplier effect in the South African context is expected to be real, rather than inflationary, in view of the high levels of unemployment and underemployment in South Africa. In order to estimate the number of extra jobs created through farm–non-farm multiplier effects, the following estimates were adopted. International evidence suggests that farm–non-farm multipliers vary from approximately 1,3 for typical sub-Saharan subsistence agriculture with low population densities and undeveloped rural infrastructure, to 2,2 in modern agricultural regions as can be found in the USA (*Haggblade et al, 1991*). Given South African conditions, a rather low income multiplier value of 1,6 was used. Given reasonable assumptions about household income shares derived from various sources, it can be shown that an income multiplier of 1,6 implies that for every farm livelihood created, 0,26 non-farm rural livelihoods will also be created.[7] Hence, a farm–non-farm livelihood multiplier of 1,26 was used to estimate non-farm livelihoods gained and lost through land redistribution.[8]

7 Assume a rural economy without rural – urban linkages in which specialized farm households exist, but no permanent labourers. Farm households derive half of household income from agriculture, a quarter from non-farm activities, and a quarter from remittances. Non-farm households receive three quarters of their income from the farm – non-farm multiplier effect, and one quarter from remittances. Set the value of the indicative incomes for farm and non-farm households at R5 000. Hence, for the farm household, the agricultural income is R2 500, non-farm income is R1 250, and remittances are R1 250. The farm–non-farm multiplier of 0,6 should be applied to the sum of agricultural income and remittances – R3 750 – to obtain a linkage effect of R2 250. Part of the latter linkage effect should be distributed over the farm household. Since it was assumed that a quarter of its income of R5 000 was derived from non-farm activities, this portion is R1 250. The non-farm household obtains the remainder, which is R1 000. Since the non-farm household was assumed to receive three quarters of its income from farm–non-farm linkages (the R1 000 just calculated) and a quarter of its income from remittances (i.e. R333,33), the non-farm household earns a total of R1333,33. But given an indicative income of R5 000, only 0,26 households could be sustained on R1333,33. Hence, one farm household effectively supports 0,26 non-farm households through the multiplier effect. This result obtains in all of the four models. Farm household's sources of income differ from the model, but the shares of non-farm and remittance income of the non-farm household are set at ,75 and ,25 in all four models. Hertz (*1994*) makes different assumptions about the non-farm household income structure and, of course, derives different multiplier values.

8 Hertz (*1994*) incorrectly states that, in the case of rural livelihoods gained, an income multiplier of 1,6 was applied to farm income and remittances in order to arrive at the additional non-farm livelihoods generated, but that, in the case of rural livelihoods lost on commercial farms, another 'wrong' multiplier is used. As explained in the previous

- **Elastic supply of labour and demand for food.** A potential constraint which is yet to be evaluated is the 'constant prices' assumption inherent in the partial equilibrium approach used here. The partial equilibrium approach assumes an elastic supply of labour and demand for food. The current high levels of under- and unemployment – in particular in rural areas – suggest that the supply of labour can reasonably be assumed to be elastic for the medium term.

  The assumption of elastic demand for food requires closer scrutiny, however. For instance, would a substantial supply reaction in, say, peri-urban vegetable production, lead to an equally substantial fall in prices? Although the extent of the downward price effect is difficult to predict, the current high rates of malnutrition – particularly in rural areas,[9] and artificially inflated consumer prices, suggest a large pent-up demand for food. Moreover, the income elasticity of demand for food among the black population is high.[10] It is clear that increases in rural and urban employment and in national income are an important condition for absorbing large supply responses. Only a general equilibrium approach would be able to shed light on this and other issues.

Results

Aggregation of the four models to their representative areas yields the following general result. Although the models are based on conservative assumptions, aggregation indicates that when land moves from large to small holdings and a more dynamic rural and peri-urban economy emerges, there is a substantial net increase in the number of livelihoods. First, this is because farming becomes more labour-intensive. Second, it is because there is a stronger linkage to labour-intensive rural or small-town production of non-farm goods and services – the multiplier effect. These two effects greatly outweigh the loss of jobs on the larger farms as they are restructured. Moreover, the cost-per-livelihood is surprisingly small. A sensitivity analysis – in which indicative household incomes were increased by 20 per cent – was also performed. Results show that

footnote, a farm–non-farm livelihood multiplier of 1,26 was used and applied consistently – both to calculate jobs gained and jobs lost.

9 Recent estimates suggest that as many as 40 per cent of black, rural children are stunted, which indicates severe malnutrition.

10 While for the most important vegetables price elasticities of demand vary between 0,12 and 0,89 (*Groenewald, 1987*), income elasticity of demand is high – 0,95 for African consumers (*Western Cape Agricultural Union, 1991*).

the number of livelihoods created is reduced by approximately the same proportion as incomes are increased.

For each of the agro-ecological zones studied, the aggregate net impact of a land redistribution process which would incrementally replace currently existing farm units in the commercial areas by the indicative models arrived at above, is now evaluated as follows:

- Availability of arable and non-arable land in the commercial sector is calculated using official statistics (Eastern Cape, summer grain, and Western Cape) or a rough estimate (in the case of peri-urban areas).
- Land redistribution is assumed to range between 10 to 50 per cent of land availability.
- Indicative farm models replace currently existing commercial farms as land is redistributed, taking into account both arable as well as non-arable land availability.
- Gross rural livelihoods created (farm and non-farm) are calculated;
- Rural livelihoods (farm and non-farm) lost in the commercial sector are calculated.
- Net rural livelihoods gained are calculated by subtracting livelihoods lost in the commercial sector from gross rural livelihoods.
- Capital costs of the establishment of farm households are calculated, representing 70 per cent of farm capital assets (land, livestock and machinery).[11]

The following results are presented in Table 17.2: (a) the gross rural livelihoods (farm and non-farm) created; (b) the net rural livelihoods created; (c) the net rural livelihoods created under the assumption that indicative household incomes are increased by 20 per cent; and, (d) the capital costs to be borne by equity or grants.

For instance, the summary results of these models convey that a land restructuring process that acquires 10 per cent of commercial farm areas in the four zones mentioned above and redistributes them to smallholdings, could establish a total of 187 556 smallholdings at indicative household income-levels of R5 000 to R12 000 per year, depending on the region. Farm sizes necessary to achieve the indicative incomes will range from 0,3 to 45 ha, and the value of farm capital (land, livestock and machinery) is about R9 803 per farm. About 30 per cent of this

11 Hertz (*1994*) fails to note the definition of 'capital costs' used here (the portion unencumbered by private debt), and proceeds to assert that a 'wrong' number was used.

capital can be rented or borrowed, but 70 per cent (R6 862) will have to come from other sources. It is assumed that the beneficiary brings in part of these capital costs as equity, and that a rural restructuring programme provides the other part as a grant. The capital costs to be borne by sources other than private farm debt are R1 287 million for the 187 556 smallholdings established under the 10 per cent scenario.

These households typically engage in varying degrees of part-time farming, depending on the region. Assuming an average household size of six, with the equivalent total of two adults from several of its members, the above redistribution process will involve 187 556 x 2 = 375 112 livelihoods in households that are engaged part-time in farming. Since these households will spend a substantial portion of their income locally, they will generate an additional number of non-farm livelihoods. Recall that it was estimated that for every four part-time farm livelihoods, about one additional non-farm livelihood is created – an employment multiplier of 1,26 is used. The gross number of livelihoods created is thus 423 877. However, if the number of farm and non-farm jobs lost on the commercial farms is deducted, the result is a net number of rural livelihoods of 360 620. Consequently, the capital costs per gross and net rural livelihood are R3 036 and R3 569 respectively.

If 50 per cent of the commercial area in the selected zones (recall that the zones cover 28 per cent of the area of the commercial sector: 14 per cent of the area in the commercial sector is transferred here) was redistributed to part-time farmers, the number of gross and net rural livelihoods created would be 2,1 million and 1,5 million, respectively. The total farm capital cost would be R 6,4 billion.

If one takes into account that, in 1991, the entire commercial agricultural sector employed around 1,17 million labourers on the total agricultural area of the RSA, the estimates presented in this section seem to hold considerable promise. (A rough estimate in Annexe 17.1 suggests that one could tentatively add another 37 per cent to the gross rural livelihoods if all agricultural land were considered.)

By far the largest numbers would arise in the peri-urban model, where a transfer of 30 per cent of the assumed 150 000 hectares of farmland could add an extra 584 000 net additional rural livelihoods, or more than half of the total of the four case studies. Transferring 30 per cent of the wine estates in the Western Cape, on the other hand, would generate only about 39 000 livelihoods. Labour intensity in this sector is already at such levels that the net employment gain from a land transfer would be more limited.

The estimated cost of a land reform programme

In the previous section, data on South African small-scale farmers was used to derive several important cost elements: (a) the area of land and the livestock and machinery small-scale farmers would have to own in order to earn a livelihood which would be sufficient to provide them with the incentive to settle or remain in rural areas and engage in farm and associated non-farm production; and, (b) the net number of additional farm and non-farm households which could be accommodated on the land base for which these models could be applicable. In this section, the estimated cost of a rural restructuring programme is calculated.

The cost estimate procedure is as follows. First, the non-land cost components are discussed, decomposed into: (a) the administrative and legal costs which would arise in both the restitution as well as the redistribution components of the programme; and, (b) the infrastructure and start-up costs to settle beneficiary households on the land.

Second, the cost and pace of a nationwide programme are discussed. A tentative extrapolation is made from the regions studied of the potential costs of a nationwide programme which would also include the land classes and regions for which the necessary data for farm modelling are not yet assembled. On the basis of this extrapolation, the order of magnitude of the costs involved in a nationwide programme is provided, and an estimate of the likely pace of implementation is given.

Third, options are discussed to reduce the fiscal costs of the restructuring programme and to defer them into the future so that they can eventually be paid out of the growth dividends which are likely to be associated with a peaceful and well-managed transition. Finally, options for financing part of the programme out of foreign resources are outlined.

Cost components

Table 17.3 shows the costs of a rural restructuring programme for the case study regions defined above. Detailed cost assumptions are given in Annexe 3. The costs are estimated for a five-year period using five different scenarios. The scenarios assume that between 10 and 50 per cent of commercial farmland in these regions is transferred to beneficiaries. In the remainder, the middle scenario – 30 per cent in five years – is chosen for illustrative purposes.

Rows (1) and (2) report the net and gross number of farm and non-farm livelihoods which are generated when between 10 per cent and 50 per cent of the commercial farmland in the four study regions (not the whole country) is transferred to beneficiaries. These numbers were

Table 17.2: Rural Livelihoods Gained and Associated Farm Capital Costs

	Proportion of Commercial Farmland Transferred over Five-year Period				
	10%	20%	30%	40%	50%
Gross rural livelihoods gained					
Peri-urban	207 333	414 666	621 999	829 332	1 036 665
Summer grain	129 744	259 489	389 233	518 978	648 722
Eastern Cape	67 356	134 711	202 067	269 422	336 778
Western Cape	19 444	38 888	58 332	77 775	97 219
Total	423 877	847 754	1 271 631	1 695 507	2 119 384
Net rural livelihoods gained					
Peri-urban	194 733	389 466	584 199	778 932	973 665
Summer grain	96 601	193 201	289 802	386 403	483 003
Eastern Cape	56 142	112 283	168 425	224 566	280 708
Western Cape	13 144	26 288	39 432	52 575	65 719
Total	360 620	721 238	1 081 858	1 442 476	1 803 095
Sensitivity analysis: indicative household income increased by 20%					
Peri-urban	160 178	320 355	480 533	640 710	800 888
Summer grain	74 977	149 953	224 930	299 906	374 883
Eastern Cape	44 916	89 831	134 747	179 663	224 578
Western Cape	9 903	19 806	29 710	39 613	49 516
Total	289 974	579 945	869 920	1 159 892	1 449 865
Capital costs (1993 R millions)					
Peri-urban	147	294	441	588	735
Summer grain	507	1 014	1 521	2 028	2 534
Eastern Cape	444	889	1 333	1 778	2 222
Western Cape	189	378	567	755	944
Total	1 287	2 575	3 862	5 149	6 435

Note: The models assume that rural livelihoods are based on part-time farming. One could roughly calculate the full-time farm equivalents by multiplying the rural livelihoods by the income share derived from farming. The farm income shares are 0,25 in the peri-urban areas; 0,50 in the summer grain areas and the Eastern Cape; and 0,75 in the Western Cape. However, such calculations would only pick up the increase in labour-intensity in farming, and not the full employment impact of a restructured rural economy.

Table 17.3: The Cost of a Land Reform Programme for the Case Study Regions

		Proportion of Commercial Farmland Transferred over Five-year Period				
		10%	20%	30%	40%	50%
(1)	Net livelihoods	360 620	721 238	1 081 858	1 442 476	1 803 095
(2)	Gross livelihoods	423 877	847 754	1 271 631	1 695 507	2 119 384
(3)	No of families	211 939	423 877	635 816	847 754	1 059 692
(4)	No of persons	1 271 631	2 543 262	3 814 893	5 086 521	6 358 152
		(1993 R millions)				
(5)	Cost of farm capital (land, livestock and machinery)	1 839	3 679	5 517	7 356	9 193
(6)	– borrowed	552	1 104	1 655	2 207	2 758
(7)	– paid by beneficiaries	368	736	1 103	1 471	1 839
(8)	– paid by grant	919	1 839	2 759	3 678	4 596
(9)	Dissemination and administration costs	944	944	944	944	944
(10)	Infrastructure	1 590	4 006	6 008	8 011	10 014
(11)	Incremental RRP	795	2 003	3 004	4 006	5 007
(12)	Health/education infrastructure	413	827	1 240	1 653	2 066
(13)	Incr. for RRP	69	138	207	276	344
(14)	Start-up grants	1 102	2 204	3 306	4 408	5 510
(15)	All capital cost	5 474	10 832	15 776	20 719	25 661
(16)	Incremental RRP	4 748	8 967	12 978	16 989	20 999
(17)	– paid by beneficiaries	368	736	1 103	1 471	1 839
(18)	– borrowed	552	1 104	1 655	2 207	2 758
(19)	– paid by government	3 829	7 128	10 220	13 311	16 402
(20)	Counting 20% of (16)	950	1 793	2 596	3 398	4 200
(21)	(19) + (20)	4 779	8 921	12 815	16 709	20 602
		(1993 R)				
(22)	Cost/gross livelihood	11 274	10 524	10 078	9 855	9 721
(23)	Cost per family	22 547	21 047	20 156	19 710	19 442
(24)	Cost/net livelihood	13 251	12 370	11 846	11 584	11 426

estimated in the previous section. In row (3) the corresponding number of families is shown, assuming two full-time equivalent workers per household. Assuming about six persons per household, there could be between 1,3 and 6,4 million beneficiaries in the case study regions alone – 3,8 million in the middle scenario of 30 per cent of land redistributed (row (4)).

In row (5) it is shown that the total farm capital (land, machines, livestock) of these households would be about R5,5 billion, of which 30 per cent would be from resources borrowed by the households (row 6). Capital costs unencumbered by private debt constitute the remainder – i.e. 70 per cent. It is assumed here that 20 per cent would come from beneficiaries' own equity contributions (row 7). For instance, such contributions would consist of animals, tools and implements the households would bring with them from the former homelands, and any contributions to the capital fund they might have made in cash or through labour contributions ('sweat equity'). The remaining 50 per cent of the farm capital costs would be financed by a grant (row 8). This implies that each household would have a debt-asset ratio not exceeding 30 per cent, so that it remains creditworthy for future borrowing of working capital and additional productive investment.

Row (9) shows the costs for dissemination and administration, including the adjudication of the programme. This finances training workshops of two and a half days for a third of the adult rural population (spread over a total of 10 000 workshops), the training of the trainers and the teaching materials in the local languages. It also includes the costs of operating a total of 250 district land committees, a national land committee, and 25 land courts for a period of five years. This cost is treated as a fixed cost, irrespective of the number of beneficiaries and, unlike the other costs discussed here, includes the nationwide cost of land adjudication and administration for both the restitution and redistribution processes, and or regularizing land which has been occupied by squatters. This cost is estimated at a little less than R1 billion. Since the existence of this system is likely to reduce the litigation costs of private parties substantially, it is not a social cost, but only a fiscal cost.

In row (10) the cost of providing the net additional households with basic infrastructure is estimated: electricity, safe-piped water supply, basic sanitation and the upgrading of the on-farm road network. This cost is about R6 billion. However, not all of these costs are chargeable to a rural restructuring programme. It is assumed that 50 per cent of the gross beneficiary households would have to be provided with this infrastructure, even if they were to remain in the former homelands or in urban areas, and therefore only 50 per cent of the infrastructure costs are charged as an incremental cost to a rural restructuring programme of R3 billion (row 11).

For health and education infrastructure (row 12), it is assumed that the households and children would have to be provided with additional classrooms and health posts, irrespective of whether they moved to new farms or stayed in the bantustans or the urban areas. It is therefore assumed that only the extra cost associated with wide dispersal is chargeable to a rural restructuring programme. This is assumed to be an additional 20 per cent of the normal costs, or R207 million.

In row (14) the cost of start-up grants is shown, covering most of the cost of feasibility studies and surveys for the farms acquired by the beneficiaries, a portion of the living costs and farm inputs during the first year, and some construction materials that the beneficiaries can use to start the construction of their housing. Both the infrastructure and the grants provided are for basic but adequate services and inputs. For example, the beneficiaries are assumed to provide their own labour for housing construction. Start-up grants are estimated at R3,3 billion for the middle scenario.

To reduce the fiscal burden while ensuring required support services, the conversion of subsidized services from usage by large to small farmers should not incur extra costs to the state. Hence, no research, extension and technical assistance costs are included because they would have to come from redirecting the expenditures of existing programmes for the large commercial farm sector to the smallholder sector. The recurrent health and education costs are also independent of the location of the beneficiaries, and the extra cost associated with greater dispersal of the population is negligible. All other recurrent costs are private costs of the beneficiaries or maintenance costs of their infrastructure for which they themselves should be responsible, rather than the state.

In row (15) it is estimated that the costs of all capital items, for a programme which would transfer 30 per cent of the land in five years, would be about R16 billion in 1993 prices, of which R13 billion would be directly chargeable to a rural restructuring programme (row 16). The beneficiaries would pay about 8,2 per cent of the incremental rural restructuring programme costs and would borrow an additional 12,3 per cent (rows 17 and 18).

This would leave a total fiscal cost of about R10 billion (row 19). All of the contingencies are added to the government cost. Contingencies are assumed to be 20 per cent of the total incremental capital cost of a rural restructuring programme (row 20). This raises the government cost to about R13 billion (row 21).

Per gross livelihood, the fiscal cost would amount to about R10 000 (row 22). Per beneficiary family, the government's upfront cost is about R20 000 (row 23). Per net livelihood created, the cost is around R12 000 (row 24). Recall that this amount includes most of the cost of productive capital, and also all of the incremental social and economic infrastructure required.

Hence, the programme is able to create a large number of livelihoods at relatively low unit cost in fiscal terms. Moreover, the resources required from the financial sector per livelihood created are also minimal. In the case study regions, the farm capital costs are only about 45 per cent of the total costs, while the land costs (at current market prices) are even less. Clearly, in this light, the concern over the cost of compensation for the land loses some of its urgency. Instead, the concern shifts to how to finance the start-up grants, infrastructure costs and administrative costs. Finally, while beneficiaries are assumed to finance about 50 per cent of the farm capital from their own equity or loans, their participation in the total costs will be about 20 per cent.

Cost and pace of a nationwide programme

Despite its low unit cost, the size of the programme could still create a substantial macroeconomic and fiscal financing problem. First of all, the calculations above are only for four farming systems in four subregions. While interest in settlement in low fertility and high risk areas such as the Karoo (excluded from the above scenario) will be limited, interest in settlement on some of the highest quality lands in South Africa (also excluded here) should be significant. The latter include the sugar sector, most of the deciduous fruit sector, the citrus sector, and the irrigated field crop sector, as well as the wheat, barley and oats growing areas. Using insights gained from the models, a rough guess was made at how many livelihoods could be created by transferring 30 per cent of the excluded areas. The assumptions are given in Annexe 17.1 and come to about 400 000 additional net livelihoods, bringing the total to about 1,5 million. We use these rough calculations to blow up the farm and non-farm cost estimates for the 30 per cent option in Table 17.3 (row 21) by factors of 2,54 and 1,37 respectively. The total five-year fiscal cost of the programme is thus R22,1 billion.[12] Farm capital costs represent 59 per cent of total costs, and beneficiaries would be responsible for about 29 per cent of total costs. This implies a total fiscal cost of roughly R15 000 per net livelihood. These figures translate into the creation of more than 600 000 additional net full-time farm employment opportunities at a fiscal cost of approximately R35 000 per additional job.

What would be the maximum pace at which such a programme could be implemented? Currently the annual rate of transfer of land in the commercial sector is about 4 per cent,[13] with a maximum rate of transfer observed of 6,7 per cent in the past. However, about half of these transactions were intra-family transactions. Hence, the land market has been able to transfer to unrelated owners close to 3,5 per cent of the land of the commercial farm sector per year. If all the measures to encourage sales discussed in Chapter 14 were implemented, it seems reasonable that this figure could be doubled. Furthermore, removal of current legal restrictions on the subdivision of land and purchase by the disenfranchised would multiply the number of potential buyers. Concurrently, the land committees and land courts would expedite the restoration process and assist in monitoring any dispute resolution for the entire process.

It would not be unreasonable, therefore, to assume that 6 per cent of the land of the commercial farm sector could be transferred in every year of the programme. This would lead to a total transfer of 30 per cent of the medium and high quality land of the commercial farm sector over five years, with a total of around 1,5 million net livelihoods created. In this calculation, it is assumed that most of the beneficiaries of restitution would be accommodated and most of the rural squatter settlements would be regularized and upgraded. This would mean that slightly more

12 Van Zyl (personal communication) suggests a consistency check following a different procedure. In 1993, the total value of farm capital in the commercial sector was estimated at R67 billion, and 30 per cent would thus represent R20 billion. The non-farm capital costs which we estimated for a nationwide programme amount to R10,2 billion. Hence, total fiscal costs (including contingencies) would amount to R26,3 billion, as compared to the R22,1 billion arrived at by extrapolating a national estimate from the four case study regions. Allowing for differences in capital intensity, these estimates seem reasonably close, and may suggest lower and upper bounds of the fiscal cost estimates for a nationwide programme.

13 At this annual rate, i.e. 4 per cent, the indicative models described in the previous section suggest that: (a) in peri-urban areas, about 83 000 gross rural livelihoods could be established; (b) in the Eastern Cape, 27 000; (c) in the summer grain area, 52 000; and, (d) in the Western Cape, 7 800. This gives a total of about 170 000 gross livelihoods or 145 000 net livelihoods per year.

than 5 per cent of the 30 per cent transferred would have been associated with judicial and administrative, rather than with market-assisted processes. It is assumed that the unit costs of these transfers would be about the same as those for the market-assisted transfers. Again, the tentative nature of all these calculations must be strongly emphasized.

Options for reducing and deferring the fiscal burden

If the government were to pay for all the above costs out of current fiscal resources, the annual fiscal burden of a rural restructuring programme which transfers 30 per cent of commercial farmland to beneficiaries would be about R4,4 billion per year. To this would be added a borrowing requirement from the banking sector of the cost borrowed by the beneficiaries – about R840 million per annum – bringing the relevant total resource requirement to about R5,2 billion per annum. It would obviously be highly desirable to reduce these costs, shift them to foreign or domestic donors, or defer them in time until growth has increased the fiscal capacity of the nation. What are the options to do so? The options are summarized in Table 17.4.

- **Sale of state land.** The first option would either be to sell all of the unoccupied arable land which is in the state sector or to use it for direct redistribution in the first years of the programme. Either way, a reduction of the fiscal cost would result. Currently, about 320 000 hectares of arable state land remain unoccupied. Valuing these at R800 per hectare, leads to an estimated sales value or fiscal savings of about R256 million, or approximately R50 million per year (row 2). Clearly, neither the fiscal nor the land redistribution problem can be solved in this way.
- **Co-payment.** The second option would be to raise the co-payment requirement of the beneficiaries. Since beneficiaries would have to borrow the amounts domestically, such a shift would neither change the social cost of the programme nor its macroeconomic impact on the capital market. This option would only reduce the fiscal cost. Moreover it would increase the debt-asset ratio of the households and enterprises created, reducing their creditworthiness for the future and making them more vulnerable to drought and price risk. Thus, the likely effect of this would be to create yet another subsidy-dependent farm sector, this time a small rather than a large farm sector. It would simply amount to a deferral of the fiscal cost to future 'bail out' operations. Moreover, there are better alternatives for shifting the budget

costs to the future. Therefore, it seems essential to maintain a debt-asset ratio of 30 per cent as originally assumed. Hence, this option is rejected (row 3).

- **Reduction of start-up grants.** A reduction of start-up grants is also not recommended (row 4). Apart from creating real difficulties for the programme participants in their first year, it is likely to reduce the quality of the feasibility studies, with an adverse impact on the environmental and economic sustainability of the resulting settlements. Most probably, an increase in debt-asset ratios above the indicative 30 per cent would be induced. Again, this will only reduce the future creditworthiness of the beneficiaries and shift the fiscal cost to the future.
- **'Sweat equity'.** A more realistic cost reduction would result from requiring the beneficiary communities to contribute for free the unskilled labour costs and local materials for building the economic and social infrastructure. For instance, in the highly successful Solidaridad programme of Mexico, beneficiaries provide between 20 and 50 per cent of the infrastructure costs in this way, depending on the project type. Requiring an average 25 per cent of 'sweat equity' from the beneficiaries would reduce the government's fiscal costs by a little more than R1 billion in total or about R200 million per year (row (5)). The disadvantage of this option is that it would reduce the labour earnings which would normally be created by an expansion in

Table 17.4: Savings on Fiscal Costs (1993 R millions)

	Options	Cost per year
(1)	Total costs payable by government without any savings option	4 420
(2)	Sale of state land	50
(3)	Increased co-payment (rejected)	—
(4)	Reduction of start-up grants (rejected)	—
(5)	Sweat equity (held in reserve)	(200)
(6)	Pension/buyout scheme	490
(7)	Grant resources	550
(8)	Elimination of current subsidies to agriculture	400
(9)	Land reform bonds (rejected)	—
(10)	Regularizing land invasions (no cost savings)	—
(11)	**Total** (1) – (2) – (6) – (7) – (8)	3 480

rural infrastructure construction. Hence, it would reduce the transitional impact of the programme on employment. A decision on whether or not to reduce the cost of the programme in this way would thus depend on the unemployment and labour income situation prevailing at the time of implementation. The option therefore is held in reserve.

- **Pension/buyout scheme.** Another cost reduction option capable of shifting the fiscal cost to the future is the pension/buyout scheme discussed in Chapter 14. As a starting point, assume that: (a) two-thirds of the value of such pensions is denominated in domestic currency but linked to the real value of government pensions, and one-third is denominated in foreign exchange; and (b) the payment of the pension is guaranteed by a third party insurer. Under these circumstances, farmers might be willing to sell their land and other agricultural assets for a present value of pension benefits substantially below the cash price of the land.

  Assuming a 20 per cent discount on about half of the farm asset cost payable by the state, this would reduce the total fiscal burden by about R75 million per year. Since beneficiaries are assumed to pay 50 per cent of the farm assets from own equity and credit, they would also accrue savings of R75 million, bringing the total to R150 million. Thus the real cost savings would not be that large. However, if the pensions were payable over 15 years on average, it would reduce the annual fiscal cost over the first five years by about R280 million (row (6) and Annexe 17.2). The pension/buyout scheme can be interpreted as a loan from the sellers to the redistribution programme.
- **Grant resources.** Additional real cost reduction, rather than deferral, can be achieved by raising grant resources for the programme from foreign and domestic donors. It should be possible to raise up to US$150 million or R500 million per year from foreign sources, and perhaps R50 million from domestic resources (row 7). Such grants could be in the form of government-to-government grants, or arise from private domestic and foreign donations to community groups and NGOs which could assist the beneficiaries in a variety of ways.
- **Elimination of current subsidies to agriculture.** Current subsidies to the commercial farm sector are about R500 million per year, excluding research and extension costs (which would be redirected to the small farm sector). It is assumed that about R400 million can be eliminated in the short run (row 8).

- **Land reform bonds.** In other countries, land reform bonds were often forced on the current owners as payment of compensation. Such bonds often had unfavourable features, such as limited transferability and low interest rates in an environment characterized by high inflationary expectations. Consequently, these bonds traded at considerable discounts or not at all. Moreover, the former owners anticipated the implementation of this device, which fuelled resistance to redistribution and increased their litigation efforts to prevent expropriation.

  Bonds which do not contain such adverse features could be used to pay the current owners, and have the same deferral effect as the pension/buyout scheme. Since they would have to be fully tradeable on the domestic capital market, unlike the pension, they would be indistinguishable from the issuing of any other government bonds in the domestic market and have the same fiscal and macroeconomic effects. However, since crowding out of other borrowing in the domestic market is to be avoided, issuing of domestic bonds to finance a rural restructuring programme is undesirable (row 9). Moreover, as mentioned above, the pension/buyout scheme can also achieve the objective of borrowing from the former owners.
- **Regularizing land invasions.** Transfer by invasion before a rural restructuring programme is initiated is unlikely to reduce fiscal costs. The owners of the invaded farms will have to be compensated, and the invaders will claim state assistance for infrastructure and other support programmes. Transfer by invasion after the initiation of a rural restructuring programme will likely result in some penalties by the state, such as denial of development and service assistance. Nevertheless, ultimately even these invasions are likely to result in compensation and infrastructure development cost.

  Finally, transfer by invasion would certainly reduce the quality and sustainability of the resulting settlements, reward the daring at the expense of the weak, increase the cost in terms of violence, and reduce investment in the remaining commercial farm sector and the economy as a whole.

In summary, the sum of fiscal savings so far considered amounts to a reduction of total fiscal cost to about R3,5 billion per year (row 11). Since this may still not be fiscally affordable, one can investigate the potential for foreign borrowing from multilateral, bilateral and commercial sources.

Foreign borrowing and grants

Multilateral loans and bilateral loans, some of the latter perhaps at partially concessional rates, should not be too hard to obtain for a programme with enormous potential impact on employment and poverty reduction. Moreover, commercial foreign borrowing could be made cheaper and more plentiful by an insurance scheme or bilaterally funded guarantees against political risks.

It seems reasonable to assume that perhaps half of the remaining annual requirement of R3,5 billion could be deferred this way, leaving a fiscal burden of about R1,75 billion or US$ 480 million per year for the initial five years of the programme. It would be ill-advised to seek further cost deferral, since it might lead to excessive foreign debt accumulation over time.

Research recommendations

The results presented in this chapter are based on available empirical work on smallholders in South Africa. However, the quality and coverage of the existing data on smallholders leaves much to be desired. More work is needed to extend the coverage of the models to additional agro-ecological zones and to confirm the hypotheses and assumptions that underlie the models presented here. More work is also needed on models of rural, non-agricultural livelihoods.

The indicative household models used above as the basis for the quantification procedure should not be taken to imply that most beneficiaries will, or should, use their land in a particular way after the reform. The fundamental operating principle of a market-assisted land reform process is the empowerment of rural households to make key economic decisions, including the decision to what extent farming should be incorporated in the household income bundle – or whether to farm at all. The models selected here should on no account be used as target or mandatory models driving the planning process of rural restructuring. On the contrary, one of the strongest messages emerging from the case studies was that local control and beneficiary participation in the process of rural restructuring are paramount to its success. All of the case studies placed great emphasis on gaining further insight into potential beneficiaries' views on future options for rural restructuring and the creation of rural livelihoods.

The above quantification is not an economic and social cost-benefit analysis – Chapter 23 provides a first attempt in this regard. A more detailed public expenditure analysis of the various options for rural restructuring is needed, indicating in more detail the financing options of the programme and their macroeconomic effects. The economy-wide effects of rural restructuring need to be addressed as a matter of urgency. The static, partial equilibrium approach used here can only give a rough indication of such effects and ignores certain price and income effects. A dynamic, general equilibrium approach would be the logical next step in the research agenda.

References

Cousins, B. 1993. Small-scale farming and agrarian reform in the Western Cape: Lessons from the MAG Experience. Photocopy. Montagu: Montagu-Ashton Gemeenskapsdiens.

De Klerk, Mike. 1993. Developing models of land use activities: agricultural development options. *Coordinator's report.* Photocopy. Johannesburg: Land and Agricultural Policy Centre.

Glover, D. 1990. Contract farming and outgrower schemes in East and Southern Africa. *Journal of Agricultural Economics,* 41(3) : 303–315.

Groenewald, J. 1987. Agriculture: a perspective of medium-term prospects. *Development Southern Africa,* 4(2) : 224–241.

Haggblade, Steven, Hammer, Jeffrey and Hazell, Peter. 1991. Modelling agricultural growth multipliers. *American Journal of Agricultural Economics,* 73(2) : 361–374.

Hertz, Tom. 1994. 1,5 million jobs in five years? Comments on the World Bank's November 1993 land reform proposal. Photocopy. Amherst, MA: Department of Economics, University of Massachusetts.

Kirsten, Johann F. 1993. Models of land use activities: the case of the summer grain region. Photocopy. Johannesburg: Land and Agricultural Policy Centre.

May, Julian. 1993. Development options for peri-urban agriculture. Photocopy. Johannesburg: Land and Agricultural Policy Centre.

McKenzie, Craig C. 1993. Farm worker equity model: mechanism to broaden the ownership base in agriculture. Photocopy. Johannesburg: Land and Agricultural Policy Centre.

Nieuwoudt, W. Lieb, 1993. Theoretical framework of appropriate levels of subsidization and cost recovery. *Background paper prepared for the Rural Restructuring Programme of the World Bank.* Photocopy. Johannesburg: Land and Agricultural Policy Centre.

Small, Janet and Winkler, Harald. 1992. Botho Sechabeng: a feeling of community. Johannesburg: National Land Committee, Community Investigation and Research Programme.

South African Agricultural Union (SAAU). 1984. *Die Finansiële Posisie van Boere in die R.S.A.* Pretoria: SAAU.

Tapson, David R. 1990. A socio-economic analysis of smallholder cattle producers in KwaZulu. *PhD dissertation.* Pretoria: Department of Business

Economics, Vista University.

Tapson, David R. 1993. Case study: livestock in the Eastern Cape. Photocopy. Johannesburg: Land and Agricultural Policy Centre.

Wentzel, Wilfred. 1993. Horticulture in the Western Cape. Photocopy. Johannesburg: Land and Agricultural Policy Centre.

Western Cape Agricultural Union. 1991. *Annual Report 1990/91*. Paarl: WCAU.

Annexe 17.1: Estimate of the possible number of gross rural livelihoods in areas not covered by the four case studies

The numbers reported in the table below are not national aggregates – they are estimated aggregates of the four areas studied, which covered 28 per cent or 24 million ha of agricultural land in the commercial sector of the Republic of South Africa. A total of 62 million ha was excluded. In order to arrive at an educated guess of the potential gross rural livelihoods to be created in the area excluded, the following rough estimates were made:

Land use	Area suitable for small-scale agriculture (ha)	Indicative smallholder farm size (ha/farm)	Potential number of smallholders (number)
Forestry on arable land (Transvaal)	300 000	25	12 000
Forestry on arable land (Natal)	400 000	25	16 000
Sugar	373 000	2	186 000
Deciduous fruit	48 000	1	48 000
Citrus	46 000	1	46 000
Sub-tropicals	22 000	1	22 000
Wheat	1 500 000	25	60 000
Sorghum, sunflower, soybeans, groundnuts	870 000	25	35 000
Oats and barley	968 000	35	27 600
Cattle	12 500 000	75	160 000
Smallstock	37 500 000	500	75 000
Total	54 527 000		687 600

Notes: (a) Total area under forestry is around 1,5 million ha, but only high-potential arable land under forestry is considered here.

(b) Vegetables (130 000 ha) are assumed to be already covered in the peri-urban case study (150 000 ha). Adding more vegetable production would result in considerable over-production.

If one assumes that 30 per cent of the above land could also become available for redistribution, the number of smallholdings created would be 206 280. This implies 466 192 gross rural livelihoods on top of the 1 271 631 already calculated in Table 17.2. The 'national' estimate of gross rural livelihoods created if 30 per cent of all land suitable for small-scale agriculture in the commercial sector is redistributed to smallholders can thus be roughly estimated at 1 737 823. This implies that an additional 37 per cent could be added to the reported estimates of gross rural livelihoods.

We roughly estimate that the farm capital (land, livestock and machinery) will be, on average, around R36 000 per farm – higher than the capital costs estimated in the four case studies. In extrapolating to the national level, it is assumed that the higher capital costs will be borne by the same debt-asset ratio of 30 per cent and the same own equity contribution of 20 per cent.

Annexe 17.2: Fiscal savings over the first five years of the RRP

Various measures can be considered which could reduce and/or defer fiscal payments over the initial period of the programme. A pension buyout option was discussed in connection with Table 17.4 in the text. Due to the complexity of the calculation of savings through the pension buyout scheme, more detail is given below.

The fiscal savings from this scheme for the first five years is equal to the total payout of farm capital through government grant (R3,5 billion), minus payouts of pension over the first five years.

The following assumptions are used:

(a) Percentage farm assets purchased under the pension buyout equal 50 per cent of total farm assets;
(b) Discount on farm assets is 20 per cent;
(c) Real interest paid on outstanding value of farm assets equals 4 per cent p.a.;
(d) Pension payment period is 15 years;
(e) Pension payouts consist of 15 equal payouts on asset value plus interest on the outstanding value of farm assets;
(f) Assets are purchased equally over five years (i.e. 20 per cent purchase p.a.).

The costing is shown on page 456.

Costing of Pension Buyout Option Savings (in million Rands)					
	Years				
Items	1	2	3	4	5
1 Pension payment on purchases (20% per annum)	20	40	60	80	100
2 Outstanding balance	653	1 259	1 818	2 331	2 797
3 Interest cost on outstanding balance	26	50	73	93	112
4 Total annual pension payment	73	144	213	280	345
5 Total grant option payment	699	699	699	699	699
6 Savings	627	556	487	420	354
Average savings per annum = R489 million					

Annexe 17.3: Cost assumptions

Total fixed overheads

These include the cost of training, training the trainers, dissemination, and administration and processing of land claims and transfers through a local/regional and national land committee system, and land courts. The assumption is that a start-up process lasting for five years will be required to allow for effective implementation. These are incremental costs which can be ascribed directly to the implementation of the RRP.

Training and dissemination costs

The purpose of training and dissemination is to ensure that communities are informed of their rights under the RRP.

- **Dissemination.** The assumption is that five person teams consisting of legal experts, agronomists, economists, community development experts and a facilitator (translator, etc) will be able to conduct two training sessions per week reaching 200–300 people per session (indicating 100 sessions over a two year period per team). The cost of a session amounts to R20 000, of which half is for professional fees, R4 000 for S&T and R6 000 for meals and accommodation for those attending. 10 000 workshops are necessary to cover 30 per cent of the adult rural population. This gives a total cost of R200 million for 100 active training teams.

- **Training the trainers.** It is proposed that each training team will undergo 30 days of intensive training. The total costs will be R12 million in salaries. In addition, each team leader will be exposed to national and international small farmer experience at a cost of R7 000 per person (R0,7m total).

The following cost items are included in overhead costs for training and dissemination:
- 5 million manuals at R3 each (R15m);
- development costs for manuals (R5m);
- additional NGO and CBO training (R50m);
- overhead management fee at 10 per cent of total training and management costs (R28,2m).

The total training and dissemination costs, therefore, amount to R310,9 million.

Land claims and transfers

- **Local land committees.** It is assumed that each of these will consist of 12 members, meeting for one day per month, or effectively 10 days per year. Professional fees are calculated at R10 000 per meeting or R100 000 per year. Each committee has three full time staff members plus a secretary. Staff costs amount to R200 000 per year. Court costs, i.e. legal plus surveyors, etc, are budgeted at R100 000 per annum. It is assumed that 250 district committees will be active throughout the country. The total costs are therefore R400 000 x 250 per year, or R500m over 5 years.
- **National Land Committee.** It is assumed that 12 members will meet one day a week for 40 weeks annually. Professional fees are calculated at R1m (12 x 40 x R2 000). Travel and related costs will amount to R1 500 per member per meeting, or R0,75m). Support staff are assumed to be drawn from the civil service and are therefore incremental. The total cost for the National Land Committee is therefore R8,75m over 5 years.
- **Land courts.** Each land court will cover 10 districts. Each court will have three professional staff (legal and land experts), plus three paralegals and two secretaries. The costs are assumed to be R1m per annum per court, i.e. R125 million over five years. The total fixed overhead costs, therefore, come to R943,75 million over five years.
- **Rural infrastructure.** This section describes the incremental costs of rural infrastructure and social services which can be directly ascribed

to the RRP. Both the infrastructure and the start-up grants will be provided to the farm and non-farm rural households under the RRP. The total number of households reached by the RRP is approximately 1,9m.

- **Electricity.** Based on the national standard, the connection costs of electricity are estimated to be R7 000 per connection. Due to the existing connection costs in rural areas, the average connection cost for RRP beneficiaries is estimated at R5 000. This cost will accommodate the upgrading of existing powerlines to accommodate the denser settlement patterns.
- **Water supply.** Water supply costs for spring protection, hand pumps, improvements to existing systems, new schemes and a stand pipe (no further than 100 metres from a residence) are estimated at R200 per capita (or R12 000/household).
- **Sanitation.** A cost of R1 000 per household is estimated.
- **Roads.** A cost of R300 per household is estimated based on standard road building costs.
- **Education.** The only incremental cost of education is assumed to be the additional costs associated with dispersion of service provision in rural areas. This amounts to capital costs of R1 800 per household and recurrent costs of R2 400 per household, based on mid-range estimates of pupil/teacher and teacher/classroom ratios.
- **Health.** The assumptions for health services provision are the same as for education. The total capital cost is estimated to be R150/household and R600/household for recurrent costs.
- **Land for community services.** Land for community services was not costed because the area is assumed to be negligible and marginal land could be used.

Implementing the Programme

PART V

18

Access to land: selecting the beneficiaries

Johan van Rooyen and Bongiwe Njobe-Mbuli

The need for, the objectives and elements of a land reform programme

Land distribution in South Africa is highly skewed. Approximately 87 per cent of agricultural land is held by almost 67 000 white farmers and accommodates a total population of 5,3 million. The remaining 71 per cent of the population, which is predominantly black, live on 13 per cent of the land in high density areas – the former homelands. Four centuries of conquest, occupation, denial, expropriation, transfers, purchases and consolidation has resulted in a land distribution pattern that is highly in favour of white occupation. It has evolved as a result of the systematic use of land allocation laws and policies aimed at controlling access and use based on race. Government support to the two sectors has been historically unequal and variable. Distortions occurred in the implementation of polices regulating markets, the provision of and access to infrastructure, agricultural credit and services, and labour (*Brand, Christodoulou, Van Rooyen & Vink, 1992; Kassier & Groenewald, 1992*).

Skewed access and ownership of land, as well as separation in the provision and quality of services has been the basis for the history of political struggle. In addition, high concentration of people in the former homeland areas has had a draining effect on the already fragile natural resource base (*Cowling, 1990*).

A necessary starting point, therefore, in the reconstruction of the rural economy, will be the need to create a system related to land allocation and land use support systems that would alleviate the 'land hunger'[1] prevalent in those communities which were systematically and

1 Land hunger almost always implies that there is existing but limited access to land and hence there is a need to increase the size of holding. In South Africa, whilst there is the physical occupation of land by blacks in squatter areas and the former homelands, it is by no means an indicator of secured access. Therefore, in this situation there is a basic need for secure access to land.

effectively prevented from gaining access on a legal basis to land and support services, i.e. those who were denied entitlement to land and support services. Furthermore, a future rural economy should ensure long term productive use of land aimed at reaching both national and household food security, creating sustainable rural livelihoods and supplying resources to the economy.[2]

Two major objectives of a land reform programme would be to effect widespread land transfers to the 'landless' disentitled and dispossessed people of South Africa, and secondly to ensure that when supported by the cost-effective provision of services, increased agricultural production, entrepreneurial activity and secure rural livelihoods will result.

The three core components of a South African land reform programme (LRP) would be, in the first place, to effect land transfers to 'landless' black people (Africans, 'coloureds' and Asians) and, in the second place, to identify the needs and aspirations, constraints and impediments facing the different social structures in black rural society. For this process, mobilization and capacity development will be a necessary ingredient for the required participation processes. In this context, it is also essential to draw attention to research pointing to important different social structures within black rural society (*Levin, Russon & Weiner, 1993*). The notion of a homogeneous black rural society should thus be replaced with a realistic description of the various socio-economic groupings in order to structure support systems based on the needs and aspirations of such groups. The reality of South African rural society does not leave scope for 'shot gun' or 'blanket' approaches to development.

The first activity essentially deals with entitlement of those who were disentitled and dispossessed through apartheid legislation. The second activity allows for empowerment through access to support services to allow productive land use. This leads to the third component: to ensure cost effective provision of required services, according to the target group, aimed at the improvement of livelihoods. It is thus clear that a multi-pronged strategy will be required to introduce and manage these three components of a LRP.

The ultimate success of a South African rural LRP should be tested against its ability to address equity in land distribution and livelihood

2 Extensive arguments about the need for land reform in South Africa and the expectations about it have been put forward by many – refer to Letsoalo (*1991*); Binswanger (*1995*); Brand, *et al* (*1992*); Van Zyl (*1994*); Van Rooyen, Mbongwa, Matsetela and Van Zyl (*1994*); Van Zyl, Van Rooyen, Kirsten and Van Schalkwyk (*1994*).

upgrading, reduction of poverty, creation of rural employment and income generating opportunities and, inter alia, raising the number of successful black agricultural producers and enhancing overall productivity whilst maintaining sustainable natural resource management and utilization.

At its heart, this programme should be aimed at redressing the impact of past wrongs; potential participants, therefore, would obviously come from black rural society which has been denied entitlements in the past in terms of apartheid ideology and laws. It is also recognized, however, that a programme of this nature cannot and will not make an agricultural producer out of every participant, although it should go a long way in creating a viable rural economy within which agriculture and the related linkages can develop.

This chapter attempts to propose a possible set of criteria for selection to participate in the land reform and agricultural activities of a LRP, which seeks to establish a democratic, non-racist, non-sexist society, a viable rural and national economy and, while focusing on the poor and those in need, at the same time ensuring broad participation of all those who have been previously disadvantaged and disentitled.

First, the relationship between access to (farm) land, welfare and poverty, production and nutrition is analysed to contextualize land hunger and demand in South Africa. Second, available land resources are discussed. Then the case for principles and criteria, managing the selection process, and a range of support programmes targeted at the different social groups, will be considered.

Access to farmland, poverty and nutrition

The concepts of 'land hunger' and demand for agricultural land need to be clarified in order to design an effective LRP. In an agricultural-based land reform programme an important issue that needs to be examined is whether increased access to land will improve the welfare position of participants. Aspects of these issues will be examined in this section.

The supply of farmland

Land hunger constitutes a real issue in South Africa. Reality, however, also dictates that the land issue is highly unlikely ever to be fully solved. The real need for land should be clarified. Indications are that the landless majority will opt for residential land use. However, it will be a mistake to underestimate the interest in land use for farming purposes.

The scope of land reform is limited by the physical availability of high-potential farmland. Roth *et al* (*1992*) calculated that redistribution of 25 per cent of land from the commercial sector to the black population would enable the settlement of 13,4 million people on 0,2 ha of arable land per resident (1,2 ha per family). This hardly provides a sound base for economic existence. If the area is increased to 4,8 ha/family, only 3,4 million people could be allocated land. If priority was given to those with farming skills, farm workers (1,0 million) would gain preference and little land would be available to others. This reality clearly calls for great realism in land reform objectives and the scope for farmer settlement. Selection of beneficiaries as well as innovative approaches to increase land supply are called for.

Land, poverty and malnutrition

An analysis by Sartorius von Bach (*1995*) using World Bank/SALDRU poverty data and discriminant analysis techniques, explored the relationship between farmland size and poverty. In addition to land size, the household food expenditure, aggregate household age, population density, average educational level and household size significantly explained why some households could be defined as poor and others as better off. As concerns the importance of land, regional differences were recorded, with land size most significant in the KwaZulu-Natal Province. This study also established a significant inverse relationship between improved access to farmland and the welfare of the poorest groups. The explanation of this is that farming constitutes a high risk activity, hence a high opportunity cost to scarce household resources. This is especially so when agricultural inputs and technology are not designed to address the basic problems and farming systems of resource-poor producers. Improved access to land per se may therefore trap these groups into economically unwise investments, affecting their welfare negatively. These findings were supported by an analysis of Carter and Zimmermann (*1993*) using parametric techniques.

A study in KwaZulu-Natal by Kirsten, Van Zyl and Sartorius von Bach (*1994*), analysed the relationship between access to land and nutritional deficiency. Farm production (crops and livestock) was used as an indicator of land access. This study was done under conditions of virtually no market-based land-use transaction such as sharecropping and rental. Access was determined by tribal custom. Land size, quality and production were thus correlated. Furthermore, participants had equal access to a set of support services under the Farmer Support Programme of the Development Bank of Southern Africa. From discriminant analyses, household cash flow (liquidity) and production (crops and livestock) were found to be the most important variables classifying rural households as nutritionally adequate or deficient. Low-producing households, with restricted access to land, fell within the deficient group. This analysis also indicated that households with access to farmer support services generally had a higher nutritional status than households without such access. (Also refer to *Singini & Van Rooyen, 1995*).

Alternative options to access farmland (and assets)

Studies by McKenzie, Van Rooyen and Matsetela (*1993*) and Van Rooyen, Nel and Van Schalkwyk (*1994*) indicate that measures to broaden the access of farm workers to equity in the farming business could enhance redistribution efforts and save considerably on a range of transactions costs such as infrastructure (to convert a large-scale commercial farm into smallholdings), mechanization costs, etc. Equity schemes could also enhance the welfare position of farm workers considerably because, effectively, they put this group in a position to accumulate welfare while the present farming business is continued, inter alia, due to the continuation of the management factor and the availability of accumulated skills. The Farm Workers Project also found that farm workers generally wanted access to farming opportunities but, at the same time, supported the concept of partnership with current farm owners (*Kirsten, 1993*). In spite of a range of potential problems associated with equity sharing, such as inter-personal complexities, (for example the master: servant status), and the low capacity of farm workers to participate actively in managerial functions (over the short run), as well as limited opportunities to expand livelihoods to non-farm workers, it must be argued that such innovative opportunities to allow farm workers (±1 million permanently employed on commercial farms) to participate as shareholders in commercially viable enterprises could significantly extend effective access to farmland, income streams and welfare accumulation on a cost-efficient basis and should be supported.

Guidelines for land reform programmes

From these studies it is clear that the relationship between access to land, production, welfare and accumulation, poverty and nutrition contextualize the land issue, and require further analysis in order to formulate clear land reform policies, programmes and models. However, the following guidelines can already be drawn:

- Access to farmland is an important although not prime factor in addressing poverty, malnutrition and accumulation of welfare.
- Access to farmland per se does not guarantee improved food production and nutrition. The required support services: extension, training, access to inputs and technology, credit, infrastructure, research, marketing, etc, are of equal importance for increased farm production.
- Land reform programmes to address the needs of the poorest and landless groups will have to be accompanied by measures to reduce farming risk, i.e. insurance schemes. However, these measures are costly in nature and require the application of cost benefit calculations. Alternative programmes such as small business, employment creation programmes, etc, will also be required to support these groups.
- Innovative and cost-effective schemes such as farm equity programmes to broaden access to farm assets, land and income streams, and schemes to increase the supply of land for land reform purposes, should be investigated to benefit farm worker groups. This will allow large numbers to participate more fully as land and asset owners and to accumulate a share of the farming profits and added value.

Beneficiaries: a case for having selection criteria

Two basic questions arise when one mentions the term criteria – namely: Why have criteria at all, and how will they be applied and used in South Africa? Both these questions are discussed below.

International evidence

Internationally, land reform and resettlement programmes have been predicated on two overlapping objectives to achieve greater social justice through redistribution to previously underprivileged groups, and to realize increased productivity of the land through the reorganization of agricultural institutions with the aim of achieving greater efficiency. Experience shows that, where such programmes have been implemented, selection criteria were derived from broad statements in national policy documents about who the prime beneficiaries in a particular programme should be. The weight given to these broad objectives is often conditioned by ideological and other policy goals of the regime in power (*World Bank, 1993*).

A trade-off was observed between having minimal criteria that could be exclusionary, or a broader-based programme that could allow for self-selection among the potential beneficiaries themselves (who would make choices about their preferred options for securing a livelihood).

International experience suggests the limitations of government administered programmes. Disadvantages of relying on free access and market mechanisms alone for the purposes of distributing land to the disenfranchised, however, point to the need to have a balance between criteria setting and self- selecting processes.

Furthermore, it can be argued that in a market driven land reform process, the market will do the selecting, generally favouring those already empowered, and leaving out the poorest and the landless of the poor (*Levin et al, 1994*). At the same time, there are imminent dangers in relying solely on bureaucratic institutions to effectively distribute land. Experiences in Mexico and more recently in Zimbabwe demonstrate that, in addition to the opportunity for corruption, it also takes a long time to carry out effective administrative land reforms.

Whilst there is no comprehensive coverage of non-administrative land reform programmes, studies which have examined characteristics of successful beneficiaries in land settlements or agricultural development programmes suggest that the following criteria emerge as most favourable in relation to land transfer performance (*Njobe, 1993*): age; education; supply of family labour; farming experience and skills; capital assets; non-farming skills; poverty status; marital status; health status; no prior criminal record, and nationality. See Annexe 18.1 for a full description.

The South African reality

The section of beneficiaries to a land reform programme will have to relate to the complex set of South African realities. Some of these are discussed below:

(a) **Converging and diverging notions.** In South Africa, as the constitutional transition advances, the positions taken by the diverse political organizations and interest groups with respect to the ownership and utilization of land could increasingly become a central point of conflict. In the examination of the existing policy positions on land reform, one point of agreement among the main political actors in South Africa is that there is a need to change the present system of land distribution and access to agricultural support services in the country to accommodate the disentitled (*Van Rooyen, Nqangweni & Njobe, 1994*). At face value, it would seem there is agreement on the end product. However, the motives for advocating change in the pattern of access to land and, in some cases, the way in which land is held and used are often different. Perhaps not surprisingly these positions vary across the political spectrum according to the perceptions of the origins and current nature of the problem and hence the motives for effecting the change. The debate on large-scale versus small-scale farming and the potential impact of market-based reform measures are manifestations of this divergence (*Christiansen, Van Rooyen & Cooper, 1993*). Consequently, there are different views on the beneficiaries, applicable models and scale of the entire exercise and the most effective policy instruments to be used. Thus, it may not be possible to arrive at a single set of criteria for determining participation based on a common political position, even within the realms of the present Government of National Unity (GNU).

However, it could clearly be said that the need for a redistributive process in South Africa, which involves the transfer of land, is broadly accepted. The controversy which then arises is in the definition of who should, would or could benefit from any land transfers and how these are to be effected. The related issue of access to services, however, is less contentious because it is recognized that a major aspect of the past policies was to deny access to services to particular racial groups and that services need to be modified (even restructured) to serve such groups in future to allow for productive land use (*Brand et al, 1992*). In view of the history of converging notions, it is proposed that these issues continue to be debated and clarified to reach a workable consensus.

(b) **A need for clear policies.** A second feature of South African reality needs to be highlighted. In the light of the differing positions advanced by the key political players, and where the motives for effecting land reform and the expectations about the scale and objectives of such a process are different, a critical issue that emerges in the development of a strategic approach is defining a framework and the degree to which the GNU will support a redistributive process. Undeniably, there will be pressure from the potential recipients as well as the current owners of land to clarify policy and procedures for land transfers. Interactive participation between stakeholders and beneficiaries will be a prerequisite for the clarification of policies and the setting of operational procedures.

(c) **Socio-economic differentiation.** A third feature of the South African reality indicates a spectrum of greatly differing social categories amongst those who have been disentitled (*Levin et al, 1994*). The notion to classify a homogeneous black rural population without attending to social class differences within categories will undermine any successful land redistribution effort. Categories describing the disentitled rural black population broadly consist of the following: victims of forced removals, emerging (even fully-fledged) entrepreneurs and salary earners; wage workers and part-time farmers; and the landless poor (*Njobe, 1993; Levin et al, 1994; DLA Workshop, 1995*). Gender issues cut across these divisions, while social class may also differ. The position of chiefs in a particular category, i.e. forced removals, for example, may differ radically from a female-headed household in the same category.

The search for target groups and 'designer' support programmes, therefore, should clearly be sensitive to the non-homogeneity of the black rural community, and account for social class differentiation within categories. The absence of such differentiation could easily lead to only certain social classes being favoured and gaining benefit from reform programmes while other may be left worse off. A market-based land redistribution strategy, in all probability, will only benefit the upper emerging commercial classes, while landless people may be deprived even more of scarce production resources due to their inherent inability to compete in the market for agricultural support and inputs.

As a possible solution to the socio-economic differentiation problem, the establishment of criteria to identify beneficiaries and appropriate support programmes must be 'tested very consistently among the affected communities' (*Njobe, 1993*). Levin *et al* (*1994*) propose that such a process will even require, as its starting point, systematic analysis of socio-economic class and gender-based differentiation, and that the following five broad socio-economic classes be recognized, 'while simultaneously acknowledging the existence of social stratums within class categories and household mobility from one category to another':

- A (small) class of salaried individuals, with substantial political influence (petty bourgeoisie);
- Petty capitalists and emerging entrepreneurs who achieved subsistence and are in the process of accumulating capital through various activities, including farming, trading, contract work, etc;
- Wage workers, including farmworkers, who may occasionally become involved in entrepreneurial activities, but for whom opportunities for accumulation remain restricted;

- Allotment holding wage workers who depend primarily on wage labour, remittances and pensions, but also engage in farming on a part-time basis on small allotments;
- Landless (or near landless) poor people with limited access and opportunities to enter and compete in the market (rural proletariat).

These five socio-economic categories will clearly require differentiated land reform options and support programmes based on diverse needs and objectives.

(d) **Farmer groupings, programmes and projects**. Empowerment entails both entitlement and creating access to support activities and opportunities. Such access should be created through a range of programmes, each with a farmer target group in mind. In general, these support activities should include the following elements (*Brand et al, 1992; Singini & Van Rooyen, 1995*):

- mobilization and capacity building;
- access to land with at least security of land use;
- access to technology and inputs;
- access to training, research and extension/counselling;
- access to market facilities;
- access to support infrastructure;
- access to financial services (which go further than credit facilities);
- advocacy and legal support to deal in land dispute matters;
- planning and management services;
- access to a lobby group to participate in the political economy.

The design of programmes and projects should occur with the full participation of the communities and individuals concerned. Certain design parameters should give direction to programmes and projects for the various groups or classes. This argument is illustrated by the following proposals (*Levin et al, 1994*):

- 'Petty bourgeoisie' should not qualify for land redistribution activities *per se*, but rather gain entitlement through access arrangements to participate in market processes (i.e. access to Agricultural Credit Board, Land Bank, etc.);
- 'Petty capitalists' should actively be supported to gain access to

land through small grants and subsidized loans within market-related support activities;

- Wage workers (including farm workers and allotment holding families) will require more substantive support to gain access to land and support mechanisms. Grants and low interest loans could be available to these groups. One particular example is where farm workers gain access to farm equity schemes through the 'land acquisition' grant scheme of the Department of Land Affairs;
- Landless people will require substantive intervention by government. Employment programmes, land grants and 'safety-net' activities will have to accompany programmes aimed at supporting this 'poorest-of-the-poor' category.

Whereas the range of required support elements will be similar for all groups or classes, the design of a support programme is also expected to differ substantially. In various research studies this argument is substantiated.

Van Zeyl (*1994*) investigated the priority elements as requested by various farming groups. Priority support elements for peri-urban farming groups ranged from legal support (with land claims); training; extension services and market access; and credit and information. Advocacy and land use planning were prioritized by communal farmers; training and micro-funding by rural subsistence groups; and funding and management support by informal settlement groups ('squatters'). Black farmers operating in the mainstream of white agriculture prioritized support to assist in the diversification of their investment in farming, to improve liaison with their white counterparts on agricultural policy matters, and to access financial services.

The evaluation of the Development Bank of Southern Africa's Small Farmer Support Programme (FSP) (*Singini & Van Rooyen, 1995*), highlighted the need to adapt a support package to local requirements. In the Phokwane project in the Lebowa area of the Northern Province, emphasis was put on training ('schools') and household food security. Access to on-farm infrastructure ('sheds') and mechanization equipment ('factors') was prioritized in programmes in the Mpumalanga Province. From an initial request to fund an agricultural project, additional facilities such as domestic water provision and infrastructure were included in a FSP in the Driekoppies dam area in Mpumalanga.

The present legal status of land and the natural resource potential will equally influence the design of programmes and projects (*Van Rooyen, Mbongwa, Matsetela & Van Zyl, 1994*). In their proposal, a matrix of options ranging from farmer settlement, the provision of support services, outgrower schemes, farm worker equity and the outright purchase of land could be linked to different types of land status and resource qualities.

While these particular studies do not fully allow for generalization, the trend is sufficient to alert planners to the need for flexibility, social stratification and participative approaches to project design, while taking into account the legal and natural resource status of particular pieces of land.

The scope for criteria

Developing a strategy to address all the above objectives and groups will require reconciling them, as far as possible, in terms of political pressure, social reality and economic feasibility. It may be argued that in South Africa, where the overwhelming number of potential beneficiaries are 'land hungry', the poor, the underprivileged and those historically discriminated against, it is not advisable to insist on a set of strict and inert criteria imposed from a 'top down' authoritarian basis. In the same vein, it may be argued that the proposed LRP aims to provide a viable framework for the development of hitherto underdeveloped rural groups, to the extent that they can improve their living standards and level of income, and contribute to the national economy. This would imply that there should be open access to a maximum (if not total) number of previously disentitled people. However, this will be economically unfeasible and will have costly implications to meet demand in terms of political expectations and available fiscal, professional and natural resources, and administrative capacity. Thus, in the South African situation, a reasonable (the only viable?) strategy would be to strike a balance between these two approaches, that is, accommodating all (the land hungry) *vis-à-vis* selecting particular individuals and groups to qualify for particular support programmes.

Essentially what is proposed is a set of critical factors that should be taken into consideration in the planning and implementation of a LRP. Thus, the criteria proposed here will be set to **facilitate** the participation of people at differing levels and in different ways within a broad rural restructuring programme.

These principles and criteria are now discussed.

Principles and criteria for beneficiary selection

As a consequence of the history of South Africa, four sets of principles[3] need to be considered and reconciled in a LRP. These are:

- The need to address inequities in the ownership of and access to land and productive assets and support services which result from disentitlement through apartheid.
- The need to address poverty issues which evolved as a consequence of the apartheid laws as well as the poor rural economy.
- The need to establish land use models which will ensure improved, efficient and sustainable use of the land under a land reform programme.
- Democratic (transparent and broad-based) participatory processes for project and beneficiary selection.

(a) **Disentitlement.** A starting point in the South Africa context is to consider those South Africans who were historically disadvantaged and victimized through disentitlement and denial of access to land and support services, to be beneficiaries in a LRP. This would justify the allocation of benefits because it is clear that apartheid was an act of segregation and disentitlement (in law). However, this is also a broad generalization that can be used to appease political demands for the recognition of lost rights to land and hence the opportunity to develop – as a result of conquest and oppression. Unless such a broad and open recognition is given, the issue of restitution is likely to be raised repeatedly in the event of dissatisfaction within the rural areas.

The pressing issue then becomes determining and categorizing the numbers of affected people and identifying how they are to benefit. Disentitlement should only be a base category for identifying beneficiaries in a land redistribution programme. Using this broad category as a basic 'entrance' criterion, it would be necessary to determine what this would imply about how particular groups of people gain access to land (particularly in the light of the objective to develop and increase efficiency in agriculture). Additionally, poverty as well as lack of access to opportunities for production should be used as a means of determining the potential participants.

3 Dispossession refers to a different aspect of land reform, i.e. restitution. This concept is not fully discussed in this article.

(b) **Poverty and need.** Poverty in South Africa, to a large degree, is structural. A reversal of the current state of affairs therefore, demands that deliberate policies targeted at the root causes of poverty be applied. There is broad agreement in South Africa about the need for non-racial democracy. This being the case, and accepting that democracy will have little meaning and may prove to be short-lived if it does not have a meaningful impact on the poor (*Jazairy, Alamgir & Panuccio; 1992*), then a key question in the feasibility of a land reform programme in a democratic South Africa is the degree to which it provides the rural poor[4] with access to opportunities for income generation, employment and self-employment.

The argument that growth in overall national output ultimately creates the capacity to reduce poverty may hold in the long run, but is not useful in South Africa for the immediate present, where poverty is primarily an outcome of disentitlement and the denial of access to opportunities through apartheid policies. In this context, the relationship between landlessness, poverty and malnutrition was examined on pages 464–465 with a clear indication that land constraints were a significant factor in household poverty, malnutrition and low rural incomes. This chapter, therefore, takes the view that any programme in South Africa aimed at rural economic development, must have a strong social orientation component, and yet not be purely welfare oriented. In the case of an LRP, this could take the form of a process of land allocations accompanied by the provision of adequate and appropriate support to enable those without resources to be able to use the land effectively to secure a livelihood.

In developing the framework of a LRP, therefore, it becomes critical to incorporate criteria which give preference to the poor and the landless amongst the victims of apartheid as beneficiaries. In addition, the programme needs to ensure their participation in the planning and implementation process. The LRP should also create opportunities for income generation (through agricultural production or non-agricultural businesses), employment schemes and self-employment activities.

The proposed principle – poverty and need – should recognize the diversity of needs that emanate from historical experience and social

4 It is assumed that, in the context of transformation of South Africa's society, the most important issue within rural policy will be the future of the rural poor – especially women.

stratification (See page 469.) Some people may require access to a more secure form of their current land holding, others may require a marketable right to holdings. A third group may need access to land and services, and a fourth only require improved services. Thus, opportunities should reflect the inclusion of poverty concerns, but may also be determined by the degree to which the beneficiaries are expected to participate successfully in viable and productive agriculture.

(c) **Productivity and sustainability.** The productive use of land emerges as a concern in the design of a restructured rural economy. Sustainable models of land use are not the subject of this paper. However, it is argued that a range of land access and land use models need to be developed to accommodate the various needs, opportunities, resource base and political circumstances (*Van Rooyen, Mbongwa, Matsetela & Van Zyl, 1994*). In developing selection criteria, it is recognized that productivity in agriculture is likely to emerge as a determining principle for participation in the LRP, particularly in relation to designing the level of support services needed to enhance the productive capacity of the beneficiaries. A broad concern for the consideration of the long-term environmental consequences of the options also needs to be emphasized. Productivity is therefore defined in terms of (long-term) sustainability.

(d) **Participative processes.** International experience shows that problems almost inevitably arise when groups are moved on a top-down basis into land-based schemes with unrealistic expectations as to what is involved in resettlement, with little clarity in what they are expected to do for themselves and what will be done for them, or how long it will take for incomes to reach certain levels. Those who are most disillusioned are usually the first to leave. Therefore, the expectations of the beneficiaries, once identified, need to be anticipated and fully integrated in the planning process, and subsequently dealt with in the implementation stage.

The basic position taken here is that local participation in planning and implementation will be a prerequisite, not only because it will produce more efficient and sustainable results, but also because it should go a long way in addressing the prioritization of particular concerns at a particular time, as well as impacting on the selection processes and enhanced productive capacity of the participants.

(e) **Case studies.** An understanding of the issues raised above points to the importance of translating principles and defining criteria through a process of direct participation with the affected people and other stakeholders. In order to achieve this goal, information from several such case studies was selected. A first case study entailed discussions in the regions of the Eastern Cape in the former Ciskei and Transkei, and the former Lebowa in the Northern Province (*Njobe, 1993*). The study areas were chosen mainly because of the mix of historical, socio-political and economic characteristics which they display. Ideally, such studies should be carried out countrywide in order to incorporate the expectations of those who, effectively, are outside the rural areas as an outcome of the past policies.

A second case study included the gathering of opinions on selection criteria from those who were already involved in farming on a range of projects and schemes (*Van Zeyl, 1994*). Groups in the urban and rural environment were interviewed. Understandably, proposals were more focused on production performance criteria such as in-depth farming knowledge, farm management ability and agricultural skills. Some criteria common with the Njobe case studies were registered, such as preference for South African citizenship, availability of labour (hard work) and positive interest in farming. What is important from the Van Zeyl case study, however, is the differences between farming groups. Urban groups insisted on 'no-criteria' for participation. Farmers in a tribal situation emphasized access exclusive to

Table 18.1: Criteria Historically Used in South Africa for Land-based Development Programmes
• Previous experience of farming
• Education
• Marital status
• Age
• Family size (linked to available labour)
• Health
• Entrepreneurial skills
• Managerial aptitude
• Farming knowledge
• Participation in local agricultural organizations
• Previous farming history
• Net worth (and positive credit record)
• Willingness to be trained
• Positive perceptions of farming

'existing members of the tribe', but were particularly unsure about how to handle returnee 'past' members. Black farmers already in the commercial mainstream and farm workers emphasized technical and managerial skills as major criteria for participation in farming.

For future use it should be noted that the criteria documented by Van Zeyl are partly 'trapped' in the policy objectives of the previous regime to establish agricultural production. A similar argument can be levelled at the criteria previously used for government supported land settlement programmes. These mainly included levels of education, proven farming experience and positive credit records. (See Table 18.1).

A third set of case studies explored the question: 'who should benefit from rural land reform?' through village-level participatory research in Mpumalanga and the Northern Province (*Levin & Weiner, 1994*). These studies which, inter alia, recognized socio-economic class differences, recommended a set of criteria somewhat similar to Njobe. Gender issues, in particular to prioritize the participation of landless female-headed households and to recognize the role of women in numerous forms of production, were emphasized. These studies in particular insisted on clarifying the need, formulation, interpretation and application of criteria through participatory processes. Table 18.2 gives a summary of the criteria suggested from the various case studies:

Table 18.2: South African Case Studies: Criteria suggested by Respondents to Interviews

- Landlessness
- Poverty
- Victims of forced removals
- Availability of labour
- No record of collaboration with the apartheid system
- Health
- Age
- Women abandoned by their husbands
- Female-headed households
- Interest in farming
- Pledge not to sell land
- South African citizenship

Making the choices: a possible option for South Africa

We now propose a list of criteria to be set at a national level, and applied at a local level. They are broken down into three broad categories, based on the principles as stated: namely, disentitlement, need, and productivity. The criteria for consideration are listed in Table 18.3 and are discussed thereafter.

Table 18.3: Selection Criteria for Participation in a South African LRP
DISENTITLEMENT
1. Disentitlement to access land and services
2. Nationality
3. Rural residency
NEED
4. Poverty
5. Number of dependants
6. Negative status in society
PRODUCTIVITY
7. Health
8. Age
9. Net financial worth
10. Education levels
11. Gender sensitivity
12. Previous experience in farming
13. Entrepreneurial skills
14. Managerial aptitude

Criteria based on disentitlement

The basis for having a rural restructuring and land reform programme is to rectify imbalances caused by the apartheid system. It follows, therefore, that at the first point of access to the programme, only those victims of apartheid who were disentitled should qualify. Preferred attributes of those disentitled would be rural residency and South African citizenship.

1. **Disentitlement:** The victims of apartheid can include a range of people and communities, both black and white. Black people (inclusive of Africans, Indians and 'coloureds'), however, were disentitled

through apartheid legislation and constitute the focus group within these criteria. This is both a moral and political statement of intent. It accepts that the need exists to make the declaration to establish a strong sense of purpose about the objectives of an LRP. It also recognizes that not all victims of apartheid need to benefit directly from an LRP. The basic benefit is that opportunities in their interest are to be created within an LRP; there will be no further denial of entitlement, and access to support services in agriculture and the rural economy will be provided for all.

2. **Nationality.** The expressed concern for the sharing of benefits with those who are not South African citizens was repeatedly put forward in the interviews conducted in the case studies. It will be difficult however to make a separation of those who are 'South African citizens' and those who are not. Similarly, there is strong sentiment about the potential loss of opportunity to South Africans should preferred access to an LRP be open for all. Notably, international experience suggests that nationality does become an important criterion, particularly where there has been a history of rural refugees.
3. **Rural residency.** We propose that there should be a preference for those who are rural residents, i.e. residents in areas where farming is a dominant economic activity. However, this does not exclude those victims of displacement who currently reside in the urban and peri-urban areas. On the contrary, due to their historical disposition and need they will still be able to exercise the option to participate in a LRP. For this purpose, 'urban farming' needs to be included as an important strategy. The difficulty will rather be with the family units which are not all rural residents. Experience in Zimbabwe points to a problem with having a strict and inert set of criteria which exclude those units who may have one or more members being non-rural residents. This should not become a limiting factor in South Africa.

Criteria based on need

Potential participants should also be evaluated in relation to their extent of need, using poverty, number of dependants and socio-economic status as criteria. Points could be allocated to determine the extent to which they qualify for awards such as special assistance in the acquisition of land and access to services, specific land and/or some form of grant/cash. However, the emphasis clearly remains to help the poor and landless rural dwellers.

It is extremely difficult to quantify need, particularly in the South

African context where so many people feel they are in need. The demand for land in South Africa is very high among the historically disenfranchised. This factor would need to be included among the criteria to satisfy political and social pressure. The approach adopted here, therefore, was to use need as an overriding principle which in practice is tested in terms of the following criteria:

- **Poverty.** To measure poverty is very difficult, and to identify and classify poor rural people is complicated. However, this factor will have to be included among other criteria. Access to land is often perceived as being the first step to poverty alleviation. As referred to on pages 464–465, research by Sartorius Von Bach (*1994*) indicates that provision of land to the very poor for farming may have a detrimental effect on family income due to the risk factor in farming. Land provision to the very poor should therefore be accompanied by a range of support services and insurance schemes. Non-farm options should also be activated.

  Poverty should not be uniquely defined in terms of income, but in terms of abilities (observable capacities and capabilities) to function (for example, being adequately clothed and sheltered, literacy and lack of political/lobby power). Inadequate abilities are not only easier to observe and to measure, but they are also closer to the mark.
- **Number of dependants.** The greater the number of dependants in a family unit, the greater the need for land (*Kirsten, Van Zyl & Sartorius von Bach, 1994*). Also, there is the availability of a larger supply of labour. Internationally it has been shown that a family labour force has a strong positive correlation with agricultural and economic performance (*Kinsey & Binswanger, 1993*). In order not to blur the understanding of family – or make assumptions about the definition, the number of dependants in the rural society could be used additionally to denote the extent of need and poverty. Poor families tend to have large-sized families with little or no sources of substantial income. If the dependants are many, then the likelihood that the family is poor is great.

  However, the caution is that claims can be made on the number and ages of the dependants depending upon the perception of benefits to be derived. One way to avoid this could be to point out that the assumption made about the number of dependants over (for example) the age of 16, indicates that there will be people to work the land – initially, hence there will be higher expectations from that unit about productivity. In group formations and or communities this could also be an indication of potential labour as well as the degree of need.

- **Status in society.** The categorization of social classes was discussed on pages 470–472. The application of such class differentiation should be considered to assist in the targeting of the poor and the selection of appropriate support systems. The position of women, however, requires special attention.
- **Gender and group issues.** Marital status – with a preference for married men – was used as a criterion by the Agricultural Credit Board and South African Land Bank on the basis that it is seen to indicate that the participant can bring stability to the farming environment and has the potential for long-term investment because of inheritance factors. This might be a useful criterion. In the broader South African sense, it can be used to indicate the need for land, the supply of labour and other socio-economic issues.[5] However, if the intention of an LRP is not to pay lip-service to enhancing the status of women in the rural areas, it would not be in their favour to lay emphasis on the fact that an applicant is married. Poor female-headed households should pertinently be added to this category.

  It is argued, therefore, that the broader status in society is used as an indicator of social and legal status – hence being a member of a particular group which has formed itself in the interest of farming could also be an important factor.

Criteria for access to land and the productive use of agricultural support services

In the interest of attaining optimum productivity in land use, and recognizing that apartheid policies resulted in structural poverty, it becomes important to offer special assistance in accessing appropriate services to the victims of apartheid. The proposed selection system is thus extended to apply to these criteria, to facilitate the determination of the level of assistance. Important attributes to be tested at this level include health; age; net worth; education levels; gender; previous experience in farming, entrepreneurial skills and managerial aptitude.

5 The assumption is that there are dependants and a sense of planning for the long-term establishment and settlement of a family unit.

- **Health.** An applicant who does not enjoy good health[6] will not be able to contribute labour and or manage a farm. For purposes of an LRP this should be tested in a manner which is not cumbersome and costly – nor intimidating as people will invariably opt out of the programme application process purely because of a deterrent such as a costly medical examination. Rather the indicator should be (as is usually done on medical forms) a record of particular illness and disability.[7]
- **Age.** International experience shows that a suitable age to get the most success is around 45 years old. Research by Botha and Lombard (*1992*) in South Africa, shows that farmers between 30–45 years have the best chances of success. When beneficiaries are too young or too old, it can have detrimental sociological effects on the age structure of the farming population and community. The consideration of age, as a criterion in this report, however, arises as a result of its interrelationship with access to labour and health – necessary qualities of potential beneficiaries. Age on its own is not a qualifying measure of one's ability to succeed in the LRP. However, it can be anticipated that where applicants are in poor health and a sufficient labour source is not available, very low performance levels could be expected. The older a person is, the less likely they are to experience long and good health. Within a thrust to develop small-scale farming, much will be dependent on the ability of the participants to work and manage their production units. On the other hand, since high output is also linked to one's ability to access labour and, in the context of small-scale production where resources are scarce, this tends to take the form of family labour.

  International experience, highlighted by Kinsey and Binswanger (*1993*), has shown that the agricultural and economic performance of households is strongly positively correlated with the number of family members able to work. The younger a person is, the less likely they are to have access and control over an adequately capable family labour source. Their own children would be of a school-going age (or younger), and they would rarely have authority over the other members of the extended family and or household. In a group or

6 A precise definition of good health will have to be developed consistent with the expectations from the primary health care service institutions. Conditions such as AIDS, physical disability and pregnancy will need to be carefully qualified so not as to discriminate against vulnerable groups including women.

7 See previous note.

community environment it then becomes important to have a fair spread of older and younger people.

Recognizing that the average age of communities in the rural areas is high, we propose that within the weighing system the age brackets be broken up as follows: (a) <17–28 yrs; (b) 28–50 yrs; and, (c) >50. In using such a breakdown of age, it is expected that in the event where the person is nearer the 50 year mark, the person will have dependants whom he/she can involve in the process of acquiring labour. Those below the age of 28 may still be more interested in pursuing opportunities in education, and the urban centres. This will obviously exclude the youth, to a large extent; however it is hoped that national reconstruction will focus on providing educational and employment opportunities for this particular category of the population. Special land programmes can also be targeted to the youth.

- **Net financial worth.** Net worth is used as a financial criterion which indicates the capital assets of individuals. Inclusion of this aspect, especially if a higher net worth is taken as a criterion, allows for the identification of the poor for selection purposes. This can also be used as a criterion to determine security and repayment ability, as well as willingness to increase size of holding and or other assets. The criterion, together with credit history, has been used by several organizations in South Africa (Land Bank, Credit Board) which lend money to commercial farmers.

  However, for purposes of determining access to financial services within an LRP, it is proposed that criteria of access do not exclude the poor. The contribution of 'sweat equity', i.e. work for free in infrastructure activities required for a project, can be considered as a substitute for net worth.

- **Education levels.** South African research by Botha and Lombard (*1992*) shows that higher levels of education are linked to better farming performance. This is backed by international experience (*Evenson, 1988; Kinsey & Binswanger, 1993*). It will be important in the case of an RRP to ensure that the definition of education goes beyond the formal systems which have been recorded in terms of certification. Whilst there is an appreciation of formal education in South Africa's rural areas, the reality is that the majority of black people have been denied access to educational institutions and particularly agriculturally related institutions of higher learning. In future, the formal education level could be an acceptable criterion; however, for the purposes of the LRP, the criterion is used in as far as it is necessary to gauge the need to approach information dissemination and technical instruction in a 'user friendly' manner.

It is important to consider literacy and numeracy levels. Lack of local research makes it impossible to indicate its influence on performance, but it should be taken into account because it is indicative of levels of disenfranchisement, need, poverty, powerlessness, etc. Ultimately the support services provided should be based on the assumption that training, information dissemination and access to services should be offered in such a way that literacy and numeracy are not inhibiting factors. Access to capacity development and literacy programmes in the rural areas will have to be provided as well.

Local work done by Botha and Lombard (*1992*) has shown that clients who are willing to be trained are more successful than their counterparts. Therefore, this could be one of the additional factors to be taken into account in applying this criterion. Another factor is the consideration of the number of any courses, formal or informal, attended. This ties in with education, but shows if a potential beneficiary has used any opportunities for self advancement in the past.

- **Gender and women criteria.** Evidence in South Africa found that males fared better than females on commercial agricultural projects (*Botha & Lombard, 1992*). However, this is contradicted by evidence in many of DBSA's farmer support programmes (*Singini & Van Rooyen, 1995*). A critical factor which will determine the success of the poverty alleviation component of an LRP will be the ability of the programme to target women as beneficiaries. The case studies also suggest a large degree of concern for the possible exclusion of women within the LRP. In South Africa, where there is growing recognition of the need to address the problems of poor rural women and female-headed households and, in order to do justice to the policy statements about the 'inclusion of women',[8] an LRP will have to target women as a particularly vulnerable group. The subordination of women, and particularly black women, was inextricably linked to the implementation of apartheid. The large numbers of black female-headed poor families present in the former homeland areas are a direct consequence of the influx control and other laws intended to keep 'separate development' alive.

8 Probably in reaction to the extreme injustices on the majority of people in South Africa, the status and participation of women in the reconstruction process has become an issue which is consistently raised at the different level discussions on political, economic and social transformation processes.

There is a clear emphasis on the demand for human rights, with an articulation on the equality of women in the proposals for legislative reform,[9] however, the reality is that women are still the subjects of this discussion. Ironically though, the bulk of the policy statements being propagated emphasize the empowerment approach.[10]

The current danger, therefore, is that extensive reconstruction programmes are being developed without the active and equal participation of affected women. This, combined with the extent to which there is a need to address a range of issues linked to the creation of a viable rural economy, could result in an increase in poverty levels for rural women. It is therefore proposed that women be given particular preference in the application of all criteria suggested. We must avoid the tendency to define merit based on the experience of male farming. Rather, those criteria which are more favourable towards the participation of men, should be changed to ensure that the quality being measured and or identified is in fact a human quality necessary for farming. One example of this is the practice of using marital status as an indicator of success.

In using marriage as a positive criterion for access to support services, it will be difficult for many South African women to qualify. Accepting perhaps that marriage and/or the presence of a family is one way of bringing stability into farming communities, it is pro-

9 The Land Manifesto of the ANC Land Commission points to the need to recognize that black women are the most disadvantaged section of the population. Consequently, attention needs to be given to their special needs and problems, and special and conscious efforts will need to be made to enable them to play an equal and full part in all aspects of life. It calls for equal rights for women in relation to land, support services and credit; full and equal participation by affected women in making decisions about land reform and rural development; and the freeing up of women's time so that they can play an active part in all aspects of social, political and economic life.

10 This approach questions the fundamental assumptions concerning the relationship between power and development that underlie the previously described approaches. It argues in terms of the capacity of women to increase their internal strength and right to determine choices in life and to influence the direction of change through the ability to gain control over crucial material and non-material resources. Longer term strategies will ultimately be needed to break down the structures of inequality between genders, classes and nations. Whilst it is a recognizably slow process, it recognizes the interaction of social and economic issues and goes beyond an emphasis on access and productivity. The irony is that it is often propagated at policy discussion forums where women are absent and no consistent attempts are made to include them in the discussion.

posed that the status of the individual be identified in such a way as to determine his/her existing commitments and/or responsibilities. In this way, whilst testing stability, the criterion does not make a moral judgement about whether or not people should be married. On the other hand, certain criteria will need to be weighted in such a way that they do not effectively exclude women or discriminate against them purely on the basis of their sex.

- **Previous experience of farming.** The use of this criterion is based on the following assumptions: the previous existence of a viable black agricultural community; the existence of (indigenous) knowledge of viable farming; farm workers who have lived and/or worked on commercial farms, through practice, have acquired some interest and knowledge of farming; and food production – activities carried out by women in the homeland areas, serves as a basis for some knowledge of farming and of formal and/or informal training in farming practices.

Whilst there are many people who can lay claim to the above, it is also true to say that the circumstances in an LRP would be different and more conducive to those who have the will and ability to farm. It is also very difficult to detect what impact previous experience of farming can have on one's own managerial and entrepreneurial ability to farm successfully. Under the productivity criteria, other criteria are proposed which would further test farming abilities (qualities). However, both local research (*Botha & Lombard, 1992*) and international experience (*Kinsey & Binswanger, 1993*) show that a background which includes successful farming experience and acquired skills is strongly predictive of good performance, whereas one with a negative experience of farming tends to drive people into other options of economic activity. It can also be expected that an understanding of the farming environment where participants would like to farm or live, could make adapting to such an environment much easier and soften the blow of resettlement with positive influences on performance.

The findings of research show positive relationships between higher levels of knowledge and farming success (*Evenson, 1989; Botha & Lombard, 1992*). The accent falls mainly on knowledge of modern farming techniques, but Indigenous Technical Knowledge (ITK) should not be disregarded, rather it should be included as well[11]

11 Indigenous Technical Knowledge is recently being given more prominence as a useful and relevant quality. It differs within the range of rural societies, and appears to be transferred to new generations through oral traditions and education and practical

within a farming systems research approach. If the above is true, in testing the previous experience of farming, it will be important to raise as diverse a range of farming situations as possible. It will also be important to give recognition to non-agricultural skills since these often contribute to the viability of the process. Business experience in particular comes to mind.

- **Entrepreneurial skills.** Research on differences between unsuccessful and successful farmers could be used to draw up several criteria for inclusion. These include farming for the purpose of making money and making a living; the cultivation of cash crops and marketing of the crops/livestock; the inclination to grow more cash crops; a desire to increase landholding; a more positive orientation towards training; employment of other people; exposure to agriculture and other information through the media; openness to the advice of fellow farmers, co-operatives and extension officers; the desire to contribute to the decision-making process; awareness of risk and the need for forward planning; a good knowledge of, and a willingness to apply modern crop and cattle farming practices; some form of previous management training, etc. International experience also shows that the above skills are strongly predictive of good performance (*Kinsey & Binswanger, 1993*).
- **Managerial aptitude.** Relatively little research has been conducted on this topic in South Africa. Burger (*1967*) developed a scale that is frequently used. However, if used, Burger's scale should be applied with great circumspection for the LRP. Other interesting observations related to managerial aptitude would need to underlie the interpretation of this criterion. For example, farmers who displayed positive and accurate perceptions about several facets of farm management were found to be more successful in commercial farming enterprises (*Botha & Lombard, 1992*).

Managing the process

In order to be clear at what point the criteria would be applied, a framework of the envisaged process is proposed (See Annexe 18.1, Table 18.4). Using this framework, it is proposed that the three levels of selection be considered as follows: the first will be the point of entry to the

demonstration. Its inclusion will demand an acceptance of its different forms within the affected communities and areas in South Africa. Consequently, there will be a need to carry out thorough research on its form and content, prior to its inclusion as part of participation criteria in the LRP.

LRP, (applying the disentitlement criteria); the second will address poverty concerns as well as the extent of need; and the third level will focus on determining the levels of assistance which could be provided to participants in order to upgrade land productivity through access to support services. The third level criteria are intended to be more specific and productivity orientated.

The management process of selecting beneficiaries requires three operational levels of interaction, namely, national, provincial and local level. The link-up between the levels needs to be clarified and co-ordinated.

A national level institution (which would also play a significant role in steering the process of rural restructuring) would need to be formed. This institution should link-up with the Reconstruction and Development Programme (RDP) as well as other national bodies concerned. It should be made up of a mix of people: those with an understanding of the complexity of the issue, but who could also be impartial to specific cases; representatives of the major interest groups; and, when a specific region is under consideration, representatives from the local area(s).

Local community level institutions should be set up to be responsible for clarifying the need and application of criteria, facilitating the collation of information on the profile of the affected communities and the selection of potential beneficiaries. These institutions should be representative with a similar mix of persons (as at the national level) and interest groups at a local level.

The role of provincial authorities is presently debated, with agricultural development a provincial competency, and land and water a national responsibility. Programme level responsibilities will have to be developed at this level to ensure the co-ordinated functioning of the programme.

Resources which will support the prompt dissemination of information to the affected communities as well as the rest of the population will need to be established in every area. In this way, one could possibly circumvent a delayed reaction when there is a crisis which goes beyond the capacity of the local committee to resolve. This could be a provincial level function.

Information dissemination will need to take into account languages, as well as different levels of proficiency in literacy and numeracy. This should also be linked to the information service to be provided as part of the technical support services.

An institutional link in the management process which will be of extreme importance, will be the creation of opportunities to enter market-based interaction. The above measures mainly appear to be public sector driven. However, this will be a mistake, especially in the South African environment with its widely developed private sector. The role of the public sector should essentially be directed to 'kick start' and facilitate access for the selected beneficiaries to enter market driven processes. The above processes are to set a framework to enable the mobilization of a range of support measures. This should inter alia entice the private sector to participate in LRP.

The monitoring of the process, preferably from a provincially-based institution to undercut favouritism and corruption, should also be introduced.

Concluding remarks

A land reform programme in South Africa will need to have, as key components, substantial land transfers, the reform and liberalization of the agricultural policy environment, effective empowerment mechanisms, and clear selection criteria to identify beneficiaries to qualify for support measures. Essentially, what is proposed in this chapter is a set of fundamental principles and critical factors that should be taken into consideration in the selection of beneficiaries insofar as providing access to land and farming opportunities are concerned. It is envisaged that, through the provision of adequate information about available options (for development opportunities), and through training and flexible but effective local management, people with particular experience and qualities will be identified, encouraged and supported to opt for one or other form of agricultural land use or rural activity. Ultimately, critical, difficult and also emotive choices will need to be made about the selection of beneficiaries and the focus of such a support programme. The inability to make timely decisions, in the long-term, may result in politically motivated and potentially 'inappropriate' criteria or the lack of effective criteria being used to establish a viable rural economy.

In drawing from international experience, it is noted that programme beneficiaries in land reform have tended to be those who occupied the land at the time of the reform. In South Africa, the extremity of landlessness and land hunger is a direct consequence of a history of dispossession, including extensive forced removals. It is therefore inevitable that land allocations to those in need will often involve the need to move out of the designated 13 per cent (black former homelands) into 'white' South Africa. This kind of activity is more typical of resettlement experiences that were aimed at relocation and the establishment of particular programmes. Such processes require that clear criteria and principles be established at the outset, in order to clarify expectations and to guide participation as well as to achieve the objectives of the programme. Innovative approaches, unique to the South African environment, may also be required to accommodate farm workers and other rural dwellers, while mechanisms to allow for the increased supply of land should be considered through inter alia private sector participation, ie. willing sellers.

This chapter suggests that the following guidelines be used in relation to beneficiary identification and selection:

- Any serious attempt at rural restructuring and land reform in South Africa will need to take into account a diverse set of long-term objectives, as well as include a substantial number of beneficiaries who will eventually become part of a viable rural economy.
- A process of facilitating participation in an LRP designed for South Africa would best meet the pressing needs if it is based on three basic levels of criteria – disentitlement, poverty (need), and productivity.
- It should seek to address the existing problems in South Africa's rural areas where land is an important socio-economic resource. The questions that need to be addressed regarding resource allocation, prioritization of issues for redress, and trade-offs, should take into account the existing infrastructure, needs and expectations of the people most adversely affected, and present production processes providing income and employment.
- Land transfers without a clear conceptualization of the target group and an appropriately designed support programme, cannot be expected to be effective. Socio-economic status, nutritional and food security needs, entrepreneurial objectives, etc, will have to be clarified.
- It is assumed that a process of restitution and adjudication through the proposed land claims process will be implemented prior to or simultaneously with this land redistribution. Whilst it is expected that most communities with a claim to specific pieces of land will have their claims addressed as a result of what is largely a politically driven process, one can also expect that they may prefer to opt out of such a process and take part in the redistribution activities of a LRP. Subsequently, it becomes necessary to identify them as those who would effectively be the first active claimants to land and services in a distributive programme.

- Clarity on the definition of disentitled victims of apartheid, poverty-stricken people, and productivity, should precede any implementation of related criteria. Any ambiguity on these issues could result in corruption and the ultimate failure of the LRP.
- A critical success factor in planning and implementation will be the participation of the different interest groups and affected communities. At best, a transparent, flexible programme with active local participation and provincial support would accommodate (to a large extent) the fears of exclusion.

The criteria proposed in this chapter will need to be tested consistently among the affected communities. There is already a danger that extensive reconstruction programmes are being developed without the active and equal participation of the affected communities, more particularly the women and the poor. This, combined with the extent to which there is a need to address a range of issues linked to the creation of a viable rural economy, could result in an increase in poverty levels and a failure in the rural restructuring. The process of developing a South African Land Reform Programme, its objectives, strategies and final implementation, will need to involve the affected, the local communities, and all major role players actively at all times. This could be seen as a way of facilitating the actual implementation. However, there should be an awareness of the possibility of historically-based conflict at a local level.

References

Botha, C.A.J. and Lombard, P.P. 1992. The effects of education and training on the managerial attributes of project farmers in the KaNgwane homeland of South Africa. In Csaki, C., Dams, Th. J., Metzger, D. and Van Zyl, J. (Eds), *Agricultural Restructuring in Southern Africa. Proceedings of IAAE Interconference Symposium.* Namibia: Swakopmund; Agrecona, Windhoek, 24–27 July.

Brand, S.S., Christodoulou, N.T., Van Rooyen, C.J. and Vink, N. 1992. Agriculture and redistribution. A growth with equity approach. In Schrire, R. (Ed), *Wealth or Poverty? Critical choices for South Africa.* Cape Town: Oxford University Press.

Burger, P.J. 1967. Agricultural progressiveness: a South African concept. Unpublished. *D. Agric (Inst Agrar) dissertation.* Pretoria: University of Pretoria.

Carter and Zimmerman. 1993. Structural evolution under imperfect markets in developing countries: a dynamic programming simulation. Typescript. USA: University of Wisconsin-Madison. December.

Christiansen, R., Van Rooyen, C.J. and Cooper, D. 1993. (Eds.) Agricultural policy experiences for South Africa. *World Development,* Vol. 21 (9) : 1447–1450, October.

Cowling, R. 1990. Options for rural land use in Southern Africa: an ecological perspective. In De Klerk, M. (Ed), *A Harvest of Discontent: The Land Question in South Africa.* Cape Town: IDASA.

Department of Land Affairs. 1995. Work session: Towards a Land Reform Policy. Warmbaths, January.

Evenson, R.E. 1989. Human capital and agricultural productivity change. In Maunder, A. and Valdés, A. (Eds), *Agriculture and governments in an interdependent world.* UK: Dartmouth. Proceedings of the 20th International Conference of Agricultural Economists, Buenos Aires, 24-31 August.

Jazairy, I. Alamgir, M. and Panuccio, T. 1992. *The State of World Rural Poverty: An Inquiry into its Causes and Consequences.* New York: IFAD, New York University Press.

Kassier, W.E. and Groenewald, J.A. 1992. Agriculture: An overview. In Schrire, R. (Ed), *Wealth or Poverty? Critical Choices for South Africa.* Cape Town: Oxford University Press.

Kinsey, B.H. and Binswanger, H.P. 1993. Characteristics and performance of resettlement programmes: a review. *World Development,* Vol. 21 (9) 1477–1494.

Kirsten, J.F. 1993. Models of land use activities: the case of the summer grain region. *World Bank Rural Restructuring Programme, Unpublished Research Document.* Johannesburg: LAPC.

Kirsten, J.F., Van Zyl, J. and Sartorius von Bach, H.J. 1994. Evaluation of the Farmer Support Programme: KwaZulu. *Unpublished Research Report.* Halfway House: DBSA.

Letsoalo, E. 1991. Land reforms – state initiatives. In De Klerk, M. (Ed), *A Harvest of Discontent: The Land Question in South Africa.* Cape Town: IDASA.

Levin, R., Russon, R. and Weiner, D. 1993. Social differentiation in South Africa's bantustans: class, gender and the politics of agrarian transition. In Levin, R. and D. Weiner (Eds), *Community Perspectives on Land and Agrarian Reform in South Africa.* Final project report prepared for John D. and Catherine T. MacArthur Foundation. USA: Chicago, Illinois.

McKenzie, C.C., Van Rooyen, C.J. and Matsetela, T. 1993. Options for land reform and funding mechanisms. *FSSA Journal*, April.

Njobe, B.N. 1993. Criteria for participation in a South African Rural Restructuring Programme. *Paper presented at the LAPC Land Redistribution Options Conference.* Johannesburg: 12–15 October.

Roth, M., Dolny, H. and Wiebe, K. 1992. Employment, efficiency and land markets in South Africa's agricultural sector: opportunities for land policy reform. *Unpublished Review Paper.* USA: Land Tenure Centre, University of Wisconsin-Madison.

Sartorius von Bach, H.J. 1995. Poverty gap assessment of rural areas in South Africa. *Draft Report*. Pretoria: University of Pretoria.

Singini, R. and Van Rooyen, C.J. (Eds), 1995. S*erving Small-scale Farmers:*

An Evaluation of the DBSA's Farmer Support Programmes. Halfway House: Development Bank of Southern Africa.

Van Rooyen, C.J., Mbongwa, M., Matsetela, T. and Van Zyl, J. 1994. Determinants of and options for agricultural land reform in South Africa. *Paper presented at the Conference of the International Association of Agricultural Economists.* Zimbabwe: Harare, August.

Van Rooyen, C.J., Nel, P. and Van Schalkwyk, H. 1994. Boerdery organisasievorme vir die toekoms: moonlikhede vir plaaswerkersaandeelhouding. *Unpublished paper.* Pretoria: University of Pretoria.

Van Rooyen, C.J., Nqangweni, S. and Njobe, B.N. 1994. The Reconstruction and Development Programme and agriculture in South Africa. *Agrekon,* Vol. 33 (4).

Van Zeyl, P. 1994. Perspectives on farmer settlement. *Unpublished Research Report.* DBSA: Halfway House.

Van Zyl, J. 1994. Market-assisted rural land reform in South Africa: efficiency, food security and land markets. *Agrekon,* Vol. 33 (4).

Van Zyl, J., Van Rooyen, C.J., Kirsten, J.F. and Van Schalkwyk, H. 1994. Land reform in South Africa: options to consider for the future. *Journal of International Development,* Vol. 6(2): 219–239.

World Bank. 1993. Options for land reform and rural restructuring in South Africa. P*aper presented at the LAPC Land Redistribution Options Conference.* Johannesburg: 12–15 October.

Annexe 18.1: International experience selection criteria: performance outcome

Age

Experience shows that those aged under 45 are generally the most successful. This positive outcome is correlated both with the physical vigour and the wider experience of individuals in this age group and with their stage/status in the household cycle. Couples of this age category tend to have children who are still resident at home, and who then assist in working the land. Some applications of age-related criteria have resulted in communities containing relatively young beneficiaries. Such communities have generally exhibited poor social cohesion and achieved dismal economic results.

Education

There is strong evidence that better-educated settlers are individually more successful than those who are less educated, and that better educated communities outperform those with lower average levels of education.

Family labour force

The agricultural and economic performance of households is strongly positively associated with the number of family members able to work, although little systematic work has been done to ascertain whether production per person is higher in larger families.

Farming experience and skills

A background which includes successful farming, business experience and acquired skills is strongly predictive of good performance. Those without skills and experience can usually eke out an existence on the land, but they seldom prosper and are vulnerable to take-over by other farmers or more powerful interests. They are also more likely to abandon their holdings.

Capital assets

There is no overall consistent evidence from international experience in favour of selecting those who already have capital or assets. It is clear, however, that different production patterns have different quantitative and temporal demands for capital, and meeting these demands is an important dimension of success.

Non-agricultural skills

Experience indicates that the more varied the mix of skills in existence, the more self-reliant and dynamic is the community. Important local multiplier effects are achieved when farmers and their families are able to purchase support services and inputs from within their communities.

Poverty

Experience shows that land-based programmes have generally been poor vehicles through which to address purely welfare objectives. If a poverty-related criterion is assigned the heaviest weighting so that uniformly poor people are selected, the outcome is likely to be the creation of highly dependant communities and the perpetuation of a paternalistic style of administration. Economic performance is also likely to be such that it fails to cover the costs of the programme.

Marital status

Married participants almost invariably outperform those who are not married. Marital status is positively related stability, social cohesion and superior access to resources, whether or not both partners in a marriage are resident on the land all the time.

No prior criminal record

There is no clear evidence of association between this criterion and outcome. Furthermore, there are obvious problems with the application of this criterion where the legal system has been employed as part of a strategy of political repression.

Nationality

No clear evidence of association with outcome.

An Envisaged Process of Land Redistribution and Provision of Farmer Support Services

LEVEL ONE: ENTRANCE LEVEL

The basis for having the Land Reform Programme (LRP) is in order to effect a restitution process. It follows, therefore, that at the first point of access to the LRP only the **victims of apartheid** who were *disentitled* and *dispossessed* should be able to qualify. Preference should be given to **South African nationals** and **rural residents**.

LEVEL TWO: PROGRAMME SELECTION

Restitution: Land Claims Court	**Redistribution: Administrative Access**	**Redistribution: Open market**
Through a legal process, the dispossessed may lay claim to specific pieces of land and also qualify for special assistance in accessing support services.	Through an administered programme (redistribution), the disentitled could be 'evaluated' and awarded points in relation to their extent of need, using poverty, number of dependants, status in society, and previous experience of farming. These points could be used to determine the extent to which they qualify for awards such as special assistance in the acquisition of land and access to services, specific land and/or some form of grant or cash. The emphasis in this phase is to help the poor.	Disentitled individuals, families, groups and/or communities may use level one support to enter the existing land market to acquire land.
Acquisition of land	Cash and/or granting of compensation	Assistance in the land market – through cash, grants or other means

LEVEL THREE: ACCESS TO SUPPORT SERVICES

Access to and provision of agricultural support services – including extension, training opportunities, credit, etc. This may happen on an individual, group or community basis. It would either be through direct access or purchase of the service, or by applying criteria such as age, health, net worth, education levels, gender, entrepreneurial skills and managerial attitude. Special assistance may be given to those in need. This could vary according to the levels of scores on a point system.

19

Providing agricultural support

Christina Golino, Stephen Hobson and Nick Vink[1]

Introduction

This chapter outlines the support services required in a land reform programme. These support services provided to beneficiaries of a land reform programme would be crucial to the success of the programme. Furthermore, the purpose of the chapter is to describe the current status with respect to support services to farmers and other rural dwellers in South Africa. In these fast-changing times, it is difficult to describe definite outcomes since these will necessarily become rapidly outdated. For this reason, the chapter starts with a brief description of the economic, political and social context of the rural areas. This is followed by a discussion of the provision of support services in the context of the Reconstruction and Development Programme (RDP) of the Government of National Unity.

The South African government is faced with the challenge not only of implementing new policies, but of simultaneously changing the institutions though which these new policies are to be implemented. Pages 508–510 describe some of the main issues that arise from this need for institutional restructuring.

Background

It is hardly necessary to describe the vast disparities in income, wealth and access to social services and economic opportunities that are found in modern South Africa. While the racial disparities receive most attention, it is arguably true that disparities between the sexes and between the different regions of the country are as large. Also, the majority of the country's poor reside in rural areas.

1 The authors are listed in alphabetical order. The views expressed are not necessarily those of the respective institutions, i.e. the Development Bank of Southern Africa or the Small Farmer Development Corporation.

Most of these inequities have been the intended and unintended result of past policies. In agriculture, a range of instruments (described in more detail in Chapter 3), were employed to support a model of full-time owner-occupied family farming for large-scale, white commercial farmers, to the exclusion of the majority of the rural population. This was done through a wide range of policy instruments aimed at excluding black (and also women) farmers from access to land; to financial services; to other support services in farm and non-farm rural enterprises; and to social and physical infrastructure.

This has resulted in a number of distortions in the production and employment structure of agriculture, the social fabric of rural areas, the rural non-farm economy and the wider agricultural-manufacturing complex, as reflected in the skewed income and asset ownership patterns. The rural poor, as a result, have had to adopt a wide range of survival strategies to cope with these adverse circumstances. These have had adverse impacts on the rural poor themselves as well as on the nation as a whole.

Two general statements can be made about the rural areas of the former homelands: first, that there is practically no viable rural economy; and second, that most rural households depend, either directly or indirectly, on the urban areas and state support for their livelihoods. These rural communities share the following demographic and socio-economic features (*Bekker & Cross, 1992 : 45*):

- Fertility rates far higher than those in the urban areas.
- Significantly low male to female ratios.
- Average population densities four times higher than rural areas outside the former homelands.
- Household incomes almost entirely dependent on exported labour, social pensions and salaries of civil servants.
- Subsistence agriculture which makes only a minor contribution to household incomes.

The unprofitability of agriculture and other rural enterprises and the unavailability of rural job opportunities, especially in the former homeland areas, forced poor households into a situation where they had to ensure access to a wide range of income sources in order to reduce risk. The primary strategies followed were to gain access to welfare payments, including social pensions, and to divert household labour to the areas of highest return which, in most cases, meant urban jobs for the able-bodied coupled with income remittances.

In more recent years, public sector employment in the administrations of the former homelands and more recently available options, such as enrolling children for supplementary feeding programmes, have substituted for some of these job opportunities. The result has also been that the country has experienced an abnormal demographic shift as reflected in high rates of rural population growth, abnormal migration patterns, and generally weak community institutions in rural areas.

The wider effects include corruption in the social welfare system, a disproportionate demand placed on urban areas and the public sector in terms of the need to create jobs, and a higher fiscal cost of social and physical infrastructure provision. A further result has been a skewed distribution of economic activity across the country. In the rural areas, this is reflected in unused and underused natural and human resources, and weak linkages between agriculture and the rural non-farm sector. Finally, the result has been a disproportionate emphasis on agriculture in the rural areas, to the detriment of other opportunities such as in small-scale processing and distribution, rural services provision and eco-tourism.

Redressing the past: the RDP

The Reconstruction and Development Programme (RDP) of the Government of National Unity has been designed to address the social and economic distortions in the country, including those in rural areas. It represents a shared vision of the people of this country, having resulted from wide consultations, first within the ranks of the erstwhile liberation movements, and later from intensive public and parliamentary debate.

One of the underlying assumptions of rural reconstruction is that equitable access to opportunities and to support is a necessary condition to address the apparent race, gender and geographic dislocations in rural areas, including the income and wealth disparities. Legal rights to equitable access have been included as a major component of the constitution and are underpinned by the RDP.

Rural economic opportunities include, inter alia, agriculture and related industries, tourism, other resource-based industry, commerce and trade. Support services include farmer and other business support, provision of physical infrastructure, social services provision and the institutions required to provide support. The design of these support services is also an important component of rural reconstruction programmes in South Africa. In the following sections we describe the current status with respect to these support services and institutions. This is informed by experience within the Development Bank of Southern Africa (DBSA), including the findings of the interim evaluation report of the long term DBSA Farmer Support Programme Evaluation (*Singini & van Rooyen, 1995*).

The role of land reform and agricultural production

The land reform programme is designed to increase income earning opportunities and decrease poverty for a number of previously excluded rural people. While the aim is to benefit as many rural households as possible, it is argued that the large majority of low income households will not directly benefit from greater access to land.

The major argument for land reform, from a food security point of view, is to achieve a shift in income earning opportunities in agriculture. It is crucial to understand that the rationale behind this argument is based on the expected increase in income, rather than on own production or increasing the physical supply of food. This implies that attention should not only be paid to the issue of land and production, but must also be focused on improving entitlements and the income earning potential of vulnerable people. Land redistribution should therefore be structured in a way that allows new entrants to increase their income levels and become financially viable.

In contrast to most other African countries, the bulk of rural household income in South Africa does not derive directly from smallholder agriculture. It is often mistakenly assumed that the self provision of food equates to security of food supply. People in these so-called rural areas cannot support themselves by subsistence farming alone. Research shows that a high percentage of rural households are in fact net consumers of food, even though many of them are engaged in food-crop agriculture. Sales of food are also highly skewed, with a small minority of households accounting for more than 80 per cent of sales (*Van Zyl & Coetzee, 1990: 112*). Food prices play a dual role in that they act as incentives to agricultural producers and form a major part of consumer expenditure. An increase in food prices will negatively affect all households which buy more food than they sell. Most poor people, under current rural conditions, will be further disadvantaged by an increase in food prices.

Remittances, pensions, salaries and wages are generally the dominant sources of livelihood in rural areas. While pensions are claimed by individuals, they are largely consumed within the household in the three-generation families in which most rural African pensioners live (*Lund,*

1993: 7). The pension pays for the household food and other consumable items, for education and health expenditures, and so forth. The reliability of pensions, and the associated access to sources of credit that they bring, further emphasizes their importance. This reinforces the view that the performance of the non-agricultural sector is as important for food security as that of the agricultural sector.

Farmer and business support services

Rural entrepreneurs require access to the following range of rights and services:

- **Legal rights of access.** The history of the Land Acts, as described in Chapter 4, shows the consequences for the affected individuals and for society as a whole of such prescribed barriers to access. The major strategies to address this injustice include the land restitution and land redistribution programmes, as described in Chapters 15 and 16; measures to allow the subdivision of agricultural land and measures to ensure security of tenure, especially in the former homeland areas. This includes proposals by the various provincial governments to vest control of state land in the local communities.
- **Input supplies.** The supply and financing of seasonal farming requisites and capital equipment to commercial farmers has largely been through private sector institutions. In most of the country, however, the immediate source of supply has often been agricultural co-operatives. These operated under a wide range of direct and indirect subsidies, and used a range of statutory and non-statutory mechanisms to exclude black farmers from access to their services. This had a particular effect in the grain-producing areas, where co-operatives were given regional monopoly powers to handle grain produce. These co-operatives were often the immediate source of credit to finance seasonal and capital inputs.

  In the former homeland areas, input supply and credit were mostly non-existent. Where these were provided, it was mostly through parastatal institutions, generally biased against female and small farmers. In addition, a large proportion of agricultural production in these areas came from state farming and state-supported project farming where input supply was arranged through public sector institutions.

  The different elements of the land reform programme (restitution, redistribution and security of tenure) all include measures to ensure access to input supplies in their design. There is a bias towards private sector suppliers where these exist in the commercial farming areas, and special measures in the former homeland areas. Specific programmes include investigations into the future role of agricultural co-operatives, provincial initiatives to address the future role of the agricultural parastatals, the co-ordination of input supplies to the beneficiaries of land reform and the wider Broadening of Access to Agriculture Thrust (BATAT) of the central Department of Agriculture. This programme, which is an evolving one, is designed to effect the policy, institutional and budgetary change required to ensure access to all support services for small farmers in South Africa.

There is also increasing recognition that the creation of rural livelihoods of sufficient quantity and quality requires support for other rurally-based economic activities. Measures to ensure that small business support programmes for the rural areas are given due attention are, for example, included in the recent draft White Paper on Small Business released by the Department of Trade and Industry, while attention has also been paid to the prospects for the eco-tourism sector.

- **Marketing.** The story of the liberalization of the highly controlled agricultural marketing system and of its effects is dealt with elsewhere in this volume. For our purposes, it is important to note that the system as it existed to the mid-1980s excluded all but the largest of the large-scale commercial farmers and, by that time, many of these supposed beneficiaries of the system had also become dissatisfied (*Kassier Report, 1992*). Farmers in the former homeland areas were often subject to an additional set of laws and regulations promulgated in those areas, as well as 'second class' access to the wider system.

  The extensive deregulation of the agricultural marketing system over the past decade has had direct and indirect beneficial effects on the commercial farming sector. However, farmers in the former homelands, at best, have benefited indirectly since the Control Boards have become increasingly unable to regulate the movement, processing and retailing of farm produce. To some extent, they have been able to exploit opportunities in the informal sector.

  A number of pronouncements on marketing policy by the ANC, and subsequently by the Government of National Unity (e.g. the draft White Paper on Agriculture and the land reform proposals) support both the further deregulation of agricultural marketing and measures to ensure more equitable access for small-scale farmers. Recently, the issue of the liberalization of international trade in terms of the GATT agreements has also received attention. A special case in international trade is the management of trade within the Southern African Customs Union, which agreement is currently being renegotiated, and in the wider Southern African Development Community. In both these cases, emphasis has been put on the creation of opportunities for small-scale farmers and processors, etc.

- **Research, extension and training.** Agricultural research in South Africa is done largely by the Agricultural Research Council (ARC), the provincial Departments of Agriculture where they have inherited the regional offices from the central department, the university faculties of agriculture, and some private sector suppliers of farm inputs. The ARC is a statutory body that has research stations in all of the major agro-climatic regions of the country. The extension function has been devolved to the provincial governments. Formal agricultural training is done at universities, the agricultural colleges in the provinces and by the agricultural development corporations. A wide range of NGOs also provide training in different aspects of agriculture.

  Agricultural research has traditionally been focused on the needs of the commercial farming sector. The extension service of the central Department of Agriculture also served the needs of these farmers. In recent years, however, the larger commercial farmers have increasingly turned to the private sector as a source of advice. The need to integrate the extension services of the former homeland areas into the new provincial structures has adversely affected what was already a weak service, at least in the short term. The result has been that there are serious institutional barriers that hinder the flow of information between the farmer, the extension service and the researchers.

Physical infrastructure

- **Bulk and reticulation infrastructure.** The provision of bulk and reticulation infrastructure in the commercial farming areas is generally adequate to service the existing large farm sector within current production patterns. The provision of reticulation of services (roads, water and sanitation, electricity and communication) to black rural dwellers is inadequate, specifically for farm labour and labour tenants. Bulk infrastructure was provided by national and provincial state institutions and utilities, and reticulation infrastructure by utilities in farm areas (at full cost recovery) and by local government authorities within rural towns.

In the former homeland areas, both bulk and reticulated infrastructure is generally inadequate, with major backlogs in roads, water, electricity and telecommunication. Where existent, physical infrastructure provision was based on population densities (often classified as displaced urbanization resulting from past policies restricting population movement), rather than on any perceived or potential production patterns. The result was inadequate infrastructure both for settlement and production needs. Infrastructure in these areas was provided by the erstwhile homeland governments and homeland utilities. Many of the latter have been or are in the process of being incorporated into the national utilities.

The institutional delivery system for infrastructure provision has changed with the new allocations of national and provincial functions, while demand will be derived from guidelines contained in a number of RDP programmes. Notably, bulk water will be handled by the central Department of Water Affairs, focusing on primary basic consumption needs. Responsibility for water reticulation rests with the local government level, and will be performed by the Department of Water Affairs where no local authority exists (as is generally the case in the traditional authority areas). The local government elections (held in November 1995), for the first time, involved all rural dwellers, including people in traditional authority areas and farm dwellers. The allocation of resources for infrastructure provision is therefore expected to be more fair, both racially and spatially, especially in terms of settlement. The lack of fiscal and human capacity in rural local authorities, however, is expected to place an additional burden on other service providers, and will be the focus of attention in the design of the fiscal transfer system in the short- to medium-term.

The land reform and restitution programmes, the impact on farm size resulting from changes to the marketing structure described above and other factors, will have a marked effect on the bulk infrastructure needs of the new evolving production systems. The commercial farm areas will arguably have higher population densities, while small-scale and more intensive production patterns will have different, and higher, infrastructure needs. The success of rural non-farm development programmes will intensify this.

In the traditional authority areas, population densities are not expected to decline (and might grow in absolute numbers), and settlement infrastructure needs will therefore remain high. Also, exploitation of the limited additional agricultural potential in these areas, as well as the promotion of non-farm entrepreneurial activities, can be expected to overstretch the resources allocated to infrastructure provision for the foreseeable future.

Little clarity exists on the future water allocation to the agricultural sector. The existing water allocations to all sectors are currently under review. With the emphasis placed by the RDP on water for human consumption, it is possible that agricultural water quotas could be cut. The impact of this on existing agricultural production, and especially on the new emerging small-scale farm sector, will have to be carefully assessed.

- **On-farm infrastructure.** The rules governing the provision and financing of on-farm infrastructure have varied according to circumstances, although it has generally been regarded as the responsibility of the farmer. Exceptions included the provision of on-farm infrastructure to plots under large-scale projects in the former homelands, and the wide range of subsidies for the erection of fixed improvements that were made available to white farmers.

  Given the changed structure of farm size that is expected to result from the land reform programme, it is expected that what has been regarded in the past as on-farm infrastructure will now, in effect, become reticulation infrastructure. However, as yet there is no clarity on the rules regarding the future financing of this, beyond the grant element included in the land reform programme.

Social Infrastructure

- **Safety-nets.** Lund (*1993*), Kruger (*1992*) and McLachlan (*1992*) provide comprehensive overviews of the South African social security system, especially in terms of welfare and pensions.

  Social security in South Africa consists of the state's welfare system, dominated by pensions; civil pensions (i.e. those to pensioned civil servants); a large private pensions industry; and occupational related social security benefits, including workmen's compensation, unemployment insurance (including maternity benefits) and provision for retirement through work-based pension and provident funds.

  Of these, only welfare has a major impact on rural dwellers. The major proportion of welfare goes to old age pensions, while disability payments, child and family care, and poor relief (which are all termed 'grants'), take up most of the rest.

In the past, recurrent droughts have resulted in ad hoc **drought relief programmes,** targeting the agricultural and mainly white commercial farm sector for assistance, to the exclusion of other rural dwellers. Up to 1993/94, the homeland areas were excluded from national drought relief programmes, leaving the homeland governments to institute their own actions. The 1993/94 drought relief programme saw the last minute inclusion of the former homeland areas, still targeted at mitigating impacts on agricultural production. The experience gained in this drought programme led to calls for improved early warning systems, and to the design of the NNSDP and the pilot public works programme instituted in 1994/95.

The programme that has been set up in recent years specifically for poverty relief is the Department of National Health's National Nutrition and Social Development Programme (NNSDP). The NNSDP came into being in 1991/92 to provide 'well targeted' relief to the poor, at the time when the General Sales Tax (which involved a large number of exemptions, especially for basic foods) was replaced by Value Added Tax. The NNSDP has had an annual budget of about R400 million. It has been a highly controversial programme, particularly in its inability to target, and in its urban bias.

The NNSDP, however, is dwarfed by the state's social security delivery system, with a current annual budget of around R6,5 billion, and also by the VAT relief on basic foods (which was conceded in 1992), valued at around R3 billion per year. The other relief programmes that affect rural dwellers are the Protein Energy Malnutrition Scheme (PEM), valued at R40 million a year, and the public works programme under the aegis of the Department of Public Works, piloted during 1994/95 under the National Economic Forum, with an initial budget of R300 million.

The importance of pensions to poor households in the rural economy cannot be over-emphasized. As well as feeding large families, pensions provide funds for all other expenditures, including education, some investment, and funds to provide subsistence for working age adults to spend long periods job-hunting. However, there are questions about the effectiveness of channelling relief to poor households through a mechanism of largely providing pensions to the aged.

The above programmes do not address the issue of the employable unemployed, by far the largest group, even after a full-scale land reform. It is argued below that there is need for a large, well-planned programme of public works. However, both land transfers and public works will need time for implementation. In the interim, both the NNSDP and the PEM require improvement; they need to be brought together, and both require the quick implementation of a surveillance system to assist in the identification of the poor and hungry.

- **Community and social infrastructure.** Schools, clinics and social amenities show much the same pattern of provision as physical infrastructure, with underprovision for black people, both in the traditional authority areas and commercial farm areas, and generally good provision for whites. Educational facilities for the children of farm labourers have been particularly poor since no secondary schools were allowed, by statutory provision, on commercial farms. In addition, farmers had to carry the full cost of the physical facilities of farm schools, with the now defunct Department of Education and Training subsidizing only the associated running cost.

  Housing delivery has also been problematic. The restrictions placed on the movement of people meant that, although very little state support was given for the provision of housing in the traditional authority areas, many households invested heavily in the stock of rural housing – given their relative lack of financial resources. In many cases, this places a financial restriction on people's future ability to move. On farms, a low subsidy was provided by the government to farmers to assist in the provision of housing for farm labour, generally resulting in very poor housing.

  Resolving issues around social infrastructure provision is central to many of the RDP programmes, as evidenced by the Homestead and Basic Needs grant component contained in the Land Reform Programme, the White Paper on Housing, the White Paper on Community Water Provision and the Extension of Municipal Services Programme, among others. Social infrastructure provision for farm labour is mostly linked to tenure reform issues and, although part of the debate, has not yet been resolved.

Special programmes

Apart from the extensive changes proposed and implemented in the education system, and the Presidential Lead Programmes within the RDP covering water, housing, literacy, capacity building, health, feeding programmes, municipal services, electrification, land and tenure reform and restitution and small farmer development, the Public Works Programme is seen as an important mechanism in urban and rural areas to address poverty and people's ability to enter into the mainstream of the economy.

A major national rural public works programme has the potential – without creating dependence – to achieve the following:

- By **absorbing labour** in constructive work, it would provide employment and income to those trapped in the cycle of poverty. It would do so in an enabling way, as opposed to a patronizing and dependency-creating approach. The estimate is that for the 1994/95 financial year, 440 000 work months of employment will be generated.
- Through its ability to draw unskilled labour into the programme, it would **disseminate valuable productive skills** and provide useful education and training to the participants. In this way, it would act as a catalyst in converting the involuntary unemployed into income-generating agents. This role can best be fulfilled if complementary policies reinforce economic growth to enhance the economy's labour absorption capacity. At present, R14 million of the current allocation to public works programmes has been made available for accredited skills training.
- The PWP could **generate socio-economic infrastructure**, both human and physical, indispensable for economic growth and development. Projects can provide infrastructure of a social nature which will contribute to an improved quality of life as well as a greater ability to work, (e.g. training and skill generation, building of houses, roads, schools and clinics; improved sanitation, water supply and electricity). Other projects supportive of economic growth include land development (erosion prevention, conservation), the extension and improvement of transportation facilities, and the establishment of industrial areas and other economic infrastructure – to give a few examples. Funds have been allocated to 512 programmes across the country, with an emphasis on rural programmes.

Implementation issues

Institutional change

Institutional change probably represents the biggest challenge facing South Africa at present. The new constitution mandates considerable institutional restructuring, and the RDP requires a concerted change management process. The agricultural sector is an integral part of this process of change, given the history of discriminatory policies and practices that have been applied in the past. These institutional change processes have to be accomplished at the same time as the restructuring of the service supply sectors described above. The following major issues have to be addressed:

- At the **central government** level, the most important issues include the presence of two ministries with responsibility over agriculture, where the latter has been described as a Schedule 6 function in the constitution (i.e. one over which the provincial governments also have competency). The two ministries are the Department of Land Affairs (responsible for the land reform programme) and the Department of Agriculture which regulates support services to farmers. In some instances, such as the provision of extension services, these powers have been devolved to the provinces, while in others, such as research and marketing, they remain a central government responsibility. There are also a range of parastatal institutions at central government level involved in agricultural and rural development that have to be restructured to meet the needs of the RDP. These include the Land Bank, the Development Bank, the utility parastatals, etc. In addition, a wide range of other ministries have a direct interest in rural issues, including the Departments of Water Affairs and Forestry, Environment, Housing, etc. The co-ordination of these functions, therefore, is an important focus of the office of the RDP where a rural development co-ordinating function has been established.
- The **provinces** have already gained control over a wide range of agricultural powers and functions in terms of the constitutional process of devolving these. However, the first order of priority has been to establish provincial administrations to deal with agricultural matters. The most intractable issues have arisen from circumstances such as the accountability of restructuring for a horizontal redeployment of powers and functions (e.g. between the former homeland administrations and the new provincial administration) as well as a vertical restructuring between the central government and the new provinces; the need for restructuring of the parastatal institutions; and the need for co-ordination with a range of new inter-governmental bodies such as the Commission for Provincial Government and the Financial and Fiscal Commission. While this process of institutional structuring at the provincial level has received much attention over the past year, it is already evident that it will remain on the agenda for some time to come.

- The process of **local government** restructuring has lagged behind central and provincial level restructuring. Rural local government has only recently started receiving sustained attention from policy makers as the country prepares for its first democratic elections at local government level. At least two issues have become critical to the success of this process. First, local government has also been defined as a Schedule 6 function in the constitution. The final structure of local governments, therefore, can be expected to differ between the provinces. For example, those provinces that incorporate the former homelands will have to accommodate tribal authorities, and will probably do so in different ways. These differences will probably also extend to details such as their sources of revenue. In the Western Cape province, for example, there is a longer history of rural local government. Draft legislation already makes provision for the reintroduction of an agricultural property tax. While such variation in outcomes is a healthy sign of experimentation, South Africa, in the past, has experienced the cost of the lack of a strategic framework to guide rural local government structuring. Second, there is considerable concern about the capacity of the majority of local authorities to manage and implement the kind of 'bottom up' development process envisaged in the RDP. While community based organizations (CBOs), in many cases, have shown their ability to assume an important role in this regard, there are concerns about their representativeness and even their ability to crowd out elected structures. This will place an additional burden on the other elements of the delivery system for RDP programmes and projects.
- A particular problem regarding the change management process refers to the **orientation, deployment and number of public sector employees**. Here the agencies of the state involved with rural development at all levels share a problem common to the rest of the country. The reallocation of functions between the various levels of government has to be accompanied by a redeployment, and often a reorientation, of staff. This process of change also has to account for the constitutional provision that guarantees job security for current employees of the state.

Programming

The success of the rural development programme is dependent on the ability of government at all levels to deliver in a programmed way, and perhaps more importantly, on the ability of rural dwellers to interact with the providers of goods and services, including government.

The service responsibilities are not only spread across various functional areas, but also between national, provincial and local levels. Planning and budgeting for a co-ordinated programme, therefore, becomes exceedingly difficult. In the Land Reform Programme, programmed co-ordination would be required between the national Departments of Land Affairs, Agriculture, Water Affairs, Education, Health and Welfare, Public Works; Eskom, and possibly two more utilities – Telkom and the appropriate Water Board; the provincial Departments of Agriculture, Local Government, Land Affairs, Rural Development (the last two functions are placed within various departments in the various provinces) and Public Works; the local government level, which currently comprises the transitional local authorities and the services boards; various private sector agencies such as co-operatives; and local service NGOs.

The above diffusion of responsibility increases the complexity of the interface from the point of view of the potential beneficiaries, the more so due to the lack of capacity, including a lack of experience in managing the development process at the local level. It is proposed that local capacity should be augmented by the appointment of rural development facilitators. The relative experience and seniority of such facilitators, however, would probably not resolve the interface problems from the beneficiaries' point of view. In addition, responsibilities are differently divided among government service providers at the local and provincial level across the country, which precludes a 'recipe' as a solution for beneficiaries.

The inclusion of women

As with any developing area, women, and especially rural women, have largely been excluded from the development process in South Africa. Even in the developed sector of society, women have largely been excluded from a meaningful role. The interim constitution emphasizes the role of women and, for the first time in the history of South Africa, guarantees legal equality for women. The historic exclusion of women, especially in the rural areas, exacerbated by the traditional authority structures, means that specific measures will have to be taken if women are to become real beneficiaries of land and rural reform. Measures to attend to this, such as selection criteria, training programmes, and the specific inclusion of women in the new local authority structures, are under discussion but have not yet been finalized.

Conclusion

Revitalizing the economy of rural areas while maintaining a sustainable household support system is likely to consist of three elements: first, the restructuring of the agricultural sector itself; second, enhancing incomes that are earned from non-agricultural activities; and third, addressing the current inadequate supply of services and facilities to the bulk of the rural population.

Land reform is an important component of this programme. At the same time, a dynamic rural economy with sufficient safety-nets is essential for the sustainability of the land reform programme.

Welfare and pensions form a vital component both of safety-nets and the land reform programme. They should not be viewed only as social safety-nets or as a political necessity, but also as a productive activity in terms of the role that they play in economic development.

The first safety-net is productive employment. Any other parts of the safety-net must be geared towards helping as many people into productive employment as possible (or at the very worst, not hinder that process by creating dependency). Many of those obtaining access to land will need preliminary assistance, for instance, until a first crop is reaped. They will also need infrastructure. These two requirements can be well married by providing the option of PWPs in newly settled areas, to build schools, clinics, roads, etc. The above recommendations on PWPs point towards a national programme, to which any organization or group could apply as it needs.

At the same time, and in recognition of the fact that many of the poorest will be unlikely to profit immediately from greater access to land, it will be necessary to support more than agricultural activity in the new areas, so that the poor may gain from increased employment in a variety of agricultural, formal and informal sector jobs.

References

Bekker, S. and Cross, C. 1992. The wretched of the earth. *Indicator SA,* 9(4).

Kassier Report. 1992. Report of the Committee of Inquiry into the Marketing Act. Pretoria: Department of Agriculture.

Kruger, J. 1992. An overview of the South African social security system. *Paper for the Workshop on Social Safety-nets.* Stellenbosch: Stellenbosch Economic Project.

Lund, F. 1993. Inserting social security between relief and development. *Paper presented at the Conference on Food Security in South Africa.* Johannesburg: National Consultative Forum on Drought.

McLachlan, M. 1992. Household food and nutrition security: how adequate is the safety-net? *Paper for the Workshop on Social Safety-nets*. Stellenbosch: Stellenbosch Economic Project.

Singini, R. and van Rooyen, J. (Eds), 1995. *Serving Small-scale Farmers – An Evaluation of the DBSA's Farmer Support Programmes*. Halfway House: Development Bank of Southern Africa.

Van Zyl, J. and Coetzee, G.K. 1990. Food security and structural adjustment: empirical evidence on the food price dilemma in Southern Africa. *Development Southern Africa*, 7(1): 105–116.

20

Restructuring rural finance and land reform financing mechanisms

Gerhard Coetzee, Masiphula Mbongwa and Kgotoki Nhlapo[1]

Introduction

The South African debate on rural finance has been put on the agenda of the new democratic order by the appointment of a Commission of Inquiry into the provision of financial services in rural areas (*Government Gazette, 1995*). In this chapter the rural finance debate will be reflected upon – specifically with respect to land reform. World-wide, the rural finance debate has changed considerably over the last two decades. It swung from a supply lending approach to a market-based approach. This view was somewhat tempered by arguments from the information economics cadres that argued for government intervention, albeit from a different departure point (*Besley, 1994*).

Similarly, the debate on land reform has been through major changes since the post-war period. Asian countries in the 40s and 50s went through the land-to-the-tiller land reform experience, whilst Latin American countries went through an experience which broke up large estates and distributed them to the landless poor, the state and small farmers (*De Janvry & Sadoulet, 1994*). The African experience of land reform was intimately linked to political independence.

The land reform debate in South Africa has also followed diverse routes. It changed from the nationalize-the-land debate of the 50s to the pragmatic perspective of the 80s which conceded justified state intervention within a market driven land reform programme (*Binswanger et al, 1993*). This view noted the complexities of land reform under South African conditions, where large commercial farms co-exist with low wage resident workers, external seasonal workers and high unemployment. Under these conditions, there could be no simple and single answer, but rather a combination of mechanisms that should be followed.

1 Although inputs in the form of ideas and discussions are acknowledged from colleagues, the responsibility lies with us. The usual disclaimers apply.

In this chapter a structure will be proposed to operationalize the demand set by the Reconstruction and Development Programme (RDP) (*ANC, 1994*) in terms of access to financial services for all. This structure will be based on certain guidelines that are discussed. The financial instruments of the land reform programme are also discussed and linked to the proposed overall structure. Land reform policy is based largely on the pilot project as proposed by the Department of Land Affairs (*1994*). Elsewhere in this book, a more detailed discussion is made of how land reform will be structured, how a market-based approach can be enhanced, etc.

Problem statement

The provision of financial support services to rural households in South Africa has to be seen against the background of past state intervention in the economy, characterized by distorted financial policies and institutional impediments. These led to a situation of extreme dualism in the rural financial sector. On the one hand, there is a highly modern and sophisticated financial system which serves the full range of financial needs of a small proportion of the South African population. On the other hand, micro-lending and an informal sector attempted to serve the majority of the population both in urban and rural areas.

These policies, such as subsidized credit, were also a major reason for the poor performance of South African rural financial markets. They distorted loan allocations by financial institutions. Low and negative real interest rate policies induced the commercial farmers to misallocate finance (*Coetzee, 1991; Mostert & Van Zyl, 1988*).

The distortions created a restrictive and unsustainable legal, financial and tax environment for micro-lenders. Legislation like the Banking Act prohibits micro-lending institutions to mobilize funding, whilst the Usury Act placed a ceiling on interest rates charged by micro-lenders with complete disregard for the high level of transaction costs associated with these institutions. Consequently micro-lenders failed to fill the void left by the formal institutions to cater for the majority of marginalized rural households. Their outreach and volume of financial services to rural people were therefore limited and small in financial terms.

The impact of these policies at an institutional level has resulted in lack of appropriate financial services in rural areas due to inappropriate delivery systems, lack of delivery systems, and inappropriate financial criteria. It resulted in the majority of rural people having limited, or no access to financial services by the formal financial sector.

Available land financing catered only for commercial farmers in the past. The Land Bank played a major role, with commercial banks financing a very small proportion of land acquisition. Another portion, also quite small, was addressed by the Agricultural Credit Board, whose efforts focused on farmers who did not qualify for assistance by the commercial banks and the Land Bank. All of these mainstream private and public financial institutions focused on white farmers. Recently, most of the public sector institutions have announced policy changes which extend their assistance to black farmers.

For its own purposes, therefore, the old order managed to make adequate and appropriate provision for the policy, legislation and institutional needs of the white rural and farming community regarding land purchases, agricultural growth and development issues. It follows that there will be problems when these old policies, institutions and legislation have to operate under a new constitutional order which requires different policies, legislation and administrative institutions to serve new rural clients. In other words, there is an urgent and pressing need for new policies, legislation and institutions to finance land reform and agricultural restructuring and development priorities of the new order.

International experience with targeted rural financing[2]

Financial intermediation in rural areas is more difficult and costly than in urban areas (for banks and clients) because of three inescapable rural characteristics:

- Spatial dispersion and the associated high information and transactions costs.
- Specialization of rural areas in a few economic activities linked directly or indirectly to agriculture which expose rural clients to the vagaries of the climate, pests, diseases and prices, and lead to co-variance of their incomes.
- Seasonality of production, with its accompanying sharp and opposite fluctuations in the demand for credit and deposit services.

If a rural bank operated in a single small area such as a group of villages, it would be able to sharply reduce the information and transaction costs problems associated with spatial dispersion. However, co-variance and seasonality would force it to operate with a large reserve ratio. This explains why traditional money-lenders in India only lent out of equity,

2 This section is based on the work by Van Zyl and Binswanger (*1994*).

rather than taking deposits. A high reserve ratio requires large intermediation margins to make such rural banking profitable.

Rural financial institutions use three ways to reduce the impact of covariance and seasonality on reserve requirements. First, they diversify their client base and loan portfolio out of agriculture into agro-processing and other rural non-farm enterprises. Second, they link their operations to an urban financial market, either through financial markets or by integrating the rural operation into a branch network which includes urban locations. Third, they set up inter-regional risk pooling mechanisms through networks or federations of individual rural financial institutions.

But inter-regional links, whether through branch banking, federations, or other risk pooling devices, still face special difficulties in supervising and monitoring operations of an individual rural branch or office. These difficulties are associated with the distance and fluctuations in branch performance that are induced by seasonality and covariance. Rescheduling of rural clients within a particular zone is occasionally required in order to tide them over years of bad crops or bad prices. This leads to opportunities for clients to collude against a single local institution, a branch or an entire system, which further increases the supervision problem.

Specialized farm credit institutions – the mechanisms of the conventional supply-led approach to rural credit – are especially poorly adapted to the difficulties associated with rural finance. Typically, they do not diversify their client base and portfolio inside the rural areas. They are not usually integrated into larger institutions with urban operations, and have limited urban diversification and risk pooling opportunities. Even with inter-regional risk pooling, they remain vulnerable to major droughts affecting an entire country or to international commodity price slumps.

Added to these inherent difficulties associated with specialized farm credit, institutions have often been used by states as conduits for carrying out agricultural and social policies, such as compensating farm sectors for excessive taxation of agriculture and for urban bias. This has led to a lack of autonomy. It is this lack of autonomy which has crippled these institutions rather than state ownership per se. In effect, financial markets were used to allocate subsidies rather than to fulfil their role of efficient allocaters of finance. Because the government has pursued social and agricultural policy objectives through rural financial institutions, these institutions have been particularly vulnerable to collusion by their politically organized clients.

Using specialized financial institutions to compensate farm sectors or pursue social objectives has mostly been futile and wasteful. The whole agricultural sector, and thus the rural areas, suffered from bad policies, bad prices and/or bad weather, and only a minority of better off clients had access to credit.

The case for targeted financial programmes

Poverty reduction objectives can justify targeted interventions and even subsidies in order to promote rural finance for particular groups of the population. However, such financial interventions must be cost-effective to achieve poverty reduction in comparison to other targeted interventions, such as small farmer extension, public works, or social safety-net programmes. Yet past subsidized and specialized credit lines through parastatal institutions have not fulfilled this condition.

Information and agency problems have been the focus of a rapidly expanding literature to justify targeted intervention (*see Graham & Von Pischke, 1995; Besley, 1994*). Much of that literature, however, applies equally to rural and urban financial intermediation. Therefore, these problems do not support a case for special attention to the rural system.

Externalities like possible risk diversification and nutrition benefits associated with the availability of improved rural savings, deposit and credit systems are another set of factors that are cited to justify intervention. Recently a rich theoretical and empirical literature on this has emerged, with lessons which point to alternatives other than intervention alone.

Infant industry considerations are well justified by the special inherent difficulties of rural, compared to urban, finance. After all, it is these special difficulties which lie behind the universally slow and painful emergence of rural banking as opposed to urban banking. The traditional policies of supporting rural financial sectors – special credit lines and subsidized interest rates which benefit the best-off rural groups – are neither appropriate nor cost-effective infant industry instruments. Specific and temporary measures need to be related to specific infant industry problems.

Special needs of institutions with strong involvement in rural areas arise from co-variance and seasonality which may require special attention from central banks, bank supervision bodies, the Department of Finance, private sector bodies such as associations of bankers, traders or producers, commodity exchanges, and bonded warehouse systems. These include adjustments of maturity lengths and grace periods to the agricultural cycles, the management of liquidity of the rural financial system, special programmes to assist in the diffusion of co-variant risks, and the management of regional or sector-wide crises associated with the climate or with prices.

The fiscal costs of rural financial policy mistakes or supervision failures have been exceptionally high in developed and developing countries alike. Such costs are usually off the budget and hidden until a crisis breaks out. In the case of the US, the Farm Credit System nearly collapsed in the early 1980s, generating fiscal costs and leading to the disappearance of nearly one-third of US farms.

Classes of rural clients

Most rural areas have rural classes which are illustrated in Figure 20.1. These groups are not clearly differentiated, and rural financial systems face great problems in differentiating between them. They are bankable large commercial farm and non-farm enterprises, commercially oriented small farms and businesses, the bankable poor, and the non-bankable poor.

Bankable commercial farm and non-farm enterprises have access to conventional collateral such as land and real estate, and movable and financial assets. They are thus served by commercial banks. Commercial small farm and non-farm enterprises have volumes of market operations which justify the use of formal financial intermediaries on the savings and the borrowing side. They usually have investment options which also justify borrowing at positive real interest rates similar to commercial bankable large farmers. The volume of their businesses, however, is too small to make banking with them profitable for commercial banks. In order to get access to financial services, these small commercial entrepreneurs often organize themselves into savings and loan associations, credit unions or co-operative credit societies. They use some form of group solidarity to reduce supervision and monitoring costs of the financial intermediary. They carry some of the costs of intermediation themselves. These organizational structures are designed for risk pooling.

The bankable poor are often in the subsistence sector with small commercial sales. The level of their savings and loan demands is very small. Banking business with them has very high transaction costs. Consequently, they are not served by commercial banks. Their access to finance is usually through informal lenders. Informal lending is characterized by very low loan amounts, short maturity periods and very high interest rates. Rarely are assets required as collateral.

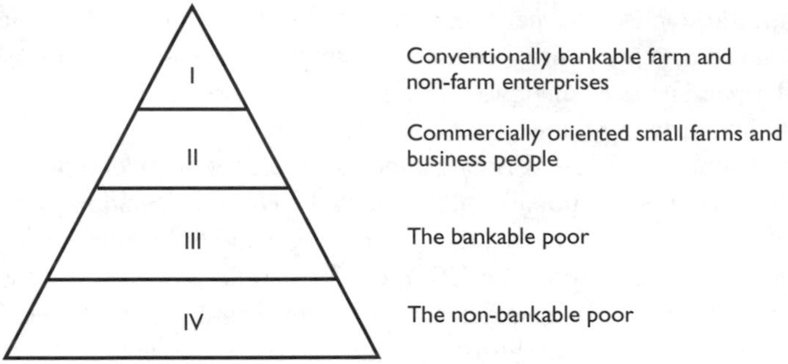

Figure 20.1: Rural Clients of Financial Institutions

Experience of the Grameen Bank (Bangladesh), the Bank Rakyat (Indonesia), and other similar schemes, show that banking facilities with a relatively low subsidy can be successfully organized for these poor groups. These financial organizations usually employ the following principles:

- Personal reputation.
- Focus on the truly poor, especially women.
- Emphasis on deposit mobilization.
- Group formation and group solidarity.
- Transparency of transaction to all members of the group.
- Group-internal assistance in case of emergencies.

The non-bankable poor include very poor farmers on marginal land who have little investment opportunities with sufficient return to make borrowing attractive, even at interest rates commercial banks offer to their good clients. Such groups need assistance on a partial or matching grant basis. Often such grants can be made to groups for group projects. They usually invest their own labour and local raw materials. Considerable capital formation can be induced in this way. Another alternative is to provide these groups with a revolving fund which can be lent to members for the purchase of inputs and is repaid to the group for use in subsequent seasons or for a group project. Interest is not usually involved; the donor or the state does not expect repayment, and members can repay their group in products or sometimes even in labour.

Objectives and instruments of government intervention

The overall objective of rural financial policy is to make banking facilities available to the rural sector at costs that are only slightly higher than corresponding urban facilities. It is also to assist in the graduation of the groups to higher levels of wealth and bankability. More specific objectives are to:

- create and supervise a set of financially viable institutions which must be capable of inter-sectorally and inter-regionally managing the seasonality problem and the co-variant risk inherent in rural banking, without periodic recourse to massive infusion of government funds;
- create enabling policies and regulations to encourage informal lending credit, so as to reduce its cost and to increase the degree of competition within the informal sector;
- foster the growth of rural commercial and co-operative banking so that more bankable rural clients have access to remunerative deposit and savings facilities, and to credit at the lowest possible intermediation margins;
- support private or non-profit institutions which serve the bankable poor with zero or low rates of subsidy dependence. These support measures should not be through interest subsidies, but through infant industry established grants and transaction cost subsidies;
- support those government, private or non-profit institutions which assist the non-bankable poor with asset creation, income transfers, resource conservation and better health and nutrition.
- take optimal advantage of complementarities between the provision of financial services and technology diffusion;
- take advantage of opportunities to foster environmentally sound investments where this can be done with the assistance of the rural financial system.

Policy instruments related to the above objectives are:

- development of a rural financial policy analysis and supervision capacity in the relevant government institutions;
- strengthening banking supervision in the rural financial sector for commercial banks, co-operative institutions, NGOs, etc;
- promotion of risk sharing in the system;
- security of contracts, enacting legislations, establishing informal and formal dispute resolution mechanisms, and a functioning court system;
- formal collateral: land, crops and stocks, moveable assets, financial assets;

- collateral substitutes: interlinking, rotating credit societies;
- promotion of appropriate maturity lengths for agriculture: seasonal, and for investment in animals, machines, fruit trees, forestry, farm real estate; promotion of savings and of graduation: through support to commercial banks, and the co-operative sector, and by improving the bankability of the poor; special programmes for the non-bankable poor;
- promotion of technology diffusion through financial institutions;
- promotion of environmental assessment capacity in financial institutions.

The criteria for government intervention in rural and agricultural credit markets are clear. Direct government credit to farmers and non-farm rural entrepreneurs should only occur when there is absolutely no chance of the private sector extending credit to these clients. However, if government find it necessary to provide such credit, the target beneficiary group should be well and narrowly defined. The duration of the loans should also be clearly specified, and conditions of the loans should not create distortions in credit markets or create disincentives for paying back the loan capital. Specifically, interest rates should be market-related and the loans should rather concentrate on reducing the transaction costs of getting the loan. In this respect, the possible role of matching grants could also be explored to address the lack of sufficient collateral.

Rules for making rules

In earlier papers (*see Coetzee et al, 1993; Coetzee et al, 1994*) on the role and activities of the public sector in agricultural finance, emphasis was placed on policy proposals, such as the state's role as a wholesale financier and how the state should apply subsidies. With the Rural Finance Commission of Inquiry underway and most financiers currently rethinking their approaches, guidelines for policy making become an important contribution to be made. These guidelines (or tests for policy) are mostly based on the work of Beghin and Fafchamps (*1994*) who propose a simple taxonomy of institutions and governance concepts applicable to the comparative analysis of the political economy of agricultural policies.

The first question to be answered is how to bring about good institutions? Good institutions will lower transaction costs and increase efficiency in economic exchange. Further, good institutions are based on clearly defined property rights and uncomplicated contract enforcement. This, in turn, is only possible if the correct information is available and the legal system is amiable. Thus two tests are evident, i.e. are property rights secured, and is the rule of law enforced? In essence, we are looking at clarity in terms of benefits of programmes and enforceability of the law related to these programmes.

Further, state programmes should be both implementable and predictable.[3] As stated earlier, the people should be clear on the rules of the programme and who the beneficiaries will be, and the programme should be implementable in practice, for example, it should be in tune with the constitution and be implementable in terms of the capacity of institutions and the training of people. Thus it should be realistic.

The willingness of government to let the market discipline work and to ensure the mobility of resources (economic openness), the contestability of political markets and public service provision, as well as the transparency of the policy-making process and the participation of pressure groups (political openness) relates to openness in general. Openness is one of the underlying concepts of good governance.

Related to some of the above concepts, for example implementability, one can also test policy based on whether it 'fits' the constitution. In practice, this also relates to whether it is implementable within the central/provincial structure and whether it addresses the demands of the RDP (*ANC, 1994*).

If our aim is to achieve efficient institutions, what could be the role of government to reach this aim given the above broad guidelines? This in essence goes to the core of two views in economics on the role of government. Some economists argue that by understanding the causes of market failure (symmetric and asymmetric information), we are led to the conclusion that there is little scope for efficiency enhancing government interventions (*Besley, 1994*). Others argue that the recognition of transaction costs and imperfect information justifies the traditional role of government, and thus intervention (*Stiglitz, 1989*).

However, in practice we have examples of the outcomes predicted by both approaches. In terms of the first view, i.e. government intervention where the market already failed, we can use the conventional approach to farm credit programmes and the failure of this approach as an example (*Coetzee, 1988*). For the alternative view and in rural financial markets,

3 Begin and Fafchamps (*1994*) use the concepts of predictability and enforcement in relation to the rule of law. Our view is that it should apply to all state actions.

only a few examples are known (e.g. Bank Rakyat in Indonesia) (*Chaves & Gonzalez-Vega, 1993*). For the latter view, a collection of experiences serves as a basis for arguing why and how governments should intervene in rural financial markets.

Further, government can intervene indirectly by facilitating transactions between different agents in the economy. In agrarian communities, local agents have superior information about the expected behaviour of community members (*Fafchamps et al, 1994*). This information allows them to reduce adverse selection and moral hazard, but they are constrained in their ability to diversify risk, and more potential exists for enforcement problems due to close relationships with local power structures. One way to solve this through government action is the linking of external and internal agents, thus solving external agents' information problems and internal agents' risk in terms of managing and enforcement problems.

Fafchamps *et al* (*1994*) provide several examples of how the state can intervene on behalf of competitiveness. These examples range from state intervention in the incentive of gains based on institutional change; intervention based on the ability to provide information superior to that of civil society; redistribution of assets and the redefining of property rights; investment in infrastructure, education and technology (*Thompson, 1994*) in order to reduce transaction costs and to oil the working of the market; the diffusion of information; and the decreasing of risk for poor families, for example, by food for work programmes.

Thus opportunities are plentiful, but these interventions can easily backfire through rent seeking and mismanagement. These interventions also need to navigate the political economy of political feasibility (*Fafchamps et al, 1994*). This in itself is a challenge to any government.

Targeted financing in South Africa and the role of the current state agricultural financiers

The current range of state financiers of agriculture (see Table 20.1 and Figure 20.2) are mainly still captured by their traditional roles. They operate within an uncertain environment with respect to the future role of specific organizations. Policy declarations and discussions, however, reflected the willingness to change. The whole system still emphasizes credit as an instrument since few state agricultural financiers provide any financial services other than credit. However, concepts like market-related interest rates are now more easily accepted by these financiers.

Table 20.1 provides a summarized overview of the role players in this regard. Although these actors targeted specific geographic areas and racial groups, the recent political changes resulted in increasing the number of state actors in the same market. Differentiated approaches result in confusion of the clients, especially with respect to levels of interest rates. At the moment, the interim constitution clashes with all agricultural financial legislation (*Olivier, 1994*). The provincial nature of agriculture does not fit comfortably with financiers who operate from a national basis as retailers. Most state financiers of agriculture were removed from their clients, and from information on their clients in the past; this is even more true now. Political changes added clients to the clientele of some financiers who did not service them in the past. This results in severe information problems. The combination of effects necessitates a restructuring of state support to agriculture.

The targeted programmes from public sector specialized credit institutions in South Africa have had negative unintended consequences on the prices of agricultural land (*World Bank, 1994*). The low interest rates were capitalized into land and manifested themselves in the artificially high prices of agricultural land. The gap between the market and the productive value of agriculture made entry by new and young farmers prohibitively expensive.[4] Those who already owned land or had greater wealth tended to buy agricultural land, especially in the 1970s. As a result, the distribution of landownership became more skewed, even among the white rural population.

The character and performance of the agricultural sector was also adversely affected by the subsidies and grants. The mechanization of the sector, which began in earnest in the 1940s, was initially accompanied by rising employment, cultivated area and output. However, this changed by the 1970s, when increasing mechanization was accompanied by sharp drops in agricultural employment. The social and economic dislocation impact of that process included the massive eviction of farm workers, and growing cases of rustling, unrest and insecurity in the commercial agricultural sector. The increasing cultivation of marginal areas worsened environmental and ecological deterioration (*World Bank, 1994; De Klerk, 1985*).

Financial institutions lapsed in their application of strict financial management discipline, whilst farmers failed to invest wisely and prudently. It is the public, however, that has borne the brunt of these adverse effects of huge subsidies and grants to the white commercial farmers. It

4 See Chapter 12 on the gap between production value and market value of farms, and the effects of the phasing out of numerous subsidies on land prices and the production and market value of agricultural land.

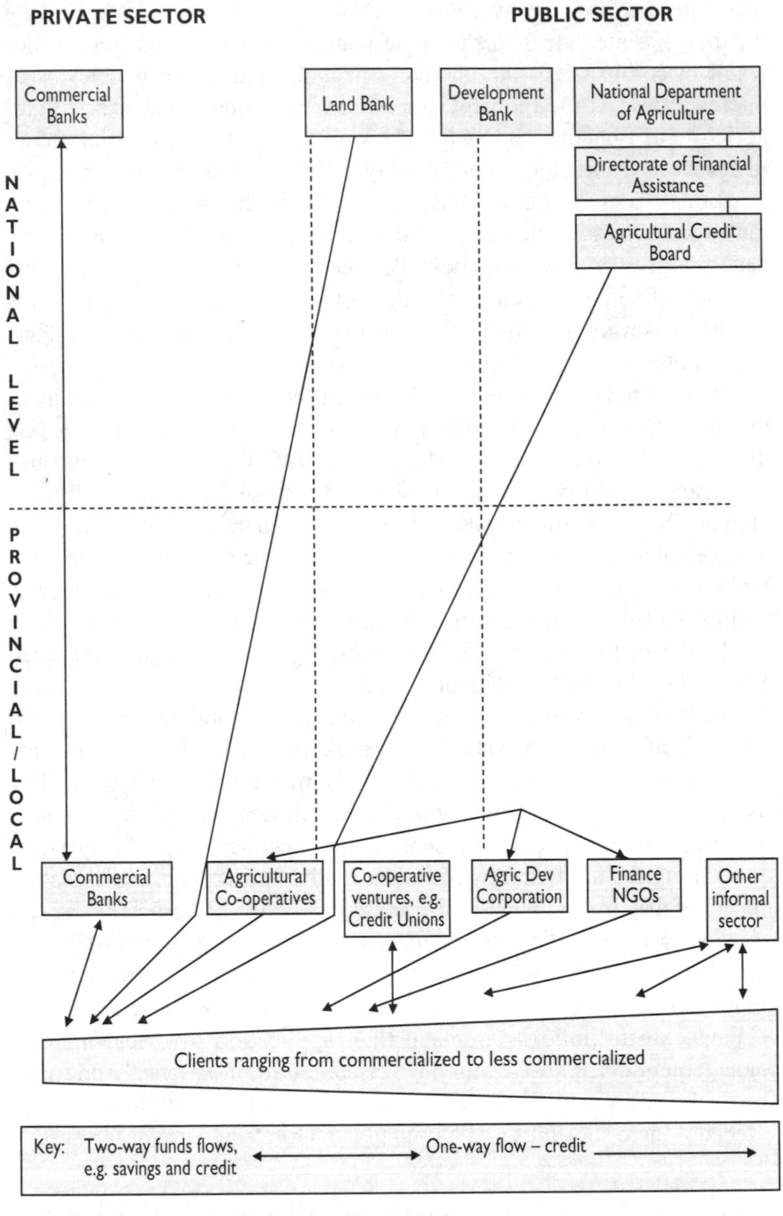

Figure 20.2: Past and Current Structure for State-supported Agricultural Finance in South Africa

is estimated that the R1,5 billion of the debt owing to the Agricultural Credit Board cost the taxpayer about R205 million in 1994 alone (*World Bank, 1994*). In fact, loan losses and provisions made for doubtful and bad debts, actually caused the taxpayer to suffer much more than R205 million. If R6 000 is the annual minimum income level for a rural household, then about 30 000 households and a total of nearly 200 000 rural people could have been sustained by the redirection of that subsidy.

Future financing structure

The problems associated with the current structure, the implementation of the interim constitution, and the lack of access to formal financial services by the majority of the rural population, necessitate a thorough study of the system and public sector policy. The Rural Finance Commission (*1995*) emphasizes four areas of investigation, i.e. demand for financial services in rural areas, institutional aspects, legal implications and land reform financing mechanisms. This Commission plans to report its findings by the end of 1995. In the interim, the land reform pilot projects are being implemented. Further, the nine provincial Departments of Agriculture and the National Department of Agriculture are busy with the restructuring of services, including financial support.

Table 20.1: Current Role Players

Organization	Source of funds	Finance function	Target
Public sector SA Land and Agricultural Bank Regional agricultural banks DBSA Departments of Agriculture Agricultural Credit Board Other Departments? Development Corporations	Public sector; National capital market; Foreign capital markets	Land; General agricultural development; Emergency aid	Specific institutions targeted specific farmer groups
Private sector Commercial banks Co-operatives NGOs Moneylenders Family and friends Other financiers	Own funds; Deposits; Investors; Loans; Levies; Service fees	Range from short- to long-term credit; Production credit	As decided by the institution

Institutions in the provinces (agricultural development corporations and agricultural banks) and at national level (e.g. the Land Bank and the Development Bank) are being studied with the aim to restructure and rationalize. Thus some broad guidelines are needed right now. These policy guidelines are needed to point all rationalization and restructuring activities in the right direction. Specific structures need to be identified to accommodate the land reform activities.

At this time the only way to reflect a view on the future structure of public sector agricultural financing is to look at the convergence of strategic views. These views are the culmination of discussions and analysis both by the state agricultural financiers and private sector financiers. This view is also based on the discussion of the guidelines in an earlier section and the problems identified with the current structure.

Constitutionally, agriculture is a responsibility of the provincial governments. In practice, all related services provided by government will also be implemented at provincial level. It therefore makes a lot of sense for financial intermediation to be a ground level private sector responsibility and, where markets are found not to exist, or market failure could be argued, a role can be considered for the state. Should enough justification be found for state intervention, it calls for a state-supported financing mechanism at provincial level. No comprehensive analysis of demand for financial services in rural areas is available. By relying on survey information, documentation of informal sector activities and experience of current state financiers, a qualitative estimation of demand indicates that the state should intervene.

The policies of intervention need to be clearly stated. These policies, based on international experience and South African evidence (*Task Team on Agricultural Financing, 1994*), call for a facilitation role for the state, rather than a direct intervention role. This translates into a state-supported financial mechanism at provincial level.

According to the new views on the method of intervention (see *Coetzee et al, 1993*, for a discussion of these views), this should take place in the form of a financial intermediary which provides a range of services (savings facilities, credit and transmission services) across a broad range of sectors. This intermediary should be both wholesale and retail. The wholesale function would accommodate the links between ground level information sources and the intermediary; a retail function with larger clients would also apply. In this way, the intermediary can manage risk by having a diverse range of clients in terms of scale and activities. Savings mobilization must be emphasized by these intermediaries, and support in the form of subsidies from government should not be applied to decrease interest rates, but rather to decrease transactions cost and enhance infrastructure and the capitalization of the institution.

In Figure 20.3, the financial mechanism (FM) is depicted at provincial and local level. The relationship with the national level is in terms of financial support flowing to the financial mechanism, and policy interaction that flows with the finance. This FM is planned as an autonomous agent of the state. Distance is needed between the FM and the state to decrease the risk of nepotism and corruption. Distance is decreased between the FM and the clients to increase client participation and information. The FM interacts with a wholesale finance function at national level.

In this way, activities can be co-ordinated horizontally: at the wholesale level nationally, and at the retail level provincially. In this regard, at the national level, institutions are needed that will fulfil the financing functions as described earlier. Part of the wholesale development support function at the national level will be a financier that cuts across sectors, financing the enabling environment creator role of the state in terms of the demands of the RDP. The specific structure at national level still needs debate and analysis in South Africa. This is under the jurisdiction of a Committee, under the control of the Deputy Minister of Finance, that will report shortly.

The important criterion at national level is co-ordination with the different activities of the state. Thus, a programmed approach to the delivery of reconstruction projects is essential for success. Without a programmed approach, the danger exists of emphasizing the dual character of previous structures. The sources of finance for the wholesale function at national level could be from the state (departmental transfers and RDP fund), from donors, and from the national and international capital markets. The wholesale function can be one institution, or a collection of institutions, as long as the programme approach is adhered to.

The illustration in Figure 20.3 needs to be qualified and further explained as follows:

- Although the illustration is agriculture specific, it can be extended to provide services to all who are not serviced by the private sector in rural areas (either by facilitation of the private sector or direct intervention in the market through the FM). Here the assumption is that servicing diverse sectors will decrease risk exposure to one

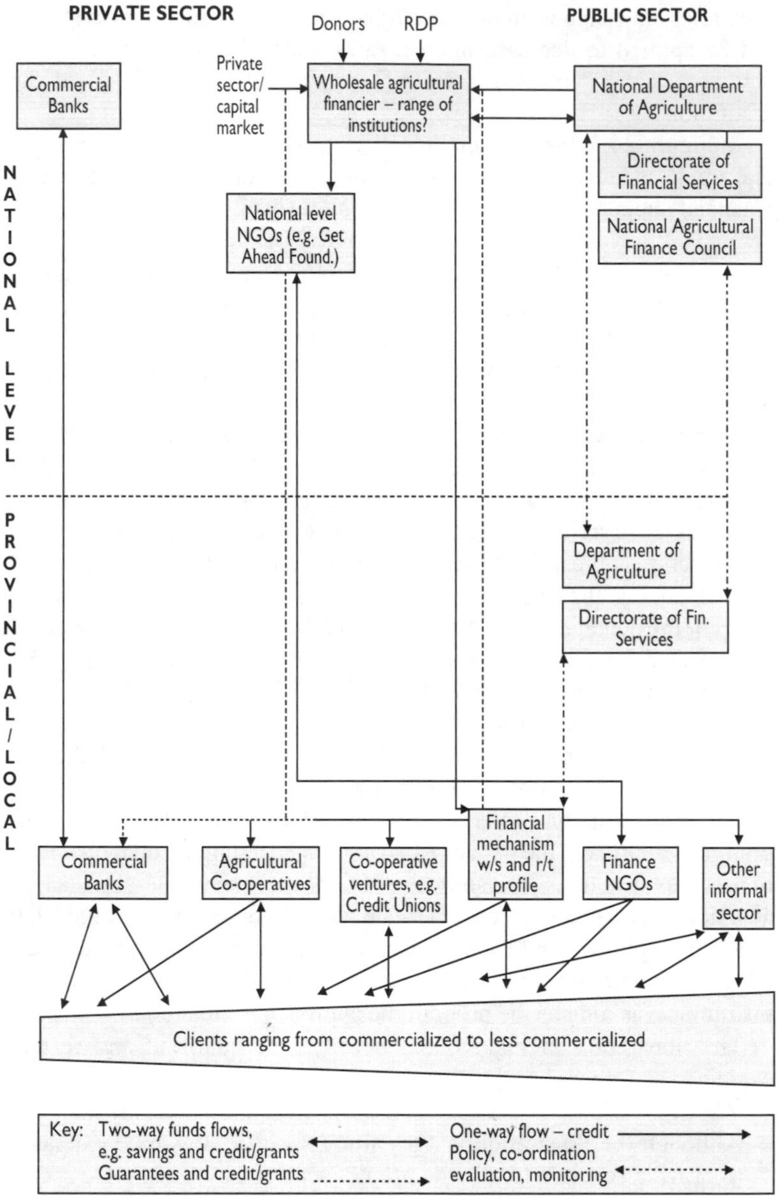

Figure 20.3: Proposed Public Sector Support to Agriculture in South Africa

sector by the ability to pool risk. This is extremely important in a country where agriculture is exposed to such a variable and hostile climate.

- It is important to note that several government departments (the RDP office, Department of Lands, Department of Agriculture, Department of Housing, Department of Trade and Industry, Department of Post, Telecommunications and Broadcasting, Department of Health, Department of Public Works, Department of Water Affairs and Forestry, to name a few), at this moment are rethinking their services and service structures in line with the RDP and the constitution. Thus, co-ordination and the programmed approach are extremely important.
- Figure 20.3 also accommodates potential political tension between National Departments and Provincial Departments. In the case of agricultural finance, this is addressed by the proposed creation of a National Agricultural Finance Council. The Council is formed by representatives from the Provincial Departments and the national department. The function of the Council is twofold. It makes national policy on state financial support to farmers, and it co-ordinates provincial policy on the same issue.
- The role of the agricultural finance sections (Directorate of Agricultural Financial Services) is to co-ordinate, monitor and evaluate at the appropriate level.
- Financial services at the ground level also include services other than credit. In Figure 20.3, this is depicted by the two-way funds flow, i.e. savings and credit, between clients and financiers. In this regard, policy change would necessitate changes to financial and agricultural legislation in South Africa.
- The clients are not depicted on any progression scale between the different financiers. Here we assume and believe that, given flows of information, the absence of policies that use financial markets as subsidy transfer mechanisms, and the autonomous operation of FMs on business principles, clients will find their way to those institutions that will best cater for them.
- An important consideration here is whether the proposed structure could be used to accommodate more than financial services alone to farmers and small business people. What about housing and land reform, for example? The issue of land reform financing will be addressed in the next section.

The application of the discussed rule here would test such a structure. Does it fit constitutionally? Is it implementable? Is the environment conducive in terms of clear laws, and economic and political openness? As stated by De Gorter and Swinnen (*1994*): Is the legal, regulatory and institutional framework of such a nature that it enables the working of government policy, and more important, the activities in the market between agents?

Land reform financing mechanism

Overview of the land reform programme

The land reform financing mechanisms discussed in this chapter are those that are explained in the policy document on pilot projects (*Department of Land Affairs, 1994*). As such, only the financial instruments highlighted will be discussed. These instruments are:

- Planning grants that will be granted by the District Officer for the purposes of planning for land acquisition, settlement and development. Planning will be executed by persons preferred by the community.
- Survey and transfer grants to cover costs related to the transfer of ownership.
- The land acquisition grant for acquisition of productive land on the basis of the Rand cost of land. This grant ensures that the poorest members of the community have access to land, while community members with their own resources can acquire more land. This grant goes to individuals. The pilot project policy document states that these grants will not work if additional credit is not available.
- The homestead basic needs grant, along the same lines as the urban housing grants.
- Access to short- and medium-term credit – important to the beneficiaries of the land reform programme.

Land reform financing mechanisms within an overall structure

Certain criteria specific to the land reform programme are needed before any proposals are made about financial mechanisms to accommodate land reform.

- The mechanisms should be linked to the proposed structure and should adhere to the same guidelines and principles.

- The mechanisms should be flexible to accommodate the pilot nature of the land reform programme.
- The proposed mechanisms should be in line with the aims of the land reform programme (*see Chapter 22; Department of Land Affairs, 1994*). In this regard, the aim concerning the testing of the most appropriate financing mechanisms for planning, land transfer and infrastructure delivery apply. The guidelines here are that the chosen mechanisms should include considerations both of macro-economic affordability and replicability, as well as sustainability and affordability at community level. The intention is 'to ensure an efficient and accountable allocation of funds in a manner that makes a significant contribution to improving rural incomes, increasing household food security, and stimulating economic growth' (*Department of Land Affairs, 1994:4*). This aim translates into providing a service in such a way that transaction costs both for provider and beneficiary are minimized, and the provider can continue to exist. At the same time, the financial mechanism must be accessible to the beneficiaries of the land reform programme.
- The financial mechanism that provides financial services to land reform beneficiaries must work in close co-ordination with Land Reform District Offices and their staff.
- Duplication of funds flowing from the state should be identified and corrected. Here, co-ordination with other departments is important. Retail financial structures supported by the state should not be duplicated.

The proposed national structure should be able to accommodate the needs of the Land Reform Pilot Project. If the proposed structure at provincial and local level is accessible to rural people, it will also be accessible to land reform beneficiaries. Over and above the grants, the land reform beneficiaries would access the normal services of the FM and other intermediaries. Social and physical infrastructure is the responsibility of the line departments in the province. It is here where a programmed approach, co-ordinating the aims of the land reform programme, the infrastructure needs, and the short- and medium-term financial needs, would be very important. This leaves two important aspects – the physical pay-out of grants and the selection of beneficiaries. With respect to the latter, the FM has to leave the selection of beneficiaries to the District Land Office. Payment of grants, therefore, would be on advice from the District Land Office. It must be ensured that the perception is not created in clients that the FM provides grant finance in general. If this becomes the case, it will threaten the existence of the FM. Land reform beneficiaries with access to land would be able to access production loans and other services like all the other clients of FMs. Even additional land financing and the financing of buildings could be accommodated by the FM. This has to be approached with care since the payment of subsidies would taint the business approach of the FM.

The question of phasing in the system while the appropriate FMs are not yet structured needs to be answered. One way to achieve this is to consider options both of an interim nature (thus to be implemented over the short-term) and which will fit in with the long-term proposals. Four possibilities exist:[5]

1. **Commercial bank funding.**
   Considering the nature of the participants, it is likely that the loan granting mechanism would need to be flexible and not require excessive collateral security. It is thus unlikely that commercial financing institutions, including the Land and Agricultural Bank as currently constituted, would be willing or able to enter the land reform programme on any substantial scale in the short- or even the medium-term. However, participants should have access to these institutions in the normal manner, and these institutions should not be prevented from doing their ordinary business with current and prospective clients.
2. **Special state support to the Land and Agricultural Bank.**
   As an interim funding mechanism, public sector funders could provide a loan or partial guarantee (covering development risk only) to the Land and Agricultural Bank for on-lending to groups and individuals under the pilot programme. Conditionalities attached to these loans to participants would have to be negotiated between the Department of Land Affairs, the lending institution and the Land and Agricultural Bank.

   Given the financial strength of the Land and Agricultural Bank, direct loan funding might not be required. However, if a direct loan is extended, a clear understanding needs to be established on repayment of the bulk loan as soon as the rural and land financing system has been finalized. The same would hold for the phasing out of a guarantee.

5 These are drawn from the work of Golino (*1995*).

The Land and Agricultural Bank criteria on an application for loan finance are essentially the same as those applied by the commercial banking sector. In contrast to the commercial banking sector, however, the Land and Agricultural Bank is prepared to take livestock, in addition to other movable assets, as collateral. Although it has a large (and very strong) loan book, the Land and Agricultural Bank has no experience in development risk, selection of farmers on criteria other than financial collateral, or issues surrounding group farming. It is also doubtful if the Land and Agricultural Bank currently has the staffing capacity to handle the additional anticipated workload which will be generated by the land reform pilot programme.

3. **The creation of a dedicated wholesaler trust fund.**
Given the problems described above in terms of providing loans or guarantees to the Land and Agricultural Bank, an alternative proposal is to create a dedicated wholesaler trust fund. This fund would be the source of a credit line for the interim phase, and could be transferred in toto (both assets and liabilities) to the appropriate institution at the appropriate time. This alternative appears to be clear of the confusion created by the first alternative in (2).

The Development Bank could provide the appraisal and monitoring capacity for such a trust fund on an agency basis on applications received from groups and individuals, and could assist in the identification of an appropriate intermediary, if that is not already in place. Examples of such intermediaries could be commercial financial institutions, provincial parastatals, NGOs or any local government organization. The requirements for such intermediaries would all be the same. They would have to be prepared to take normal business risks and to operate within normal commercial interest rate criteria. It is important to note that, from the point of view of the retailer of funds, duplicate systems with different conditions attached become impossible to handle. It has to be stressed that this option will require intensive negotiation between the community and the intermediary on issues surrounding non-payment.

The drawback of this proposal, however, is that direct applications from communities, where appropriate intermediaries cannot be found, will not be able to be entertained.

4. **The creation of a dedicated retail trust fund.**
The fourth and probably the easiest option for the short-term, would be the creation of a trust fund as above, but to act as a direct retailer to beneficiaries. The fund would carry the risk, and could contract a

commercial bank to provide the banking services. In this case, the issues around non-payment would need to be conditional to the title deed.

Appraisal and, if necessary, monitoring capacity can be implemented in the same way as for the wholesale trust fund proposal.

Financing functions of the private sector

In principle, the public sector cannot directly instruct the private sector where, how and who to finance. By providing an incentive structure, however, the state can influence the private sector to provide services in circumstances where, without state assistance, the private sector normally would not be willing to intermediate, due to perceived risks. The state can achieve this, for example, by providing equity financing mechanisms and guarantee schemes. This should be approached cautiously, as inadequate structuring of schemes may lead to inefficiencies.

In essence, commercial banks state the prerequisite of the profit objective in a legal environment. They acknowledge their role in socio-economic development, but it should still adhere to the profit motive. A careful analysis of the financial behaviour of rural dwellers (mostly black rural dwellers)[6] indicates that a small segment is already in a position to qualify as commercial bank clients, given their stated criteria. It is perceived that commercial banks themselves do not possess enough information on this segment to structure and provide services. The state, in its role as the creator of the conducive environment, should address this information need of the private sector in its policy planning. This could be achieved by using mechanisms such as guarantee systems.

Conclusion

In this chapter, a state financial assistance structure to ensure access to financial services in rural areas is proposed. The land reform mechanism should link into these mechanisms. No justification can be found for a separate financing structure just for the land reform programme and the land reform beneficiaries. Only in the case of the disbursement of grants may there be a temporary role for District Offices to act as a financial mechanism. Programming development activities at the local level is important, and the importance of programming macro-level activities has also been argued. What has not been discussed in this

6 Done by Charles Simkins of the University of the Witwatersrand for the Community Bank.

chapter is detail on how the FM should operate. Discussions on innovative approaches to be followed by these institutions in order to survive while providing much-needed, costly services also need to be addressed during the structuring (or restructuring) of these institutions. Proposals were made for an interim structure to accommodate the urgent needs of the land reform programme.

The role of the informal sector and savings mobilization are extremely important in this regard. The importance of access to information cannot be emphasized enough. Land reform beneficiaries must have information on their rights and what is being offered in the programme. Financial intermediaries must have information on which to base decisions. Without information, the market-based approach would not be implementable.

References

ANC. 1994. *The Reconstruction and Development Programme – A Policy Framework.* Published for the African National Congress. Johannesburg: Umanyanou Publishers.

Beghin, John and Fafchamps, Marcel. 1994. Constitutions, institutions and political economy of farm policies: What empirical content? *Invited paper read at the International Conference of Agricultural Economists.* Zimbabwe: Harare, August 22–29.

Besley, Timothy. 1994. How do market failures justify interventions in rural credit markets? *World Bank Research Observer,* 9:27–47.

Binswanger, H.P., Deininger, K. and Feder, G. 1993. Power, distortions, revolt and reform in agricultural land relations. In T.N. Srinivasan and J. Behrman (Eds), *Handbook of Development Economics,* Vol. III.

Chaves, Rodrigo and Gonzalez-Vega, Claudio. 1993. The design of successful rural financial intermediaries: Evidence from Indonesia. Washington D.C.: Finance 2000 Conference, 25–26 May.

Coetzee, G.K. 1988. The financing of small farmer agricultural production. *Unpublished MSc dissertation.* Stellenbosch: University of Stellenbosch.

Coetzee, G.K. 1991. Financing agriculture – challenges to financial institutions. *Agrekon,* Vol 30 (4): 203–209.

Coetzee, G.K., Breytenbach, S. and Stander, J. 1994. An agricultural finance servicing structure for the Western Cape from a government point of view. *Proposal prepared for the Working Group on the Restructuring of the Department of Agriculture in the Western Cape.* Elsenburg.

Coetzee, G.K., Kirsten, M.A. and Christodoulou, N.T. 1993. Financing entrepreneurs in rural areas: a new approach. *Paper delivered at the Annual conference of the Agricultural Economics Association of South Africa.* Cape Town.

De Gorter, Harry and Swinnen, Jo. 1994. Political and institutional determinants of agricultural policy. *Paper read at the International Conference of Agricultural Economists.* Harare: 22–29 August.

De Janvry, Alain and Sadoulet, Elizabeth. 1984. Transaction costs, market failures, competitiveness and the state. *Paper read at the International Conference of Agricultural Economists.* Harare: 22–29 August.

De Klerk, Michael. 1984. Technological change and employment in South African agriculture: the case of maize harvesting in the Western Transvaal, 1968–1981. *Unpublished MA Economics dissertation.* Cape Town: University of Cape Town.

Department of Land Affairs. 1994. Land reform pilot programme: a project of the Reconstruction and Development Programme – Programme Overview. Pretoria.

Fafchamps, Marcel, De Janvry, Alain and Sadoulet, Elizabeth. 1994. Transaction costs, market failures, competitiveness and the state. *Paper read at the International Conference of Agricultural Economists.* Harare: 22–29 August.

Golino, Christina. 1995. Proposals towards interim land financing mechanisms. *Working Document.* DBSA.

Government Gazette. 1995. Government Gazette of the Republic of South Africa, Vol 355, No 16235 (No3 of 1995). Pretoria: Government Printer.

McKenzie, C.C. 1994. Farm worker equity model: broadening the ownership base in agriculture. *Agricultural Policy Conference.* Johannesburg: 25–27 October.

Mostert, Chris and Van Zyl, Johan. 1988. The evaluation of support strategies for farmers with serious liquidity problems (in Afrikaans). *Annual Conference of the Association of Agricultural Economists of South Africa.* Stellenbosch: 26–27 September.

Olivier, D. 1994. Constraints on fundamental rights in agricultural legislation (Afrikaans – Beperking van fundamentele regte en landbouwetgewing). *De Rebus.* October: 753–756.

Rural Finance Commission. 1995. *Press release published in Farmer's Weekly, Business Day.*

Stiglitz, Joseph. 1989. *The economic role of the state.* Cambridge: Blackwells.

Task team on agricultural financing. 1994. *Draft minutes of the meeting of the Task Team on Agricultural Finance.* Pretoria: Secosaf, 17 May.

Thompson, Robert. 1994. *Summary and final comments on the 1994 IAAE conference.* Harare: 22–29 August.

Van Zyl, Johan. 1994. Farm size efficiency, food security and market-assisted rural land reform in South Africa. *Agrekon,* Vol 33, No 4.

Van Zyl, Johan and Binswanger, Hans. 1994. South Africa: Agricultural finance and the role of the government: managing the agricultural debt. Mimeo. The World Bank.

World Bank. 1994. South African agriculture: structure, performance and options for the future. *Discussion Paper 6.* Washington, D.C: The World Bank Southern Africa Department.

Managing the Process

PART VI

21

Notes on the process

Robert Christiansen and Katrina Treu

Introduction

The rapid and dramatic political changes in South Africa between February 1990 and April 1994 provided the international community with a preliminary, but critical opportunity for supporting a transition from white minority rule to a multiracial democracy. In South Africa, however, many groups both on the political left and right had serious reservations about relying on external advice in formulating future policy options. These reservations stemmed, in part, from a view that South Africa's circumstances were unique and that South Africans were best equipped to understand and develop policies for those circumstances. There was also a concern that South Africa could well travel the same path of economic mismanagement and stagnation that has been followed by many of its regional neighbours. The latter outcome is often attributed, in large measure, to the policy advice and intervention provided by the international community. This perception, combined with the view that reliance on outside assistance is an indicator of failure, defined a significant gap between donors and many groups in South Africa. Hence, establishing a working relationship and dialogue with the various groups in South Africa was critical.

From a South African perspective, the period of transition to a multiparty democracy represented an opportunity for creating new approaches to development policy. Although those who opposed the apartheid government had some experience with working in rural areas, there had been very little involvement with government. In this context, developing policy for a new government was a formidable challenge. How does one engage government while, at the same time, oppose its existing policies? In the rural sector, it seemed impossible to develop sufficient agreement on an approach to allow policy development to progress without the involvement of external actors. In this scenario, the World Bank was able to play a role as facilitator, bringing together players from the different political groupings and from government, to enter into a policy dialogue. At the same time, a South Africa cut off from international experience by sanctions had much to gain from engaging with the World Bank sooner rather than later.

Fostering engagement and dialogue

The World Bank, however, was not an automatic choice to act as initial facilitator. There was considerable scepticism of the World Bank because many South Africans understood that the focus of the Bank programmes in other countries was the need to lend and to dictate policy changes as a precondition to lending. In light of this concern, and as discussions with a variety of groups in South Africa progressed, it became clear that the typical product blend of Bank-managed sector work followed by lending was not a viable approach. Instead, the starting point for Bank involvement – whatever that might constitute – in South Africa, was the need to establish a dialogue with a wide range of groups in such a way that these groups managed the Bank's activities. Hence, the role of facilitator had to be well-defined and accepted by all interested parties.

In the agricultural sector, the Bank's direct involvement with South Africa began in late 1991. Due to the political climate at the time, it was critical that the Bank establish a working relationship and dialogue with the various groups in South Africa which represented the full range of political interests. Four principles guided this process:

- Relying on extensive collaboration in defining the content of the products that were to be prepared for the client (i.e. groups in South Africa).
- Ensuring that the content of the work programme was well-defined and of the highest quality.
- Ensuring that the process of preparing the outputs was fully transparent and open to discussion.
- Relying on South African expertise wherever possible.

After consulting with several groups, but in particular the ANC and the government, it was agreed that the first product would be an agricultural sector report that would contain an analysis of the rural economy. In keeping with the principle that South African expertise would be used wherever possible, international experts were paired with South African counterparts. In several cases, these pairings resulted in co-authored background papers for the sector report – a result that further enhanced ownership. To ensure that the preparation process was fully transparent and consultative, the drafts of the report were circulated for comments at several stages to various groups including the Department of Agriculture, the ANC, and the Development Bank of Southern Africa (DBSA). In addition, the preliminary results of the report were discussed in separate meetings with these groups. These results indicated that the South African rural economy was plagued by numerous distortions that would seriously hinder the ability of a new government to address rural poverty and unemployment. The report concluded that the policy environment which gave rise to these distortions would need to be substantially reformed and a major restructuring of the rural economy attempted, if a new government was to address these problems of poverty and unemployment.

Consequently, it was agreed that the Bank (in collaboration with UNDP, and in consultation with groups in South Africa) would organize a workshop at which international policy experience in several areas (e.g. marketing reform and land issues) would be examined. The areas examined in the workshop corresponded to those areas in South Africa that were characterized by the greatest distortions. Up to this point, the extent of separation and suspicion among many parties in South Africa was such that the Bank tended to work bilaterally with most of these groups. The workshop, however, presented an opportunity to bring the groups together to discuss topics of mutual concern and interest. In addition to representatives of the Departments of Agriculture and Land Affairs, the ANC and the DBSA, a wider range of groups also participated. These included representatives from other extra-parliamentary groups (e.g. the Pan Africanist Congress and Inkatha Freedom Party) as well as the South African Agricultural Union, the National African Farmers' Union, universities, and other interested parties.

Because the purpose of the workshop was to provide lessons of experience and not provide prescriptions, the authors of the papers presented were careful not to make recommendations or prescriptions for South Africa. However, the salient theme that emerged from the workshop, both in the presentations and the plenary discussions, was the need to restructure the rural economy, beginning with a land reform programme. Although the definition and extent of this programme was debated, there was an interest in exploring options for a land reform programme that could be considered by a new government. The Bank's participation in this effort would be as one of several groups exploring alternatives that would be presented at a subsequent workshop on land reform options.

Preparing options

The principle that guided the preparation of the land reform options was that 'the preparation process is as important as the product'. In the politics of opposing apartheid, and from the existing South African development experience, there were a host of land reform approaches already being prepared. These were based on several needs, including local involvement and participation in the design of activities, community participation in decision-making at the policy and implementation level, and a strong role for the public sector. Given the highly participatory nature of the extra-parliamentary groups, it was of central importance to the South Africans involved in the preparation of additional options for a rural restructuring programme, that the starting point should be a blend of international and South African experiences. In addition to enhancing the quality of the options, this approach greatly strengthened South African ownership of the options.

In order to ensure that the process was locally driven, several teams of South African experts were recruited to prepare background reports on various aspects of land redistribution.[1] The draft terms of reference for the teams were prepared by the Bank and subsequently revised by the South African teams. In order to ensure that experience from other countries was reflected in the final reports, each of the teams had access to international experts. Finally, an independent Advisory Group – of which the Bank was not a member – was formed to guide the process of report preparation and facilitate the transfer of ownership. The Advisory Group quickly took the lead in guiding the preparation work, and accepted responsibility for managing the Bank's role in the preparation activities.

In total, about 120 South Africans were involved in the preparation of the background reports. The final report, produced by a Bank team working in collaboration with the South African teams, was presented to a workshop on 'Land Reform Options' in October 1993. As was the case

1 The subjects of the individual reports were: i) Required Elements of the Constitution and the Land Act; ii) Legal Aspects of Land Titling and Transfer; iii) Agricultural Pricing and Marketing; iv) Land Market and Land Price Analysis; v) Administrative Requirements for a Rural Restructuring Programme; vi) Community Group Structures and Requirements; vii) Criteria for Group and Individual Participation; viii) Review of South African and International Experience with Rural Resettlement; ix) Agricultural Livelihood Options; x) Advisory Services for Programme Participants; xi) Mechanisms for Support Services for Programme Participants; and xii) Financing a Programme for Restructuring the Rural Economy.

with earlier reports, the report on rural restructuring options was characterized as a starting point for discussions among South Africans and not as a set of recommendations.

In the months following the workshop, and in a process directed by the Advisory Group, the various options were presented and discussed with communities and interest groups. This exercise was intended to provide information on the options and to encourage these groups to think of how the basic model could be refined and customized to meet their needs.

After the elections in April 1994, the newly elected government moved quickly to assume direct control of the rural restructuring programme, with the Ministry of Land Affairs assigned responsibility for land reform. The initial focus of the Ministry's efforts was to design pilot land reform projects based on the options developed in the various workshops and in consultation with the intended beneficiaries of land reform.

Conclusion

Between 1990 and 1994 in South Africa, the government was not a suitable sole contact point for international agencies and, therefore, it was necessary to work with a wide range of groups. These circumstances, and the resulting process, provided important lessons both to donors and South Africans. The most important centre on: (a) focusing on developing working relationships in which the Bank works for the client; (b) needing to clearly define the range of Bank products that are available to the client; (c) willingness to listen to the needs of the client and requiring active client participation in defining and completing the work programme; (d) transparency and openness in the formulation and completion of a work programme; and – perhaps most importantly, (e) the willingness and ability of the client to define and manage the relationship with donors.

In this way, it was possible to engage a wide variety of groups in South Africa in a substantive dialogue about the policy and investment options that were likely to face a post-apartheid South Africa. As a result of this engagement, there emerged a significant (although certainly not complete) consensus on the outline of a development agenda to be pursued in the rural sector. This agenda included the need to address rural poverty which is deeply entrenched in South Africa, along with issues of social justice such as land restoration and economic development within an integrated policy framework (e.g. including legal and constitutional issues and local administration). The process of preparing this agenda has ensured broad South African ownership and a framework within which donors can operate.

While acknowledging the uniqueness of the circumstances in South Africa during this period, it is still possible to replicate the working relationship developed between the Bank and groups in South Africa in other countries in the region. This is particularly important at a time when the Bank is concerned about improving its ability to respond to the needs of clients. Whether or not this change takes place in other countries where indicated below depends on the willingness of the Bank to change its mode of operations (including its definition of products and clients) significantly, and on the willingness of governments to lead the relationship with the Bank and other donors.

22

An overview of the Land Reform Pilot Programme

Susan Lund

Introduction

Land reform is identified by the Reconstruction and Development Programme (RDP) as 'the central and driving force of a rural development programme'. The RDP notes the three key elements of a land reform programme as being *restitution* of land to victims of forced removal; *redistribution* of land to landless people; and *tenure reform* that would provide security of tenure to all South Africans.

The *restitution* process is to be addressed through the establishment of a Land Claims Court and Commission to process land claims from victims of forced removals (see Chapter 15). *Tenure reform* is being tackled through a review and reformulation of present policy and legislation. The initial phases of *land redistribution* are being developed and expanded over a two-year period in each of the nine provinces through a Land Reform Pilot Programme.

The Land Reform Pilot Programme is an effort to translate the outcomes of an extensive South African debate on land reform policy options into an implementable strategy for redistribution. The principle is to test a mechanism of delivery which places decision-making on the use of the budgeted resources at a district level. The piloting process is one of learning by doing, while establishing a facilitative role for the state in a manner that is both replicable and affordable. The Pilot Programme places strong emphasis on shaping the nature of government intervention for land reform, and on building rural capacity to plan and manage the expenditure of state resources for land purchase according to locally devised and negotiated solutions.

The essence of a Pilot, of course, is that it is not written in stone. As events unfold, elements will change. By the time this chapter reaches print, much will have been adapted in the course of implementation. Much that may appear unworkable in the pages that follow may have been proven so; while other, odd-looking aspects, may just prove appropriate and replicable in the South African circumstances.

This chapter provides an overview of the Land Reform Pilot Programme which is currently being implemented by the Department of Land Affairs. Firstly, it provides some background on the Department, its priorities for land reform, and the allocation by the RDP of the budget for the Pilot Programme; secondly, it outlines the design of the Pilot Programme; and thirdly, it sets out management and time plans. The chapter concludes with a discussion on the expansion of the Pilot phase into a national programme.

Developing a national land reform programme

In April 1994, the new Minister of Land Affairs inherited the former Department of Regional and Land Affairs in which the land component consisted mainly of the Deeds Registry and the Surveyor General's office. 'Land Reform' had been established as a Chief Directorate in the Department following the abolition of the 1913 and 1936 Land Acts in 1991. It received its first budget in 1993, and although its work expanded enormously during that year, its budget and staff allocations in the 1994/1995 financial year remained insignificant in the face of the land reform priority of the newly-elected government. The work done by this small Chief Directorate included implementation of restitution decisions made by the Commission on Land Allocation (since disbanded); implementation of tenure 'upgrading'; implementation of subsidized access to land (through 1993 legislation providing for state assistance in land purchase); and the winding up of the estate of the former South African Development Trust.

The reorientation of the Department towards RDP objectives entails restructuring. To this end, the Public Service Commission has approved a re-vamped departmental structure, and personnel are being recruited to fill new posts. However, further institutional change is anticipated as the respective roles of national and provincial departments in the implementation of land reform are negotiated and resolved. The Minister of Land Affairs meets regularly with the Provincial Members of Executive Councils designated with land responsibilities, in order to establish collaboration in land reform delivery. While land is likely to remain a national function under the constitution, it is probable that the provincial governments will carry major responsibility for delivery within a policy framework set and facilitated through national agreement, and monitored by the Department of Land Affairs.

In the meantime, the current legislative programme of the Department points, in part, to the extent of change which is presently underway as a start to land reform. Firstly, legislation was passed by parliament in late 1994 providing for the establishment of a Land Claims Court and Land Claims Commission to deal with *restitution* (see Chapter 15). Commissioners have subsequently been appointed, and restitution claims are being lodged. Secondly, the unravelling of the complex web of apartheid-designed *land administration* systems has begun, and a Land Administration Bill is before parliament to allow delegation of responsibilities in this regard to the new provincial governments. Thirdly, reform to facilitate speedy *allocation of land for development* purposes is being prepared in the form of the Development Facilitation Bill which was passed by parliament in 1995. Fourthly, the work being done by a joint departmental/ non-government task team on tenure reform has proposed new legislation during the course of the year. Finally, amendments to, or re-formulation of the legal facility contained in the Provision of Certain Land for Settlement Act (Act 126 of 1993) is under consideration in the process of implementing the Pilot Land Reform Programme, to ensure sustainable and equitable *state assistance for land purchase* as part of a land redistribution initiative.

In addition, the Minister has appointed a Commission of Inquiry into the Provision of Rural Financial Services, through which the roles and systems of the existing *credit* institutions come under scrutiny (see also Chapter 20). The Commission's recommendations are due by the end of 1995, and are likely to include institutional, legislative and regulatory change to remove obstacles preventing poor and entrepreneurial rural dwellers from accessing credit on the market. The Pilot Programme had identified an urgent need to secure credit access by beneficiaries of the Programme, and early investigations recommended a fast-track facility as part of the Commission's work.

Increasing numbers of applications, demands, claims and requests for access to production land reach the Department from communities, farmers' associations and individuals. If the precise extent of demand for rural production land is as yet unclear, national expectations remain high that a redistribution plan will rapidly shift racial patterns of ownership and access.

The need to establish delivery mechanisms by which the state will respond to such demand is thus a priority for the Department of Land Affairs. At the same time, the capacity to deliver is constrained not only by a fledgling policy framework for land redistribution, but by the politically difficult establishment of institutions for rural development and land reform, and by severely limited budgets inherited from previous government designations. It was in this context that the Department approached the RDP Fund for the allocation of a kick-start budget for a redistribution Pilot Programme that could begin to establish mechanisms of delivery in the period 1995/96 when institutional and financial restructuring was taking its course.

By mid-1994, the RDP had earmarked an initial round of resources for allocation to a number of Presidential Lead Projects located in a range of sectors for the purposes of restructuring priorities and budgets. The selection of these Projects was to be determined by the extent to which they set out to reorientate government expenditure and operations to meeting the goals of the RDP while, at the same time, bringing tangible development outcomes for the intended beneficiaries. The Land Reform Pilot Programme met these criteria.

Part of the RDP's reorientation drive has entailed an insistence on the drafting of business plans for the Presidential Projects, with the aim of introducing objective setting, plus time and cash flow planning and reporting as a norm for future departmental programme management. A Core Business Plan for the Land Reform Pilot Programme was therefore formulated by the Department, and approved by the RDP, effectively securing a sum of R315 810 000 for expenditure according to plan over a two-year period in each of the nine provinces of the country. R45 million of this sum is to be contributed by international donors.

The design of the Pilot Programme

The Pilot Programme is based on the understanding that land redistribution is part of a rural development strategy to address poverty, and to introduce equitable access to the land market. The aim of the Programme is to develop efficient, equitable and sustainable mechanisms of land redistribution in rural areas, as a kick-start to the redistribution element of a national land reform programme.

The Pilot is being undertaken on a limited scale (one selected district per province), in response to a diversity of rural land needs. It intends to ensure that lessons are learnt quickly, and that policy is consolidated in the course of the Programme for expansion to a national initiative within the two-year period.

In identifying a starting point for this endeavour, it is well recognized that, internationally, land reform does not reflect great or easy success. Considerable time and effort has been expended on ensuring the participation of a wide range of groupings both inside and outside government in the design of the Pilot Programme. In particular, the expressed land problems and demands of rural communities, and the solutions that they have suggested over many years of land struggles, have been taken into account.

The emerging Pilot design for the use of state resources intends to ensure that:

- access to land will assist the poor to break poverty;
- the intended land use will be sustainable and productive;
- secure forms of tenure are instituted;
- access to credit, information and technology is made available to all;
- land prices are fairly negotiated between the beneficiaries of state assistance and the seller;
- all local stakeholders contribute to solving land problems;
- local decision-making capacity around the application of limited resources is strengthened;
- disputes over access to land are mediated;
- the resources applied are affordable to the state and have positive returns to the national economy.

Within these priorities, land and land use is identified through a planning process run by beneficiaries themselves. Plans are devised in terms of a Framework for Planning set out in the Core Business Plan. Expenditure is within broadly designated but fixed budgets. State assistance for the purchase of land is allocated within budget on approval of plans.

The state thus acts as facilitator in the Pilot Programme by providing:

- guidelines, norms, standards and policy for the use of the budgeted resources – within which people can make choices according to local conditions and requirements;
- grant finance for land purchase and basic needs provision, within national housing subsidy limits, to assist people to purchase land and settle in terms of locally designed plans;
- grant finance and planning facilitation so that people who want land can assess the conditions and make informed choices on the use of available resources;
- access to credit to contribute to land purchase;
- mediation resources to enable people to resolve conflict over access to land at local level;

- training and capacity building to increase skills in planning, management and resource use.

Table 22.1: Budget Line Items of the Pilot Programme and their Application Provision	
1. District management and facilitation (one set sum per district to be budgeted by the Land Reform Steering Committee in each province)	Overheads: staff, transport, administration, community workshops
2. Planning Grants (one set sum per district to be budgeted by the Land Reform Steering Committee in each province)	i) District-level planning; ii) Project-level planning for each beneficiary group within the district; iii) Expansion planning (one/two additional district-level planning per province)
3. Settlement subsidy to a maximum of R15 000 per household	For use in terms of beneficiary plans for land purchase, survey and title transfer costs and basic needs provisison. The use of the grant must: i) reach the poor directly; ii) increase household incomes; iii) ensure tenure security; iv) lever both credit and own/other contributions to land acquisition

The Pilot design may be elaborated as follows:

District selection

The nine pilot districts were selected through provincial-level investigation and negotiation (including both government and non-government sectors) towards the end of 1994. Each district was to include as many of the following conditions as feasible – with the inclusion of state land being a prerequisite:

- state-owned land;
- private land for acquisition;
- farm workers and labour tenants requiring tenure security;
- rural, peri-urban, small town conditions;

- high rates of poverty;
- small groups and larger groups/communities as potential beneficiaries;
- potential for leasehold land use;
- communities seeking restitution;
- organized/unorganized communities/groups with land needs.

The selected districts include some of the most pressing current land needs and demands, many involving decades of conflict over land access, and many having experienced a litany of interventions by government.

Facilitation, training, mediation

The Pilot design provides for mechanisms of land reform facilitation that should, in time, become integral to rural local government and contribute to sound systems of land management and infrastructure delivery at local level. Local government is unlikely to be operational for such work within the next few years. In the meanwhile, district-level facilitation for rural organization development is essential. To this end, a Pilot facilitation team has been appointed by the province to assist the communities in the Pilot District to establish the necessary organizational structures for land access planning and basic needs provision planning. The district team is also responsible for district co-ordination of training and mediation resources that are available throughout the programme.

Donor allocations (over and above the Pilot budget total) are being finalized for a substantial training and technical assistance programme that will support the Pilot initiative. Training will be managed by the Department of Land Affairs but will be provided by a range of non-government, university and private institutions as appropriate. Topics will include, inter alia, participatory planning, financial management, administration, mediation, the establishment of legal entities and tenure options. It will be targeted at a number of levels: beneficiaries, facilitators, planners, government departments and the Land Reform Steering Committees.

Planning

Potential beneficiaries of settlement subsidies qualify for such financial assistance on the basis of their engagement, through organization, in a planning process.

Planning is to be undertaken through participatory methods, using planning agents (from any sector – non-government, public or private) chosen by the potential beneficiaries. In the first instance, planning will be at district level, involving all local stakeholders (not only the direct beneficiaries), to identify, through negotiation, the beneficiaries, land availability, resource use and resource constraint. The Framework for Planning sets out the parameters that are to be covered in the exercise. In the second instance, planning will occur at beneficiary or Project level (there may be a number of these planning undertakings within each Pilot area). The Framework for Planning specifies the broad requirements of such Project planning, notably those for environmental sustainability, productive land use, basic needs provision, maintenance of infrastructure, and overtly beneficial impact for women and the poor.

For the approval of plans, the Pilot Programme establishes a Land Reform Steering Committee in each province, made up of inter-departmental representatives and the organized non-statutory sector. This body will oversee the Programme, appraise plans made by communities, and recommend the allocation of finance accordingly. No plans will be approved if they do not have the backing of the appropriate government department (e.g. for water supply) and if they do not make provision for long-term maintenance of infrastructure. This inter-departmental approach to land reform responsibility is intended to ensure that departmental budgets and personnel will be allocated in due course to land reform areas as a matter of priority, and that planning is within provincial norms and standards. Leverage of additional resources to the Pilot areas is also intended through the planning undertaking.

On approval of plan, the beneficiaries will be given land acquisition assistance finance in terms of their plan, and will be able to begin basic needs infrastructure development according to plan.

Land acquisition assistance

A lump sum budget is made available to each geographically designated Pilot District for land purchase by the people in the area who have previously been denied access to the land market, and for whom land purchase without state assistance would be impossible. From this budget, a Settlement Subsidy may be allocated through the planning process at a maximum of R15 000 per household. Plans must show that the beneficiaries' basic needs have been catered for within this amount. Allocation of the subsidy will be registered on the same national data base as the Housing Subsidy and a household will not qualify for both.

The subsidy is thus a once-off, umbrella subsidy in the sense that any government assistance for either land purchase or basic needs provision will be debited against the R15 000 per household maximum. The grant will be applied flexibly for a range of basic needs, likely to include land for residential purposes, water provision, sanitation, fencing, homestead improvement, internal roads and, where possible, broader social and economic infrastructure. The costs of bulk infrastructure are not included in the grant, and will have to be sought during the planning phase from other sources. The intention is to secure an initial contribution to meeting basic needs in such a way that local priorities are addressed; the beneficiary is given real choice over the use of the standard subsidy; and leverage is obtained for ongoing sector-specific contributions.

The Programme is essentially designed to encourage locally negotiated trade (within the settlement subsidy and available credit limits) in terms of local demand, local supply, and the productive potential of the land as planned by the beneficiaries themselves. Within this, the grant element is intended to reach the poor directly, while leveraging credit and equity towards purchase amongst those who can afford to repay loans from production. While the national constitution provides for land expropriation by the state (with market rate compensation), it is unlikely that this will be necessary or desirable within the Pilot Programme. On the whole, there is ample land supply for purchase on the market.

Thus, in the course of planning, the stakeholders in the Pilot area will have to ensure, firstly, that the poor are identified (they are usually the least organized) and, secondly, will benefit directly; that land use planning shows projected household incomes increase (household income need not be measured by commercial profit margin alone – household food security is as important in the assessment of plans); and that tenure security is negotiated and resolved. In addition, credit is a required component, and own contributions are necessary. The extent of these requirements is specifically not stated, but left to local negotiation and creative solution-finding. Plans cannot be approved unless these four criteria are met.

Credit reform, tenure reform, and reforms to farm subdivision restrictions, will be critical to successful land redistribution. A basic element of the land acquisition approach is that a variety of tenure forms should be possible, and that tenure should be planned by beneficiaries in terms of their needs and preferences.

Conflict is inevitable with the delegation of responsibility to potential beneficiaries for deciding the use of a state budget for land purchase. Provision is thus made for mediation services throughout the Programme.

The design is nevertheless based on the notion that locally devised solutions to conflict stand a greater chance of sustainability than those engineered at a national or even provincial level concerning the nature of the beneficiary or the scale at which any beneficiary is assisted. Conditions vary greatly from area to area, and the 'ideal' beneficiary in one area would not be viewed as deserving in another. Furthermore, it is precisely in the process of negotiation – the trading of the needs of the poor with the needs of the farming entrepreneur – for access to limited state finance, that equitable solutions may be found. Thus, the potential beneficiaries will not only be negotiating land prices with the sellers, but will be negotiating amongst themselves to establish the most suitable allocation and contributions of grant, credit and equity in what are highly differentiated communities. Mediation, facilitation, planning support and training are provided to the greatest extent possible to bolster the endeavour.

This delegation of choice to the potential beneficiary communities themselves, constitutes the pivotal mechanism of delivery design that is being tested within the Pilot.

Planning towards Programme expansion

Programme expansion is to be embarked on at the outset, by stretching the planning budgets to cover district-level planning in one or two additional districts in each province. Choice of expansion will rest with the Land Reform Steering Committee, based on identified pressing demand.

Monitoring and evaluation

A monitoring and evaluation system that will generate the information necessary to assess progress has been designed by the independent Land and Agriculture Policy Centre, and is being implemented by the Department of Land Affairs from the inception of the Pilot. The information generated is available to the public throughout the process of implementation.

Agricultural production support

The Pilot Land Reform Programme is a state-assisted land acquisition initiative that also provides funding for planning, and some basic needs delivery. Agricultural development opportunities within the Pilot District form part of this facilitated planning process, and the active involvement of both national and provincial Departments of Agriculture in the Programme is intended to provide ready access to appropriate support from the outset. The Pilot itself does not provide finance for a full-scale farmer development programme.

Pilot management

The Pilot Programme is the responsibility of the Department of Land Affairs whose Director-General is accountable to the Department of State Expenditure, the RDP and donors, for expenditure on the Programme. Its executive is headed by a Pilot Programme Manager who recommends policy development to a National Pilot Land Reform Task Force. This National Task Force is made up of officials from all nine provincial governments; four national government departments (Water Affairs and Forestry, Agriculture, Housing, and Land Affairs); and three national non-government organizations from the land and rural development sector.

At provincial level, Land Reform Steering Committees are being established to drive the Pilot Programme. The composition of these Steering Committees varies from province to province, depending on departmental responsibilities. In all, there is inter-departmental representation (usually including departments of agriculture, local government, housing, planning, environment) plus representation from the organized non-statutory sector in the Province (whose composition also varies widely). The national Department of Land Affairs is represented on the Steering Committees. This body is responsible for devising cash flow and time plans within budget; for selecting the district management and facilitation team; for appraising planning proposals from the district and from beneficiary groups within the district; for appraising both district- and Project-level plans; and for recommending the payment of finance accordingly. It is also responsible for initiating Programme expansion within the province. For the purposes of the Pilot, this structure is non-statutory. Its formalization as an institution of land reform will depend upon its performance during the Pilot Programme.

District budgets (dedicated funds) are transferred from the Department of Land Affairs in tranches to the provincial governments for implementation in terms of cash flow plans devised by the Land Reform Steering Committee in each province. Payments are made by the province in terms of approved beneficiary plans, on the recommendation of the Steering Committee. Financial and progress reporting procedures have been agreed between the provinces, the Department of Land Affairs and the RDP for the implementation of the Pilot Programme.

At district level, the Pilot has established a District Forum which

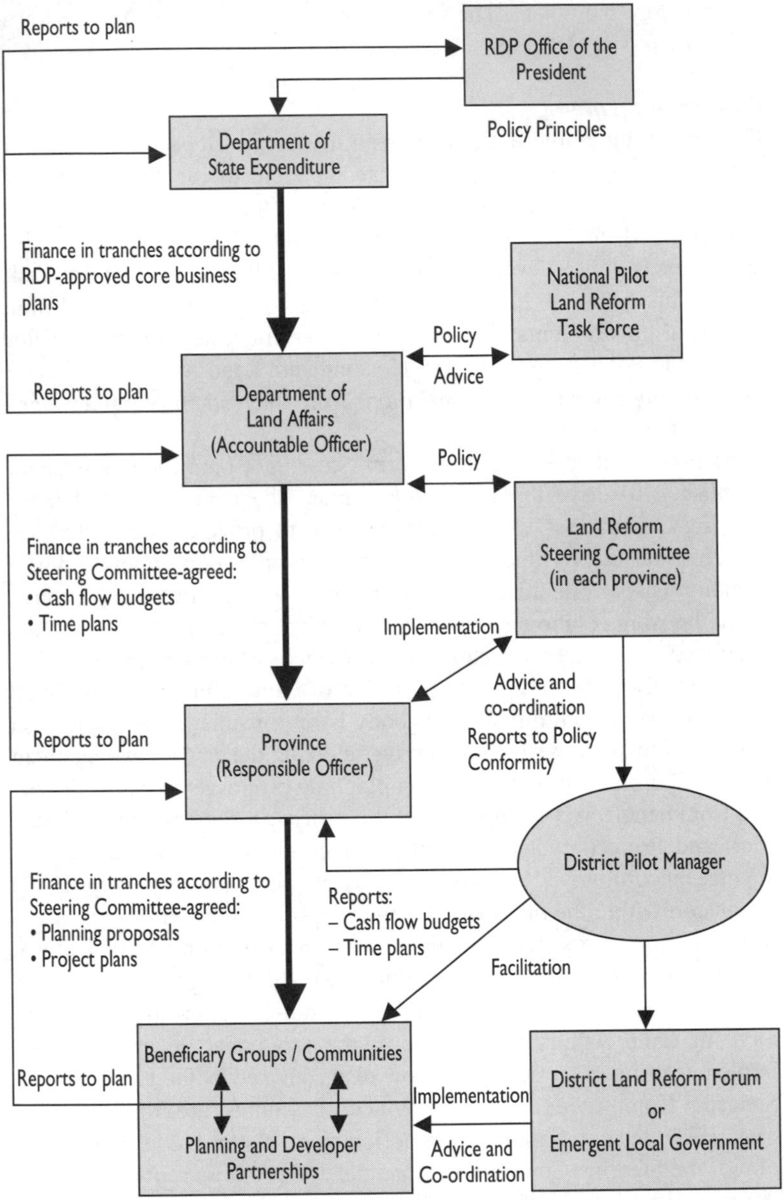

Figure 22.1: Pilot Land Reform Programme Management Flow Chart

oversees district-level planning and constitutes the main negotiating institution for district prioritizing and beneficiary selection. The relationship of such forums to emergent local government will have to be worked out in due course.

Programme expansion

The Pilot Land Reform Programme is a step in a much larger national land reform programme. It is intended to provide the means through which to develop effective and sustainable mechanisms for the financing and administration of land redistribution. Expansion – on the basis of lessons learnt – is therefore part of Pilot design. The intention is to begin expansion planning in additional districts as early as possible in the Pilot timeframe. It is hoped that, as this expansion begins, both the demarcation and form of rural district and local government will be clarified, to provide a framework for sound budgeting for land acquisition assistance within which to establish local facilitation capacity.

As Pilot lessons are learnt, it will be possible to scale up budgets for land redistribution across the country over time, with calculated and arguably justified fiscal impact nationally. Budgeting for land acquisition assistance grants would need to be done on the basis of locality-

Table 22.2: Time Plan for Implementation of the Pilot Programme

	1994	1995				1996				1997
Element	4th	1st	2nd	3rd	4th	1st	2nd	3rd	4th	1st
Establish National Task Force	x									
Select 'Districts'	x	x								
Establish Steering Committees		x								
Appoint District Managers		x	x							
Planning proposals		x	x	x						
District planning		x	x	x	x					
Project planning		x	x	x	x	x	x			
Plan appraisals				x	x	x	x	x	x	
Land acquisition				x	x	x	x	x	x	x
Basic needs delivery				x	x	x	x	x	x	x
Training and mediation			x	x	x	x	x	x	x	x
Monitoring and evaluation			x	x	x	x	x	x	x	x
Expansion (on receipt of further funds)						x	x	x	x	x

specific land demand, poverty indicators, and agro-ecological production potential to establish an appropriate balance for grant, equity and credit sources for land purchase.

The potential to scale up facilitation capacity is perhaps more difficult. The Pilot is based most essentially on a facilitative role for the state – particularly on the capacity to provide appropriate support to local decision-making. This 'facilitation factor' may just prove to be the Achilles' heel to rapid replicability. It may already be indicating a training need in facilitation on a scale not planned hitherto, but doubtless pivotal to sustainable land reform.

Annexe 22.1: Elements of the Land Reform Pilot Programme

(Extract from: *Land Reform Pilot Programme Core Business Plan, Department of Land Affairs, November 1994*)

The elements of the Land Reform Programme are applicable to each province. However, they may not necessarily apply in the order in which they appear, and many of the listed elements may run concurrently.

1. Establishment of a National Pilot Land Reform Task Force, made up of representatives from the Provincial Governments; the national Departments of Housing, Agriculture, Water Affairs and Forestry, and Land Affairs; the RDP Office of the President; and three national non-government organizations: the National Land Committee, the National Rural Development Forum, and the Land and Agriculture Policy Centre.
2. Identification of a Pilot District in each province in terms of criteria set by the National Task Force.
3. Establishment of a Land Reform Steering Committee in each province (between the Provincial Government and the Department of Land Affairs) to oversee implementation and to report to the accountable and responsible officers on compliance with this Core Business Plan.
4. Agreement on whether the Provincial Government or the Department of Land Affairs will act as the responsible officer for the programme.
5. Selection by the Land Reform Steering Committee of a Pilot District

Manager to facilitate the planning and development processes in the chosen Pilot District.

6. Establishment of a District Pilot Office.
7. Compilation of Time Plan and Cash Flow Plan for the Pilot District by the responsible officer for approval by the accountable officer.
8. Formation of a District Land Reform Forum or Co-ordinating Committee or emergent local government structure/s, representing all relevant stakeholders in the Pilot District.
9. Identification of beneficiary groups within the pilot district.
10. Identification of further districts for programme expansion planning.
11. Receipt by the Land Reform Steering Committee of District Planning Proposals (for Pilot District and expansion District/s Planning) in terms of the Framework for Planning.
12. Approval by the Land Reform Steering Committee of District Planners (for Pilot District and expansion District/s Planning), and awards of District Planning Grants by the responsible officer.
13. Establishment of legal entities representing the Pilot District's beneficiary groups, capable of forming Partnership agreements for the planning processes. A bank account must be opened by each group.
14. Selection by each beneficiary group of a planning agency (government or non-government or private).
15. Establishment of Partnership contracts between each beneficiary group and their chosen planning agency to undertake the Project Planning process.
16. The design of a Project Planning Proposal by the Partnership in terms of the Framework for Planning. Additional planning criteria specific to the communities involved may be set by the Land Reform Steering Committee.
17. Approval of the Project Planning Proposal by the Land Reform Steering Committee, in consultation with relevant expertise and authorities as necessary.
18. The allocation of a Planning Grant to the beneficiary group, by the responsible officer.
19. Formulation, by each Partnership, of a project plan for (inter alia) land acquisition and basic needs delivery, in terms of an approved Project Planning Proposal.
20. Appraisal of project plans by the Land Reform Steering Committee in terms of the Framework for Planning, the approved Project Planning Proposal, additional criteria that may have been set (see 16

above), and the Core Business Plan. Relevant expertise and authorities will be consulted during plan appraisal.

21. Upon approval of the project plan (or approved elements of the plan), the establishment of a Developer Partnership between the beneficiary group and its chosen management agency (government, non-government or private). This Developer Partnership may be the same partnership as established for planning purposes under 15.
22. Allocation, by the responsible officer, of Land Acquisition Grants to the beneficiaries in terms of approved project plans.
23. Allocation, by the responsible officer, of Basic Needs Grants in terms of approved project plans.
24. Land acquisition by beneficiaries in terms of approved project plans.
25. Implementation of basic needs delivery in terms of approved project plans.
26. *Throughout:* Training of managers, planners, and community leaders in participatory planning methods.
27. *Throughout:* Monitoring and evaluation.
28. *Throughout:* Facilitation of access to credit, training and farmer support through other departments and agencies.
29. *Throughout:* Mediation services facilitated by the District Manager.
30. Expansion of the Programme on the basis of approved district plans, funds permitting.

23

Benefit-cost analysis of land redistribution: conceptual issues

Michael Aliber

Introduction

A benefit-cost analysis of land redistribution helps to answer the broad question of whether land redistribution is in society's best interest. Simultaneously, it may help to answer more specific but equally important questions: (a) Who gains from land redistribution, who loses, and by how much? (b) Is the target beneficiary group really made better off? (c) What is the cost to the taxpayer? (d) What are the unseen social effects (benefits and costs) that accrue to third parties? In all respects, however, it only 'helps' to answer such questions; full answers must also take into account the political and social impetus for land reform which the benefit-cost analysis touches upon only tangentially.

In order to begin to enumerate the benefits and costs of South Africa's nascent land redistribution effort, it is useful to think in terms of a stylized scenario, each piece of which suggests some critical questions:

1. A white commercial farmer sells his 1 000 hectares of land to, say, 20 black households; they purchase it with a combination of their own resources, loans and government grants.

    - What are the welfare consequences for families buying the land?
    - What are the consequences for the farmer selling the land?
    - What happens to the total output from the land that is sold?
    - What are the changes in the use of labour, capital and other resources on the land?
    - Do the new owners take better care of the land, or worse, i.e. what is the net environmental impact on the land?

2. The 20 families moving onto the land previously lived in another community. They had either cultivated small parcels, grazed some livestock, engaged in non-farm enterprises, sold their labour, been unemployed, or some combination of these.

- What amount of farm production from the old locale is lost when these families relocate?
- What happens to their non-farm productive activities? (Are they abandoned? Are they brought along to the new locale?)
- What happens to the households and land resources that are left behind?

3. As the 20 families take up residence on their new land, they establish patterns of input sourcing and output marketing, with some of these outputs being intermediate goods for other enterprises (perhaps their own). They also spend their incomes on whatever goods and services are available to them in their new locale. The previous owner, meanwhile, ceases doing business with former suppliers, and no longer provides intermediate goods to the economy. (Presumably his/her consumption habits do not change, although they may be relocated.)

   - What are the net off-farm effects due to the replacement of old farm–off-farm linkages with new ones?[1]

4. Suppose many such land sales have taken place in the area. This may mean that there are regional or even national production changes, possibly involving an alteration in the output mix.

   - Will this lead to general or relative price changes of foodstuffs?
   - Who will be affected?

As the grossly simplified scenario above begins to suggest, the goal of the benefit-cost analysis is to identify all of the parties affected, while keeping in mind the intricate trade-offs that occur when resources are added to and/or moved between sectors of the economy. Secondary and tertiary effects must also be accounted for. The benefit-cost analysis also seeks to quantify all or nearly all of these effects, in such a fashion that the overall balance of benefits and costs from land redistribution can be determined.

The primary purpose of this chapter is to lay out conceptually the main ingredients of an ex ante benefit-cost analysis for South Africa's

1 In this chapter a distinction is made between 'non-farm' and 'off-farm'. The former denotes those non-agricultural income earning activities undertaken by a household receiving land; the latter concerns non-agricultural pursuits of other households.

land redistribution, in such a way as to enable the reader (and policy maker) to appreciate the layers of complexity that ultimately determine the economic desirability of this redistribution. The secondary purpose is to outline a preliminary benefit-cost analysis for the Land Reform Pilot Programme. This helps to illustrate more tangibly the principles of benefit-cost analysis, as well as provides an approximate sense of the economic benefits of land reform generally.

The discussion that follows is divided into four parts. The first part is an overview of how benefit-cost principles can be applied to South Africa's land redistribution, and also provides a preview of some of the methodological considerations elaborated upon later. The second part presents a fuller survey of the benefits and costs to be expected with land redistribution, highlighting those that are either of particular importance or that are characterized by especially thorny measurement problems. The third part comprises the benefit-cost analysis of the Pilot Programme, and the fourth part concludes.

Principles of benefit-cost analysis for land redistribution

The idea of assessing the benefits and costs of land redistribution is intuitive enough but, in contemplating the execution of such a task, one quickly encounters a variety of analytical and methodological obstacles. Some of these are part and parcel of benefit-cost analysis, and some pertain especially to the South African context and its land reform:

- How can one predict the difference in agricultural production between typically large white farms and small black farms, given the virtual absence of the latter in recent South African history?[2]
- How can one compare or summarize benefits and costs that occur at different points in time?
- How can one summarize effects on areas that are very different in terms of farm type?
- When would one expect the redistribution to affect output prices, and who would be affected?
- How can one quantify phenomena such as 'quality of life', environmental impacts, etc?
- How does one measure the (net) effects of redistribution on non-farm

2 The exception, of course, is farms in former homeland territories; however, these serve as a dubious basis for comparison given their general over-population and poor service provision.

industries, such as those providing farm inputs, those making use of intermediate outputs, and those providing consumer goods?

The list is not exhaustive, but furnishes a good start from which to consider a strategy for assessing the merits of redistribution.

Assumptions and international experience

Where relevant empirical information is scarce – such as on the relative productivity in South Africa of large, white, capital-intensive farms and small, black, capital-poor farms – one must extrapolate from what information is available and/or make assumptions based on existing literature from around the world. The problem for South Africa is particularly severe, given the effects of the Group Areas Act and other legislation in creating a highly distorted economic environment which, inter alia, has left policy makers with a limited fund of relevant knowledge from which to predict the course of the post-reform economy. The most significant gaps in knowledge naturally concern the future participation of rural blacks, who face new opportunities in a relatively fluid policy environment. The sorts of farms they establish, the intensity with which they use land and other resources, the efficiency of their production processes, etc, are largely matters of speculation.

At a later stage, when land redistribution has been under way for some years and relevant data has been gathered, the benefit-cost analysis can be repeated and refined on a more solid empirical foundation. In the meantime, one makes reasonable, conservative assumptions and educated guesses, and then performs 'sensitivity analysis', i.e. one ascertains how much the end result of the analysis will vary as these assumptions are modified. The less the sensitivity to initial assumptions, the more confidence one has in the conclusions of the exercise.

Discounting and present values

Benefits and costs typically occur at different points in time. For instance, much of the cost of land redistribution is likely to be borne early on (initial productivity losses, infrastructure investments, moving and set-up, administrative costs, etc), while many of the benefits will be spread out long into the future (nutrition benefits, risk diffusion, future productivity enhancements, etc). In order to form a comparison between benefits and costs, as well as to sum up benefits or costs that accrue in different periods, it is analytically necessary to translate all of them to a common point in time. For convenience, the common point is usually chosen to be the present.

This procedure, called 'discounting', adjusts for the fact that, say, a cost incurred this year is more onerous than the same cost imposed next year (adjusted for inflation), because the fact of having to pay now deprives one of the opportunity to invest that money in the interim to generate a positive return. An analogous argument can be made about one's preference for receiving benefits sooner rather than later. Discounting all future benefits and costs and summing up, results in the project's 'net present value' or 'NPV'.

Regional differences and the question of aggregation

The benefits and costs of redistribution are likely to differ between, say, a capital intensive horticulture operation in the Western Cape or Eastern Transvaal, and an extensive sheep farm in the Karoo. To conduct a single benefit-cost analysis for a national land redistribution programme, one needs to consider the typical benefits and costs associated with each farm (and market) type, then forecast how many such farms (or how many hectares) from each distinctive zone are likely to be involved, and then aggregate appropriately.

Another approach is to perform separate benefit-cost analyses for each distinctive zone/farm type, attempting no aggregation at all. The advantage of this second approach is its simplicity, particularly in that it avoids the need to speculate on how much redistribution is likely to take place in each zone. The disadvantage is that if some of the farm types reveal favourable benefit-cost ratios and others unfavourable, then the overall result remains uncertain. Aggregation may also be an important step in evaluating the probability of endogenous price effects, as discussed in the next section.

A similar issue is that of heterogeneity among beneficiary households within the same region or community. Some households may have a strong ambition for engaging in agricultural production, while others may lack either the ambition or the resources to realize it (labour, capital, etc). Two tacks can be essayed: either one can elaborate a model of the most 'typical' or 'average' household as a basis for predicting farm and non-farm household effects, or one can conduct the analysis separately for different household 'types', including one that *a priori* would seem the least promising as a land redistribution beneficiary (e.g. a poor household with scarcely any labour power to draw upon).

Price effects

If land redistribution results in significant changes in aggregate agricultural production then, depending on the commodity and on the marketing and trade regimes, there may be consequences for domestic producers and consumers, as well as the government. For instance, if the commodity is a non-tradable, e.g. many horticultural products destined for urban markets, and there is a significant increase in production, then, depending upon the magnitude of the price elasticity of demand, a significant price drop may result. This will benefit consumers, and either hurt producers or limit the gains that they might otherwise expect to enjoy from increased output. A full accounting of the effects of the production increase would have to adjust for whatever changes in resource use allowed for the production surge in the first place.

For a tradable, no such direct price effects would be generated since the domestic price relates to the international price, adjusted for whatever trade and other farmer support policies are in place. However, a domestic production increase, apart from its benefit to domestic producers, would result in a loss of tariff revenues if the commodity is import-competing, as are wheat and maize in certain years.

Quantifying non-monetary effects

Quantifying policy or programme effects that are not naturally denominated in money units is notoriously difficult and controversial, but it is their inclusion that lends benefit-cost analysis much of its usefulness. This is especially true for the case at hand, since a disproportionate number of the benefits of land redistribution are not to be directly realized in cash terms. First, nutritional improvements accruing to land reform beneficiaries bestow a benefit quite apart from any additional income. The 'money equivalent' of the nutritional benefit, while highly dependent upon the choice of methodology used to establish it, is nonetheless significant. Second, the role of additional agricultural resources in diversifying a household's income base confers a risk diffusion benefit that, again, improves the quality of life more than can be appreciated by looking at additional income alone. These two effects together comprise the major 'quality of life' effects that the benefit-cost analysis outlined here considers. Additionally, net environmental effects due to land transfers must be considered. These are difficult to evaluate economically for several reasons, not least of which is that they often manifest themselves elsewhere than on the farms where they originate.

Whatever the specific difficulties and strategies chosen to cope with them, the goal is not only to recognize these other costs and benefits, but to attempt to denominate them in a common unit (e.g. Rand) so that they figure in the larger analysis. Where this is simply infeasible, as with certain social effects, then the analysis can be supplemented by verbal explanations.

Net off-farm linkage effects

A linkage effect, or 'multiplier effect' occurs when an autonomous income increase in one sector inspires income increases in other sectors. The same is true in reverse: an autonomous income drop will have negative repercussions in other sectors.

For instance, when a productive hectare of commercial farmland is redistributed, the economic effect is felt among input suppliers and other producers who had relied upon the farm's outputs as intermediate inputs (production linkages), and on the trickle down effects from the farmer's spending on either inputs or consumer goods (consumption linkages). When this same hectare is put to productive use by a land reform beneficiary, a countervailing economic boost is realized in these linked markets. The question is to what extent the losses due to the disruption of the linkage effects associated with the commercial farm are offset by gains associated with the linkages to the new land use.

Since the transmission mechanism is multi-faceted and complex, a correct empirical measurement of the 'multipliers' requires a careful study of its own. Moreover, there is no consensus as to what methodologies are best for estimating the multiplier (*Haggblade et al, 1991*). Finally, while some empirical estimates for homeland areas have been made (*Black et al, 1991*), again one must ask if they serve as reliable guides to multiplier magnitudes in the post-reform era. Nonetheless, suppositions as to the probable sign of the net effect will be offered in the next section.

Elements of the analysis

The matrix on page 570 sets out the principal elements of a benefit-cost analysis for land redistribution. Rather than creating separate lists for benefits and costs, all the items are listed together, with '+' signs indicating a probable or certain benefit, a '–' sign indicating a probable or certain cost, a '?' indicating that the element's overall effect remains relatively uncertain, and a '0' meaning either that there is no net effect or that the category is not applicable for the case at hand. The '*' next to 'employment impact' and 'social stability effects' indicates that these items are not to be rendered in monetary terms, but rather are included in some other manner to lend additional perspective to the effects of the redistribution.

Table 23.1: Principle Elements of a Benefit-cost Analysis for Land Redistribution

Item	Private	Social	Gov't
FARM:			
Farm output	?	?	?
Purchased inputs and labour	+	+	0
Private capital	+	+	0
Land purchase	–	0	–
Farm development (irrigation, marketing and storage)	–	–	–
Family labour	–	–	0
NON–FARM:			
Non–farm output	+	+	+
Purchased inputs and labour	–	–	0
Private capital	–	–	0
Family labour	–	–	0
OTHER:			
Off–farm linkage effects	0	?	?
Nutrition effects	+	+	+
Risk diffusion effects	+	+	0
Environmental effects	?	?	?
Employment impact (*)	+	+	+
Social stability effects (*)	0	+	+
Infrastructure (construction and maintenance)	–	–	–
Preparation and planning	0	–	–
Moving and set–up	–	–	?
Programme management	0	–	–
Agricultural services	0	?	?
Assistance and compensation to displaced persons	0	?	?

The three columns to the right provide a rough breakdown of the incidence of benefits and costs. 'Private', in this context, is taken to mean those effects that are experienced by either the direct beneficiaries or the farmers that are being displaced by them. The 'social' category includes these groups, but in addition any other groups in society that are affected. 'Government' includes effects felt by local, provincial and national levels of government.

Virtually all of the items listed above are net effects in one or two respects. 'Farm output', for example, is the level of production on redistributed land, net of what would have been produced on it had it not been redistributed, and net also of any production lost when those acquiring the land left their previous circumstances. Nutrition effects comprise the value of improved nutrition among beneficiary households, less whatever nutritional improvements can be ascribed to other programmes or interventions, (e.g. nutrition education).

Some of the items can be measured in a variety of ways. Of special interest are the capital items, for which one can use either the (present value of) the purchase cost, or alternatively the 'service cost' or imputed rental cost. The advantage of the latter option is twofold. First, it more easily and accurately treats assets that have already been in use, such as the farm implements owned by the existing white farmer. Secondly, it allows one to combine several of the categories towards the top of the table into a single category that is roughly equivalent to net farm income (gross income less cash costs less imputed rental on capital). If one chooses to employ hectares as the numeraire of the analysis, then most of the benefits and costs related to farming can be summarized as the difference between net farm income per hectare (NFI/ha) on black farms versus that on white farms; in other words, 'net NFI/ha'. This facilitates the task of sensitivity analysis, when one would like to investigate how critical are various assumptions about the future relative efficiency of black versus white farms.

Change in net farm income

Of primary importance is the comparison of the productive use of a parcel of land before and after its transfer. As suggested above, this can be summarized by comparing the per-hectare net farm incomes of the past and future uses (or more precisely the present value of these per hectare net farm incomes forever). This comparison implicitly takes into account differences in capital use, hired labour use and expenditure on other current inputs. It also reflects technology differences between former and future users, as well as differences in technical, allocative and scale efficiency.

Using current data from commercial agriculture and former homeland areas is a poor basis for comparing yields of beneficiary farms to those of the commercial farms which they will tend to occupy. What empirical studies from the former homeland areas demonstrate is that, despite their comparatively poor resource base, agriculturists in these areas were efficient and often demonstrated higher growth rates than their commercial counterparts (see Chapter 11). Moreover, international experience suggests that, when given access to comparable support and credit services, small farms perform or outperform larger farms (*Binswanger et al, 1993*).

A conservative approach would be to assume that in the medium to long term, NFI/ha is the same as or slightly higher than before the transfer, but that in the short term some production loss occurs. It is reasonable to expect such a period of adjustment, given that a considerable number of land redistribution beneficiaries may lack much agricultural experience, while those that do have experience, e.g. former farm labourers, may initially lack managerial skills.

Another consideration is the difference between a beneficiary household's productive activities subsequent to the transfer versus before. Given the pervasive conditions of under- and unemployment that have prevailed, particularly in the former homeland areas, this factor is arguably not of great importance since relatively little previous economic activity is displaced on account of the transfer; rather it is supplemented.

Land purchase

The notable feature about land purchase costs is that they are incurred by the beneficiaries and the government, but not by 'society'. This may seem peculiar, given that the cost of land would appear to be the most significant cost of redistribution overall; and from a fiscal point of view, that may be. However, the purchase itself involves a transfer from the beneficiaries and the government to the farmers selling the land (or their creditors). Thus, excepting any distortionary fiscal effects, the cost to society is zero. This is so whether the sale is voluntary, or whether the land is expropriated with a level of compensation below what would have been the white farmer's reservation price. However, in these latter cases, the benefit-cost analysis would have to contemplate whatever additional assistance or compensation would be necessary to make those whose land is expropriated just as well off, and reckon this amount as a social (and private) cost. In practice this would be very difficult to gauge, but otherwise the analysis fails to take into account all effects on social welfare.

Change in non-farm output

This concerns beneficiary households' non-agricultural activities before and after transfer. For instance, if beneficiaries entirely forfeit one income generating pursuit merely to pursue another, then it is not clear that either they or society have gained. The idea is that household members possess a finite amount of time, such that additional time spent in agriculture would be at the expense of other value-generating activities.

It is very likely, however, that for the majority of those acquiring land through land redistribution, agriculture will continue to comprise only a share of total income and/or time available for income earning. Moreover, the extent of under- and unemployment in most communities eligible for participation in the land redistribution is such that non-farm output need change little, if at all.

Off-farm linkage effects

Although there can be no doubt that these linkage effects are critical, there is little that is known with certainty about their relative magnitudes. There is reason to believe that, in the long run, both components of the overall multiplier – production and consumption linkages – will be greater where land has been transferred than where it has not. On the production side, this is for technological reasons since, with the tendency to substitute labour for mechanical and chemical inputs, there will be fewer leakages to input industries with relatively low value-added. On the consumption side, the greater use of labour per unit area with higher remunerations, together with the fact that these individuals tend to have a relatively high propensity to spend on locally and/or domestically produced goods relative to their commercial farmer counterparts, also suggests stronger linkage effects.[3]

Multiplying the per hectare farm income for small, black farms by the appropriate multiplier, then subtracting the product of the per hectare farm income for white commercial farms by the corresponding commercial agriculture multiplier, one obtains a per hectare measure of the 'net gross off-farm linkage effects'. It is 'net' in the sense that it takes into account the difference between the multiplier effects of the post- and pre-redistribution scenarios. But it is also 'gross' because it does not subtract the respective resource cost changes in the off-farm sector that result from autonomous income changes within agriculture. Once these are subtracted, the 'resource cost-adjusted net off-farm linkage effect' can be obtained but, in order that it be entered into the final analysis, it must still be appropriately discounted in cognizance of the time it takes for multiplier effects to manifest themselves fully.

3 The counter-argument would be that lesser-developed economies – in particular those in sub-Saharan Africa – tend to have relatively low farm–off-farm multipliers (*Haggblade et al, 1989*).

The overall sign of this final effect remains uncertain in the short run when it is supposed that per hectare farm incomes (as well as per hectare *net* farm incomes) are likely to be lower on newly redistributed lands. In the longer term, however, it is probable that the net effect will be positive.

Nutrition effects

At very low incomes, food expenditures represent a large share of total household expenditures. Given uncertain income streams, this means that many household members, particularly children, are vulnerable to malnutrition. The costs to society of malnutrition are far-reaching, and include lost productivity, increased mortality and health costs, and of course private suffering (*Levin, 1986*). On the other hand, increases in household food yields have been shown to have a strong association with improved household nutrition (*Van Zyl et al, 1994*).

One way of quantifying the benefits of improved nutrition *vis-à-vis* better access to productive resources, is to consider the cost of nutrition programmes that would have similar nutritional benefits. Thus, the nutrition benefit can be conceptualized as the cost savings that would accrue to the government by not having to include some households in the programme.

At least two approaches can be taken in this regard. One is to look at cost-effective health/nutrition interventions overall. International experience suggests that effective programmes can be established for 0,5 per cent to 2 per cent of per capita GDP (*Gwatkin et al, 1980*).

Another approach is to use recent South African experience in nutrition interventions, one of the broadest of which is the Department of Health's National Nutrition and Social Development Programme which, inter alia, distributes food parcels to needful individuals who are not recipients of any other government assistance. These parcels cost R20–R30 apiece, are distributed on a per individual basis once a month, and are designed to meet one third of that individual's nutritional needs. The range in costs is associated with several factors, one of which is the cost of transport to less central areas. One advantage of this approach is that it does not factor in the overhead costs of the nutrition intervention programme, which is appropriate.

Risk diffusion effects

An important feature of land reform is that it will assist a large number of households to diversify their income sources, again on the presumption that most participating households are not going to pursue agriculture on a full-time basis. Additionally, land reform will assist families to augment their capital base. These facts imply a reduction in income variability, that is to say, in riskiness. This renders a psychological benefit to participants quite apart from their improved average income levels, because less income variability allows beneficiaries to smooth out their consumption over time, thus averting crises that might otherwise have been unavoidable.

The riskiness of agriculture notwithstanding, one would expect the risk diffusion benefit due to expanded agricultural opportunities to be substantial, given an environment characterized by high overall unemployment, high dependence on casual employment, a high degree of dependence on household members' pensions, and a scarcity of agricultural resources with which to meet subsistence needs. Add to this the inaccessibility to poor, rural South Africans of other means of coping with risk, such as insurance markets or affordable credit. The loss of employment or the death of a pensioner are very real possibilities, against which there are sometimes few safety-nets.

The introduction of additional land with which to meet subsistence needs, and perhaps produce a marketable surplus, thus serves an important income/consumption smoothing function. This smoothing function, which will be of particular importance to those living close to the poverty line, can be conceptualized as a partial insurance policy, the value to the beneficiaries of which would be the insurance premium they would be willing to pay to avoid especially bad years.

Unfortunately, in the absence of insurance markets or short-term consumption credit, such insurance premia cannot be directly observed but must be inferred, say, through sophisticated panel studies. For example, using data from rural India, Rosenzweig and Wolpin (*1993*) observed that when times are favourable, farmers accumulate more bullocks than are necessary for productive purposes. The reason is that bullocks are easy to sell off when times are bad, most notably when poor weather results in poor agricultural performance. Also, unlike land assets, bullocks can be sold to people in other areas who may not have been hit by the same adverse conditions.

The accumulation of bullocks in good times is at a cost, however, since they must be fed and their purchase may preclude other expenditures, including those on pumps and other agricultural investments. This direct and indirect foregone consumption is essentially the same as a risk premium that farmers pay in order to cope with risk. Rosenzweig and Wolpin model farmer preferences by means of estimating a profit function econometrically, and then use simulations to draw inferences about farmer decision-making under random weather conditions. On the basis of these simulations, they estimate that subsistence level farming households pay a premium equal to about 30 per cent of the average loss they stand to suffer in the event of a negative shock, if that shock occurs roughly 30 per cent of the time.

Environmental effects

Agriculture generates both on-farm and off-farm environmental damages, much of which either diminishes the inherent productivity of the land or imposes costs on other segments of society. For instance, water erosion displaces topsoil wherein much of the soil's fertility resides; but this erosion also leads to siltation of dams, reducing their water retention capacity.

There is very little one can say with certainty about the net environmental effects of passing a parcel of land from the commercial sector to the small scale, part-time farming households envisioned with land redistribution. This is particularly so since the experience of the former homelands cannot be used as a guide. Arable land per rural resident in former homelands has tended to be about one tenth or less than that of other areas (*DBSA, 1990*).

There are three problems with the measurement of environmental effects. First, many of these effects manifest themselves as externalities, and thus their costs are borne not by those who own the degraded land, but by their neighbours near and far, or by municipalities who must pay to remove silt from dams. Second, even the internalized costs are difficult to measure since they involve a complex chain of cause and effect from, say, soil erosion, to lost productivity, to lost profitability, year after year. Third, and most seriously, it is the net effects of land redistribution on the environment that are sought, meaning that one is explicitly comparing the environmental costs on land transferred to what the cost would have been if that land had not changed hands.

On the whole, one might expect a net reduction in environmental destruction for two broad reasons. First, labour-intensive technologies tend to be more environmentally sound than capital- and chemical-intensive ones. Second, the land reform process will somewhat ease the crowding problem in the former homeland territories, which may thus reduce cultivation of marginal land and over-stocking.

Employment impact

Increased employment is socially desirable, particularly insofar as it enhances equity and contributes to social stability. However, from a benefit-cost perspective, increased employment is not strictly speaking a benefit of land redistribution; it does not itself constitute social surplus that can be assigned a monetary value. On the other hand, if those newly employed had been previously involuntarily unemployed, such that neither the productive use of their labour nor leisure were disrupted, then their employment does not constitute a social cost.

Since increased employment is nonetheless an important aspect of land reform, it can be indicated outside of the main analysis, expressed simply in raw numbers of livelihoods created rather than aggregated with other benefits and costs. The net employment effect may be conceptualized as the gross employment on new farms, less the employment displaced from that land under the previous owner, less whatever employment the new tenants gave up when relocating, plus the additional employment of individuals left behind, plus whatever additional, off-farm employment is generated through the multiplier effects.

The effect on employment levels is almost certainly positive due to the fact that beneficiary households are likely to use more labour-intensive methods than their predecessors (probably by a factor of two or three), while the amount of employment lost due to relocation is likely to be low given typical unemployment levels. Moreover, so long as the overall multiplier effect is positive, the employment effects associated with it are likely to be positive, although careful comparison of the labour intensity of the forwardly and backwardly linked industries before and after redistribution would be necessary to be certain.

Social stability effects

These fall into the category of non-quantifiable benefits but, like the employment effects, are nonetheless important to consider in assessing the overall desirability of land redistribution. The following is a brief summary of some of these expected benefits:

- **Reduced crime:** as families are better able to provide for themselves, become property owners, and are removed from overcrowded, poorly serviced areas, the incentive to engage in illegal activities will decline.
- **Better quality of life:** as overcrowding diminishes, and as individuals' needs are better met, the tendency towards intra- and inter-community conflicts will decline.
- **Increased capital accumulation:** households will invest more in

their individual residences, community infrastructures, and productive potentials as they sense improved security in their tenure rights, and as their economic stake in their present circumstances are augmented.

- **Improved household stability:** households are more likely to stay together when household members have less need to seek employment far away.
- **Positive impact on urban areas:** better local income earning opportunities and greater family integrity in rural areas mean fewer rural-based individuals will be compelled to compete for urban jobs and accommodation.
- **More access to grazing land:** both communities which move and those which do not will have access to more grazing land, meaning many individuals who were previously unable to pursue this traditional way of life will now be able to do so.

Infrastructure costs

The costs of the new infrastructure investments necessary to accommodate households and communities relocating are likely to be very high. This is so because only a limited amount of the infrastructure needs of those acquiring land can be obtained by simply adapting whatever former owners leave behind. There is also much uncertainty about infrastructure costs, since much depends upon how households choose to co-ordinate their efforts. For instance, setting up an infrastructure to serve scattered homes costs far more than to serve the same number of closely spaced homes. Also, given the RDP's commitment to upgrading service provision and infrastructure to all disadvantaged communities, it is unclear whether and to what degree (non-agricultural) infrastructure costs incurred in association with land redistribution beneficiaries should be attributed to the land redistribution itself.

Estimating benefits and costs of the Pilot Programme

This section presents a preliminary benefit-cost analysis for the Land Reform Pilot Programme. Although the Land Reform Pilot Programme (LRPP) represents only one route through which land reform will take place, it serves as a fairly generalizable model for land reform in its economic implications for beneficiaries and for society at large.

The approach taken is non-aggregative in the sense that the analysis is conducted independently for different regions – thus different kinds of farms – without attempting to sum up to the national level.[4] Also, the analysis takes farm areas as its numeraire, where the areas are those for 'typical' beneficiary household farms for each region. These, in turn, are determined by considering what an LRPP beneficiary household can purchase with a land acquisition grant of R6 000.[5] These farm sizes are 14,6 hectares for the Eastern Cape (of which 2,6 are arable and 12 grazing), 11 hectares for the Summer Grain Area (SGA) (of which 1 is arable and 10 grazing), and one third of an irrigated hectare for the Western Cape.

The principal assumptions are that: (a) there are no price effects; (b) all redistributed land is sold voluntarily by commercial farmers; (c) non-farm output (by which is meant the product of beneficiary households' non-agricultural activities) is not affected by land redistribution; (d) the discount rate is 10 per cent, representing a healthy real rate of return on agricultural capital for South Africa; and, (e) the cost of agricultural services (extension) is assumed not to change with the redistribution; instead, these services are reoriented.

Assumptions of a more specific nature are discussed below. Since the effect of redistribution on land productivity remains one of the largest unknowns, and since it underpins many of the other elements of the analysis, different scenarios are treated by way of sensitivity analysis. The same is done for linkage effects.

For the sake of brevity and clarity, the results are presented only for the full social effects of the LRPP, rather than for the private and governmental effects as well.

Change in net farm income

For each of the three regions treated, the following interim loss in income is assumed where NFI/ha represents the baseline potential of the commercial farms for the respective region:

- Year 1: 40 per cent of NFI/ha
- Year 2: 30 per cent of NFI/ha
- Year 3: 20 per cent of NFI/ha
- Year 4: 10 per cent of NFI/ha
- Year 5+: no gap

4 Underlying farm models are taken from De Klerk *et al*, 1993.

5 The level and type of financing for the LRPP are not, as of this writing, certain. Moreover, regardless of the amount of direct Programme assistance, households may take out (additional) loans in order to purchase additional land. This alters very little for the purposes of the present analysis.

These percentage losses are then applied to the baseline long term NFI/ha. Thereafter, the loss per hectare is translated into a loss per farm (based on the stylized beneficiary farm size as mentioned above), and then future losses are discounted back to the present and summed. Alternative loss patterns are treated later.

Net off-farm linkage (multiplier) effects

For the Eastern Cape and SGA models, the production linkage component is initially assumed to increase by 10 per cent. For the Western Cape, it is assumed not to change since technology and land use – and thus the input and output mixes – are likely to stay the same.

On the consumption side, the linkage is assumed to increase for all areas by 20 per cent, a conservative guess in light of the consequent doubling or trebling of employment per hectare for the Eastern Cape and SGA, and the higher income elasticity of demand overall.

The consensus from international evidence is that consumption linkages comprise a larger share of the overall multiplier than production linkages, ranging between 60 per cent for Oklahoma and 80 per cent–90 per cent for parts of Malaysia and Sierra Leone (*Haggblade et al, 1991*). Assuming an intermediate share for South Africa of 75 per cent, and a base multiplier of 1,6 (*World Bank, 1993*), the new multiplier for the Eastern Cape and SGA cases is 1,71, and for the Western Cape 1,69.[6]

Nutrition effects

Assuming R26 per parcel and 6 individuals per household, the present value of cost to the government forever per eligible family would be:

(R26 x 6 x 12 months) / 0,10 = R18 720

Assuming the same proportion of households that will benefit directly from land reform can be classified as malnourished as suffer from malnutrition nationally, the *average* benefit per household (taking well-nourished and malnourished households together) would be something

6 For example, for the Eastern Cape and SGA, the new multipliers may be calculated as:

(1,6–1) x 20% x 110% (for production side)
+ (1,6–1) x 80% x 120% (for consumption side)
+ 1 = 1,708.
So the present value of the gross linkage effect would be
0,708 x 60% x NFI/ha – 0,6 x NFI/ha

in the order of R2 800, given an incidence of malnutrition of about 15 per cent.[7]

This can be considered a very conservative estimate, because it uses as a basis food parcels that meet only one third of an individual's needs, while access to productive land, in most cases, would allow for a far greater share than this (*Tapson, 1993*). Additionally, it does not attempt to account for the nutrition benefits associated with land transfers that would manifest themselves *indirectly,* i.e. through the income-augmenting linkage effects.

Risk diffusion benefits

The household risk benefit is tentatively based on the empirical results of Rosenzweig and Wolpin (*1993*), discussed above. Translated into the South African context, this would imply a household risk diffusion benefit of more than R900 per year, corresponding to a present value of R9 000 (at a discount rate of 10 per cent) for all years. For the sake of being conservative, (particularly in view of the danger of borrowing empirical results from another part of the world), this is reduced by an arbitrary 30 per cent, thus R6 300 is used.

Environmental effects

The consequence of taking water erosion into account is to lower the calculated present value of income streams associated with agricultural production, since water erosion diminishes the productive value of the land. In the absence of quantifiably intelligent guesses as to whether this adjustment must be more or less severe in the case of commercial production than under the land's new use, the present exercise assumes that the effect on land quality is the same. However, this means that the initial 'interim gap' between NFI/ha's for commercial and new farms must

$+ \quad [0{,}708 \text{ x } 70\% \text{ x NFI/ha} - 0{,}6 \text{ x NFI/ha}] / (1 + d)$

$+ \quad [0{,}708 \text{ x } 80\% \text{ x NFI/ha} - 0{,}6 \text{ x NFI/ha}] / (1 + d)^2$

$+ \quad [0{,}708 \text{ x } 90\% \text{ x NFI/ha} - 0{,}6 \text{ x NFI/ha}] / (1 + d)^3$

$+ \quad \{[(0{,}708 - 0{,}6) \text{ x NFI/ha}] / d\} / (1 + d)^4,$

where d is the discount rate. This is then multiplied, say, by 0,10 to capture the net effect of changes in resource costs.

7 A firm estimate of the percentage of people malnourished is impossible to come by, not least because the concept malnourishment is not amenable to a simple, universally applicable definition. The 15 per cent figure taken here is considered conservative, given estimates of 'vulnerability' among South Africa's rural children of about one third (*May et al, 1994*).

be adjusted down, since the cost of environmental degradation is proportional to the earning potential of the land at any given time.

The extent of soil compaction may well be different between old and new land uses because of different technologies. As labour is substituted for heavy machinery, there will be less soil compacting traffic on the soil. Areas that are vulnerable to compaction, and would have succumbed to it under continued commercial agricultural technologies, may thus be spared.

(a) **Water erosion adjustment.** The countervailing adjustment due to water erosion would start from the second year, being a function of the annual percentage decrease in NFI/ha, where NFI/ha here is specifically for exploitation of arable land.[8] For commercial arable areas as a whole, 22 per cent of the land is vulnerable to moderate erosion, and 9 per cent to severe erosion. On the assumption that transferred land will fall into these categories with these likelihoods (as opposed to a bias towards, say, transferring the most susceptible land), and further that these percentages apply roughly to the arable lands of the three regions, the total adjustment per farm works out to R19, R24 and R273 for the Eastern Cape, SGA and Western Cape respectively.[9]

(b) **Soil compaction.** The soil compaction benefit associated with a change towards more labour-intensive techniques is held to accrue to the Eastern Cape and SGA case. Despite their otherwise more degraded resource bases, the incidence of compaction in former homelands is about 19 per cent, as opposed to 27 per cent for commercial areas. A reduction in compaction is not assumed to occur for the Western Cape, where the assumption of change of technique is less justified.

8 That is, the adjustment for land undergoing degradation would be

$$(30\% \text{ x NFI/ha}) \text{ x } (1-e)/(1+d)$$
$$+ (20\% \text{ x NFI/ha}) \text{ x } (1-e)^2/(1+d)^2$$
$$+ (10\% \text{ x NFI/ha}) \text{ x } (1-e)^3/(1+d)^3,$$

where e is the annual percentage decrease in productivity, set equal to 1,35% and 3,6% for moderate and severe erosion respectively (*McKenzie, 1994*).

9 That is,

$$22\% \text{ x } [(30\% \text{ x NFI/ha}) \text{ x } (1-0{,}0135)/(1+d)$$
$$+ (20\% \text{ x NFI/ha}) \text{ x } (1-0{,}0135)^2/(1+d)^2$$
$$+ (10\% \text{ x NFI/ha}) \text{ x } (1-0{,}0135)^3/(1+d)^3\ ]$$

If compacted soil leads to a 45 per cent reduction in NFI/ha, 2 per cent of the vulnerable area is compacted per year under present technologies, 27 per cent can be taken as the degree of vulnerability associated with typical commercial agricultural techniques and 19 per cent as that for more labour-intensive techniques, then the present value of losses associated with compaction is R13 and R16 respectively for the Eastern Cape and SGA.[10]

Infrastructure costs

Infrastructure costs naturally vary enormously depending upon the circumstances. The approach taken here is to focus on water costs. If the density of settlement conforms to the Transvaal Provincial Administration's rural housing plot standard of 40 x 50 metres, and given a per hectare bulk water supply cost of about R10 000,[11] the cost per household works out to R2 000.

Other significant infrastructure expenses might include those for roads and electricity. New roads will only sometimes be necessary, because beneficiaries will be settling on land that has been relinquished by white farmers, and thus will already have access to local routes as well as have some service roads. Electricity hook-up is another matter, since not all portions of a commercial farm are likely to be linked to the grid. At any rate, the cost of extending electricity to such areas will be modest relative to that for water. A conservative approach in the absence of realistic, detailed information, is to account for these other types of infrastructure by taking 150 per cent of the water costs estimated above. In addition, assuming maintenance costs in the order of 5 per cent of initial capital costs per annum forever, the total PV of infrastructure plus

$$
\begin{aligned}
+ 9\% \times [\ & (30\% \times \text{NFI/ha}) \times (1-0{,}036) / (1+d) \\
+ & (20\% \times \text{NFI/ha}) \times (1-0{,}036)^2 / (1+d)^2 \\
+ & (10\% \times \text{NFI/ha}) \times (1-0{,}036)^3 / (1+d)^3\ ].
\end{aligned}
$$

10 That is,

$$
\begin{aligned}
& 45\% \times 2\% \times \text{NFI/ha} / d && \text{(PV of loss due to compaction in year 1)} \\
+ & [45\% \times 2\% \times \text{NFI/ha} / d] / (1+d) && \text{(PV of loss due to compaction in year 2)} \\
+ & [45\% \times 2\% \times \text{NFI/ha} / d] / (1+d)^2 && \text{(PV of loss due to compaction in year 3)} \\
+ & \ldots
\end{aligned}
$$

$$= [45\% \times 2\% \times \text{NFI/ha} / d] / d$$

$= [45\% \times 2\% \times \text{NFI/ha}] / d^2$, is the PV per affected hectare.

Then $(0{,}27-0{,}19) \times [45\% \times 2\% \times \text{NFI/ha}] / d^2$ is the PV per 'average' hectare, which must then be adjusted by farm sizes.

11 This is taken from recent resettlement schemes of the DBSA.

maintenance is R4 500 per household. Given the need to redress the poor state of infrastructure in most former homeland areas, many if not all of these expenses should not be ascribed to the land reform process. For the purposes of this exercise, the conservative assumption is made that half of these infrastructure costs are ascribable to the LRPP.

Programme management

Table 23.2: Programme Management Costs

Year	National Pilot Office	District Pilot Offices	Planning and preparation	Survey and transfer
1	R600 000	R1 800 000	R3 600 000	R3 420 000
2	R660 000	R8 100 000	R20 000 000	R5 840 000
3	R726 000	R8 100 000	R3 400 000	R1 000 000
PV sum	R1 800 000	R15 857 851	R24 591 736	R9 555 537
PV/HH	R133	R1 174	R1 822	R708

Programme management costs include those for the national pilot office at the Department of Land Affairs, those for the district pilot offices, planning and preparation costs, and survey and transfer costs. The latter three categories have already been explicitly budgeted for, while the first category is estimated on the assumption that there will be seven full time staff members, four of whom incur substantial travel expenses.

The table above shows the year by year budgeting, the sum of the present values, and the PV per household, assuming 13 500 beneficiary households.

The total per household administrative cost is thus R3 837.

Results

Table 23.3 shows results on the basis of the assumptions explained above. These results will serve as our 'baseline' against which to judge the results based upon alternative assumptions. All of the figures represent the full social benefits and costs of the redistribution, inclusive of private or governmental benefits and costs that are not netted out through pure transfers.

Table 23.3: Results of the Benefits and Costs Estimations

Item	Eastern Cape	SGA	Western Cape
Net farm income	–747	–1 808	–1 881
Linkage effects	44	107	–100
Nutrition effects	2 808	2 808	2 808
Risk diffusion effects	6 300	6 300	6 300
Environmental effects	32	40	273
Infrastructure	–2 250	–2 250	–2 250
Programme management	–3 837	–3 837	–3 837
NPV (Total)	R2 350	R1 359	R1 312
Benefit–cost ratio	1,34	1,17	1,16
Internal rate of return	16%	13%	12%

Several observations are in order. First, the overall result is overwhelmingly positive, particularly when taking into account the magnitude of the total relative to any of the constituent items. Second, the less palpable benefits – namely nutrition and risk diffusion effects – are indeed the most significant of the benefits. Third, the programme management costs comprise a large share of the overall costs, but it is reasonable to assume that this is one area in which the LRPP may not be indicative of the broader land reform effort since the state will most likely benefit from some economies of size in administration. Finally, net off-farm linkage effects turn out to be relatively small, apparently contradicting the claims commonly made as to their importance in land reform.[12]

Sensitivity analysis is performed with respect to assumptions on relative productivity and the nature of the linkage effects. Three additional scenarios are considered: the case where the overall multiplier decreases by 10 per cent; the case where rather than reaching productive parity after the fourth year, beneficiary farms' NFI/ha is 10 per cent below that of their commercial counterparts from the fourth year on; and the case where parity is reached in the fifth year, and thereafter beneficiary farms outperform their commercial counterparts by 10 per cent per hectare. None of these scenarios is entirely implausible, although the accumulated international evidence on the relative efficiency of small farms –

12 However, these claims typically concern the importance of linkage effects in generating additional employment which is quite different from the money metric used here.

based as they are on the use of family labour – would suggest that the last scenario is the most likely.

Table 23.4 reports the social NPV per beneficiary farm household for the baseline as well as these three other cases.

The analysis' overall results are indeed sensitive to the assumptions regarding the multiplier and the relative productivity of land reform beneficiaries. The biggest danger seems to be a sustained drop in productivity, which could result in a negative NPV. On the other hand, a sustained productivity increase could make the land reform highly successful in benefit-cost terms and perhaps, as suggested earlier, this productivity increase is a more realistic expectation. Thus, even though there is much more to land reform than agricultural production, these results highlight the importance of an environment that promotes the effective use of the land, meaning, inter alia, that beneficiaries must have access to effective agricultural extension services and input supplies.

Table 23.4: Social NPV per Beneficiary Farm Household for Baseline and three Other Cases

Item	Eastern Cape	SGA	Western Cape
'BASELINE'			
NPV	R2 350	R1 359	R1 312
Benefit-cost ratio	1,34	1,17	1,16
Internal rate of return	16%	13%	12%
MULTIPLIER DOWN BY 10%			
NPV	R2 204	R984	R1 143
Benefit-cost ratio	1,32	1,12	1,14
Internal rate of return	15%	11%	11%
NFI/HA DOWN BY 10%			
NPV	R1 664	-R269	-R590
Benefit-cost ratio	1,22	0,97	0,94
Internal rate of return	15%	11%	7%
NFI/HA UP BY 10%			
NPV	R2 922	R2 775	R2 438
Benefit-cost ratio	1,47	1,42	1,37
Internal rate of return	19%	18%	15%

Also of interest is the fact that assumptions about the multiplier matter less than those about relative productivity. This is not to suggest that farm–off-farm linkage effects are unimportant – multiplier effects will certainly be critical to the welfare of beneficiaries and rural non-beneficiaries alike – but it does imply that the overall success of land reform does not greatly hinge upon the exact magnitude of these linkages, about which we know so little.

Conclusion

This conceptual and numerical exercise reveals, among other things, the tentative nature of much that must enter into a benefit-cost analysis of land redistribution. It must be conceded that many of the valuation strategies employed above owe as much to art as to science. However, the power of the analysis lies not in its claim to precision – it does not make one – but in its ability to conjure rough orders of magnitude, to discern how sensitive the results are with respect to particular assumptions and, most of all, to encourage analysts and policy makers to perform a full accounting of the economic effects that can be anticipated from a given policy. This is no less true for the case of land redistribution: only with a careful enumeration of the wide variety of its effects and the agents that enjoy or bear them, can informed planning be pursued.

The numerical part of the exercise, for instance, reveals the overriding importance of the effectiveness with which land reform beneficiaries use the land. Under conservative assumptions about relative productivity, the land reform promises to be decidedly in society's economic interest. However, given a complementary set of policies which further assist emergent agriculturists – whether effective agricultural extension or improved access to credit – the land reform would appear to promise an impressive success.

References

Binswanger, Hans, Deininger, Klaus and Feder, Gershon. Power distortions, revolt and reform in agricultural land relations. In T.N. Srinivasan and J. Behrman. (Eds), *Handbook of Development Economics,* Vol. III. Amsterdam: North-Holland. Forthcoming.

Black, P.A., Siebrits, F.K. and van Papendorp, D.H. 1991. Homeland multipliers and the decentralization policy. *The South African Journal of Economics*, 59(1) :36–44, March.

DBSA. 1990. *A Regional Profile of the Southern African Population.* South Africa: Halfway House.

De Klerk, Michael, May, Julian, Nyamande, Agnes, Tapson, David, van den Brink, Rogier and Wentzel, Wilfred. 1993. Developing models of land use activities: agricultural development options. *LAPC Research Paper No. 8*. September.

Gwatkin, D.R., Wilcox, J.R. and Wray, H.D. 1980. Can Health and Nutrition Intervention Make a Difference? *Overseas Development Council Monograph No. 13*. February.

Haggblade, Steven, Hammer, Jeffrey and Hazell, Peter. 1991. Modelling agricultural growth multipliers. *American Journal of Agricultural Economics,* 73(2) : 361-374. May.

Haggblade, Steven, Hazell, Peter and Brown, James. 1989. Farm–non-farm linkages in rural sub-Saharan Africa. *World Development,* 17(8) : 1173–1201. August.

Levin, Henry M. 1986. A benefit-cost analysis of nutritional programs for anemia reduction. *The World Bank Research Observer,* 1(2) : 219–245. July.

May, Julian, Persad, Ranveer, Attwood, Heidi and Posel, Dori. 1994. Rural poverty and institutions. *LAPC Policy Paper No. 7*. October.

McKenzie, Craig. 1994. Degradation of arable land resources: policy options and considerations within the context of rural restructuring in South Africa. *Paper prepared for the LAPC Workshop,* Johannesburg: March.

Rosenzweig, Mark R. and Wolpin, Kenneth I. 1993. Credit market constraints, consumption smoothing, and the accumulation of durable production assets in low-income countries: investments in bullocks in India. *Journal of Political Economy,* 101(2) : 223–244.

Tapson, D.R. 1993. Models of land use activities: case study of livestock in the Eastern Cape. *Land and Agriculture Policy Centre Research Paper 8a*. September.

Van Zyl, J., Kirsten, J.F. and Sartorius von Bach, H.J. 1994. Poverty, household food security and agricultural poduction: evidence from South Africa's communal areas in a period of drought. *Unpublished typescript.* Pretoria: Department of Agricultural Economics, University of Pretoria.

World Bank. 1993. Options for land reform and rural restructuring in South Africa. *Paper presented at the LAPC Land Redistribution Options Conference.* Johannesburg: 12–13 October.

24

Land reform and management of environmental impact

David Cooper, Mohamed Saliem Fakir and Daniel Bromley

Introduction

The national land reform programme in South Africa is aimed at contributing to rural development as an integral part of the Reconstruction and Development Programme (RDP).[1] The land reform programme, through restitution, aims to address the injustices of forced removal and, through redistribution, aims to allow more people access to land. Coupled to these aims is the objective to ensure security of tenure for rural dwellers utilizing various forms of tenure. Land reform can also be seen as a way of experimenting with new property regimes through the transfer of resources to individuals or groups previously disadvantaged by existing land rights systems which are either inadequate or further disadvantage the poor. The intent of the RDP with regard to land reform is to create new livelihood opportunities for the rural poor and, in so doing, to alleviate pressure on the natural resource base.

Although most of those responsible for the design of the land reform programme would subscribe to the view that sustainability is a central concern, there have been few steps to concretize what this means in practice. For example, no environmental criteria are mentioned by the Land Claims Commission when return of land is considered.

On the other hand, environmental considerations are frequently mentioned by those opposed to land reform. The threat of damage to the land is presumed by opponents to be sufficient justification for refusing to contemplate land reform. Perhaps this in itself explains why proponents of land reform are reluctant to take environmental concerns into account in their planning.

In fact, the environmental concerns around land reform are highly complex. South Africa's rural history is one of removing people from the

1 The Pilot Land Reform projects received R300 million to establish nine district pilot studies in the nine provinces.

land in the name of conservation. South Africa is characterized by a skewed distribution of natural resources. The essence of apartheid was to bestow exclusive rights of key resources to a few – those individuals deemed to be important to the success of the political system. The formulation of the bantustan policy, and the enclosure of the majority of South Africans in limited land space, is a case in point. In some cases, the fear associated with land reform is an economic one which is couched in terms of environmental degradation by changing the character of land ownership. Land reform gives the potential of creating new claimants and economically active agents competing with individuals who, in the past, have enjoyed state patronage and protection.

A government commission of inquiry in the 1970s identified larger farm size with better land use practices, and entrenched in the minds of South Africa's white bureaucrats the notion that small-scale farmers are more destructive than large-scale ones. The Subdivision of Agricultural Land Act codified this notion into law. People point out that the evidence is overwhelming – you just have to look at the difference between white farms and the former homelands. The recent work of the World Bank on agriculture was the first to seriously challenge this notion (*World Bank, 1994*). Using international evidence, this work pointed out that there was no evidence that small-scale producers are more destructive land users than large-scale producers. In South Africa, the record of large-scale producers is not good – large-scale soil loss results from practices like monocropping, poor irrigation and intensive mechanization.

Clearly, no one would want the consequence of land reform to be a deterioration of the environment. It is quite evident that overcrowding and lack of investment in the former homelands have led to widescale soil erosion. There is a serious level of deforestation in some areas, and streams have been denuded. But there is also a healthier biosystem because of the absence of monocultures and minimal use of pesticides.

At one level, land reform has the intention of reducing environmental destruction in the former homelands by reducing population pressure on the land there. But how can degradation on new lands be prevented, or the damage limited?

McKenzie (*1994*) argues that it is expensive to reclaim degraded lands, and far cheaper to prevent degradation. He adds further, that most of the costs of degradation are borne off-site-in the silted rivers in particular. Hence the state, in the name of those who are affected off-site, should contribute to the costs of preventing land erosion. This effort can be assisted further by the giving of incentives to the landowner to farm using sustainable practices. In order to do so, a number of issues have to be dealt with:

- security of tenure;
- education and extension services for conservation;
- sound land use planning using environmental guidelines;
- the provision of infrastructural support in terms of markets, access to credit, etc.

This chapter, while conceding that land reform has an important developmental component, argues that environmental impacts are best managed through coherent policy that establishes properly legitimated rule systems and incentives. That policy applied to property regimes as well as other economic incentives and sanctions can induce behavioural patterns that encourage sound and efficient use of resources. Environmental impact is best managed when the basic needs of people are met, i.e. when addressing the problem of environmental scarcity.

Environmental scarcity

The scarcity of environmental resources can lead to violent conflict between different ethnic, racial and class groups. Environmental scarcity is a result of three major factors (*Homer-Dixon, 1994*):

1. The result of property regimes which lock environmental resources in the hands of a few. Environmental scarcity occurs because there is unequal access to resources. In South Africa, conflict over resources such as water and land is a result of unequal property regimes (the right to a specific stream of benefits) that privilege some and exclude others. About 60 per cent of water is in private hands due to private water rights or riparian rights linked to land rights.
2. The result of too many people in limited space. With the increase in population, the demand for resources exceeds supply. In the former homelands, degradation of grazing land and denudation of indigenous forest is attributed to high population densities. The policy of betterment schemes applied during the apartheid era heightened this problem because it collectivized agricultural activity, spread over a larger land space, into a narrower domain. For instance, former homelands such as QwaQwa and Ciskei have 9 or 10 times more inhabitants than the areas can accommodate.
3. The result of over-utilization of renewable resources. This can be

attributed to both points 1 and 2. In the first instance, where a system of 'open access' or non-property exists, resource depletion can be extensive. The non-existence of rules means that enforcement of user-limits is not workable. Individuals then seek maximum benefit by over-exploiting the resource since there is no sanction or incentive to conserve the resource. Over-utilization of resources affects both the quality and quantity of the resource. The over-utilization can occur equally in different forms of property regimes; it is not exclusive to any particular kind of tenurial relation.

Environmental scarcity is a result of political, socio-economic and biological causes. The implementation of an environmental management system will have to take into consideration a host of key indicators ranging from political, economic, legal, social and biological systems to account for environmental degradation. These are inextricably linked. Resource scarcity is, in part, subjective; it is determined not just by absolute physical limits, but also, preferences, beliefs and norms (*Homer-Dixon, 1994*).

Land reform can potentially lead to environmental sustainability, by ensuring social sustainability, if the programme has institutional and infrastructural support.[2]

Land use and degradation[3]

Change in land use patterns impact not only on soil quality but also on the quality of water. South African soils are by nature fragile; they have low organic matter[4] and are susceptible to high rates of erosion. The total land area of South Africa is 1,1 million hectares and the predominant forms of land use in rural areas are agriculture (i.e. under arable and non-arable use), forestry and nature conservation.

2 Environmental sustainability is the ability of the natural environment to provide benefit to future generations. Social sustainability is human institutional systems that are in place which manage the resource base in an effective and efficient manner. The reproduction of social institutions is linked to material production. The net economic value of the resource is an important criteria in the management of the resource. As long as the value of the resource is seen through net social welfare, the more likely is the resource to be conserved.

3 A large part of this section of the work has been drawn from the draft paper by Daniel W. Bromley. *1994*. Environmental Issues in South Africa: Development Options and Policy Concerns for the Land and Agriculture Policy Centre (LAPC).

4 About 60 per cent or more of South African soils have very low organic matter content.

About 16 per cent of this land area is considered arable, of which 3 per cent is high-potential agricultural land. Most of the arable land is found on the coastal plains of South Africa where the average rainfall is above 500 mm. Approximately 13 per cent or 13 million hectares is under commercial farming and 2,5 million hectares are small-scale farming areas of the former homelands.

In general the South African climate is arid and unsuitable for intensive farming. About 65 per cent of South Africa has an average annual rainfall of less than 500 mm. This rainfall is erratic and sparse. South Africa has poor irrigation potential which is further constrained by the limited and unpredictable supply of water. Major degradation problems associated with arable land production are compaction, salinization (due to improper irrigation of land) and acidification (*McKenzie, 1994*).

Agriculture has been an important instrument of the state to control the economic and political tenor of rural areas. In the state's past drive to secure the welfare of the white farmer and also attain food self-sufficiency, several policy instruments for the agricultural sector were devised. Input subsidies, tax policies, and agricultural output policies distorted the true cost of agricultural production. These distortions encouraged cultivation of crops on marginal land, over-use of surface and groundwater, and excessive use of agricultural chemicals which impacted on the quality of land and water resources. This contributed to serious soil erosion of fragile land. Tax incentives and low interest finance encouraged large-scale mechanization of agriculture with attendant soil compaction. In the former homelands, the problems are even worse and are attributed to environmental scarcity and overpopulation.

New environmental policy concerning the use of rural land will need to better understand existing agricultural policies in the commercial and traditional agricultural areas of the former homelands which have led to various unsustainable land use patterns. Only in this way can alternative agricultural reforms be introduced to change undesirable land use practice.

Extensive land use in South Africa is undertaken by three major users. About 300 000 km^2 are dedicated to rangeland farming. The bulk of this land is under the control of freehold white farmers. The second major user of extensive lands are the tribal authorities; their land covers 100 000 km^2 and is used for communal grazing. The third major user is people involved in a variety of nature conservation purposes, ranging from the state to private individuals; this land comprises approximately 100 000 km^2. It is important to note that about 8,1 million hectares of rangeland in white commercial areas is regarded as being seriously degraded, while an additional 21,1 million is said to be moderately degraded. The extent of degradation in the former homelands is not clearly quantifiable, but estimations are that about 4 million ha is seriously degraded.

The total land surface used for plantation forestry is about 1,4 million hectares. It is also estimated that an additional 500 000 hectares need to be afforested in order to meet the shortfall of supply for the industry in the coming years. Presently, 80 per cent of forestry plantations exceed 1 000 hectares in size, and the policy is directed towards monocultural afforestation.

The major impacts associated with afforestation are reduction in water flow which is regulated by the Afforestation Permit System (APS). The APS system is based on the Mean Annual Reductions (MAR) calculated in 1972, but does not take into account new developments in silvicultural practice that are aimed at improving yield per hectare of trees planted. The reduction of water flow at lower parts of the catchment areas means that the off-site cost of afforestation is borne by other competing land users. Soil compaction through the use of heavy machinery, loaders and vehicles is also of concern to the forestry industry. Other environmental concerns associated with the forestry industry are the use of insecticides and herbicides which affect workers and water quality (*Gandar & Forster, 1994*).

The rate of soil formation in South Africa is about 1 mm every 40 years. Current research indicates that the rate of soil loss is at 10 times the rate of formation, about 3 tons per hectare per year. However, the rate of soil loss varies from region to region and according to different land use practices. The annual total loss of soil is estimated to be about 400 million tons per annum (*Verster et al, 1992*).

> Soil, one of South Africa's most basic resources, is being lost to erosion at a frightening rate; this may well be the greatest environmental problem facing South Africa, yet the South African population appears to be the most complacent about it.
>
> (*Verster et al, 1992: 181*)

Reform of environmental policy

The challenge facing the new government is to transform environmental policy designed for the apartheid state by devising new rules, laws and organizational structures that represent the collective will as a whole.

Public policy must pay attention to three essential components: (a) the intentions of the collective will; (b) associated incentives and sanctions that will make the intentions effective; (c) the necessary enforcement structures and mechanisms to ensure that policy intents are carried out.

Currently, environmental legislation is fragmentary, ad hoc, weakly enforced and reluctant to involve wide public participation in policy. While reform in the legislation is required in some circumstances, as a whole, radical changes must occur in the way policy debate and content is gathered. In general, environmental policy must work out new rules which will reward individual initiatives and, at the same time, protect environmental resources by penalizing offenders.

Legal reforms

The repeal or introduction of specific pieces of legislation such as the Subdivision of Agricultural Land Act and the Development Facilitation Act (DFA) which will affect land use planning and management from a developmental perspective are also likely to have environmental implications.

The Subdivision of Agricultural Land Act 70 of 1970 was designed to prevent the creation of smaller 'uneconomical' farming plots and increase white farmer control over larger agricultural land. The Act also had the premise of preserving large areas of agricultural land, i.e. in environmental terms. However, the particular model of agricultural production, which relied on monocropping and large-scale mechanization had adverse effects to those intended. This piece of legislation, in the next parliamentary session, is likely to be reformed or repealed because the Act has outlived its purpose and context. It is uncertain what impact the division of land into smaller holdings will have; nonetheless, it is clear that land use practice should be the target of policy to ensure better management of resources.

The DFA is aimed at fast-track development of land that allows the rationalizing of institutional and technical processes to ensure the speedy release of land for development. It incorporates environmental impact assessment as part of the planning phase of developmental projects. However, the possibilities of institutional devolution of Integrated Environmental Management, which requires Environmental Impact Assessments (EIAs) from the national Environmental Ministry to provincial governments, increases the likelihood that this aspect of the DFA will not be implemented if the provincial governments do not have the capacity to undertake such assessments. Most provincial departments are poorly financed and ill-equipped to undertake adequate environmental management programmes. In this sense, the clause is superfluous. The clause also does not make explicit the need for social impact studies of major development projects.

Tenurial relations

Property rights secure an exclusive claim over a stream of benefits that arise from the use of resources by excluding others from the use of such resources. Property rights bestow on the beholder the collective power to protect his/her claim to a benefit stream. The system of rights only work if there is an authority system in existence that can enforce those rights. Broadly, there are four major types of property regimes: state property, private property, common property and non-property. In South Africa, natural resources are managed under one or the other of these types of property regimes.

There is a predominant view amongst economists and policy makers that freehold tenure, which provides exclusive individual rights over the resource base, establishes the correct incentives for efficient use of resources. In the former homelands, resource degradation was blamed on communal tenure systems. The nature of communal land rights was ill-understood by policy makers.[5] It was largely thought at the time that communal arrangements were 'primitive', hence policy interventions such as betterment schemes and the Trust Land system were introduced into traditional institutions. In general, these policies have failed due to their rigidity and the concurrent breakdown of traditional institutions. Property regimes function best when the authority system is intact. When these breakdown, there is a collapse in coherent management over environmental resources.

Cousins states that the premise for the argument for communal land rights is the idea that they are potentially: (a) equitable; (b) economically efficient; and, (c) ecologically appropriate and sustainable. Firstly, communal land rights reduce transaction costs of people who live in relative poverty. Secondly, they minimize individual risks associated with natural resources which vary widely over space and time. Thirdly, when resources are thinly spread over space and time, the possibility of

5 Common property regime will consist of a well-defined group of authorized users, a well-defined resource that the group will manage and use, and a set of institutional arrangements that define each of the above, as well as the rules of use for the resource in question. (Bromley, D. W.)

skewed and unequal resource endowments increases. Resources held in common can remove inequalities through common user rights, and hence prevent social destabilization. Social destabilization often leads to the breakdown of efficient resource use. Cousins recommends that the potential of this tenure, in its traditional or modified form, needs to be assessed.

Policy makers may want to consider which types of tenurial relations best secure the interest of individuals and, in turn, lead to sustainable utilization of resources. It may be that in some parts of the country freehold tenure works best and, in others parts like arid regions, common and state property regimes may need to be considered.

Role of integrated land use planning

The conservation of agricultural resources is administered by the Department of Agriculture through the Conservation of Agricultural Resources Act 43 of 1983. There are also the Physical Planning Act and the Environmental Conservation Act which, in one way or another, regulate land use planning. While these pieces of legislation boldly lay out integration of environmental management into land use planning, at the administrative level land use planning remains sectoral and fragmented.

The problem of integrating land use planning seems to be an administrative and planning problem, rather than one of legislation. The reorganization of state departments in terms of different tiers of implementation is best facilitated according to the practicable means and appropriateness of the line functions at these levels. In other words, national norms and standards are best performed at one level of government, and planning and enforcement at another level of government. The exact conceptualization of the implementation of integrated land-use planning can only be worked out once there is inter-departmental co-ordination of planning and policy.

Integrated land use planning ensures that there is efficient use of land and water resources according to different agro-ecological and climatic zones.

The role of monitoring and evaluation of environmental impacts

The Land and Agriculture Policy Centre (LAPC) has embarked on the design of the Monitoring and Evaluation System (M & E) for land reform which is to be implemented by the Department of Land Affairs (DLA). The design phase is to be completed by March 1995. This is the first time in the world that an M & E system is being done in conjunction with a land reform programme. The focus of the M & E system will be the access of agricultural land, and therefore agricultural land use will be a major concern of the M & E system.

Monitoring is the frequent collection of key data to indicate whether the implementation process designed to bring about change is working. Monitoring provides valuable information to managers during the implementation process, allowing quick interventions to be made in order to keep the process intact.

Evaluation is a way of assessing whether the project has been successful or not in terms of the objectives and targets laid out by the managers. Evaluation is done after the implementation phase and is based on the analysis of data acquired during the monitoring phase. The M &E system is being designed to give quick and easy information relating to land reform.

A key component of the M & E system is to establish suitable indicators for environmental impacts associated with land use activity. This aspect of the work is still in the proposal phase. Initially these indicators will be biophysical indicators based on quick reconnaissance made by physical observations, followed by village interviews to investigate people's perceptions of change to the environment. This reliance on the gathering of data through local people's perceptions of environmental data is the central focus of the proposal. The task of experts would be to analyse the data or further investigate environmental impacts in detail if there is a need to do so. Information gathered will determine the extent of environmental deterioration, and will serve as an early warning system to managers so that real environmental damage is prevented.

Monitoring and evaluation cannot address the problem of environmental impacts; it can only establish suitable indicators that will show whether environmental and economic policy is working or not. Monitoring and evaluation merely provide a feedback loop so as to improve management of natural resources and net social benefit. Indicators also provide a relative assessment of the impacts of various land use practices, and hence inform decision-making in the future. It is important to note that, at present, environmental planning is not an integral part of the development of the Land Reform Pilot Programme. The mechanism of integrating environmental policy into the planning of the pilot projects will evolve with time as the monitoring and evaluation system is able to establish baseline indicators for environmental impacts.

Indicators should establish the viability of the natural resource base through various land use activities and their effects on other ecosystems as a whole (*Standing Committee on Agriculture and Resource Management, 1993*), as well as the social costs associated with the impairment of the natural resource viability. These indicators should be a set of on-site and off-site indicators determining sustainability of agricultural production. On-site indicators would be items such as quantity and quality of water and land, climate, labour inputs, quantity and price of inputs and outputs, management skills, technology, information and capital investment and major factors which impact on sustainability. A set of off-site indicators would be ecological, social and financial impacts. These indicators will establish whether agricultural activity is of such a nature that the extent to which it is undertaken causes a decline in the quality of the resource base over time. Some of these measures can easily be extended to other land use types.

The frequency of assessment will be done on an annual basis to identify trends in the resource base. Changes take a long time to emerge and make their impact. The proposal to develop environmental indicators for the M & E system also envisages the training of members of local communities to identify changes requiring very few scientific or technical skills.

Environmental education

Environmental practices vary enormously between users, even from the same educational and economic background. Land reform programmes must try and develop best use practices among users, and one of the main tools for doing so is education. This education should be participatory and based on continuous learning, not top-down messages. As part of the development of environmental education packages, it will be necessary to investigate 'best practices' which can be used as models. Farmers teaching farmers is an approach which can be followed to good effect.

Extension staff and specialist environmental staff will need retraining in meeting the education needs of land reform participants, both at the level of content about what constitutes good environmental management, and at the level of pedagogy – how to transfer this knowledge. However, the returns in terms of environmental impact can be dramatic.

Environmental planning

Planning is central to the conceptualization of the land reform programme. The Pilot Programme intends to provide funding to land reform beneficiary groups for them to undertake participatory planning, using outside experts. Groups will have to submit plans to be able to draw on grant funding. Clearly, environmental impact can be one of the planning tools used to ensure that rural communities benefit from land reform in the long-term. A particular need is planning tools which do not seek to restrict communities in their use of natural resources, except in the case of sensitive ecosystems, but work as mechanisms which encourage investment in the environment. For example, what incentives would there be for building small dams or erecting contours, and how could such investments yield both improved livelihoods and better environments? Planners need to consider environmentally friendly land use as a tool for encouraging investment, for example, through eco-tourism and agricultural projects which encourage the preservation or planting of indigenous species.

At the same time, planners must take special environmental resources into account and plan for their protection, for example, wetlands, protecting river banks and indigenous bush. In the first instance, this involves drawing up inventories of environmental resources, learning their value with communities, and planning ways both of protecting them and producing incomes from their use. Planners should also concentrate on diversifying land use as much as possible, since this is more likely to lead to sustainability than monocultures.

Conclusion

The objective of the land reform programme is clearly to enhance environmental sustainability in rural areas. This will only be achieved if environmental considerations become a specific part of the planning for land reform at both a macro- and micro- level. For example, removing incentives for poor land use is an essential prerequisite for sustainability.

In addition, more intensive land use must be expected from beneficiaries of land reform. Planners need to assist communities to find viable livelihood strategies that are not land destructive, for example, the use of non-farm income generating activity such as tourism and small-business development.

Poverty is most likely to lead to environmental destruction, so enabling communities to grow out of poverty will result in long-term sustainability. Although short-term intensive use of the environment may seem to be destructive, if it leads to long-term poverty alleviation, it is likely to encourage sustainability.

Investment in land improvement also needs to be considered –

whether through public works programmes or other means. Investment in the environment through public works programmes should receive priority attention from planners.

Environmentally sensitive areas need to be defined and protected, especially through community co-operation and education. Regulation, even policing, may be needed as short-term tactics but, in the long-term, the strategy must be one of stressing community ownership and seeking positive action rather than regulation.

Officials also need re-orientation that small-scale (and black) farming is not more destructive than large-scale (and white) farming. All of this is not easy to achieve, and will not be achieved unless positive and active steps are taken.

The perception that the rural poor are responsible for environmental degradation is tenuous and debatable, given the fact that the rural poor had no support from the state. Rather, it is inadequate policies and prejudice against the poor that have led to the unsavoury state of affairs in rural areas.

References

Cousins, B. 1993. Common property institutions in land redistribution programmes in South Africa. *Background paper for Rural Restructuring Programming*. LAPC/World Bank.

Gandar, M. and Forster, S. 1994. The impact of commercial afforestation on the rural areas of South Africa. *Unpublished Research Report.* Land and Agriculture Policy Centre and Water Research Commission.

Homer-Dixon, T.F. 1994. Environmental scarcities and violent conflict: evidence from cases. *International Security,* Vol.19, No.1: 5–40.

McKenzie, C. 1994. Degradation of arable land resources: policy options and considerations within the context of rural restructuring in South Africa. *Paper presented at LAPC workshop on Natural Resource Management*. Broederstroom: October.

Standing Committee on Agriculture and Resource Management. 1993. Sustainable agriculture: tracking the indicators for Australia and New Zealand. *Report No.51.* Australia.

Verster, E. *et al* 1992. Soil. In Fuggle, R. A and Rabie, M.A. (Eds), *Environmental Management in South Africa*. Cape Town: Juta and Co. Ltd.

World Bank. 1994. South African agriculture: structure, performance and options for the future. *Discussion paper no. 6.* Washington, D.C.: Southern Africa Department, World Bank.

25

A new vision for agriculture and management of the transition

Johan van Zyl

Both the political and economic management of the land reform process are of crucial importance. However, this cannot take place in a void and there is a need for a common vision which will focus the different stakeholders, diverse approaches and limited resources. Since much of this has to do with agriculture, the vision articulated here centres on agriculture. However, there are a host of other players and issues involved.

A vision for agriculture

Commercial agriculture in South Africa has been widely regarded as highly sophisticated and successful. However, detailed analysis showed that the development pattern was inequitable, inefficient and centralized – therefore unsustainable. Commercial agriculture in general is going through a period of crisis associated with change. Many of the privileges to large-scale commercial agriculture, such as taxation benefits, subsidized credit and price supports have been withdrawn. Combined with the effects of increasing international agricultural competition under the GATT agreement, this will place agricultural profits and land prices under severe pressure, especially in the grain and livestock sub-sectors.

Considerable achievements in addressing the problems facing South African agriculture have either been realized, in particular over the past five years, or are planned for the immediate future, most notably:

- the elimination of many policy distortions at both the macro and sector level;
- market liberalization across a wide range of commodities and services;
- broadened access to agricultural services for all farmers, large and small (particularly under the BATAT);
- decentralization of agricultural services under the new constitution;
- implementation of land restitution through the Restitution of Land Rights Act, and land reform under a Pilot Programme.

These actions are important to place South African agriculture on an efficient and sustainable development path which, following international experience in both the developed and developing world, would be based on a family farm structure. A family-type farm structure is characterized by: the majority of produce being marketed (non-subsistence); the predominant use of family labour; the hiring of some labour; the frequent additional off-farm sources of income of households; the enormous differences in size of these farm enterprises. The major determinants of these widely varying farm sizes are: climate; land quality; capital availability; distance to markets; commodities cultivated; family size; and the management capacity of the household. Profitability is not determined by the absolute size of farms in hectares, but by the on-farm organization of the means of production.[1] In addition, international experience shows that owner-operated farm enterprises have clear advantages over other forms of operation and tenure arrangements when the majority of goods are produced for the market, particularly with respect to credit arrangements.

Recent research on South African agriculture, some of which is reported on earlier in this book, firmly supports the above international findings. It demonstrates that efficiency and employment would increase if average farm size were to decrease in the commercial sector and increase in the former homelands.

A common vision for agriculture derives from these observations. South Africa's agricultural sector vision should be to **increase efficiency, equity and employment-intensity by moving towards a more diversified farm structure centered around competitive commercial, owner-operated, family farms. These farms should not be dependent on subsidies and government support for their sustainability, but primarily should be supported and serviced by the private sector. The role of government should be to establish a comprehensive legal, institutional and policy framework which will ensure a level playing field for all players. This framework will include increased reliance on markets, privatization, deconcentration and decentralization.**

Realizing the vision

To attain this vision, three complementary approaches exist:

1 For example, average farm sizes in other countries or states which are regarded as highly efficient (*1990*) are: Holland (16,1 ha); France (28,1 ha); Indiana, USA (102,8 ha); Wisconsin, USA (138,4 ha); and China (0,5 ha).

1. Accelerating rural development.
2. Increasing access to land for the poor.
3. Increasing access to land for small- to medium-sized commercial farmers.

1. **Accelerating rural development.** This option is largely the one presently advocated and followed by the Department of Agriculture, and consists of: (a) a level playing field where many of the distortions and policies favouring large-scale white farmers are removed – much progress has already been made in this respect, although subsidized credit still distorts markets and crowds out further private sector involvement; (b) rural development (RD) in the former homeland areas – many of the ideas and plans still have to be executed; (c) services with cost recovery in the commercial farm sector – good intentions still have to be put into practice; and, (d) labour reform in the commercial farm sector – efforts already underway have to be expanded and implemented. While all these actions are necessary, many of the good intentions still have to be translated into actions. Moreover, these actions are not enough to address the issues of inequity, inefficiency and unsustainability:

   - Duality and poverty will continue to be a major feature of South African agriculture.
   - The issue of farm bankruptcies, loss of employment on existing farms, and inefficiencies in the commercial farm sector will persist.

2. **Increasing access to land for the poor.** This option is the one followed by the Department of Land Affairs. It involves support of all the above, plus the additional issues of: (a) restitution – under the recently passed Restitution Act; (b) distribution of state land to the poor; and, (c) acquisition of additional land to redistribute, essentially, to the poor under the current Pilot Programme and grant structure proposals. While these actions are very positive and should be encouraged, they do not address: the method for redistribution of currently technically bankrupt commercial farms; the present inefficiencies in the farm structure, particularly with respect to size; and market-assisted acquisition of farms by less poor households. In isolation, these actions will neither attain the government's redistribution target of 30 per cent of land in five years as reiterated in the RDP,

nor substantially improve the overall farm structure in the direction of family farms.

3. **Increasing access to land for small- to medium-sized commercial farmers.** Together with the previous options, this option has a pivotal role to play in addressing the shortcomings of South African agriculture and attaining the common vision. Increased efforts should concentrate on owner-operators, whether they be individuals or communities. The five categories mentioned below are relevant:

   - *Land rental markets* (including sharecropping). While a more flexible rental market would improve opportunities for households with some farm resources to engage in agriculture – and hence should be encouraged, land rentals should not become a dominant aspect of the farm structure because of the inherent problems of underinvestment, environmental costs and collateral limitations for credit.
   - *Joint ventures of workers and owners.* Joint ventures such as envisaged under the 'DBSA equity sharing option' will contribute to broadening the ownership base in agriculture. However, the joint venture option may have limited replicability, requiring special conditions to be successful. There may be high transaction costs involved due to the unique characteristics of each case; success may crucially depend on the goodwill of the owners, and the schemes may only be applicable to high value products.
   - *Contract farming between a processing or marketing firm and owner-operators.* Contract farming is particularly appropriate for commodity production which involves a central processing plant or packing house, requiring large capital outlays and benefiting from economies of scale. Examples are sugarcane and cotton farming.
   - *Land redistribution to communities or groups of owner-operators.* This approach is piloted in the Land Reform Programme of the Ministry of Lands. It has a strong poverty alleviation orientation.
   - *Market-assisted land redistribution to individual, small- to medium-scale owner-operators.* At this point, no particular programme has been implemented.

Facilitating the process

Facilitating access to land for small- to medium-scale owner-operators requires close co-operation and harmonization of efforts between the

Departments of Agriculture, Land Affairs and also Finance, as well as the active participation of the private sector. The following elements are crucial:

- grants;
- a viable credit system;
- substantial private sector involvement;
- selective government provided/funded support services.

1. **Grants.** The function of grants is to provide equity to beneficiaries who do not have it, in order for them to access the credit system and to operate within acceptable financial parameters. While these ratios vary widely with different types of farming, a rule of thumb is that a debt-asset ratio of higher than 0,35 is unviable in grain production. The size of the grant depends on the size of the farm operation and the financial requirements, as well as the contribution the beneficiary (or group) can make from their own sources.
2. **Viable credit system.** A viable credit system involves credit for the purchase of land and for operating capital. The criteria for government intervention in agricultural credit markets are clear (see Chapter 20). Direct government credit to farmers should only occur when there is absolutely no chance of the private sector extending credit to these farmers. However, if government finds it necessary to provide such credit, the target beneficiary group should be well and narrowly defined. The duration of the loans should also be clearly specified, and conditions of the loans should not create distortions in credit markets or create disincentives for paying back the loan capital. Specifically, interest rates should be market-related and the loans should rather concentrate on reducing the transaction costs of getting the loan. In this respect, the role of grants is important to address the lack of sufficient collateral.
3. **Substantial private sector involvement.** The involvement of the private sector hinges on the creation of economically and financially viable farm enterprises, small and large. Without this condition, there will at best be very small contributions from the private sector. However, without the private sector, many of the goals set for agriculture will not be achieved. The following are crucial and should be involved: agro-industry; developers; technical advisory services; and the co-operative system for inputs, marketing, credit and extension.
4. **Selective government financed/provided/funded support services.**

The role of government financed/provided/funded support services should concentrate on the following:

- strategic research;
- public infrastructure;
- adaptive research and extension, particularly with respect to: natural resource management; small farms; integrated pest management;
- control of epidemics;
- standards and grading, and other measures which make markets operate better;
- regulation of pesticides, environmental issues, etc.

Management of the transition

The transition from the present situation in agriculture to one that will embrace the principles set out in the vision above will, without a doubt, require careful management both of political and economic aspects. As illustrated in the previous chapters, there are several actors, all with some vested interest, involved in the process. Some will be winners and others will be worse off, given these changes in direction. In addition, land reform according to the principles set out in this book will cost a substantial amount, the major portion of which will have to come directly from the redirection of limited public funds to which other worthy causes also lay claim. But, as is argued, there are few other options open to achieve the stated goals of greater equity and efficiency within a relatively short period of time.

The two fundamental political hurdles to land reform and change of the required magnitude are:

- to design an approach which is acceptable to a sufficiently broad spectrum of key interest groups which will allow them to form a broad coalition for change and reform;
- to design a land reform programme so that a sustained coalition emerges which, year after year, will insist on sufficient appropriation in the budget process to carry out the land reform and resettlement for the many years required to implement it.

It is believed that the approach articulated in this book satisfies both these criteria. However, within these broad parameters, it is of critical importance that there is a strong and sustained political will and commitment to achieve these goals – even against tremendous odds and with frequent set-backs. This implies persistent and continued backing not only by the highest hierarchy of politicians, but also across a wide front.

Within this context, the packaging of land reform and the associated publicity, information and consensus-building campaigns are important. On an issue as politically loaded as land reform, a packaging strategy should accentuate conservatizing forces, thus making it easier in general – and more attractive to opponents of the initiative and their lobby – to support the land reform process.

In addition, it is important that the political leadership keep in touch with grassroots aspirations, expectations and perceptions. In general, lobby groups are vocal but their messages often deviate or fail to address the real issues and desires of those very people they represent.

INDEX

J

K

T